T0213550

Challenges in Physics Education

Series Editor

Marisa Michelini, Dipartimento di Fisica, University of Udine, Udine, Italy

This book series covers the many facets of physics teaching and learning at all educational levels and in all learning environments. The respective volumes address a wide range of topics, including (but not limited to) innovative approaches and pedagogical strategies for physics education; the development of effective methods to integrate multimedia into physics education or teaching/learning; innovative lab experiments; and the use of web-based interactive activities. Both research and experienced practice will feature prominently throughout.

The series is published in cooperation with GIREP, the International Research Group on Physics Teaching, and will include selected papers from internationally renowned experts, as well as monographs. Book proposals from other sources are entirely welcome.

Challenges in Physics Education addresses professionals, teachers, researchers, instructors and curriculum developers alike, with the aim of improving physics teaching and learning, and thereby the overall standing of physics in society.

Book proposals for this series my be submitted to the Publishing Editor: Marina Forlizzi; email: Marina.Forlizzi@springer.com

More information about this series at https://link.springer.com/bookseries/16575

Hans Ernst Fischer · Raimund Girwidz
Editors

Physics Education

 Springer

Editors
Hans Ernst Fischer
Universität Duisburg-Essen
Essen, Germany

Raimund Girwidz
Ludwig-Maximilian-Universität
München, Germany

ISSN 2662-8422   ISSN 2662-8430 (electronic)
Challenges in Physics Education
ISBN 978-3-030-87393-6   ISBN 978-3-030-87391-2 (eBook)
https://doi.org/10.1007/978-3-030-87391-2

This Springer imprint is published by the registered company Springer Nature Switzerland AG
The registered company address is: Gewerbestrasse 11, 6330 Cham, Switzerland

The ideas for publishing Physics Education developed from numerous discussions with colleagues in our science education community.

With Norman G. Lederman, I discussed many aspects of science education when we met at conferences. He contributed many ideas to this book until he passed away in early 2021. We tried to implement these ideas in our own contributions. Norm greatly encouraged me to make the European view of physics education internationally visible, at least to some extent, as possible within the limitations of such a book.

I have always been deeply impressed by his theoretically sound but practice-oriented approach, which is also evident in his contribution in Chap. 5 entitled "Nature of Scientific Knowledge".

We miss Norm very much. We think that his contributions and great ideas for science education have always moved us forward! We dedicate this book to Norm.

Hans E. Fischer

Preface

This book on physics education built on its German edition[1] is addressed to researchers, teacher educators and teachers in pre-service and in-service training. The chapter authors are experts in physics education who represent the subject in research and teaching in their universities. Each chapter deals with a respective topic by integrating current research and examples from practice. The chapters have been peer-reviewed by colleagues from different disciplines as the content of the chapters covers different aspects of learning psychology, pedagogy or subject-specific education.

Physics education deals with the questions of what, why and how physics should be taught and learned in lessons and therefore should always be regarded as a very complex and interconnected social field of activity. This is especially true for the natural sciences as they play an increasingly important role in our modern societies in terms of content and methodology. Our interaction with nature can only be purposeful if we investigate nature empirically and if we understand the methods with which the investigation is carried out. Science education already starts in kindergarten and continues for all students at school; it is not limited to those who later become biologists, chemists or physicists.

Since citizens can only respect and appreciate what they know about science, at least to some extent, if they can assess its connections with the development of their respective society. Science teachers therefore have to fulfil responsible and demanding tasks of teaching science. Professional teacher education therefore also has to be oriented towards these demands. Teaching and learning strategies also have to be adapted to the target groups and the fields of education. We have chosen to focus on physics teaching rather than addressing teaching science in general, as the content of the lessons naturally plays a central role in teaching physics. Content-related research, teacher education and the design of biology and chemistry lessons should therefore be presented in other volumes. Many problems of learning theories and teaching methods, the motivation of teachers and students and, last but not least,

---

[1] Kircher, E., Girwidz, R., & Fischer, H. E. (Eds.) (2020) *Physikdidaktik—Grundlagen*, Berlin: Springer Spektrum.

the relevance of the subjects to society must be discussed and described for each subject area.

We start with the general aspects of physics and physics education to relate historical, learning-theoretical and general pedagogical basics to the subject and the professionalism of physics teachers. It becomes clear that teachers bear a great responsibility, which can be compared to that of medical professionals or lawyers. The professionalism of teachers therefore refers not only to subject matter expertise but also to psychological and educational expertise and even the maintenance of their own physical and mental health. Accordingly, we describe the conditions for teacher education, methodological principles for structuring lessons as well as the content-related and structural basics of physics teaching. In order for future physics teachers to be able to classify new research results on teaching and learning physics, we describe the basics of empirical research at the end of this book.

The theoretical approaches and research findings presented also characterise the basics for future research in physics education. However, they only make sense if they are integrated into teacher training, the practice of teaching and learning physics under the given social and school conditions. This book alone cannot achieve the connection of theory and research to teaching practice, but it can be supportive. We, as teacher educators, should make even more effort to link empirical research and teaching practice. Our discipline of teacher education is still rather young. However, we are not always effective in communicating our research findings more widely and in using them to improve teaching and learning in schools via teacher education. Therefore, it is important that this topic has a fixed place in our professional conference programmes.

Professional knowledge of teachers is usually defined as pedagogical knowledge, content knowledge and pedagogical expertise. In this book, professional competence is used as an overarching term for extending the professional knowledge of physics teachers to other areas of teachers' professionalism, such as their professional beliefs in general, their beliefs about the subject matter or content in particular, as well as their values and attitudes towards their profession as teachers.

We hope we can offer you at least some new insights or points of discussion to improve teaching and learning of physics.

Essen, Germany   Hans Ernst Fischer
München, Germany   Raimund Girwidz

Acknowledgements

This book *Physics Education* would not have been possible without the enormous commitment, enthusiasm and competency of the colleagues whose contributions have allowed its publication. There is however a lot of more work that needs to be done in addition to writing the chapters. The technical editing of Chaps. 1, 2, 4, 8–15 and 17 was done by Chi-Yan Tsui (Auckland, New Zealand). We appreciate his extremely thorough and conscientious work that was more than just improving the English and checking the references. Chi-Yan's work also included improving the comprehensibility of the text, the argumentation and the logical flow of thoughts. Many thanks to Chi-Yan!

Hans Ernst Fischer
Raimund Girwidz

Contents

Chapter 1
Topics of Physics Education and Connections to Other Sciences

Hans E. Fischer, Raimund Girwidz, and David F. Treagust

Abstract The contributions of research presented in *Physics Education* are genuinely necessary for content and quality of physics teacher education and for developing new lessons based on research. Teaching and learning in a subject, physics in this case, can only be successful if it is informed by research in the respective areas of educational activities. Nevertheless, to improve the quality of physics teaching and learning from kindergarten to university, many references to other disciplines are needed. This chapter of *Physics Education* brings teaching and research of physics education in line with academic physics, psychology and pedagogy to enable student teachers of physics and in-service teachers to understand research findings and provide them with opportunities to transfer theoretical approaches and empirical results to their own practices. *Physics Education* is about research, teaching physics and learning physics and lesson design on all levels of the respective educational systems. Therefore, the area of research, teaching and developing learning environments covers not only processes of teaching and learning in pre-schools, all kind of schools, universities and state-dependent seminars but also non-school institutions such as museums, science centres or learning laboratories.

H. E. Fischer (✉)
Universität Duisburg-Essen, Essen, Germany
e-mail: hans.fischer@uni-due.de

R. Girwidz
Ludwig-Maximilians-Universität München, München, Germany
e-mail: girwidz@lmu.de

D. F. Treagust
Curtin University, Perth, Australia
e-mail: D.Treagust@curtin.edu.au

© Springer Nature Switzerland AG 2021
H. E. Fischer and R. Girwidz (eds.), *Physics Education*, Challenges in Physics Education,
https://doi.org/10.1007/978-3-030-87391-2_1

1.1 Determinants and Reference Disciplines of Physics Education

During the last four decades, a field of research on teaching and teacher education related to different subjects has developed rapidly all over the world. In particular, mathematics education, biology education, chemistry education and physics education have developed as distinct sciences for research on subject-dependent teaching and learning processes and for developing empirically tested learning environments and teaching methods for educational institutions such as schools and universities but also out-of-school educational settings.

At the same time, educational research has become more and more involved in teacher education at university. It is usual in universities all over the world that research and teaching are closely connected. One principle of academic education is that only content which can be scientifically confirmed should be taught. The parties involved in educational research are teachers and students of all ages and fields of education, not only at school but also, for example, in pre-schools, universities, teacher seminars and out-of-school institutions such as museums, science centres and even different kinds of private coaching schools. General conditions of teaching and teacher-dependent prerequisites and influences on teaching in institutions that might be expected should be addressed systematically in teacher education. In principle, such influences can be divided into those which are independent of teachers and those which may be influenced by teachers' professional competencies such as abilities, motivations and beliefs (see Fig. 1.1).

As shown in Fig. 1.1, learning environments at schools refer to educational, cultural and social conditions and requirements, as well as to certain physical and psychological settings. The assumed impacts require specific professional competences for teaching in general and additional competencies related to the respective subject in particular. In addition, teachers also need to know how to perform professionally in school consultations with parents or students, and undertake excursions and school trips. Last but not least, they should know how to maintain their mental and physical conditions to reach and uphold high-quality teaching (see Chap. 2 on professional competences and Chap. 3 on teacher education).

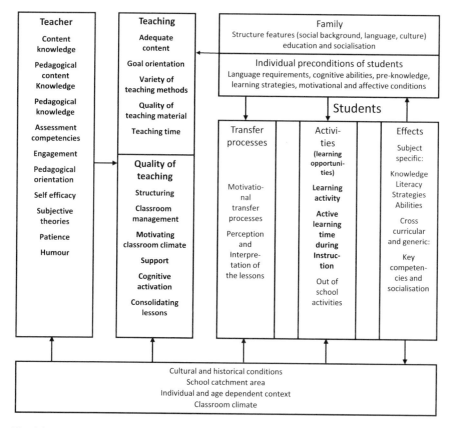

Fig. 1.1 Utilisation-of-learning-opportunities model (translated and adapted from p. 73)

1.1.1 The Main Reference Disciplines of Physics Education

To consider the determining variables in Fig. 1.1 and to meet the requirements for optimising learning processes, researchers of physics education and educators of physics must apply physics education theories and methods from other sciences (see Fig. 1.2), consider the findings of related research areas and cooperate with the respective scientists (Reusser 2008).

The extent to which the content to be taught at educational institutions depends on content type and on the age of students. Blue ovals in Fig. 1.2 indicate disciplines that should necessarily be part of teacher education; green ovals indicate disciplines that should be taught depending on student teachers' and teachers' interest and availability of resources during teachers' education study and their in-service training.

Therefore, research and teaching in physics education at university has a strong relation to physics and mathematics, epistemology, pedagogy of teaching and lesson design and psychology of learning and teaching. These disciplines are outlined in

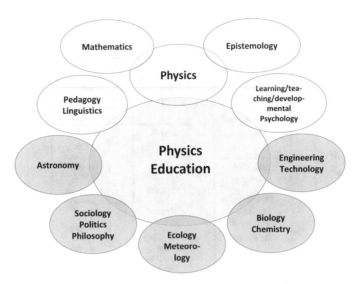

Fig. 1.2 Important reference disciplines for physics teaching and physics teacher education

the following sections. Furthermore, neurobiology, educational governance, development of educational institutions or sociology sometimes should be considered in order to obtain insightful research results and information about the functions of thinking, the influence of school organisation on classes and the effect of different social environments on teaching and learning. Applying related research findings to physics education can help to provide teachers, and also policy makers and heads of institutions, with trustworthy information to support their own teaching and organisation of general system conditions for good teaching and learning. Criteria of trustworthiness of research in science education are explained in Chaps. 16 and 17, and the transfer of research findings to professional competence of teachers in Chaps. 2 and 3.

1.1.2 Interdisciplinary Approach to Teacher Education

Besides other goals, the goal of physics teaching in schools is to provide an understanding of systematic and empirically based relations to the natural, social and technical world. In order to perform teaching to meet this goal, students of different age and with individual cognitive and developmental conditions should be presented with physics content and concepts, including different content areas related to physics (see Fig. 1.2). For this reason, physics teacher education should always follow an interdisciplinary approach. Therefore, in addition to academic physics, which must be included as a subject in physics education, aspects of epistemology, pedagogy and psychology should also be included. To organise learning environments and to assess students' progress, student teachers of physics have to learn about cognitive processes for generating knowledge of physics concepts (see Chaps. 6, 14 and 15). Knowledge and concepts to solve tasks and problems should also be offered (see Chaps. 8 and 9).

Recently, students' development of language and argumentation becomes increasingly recognised as more important for physics teaching; and therefore, findings of research on linguistics and communication theories should also be a focus of physics education research and teaching as explained in Chap. 13. Considering examples of research studies and their relevance to student learning, teacher educators and teachers should also know about any relationships of physics education with arts and music, economics, ecology, geology, history, medicine and other disciplines (Crease and Pesic 2014; Karlqvist 1999; van der Veen 2013) because they all have relations to physics.

Researchers in physics education can apply and adapt findings from other sciences to investigate the specific conditions for successful physics teaching and learning, beyond developing their own theoretical attempts and empirical studies to obtain useful results in research for improving teaching and learning. This process generates newly designed learning environments and insights, rationales and knowledge, which would not come into being without physics education research.

1.2 Physics Education

The term *education* covers teaching and learning, didactics, all thinkable prerequisites inside and outside of all kinds of institutions and historical, as well as political, individual and social requirements and conditions in general (see Fig. 1.1). *Education* also includes describing and reflecting upon the historical development of the educational system or conceptualising new designs for learning at school because of changes of societies through political or technical developments. During the last century, teaching and learning in institutions in many countries became increasingly oriented to learning processes and physics content changed radically through *relativity theory* and findings in *astronomy* and *particle physics*.

Research on education aims to describe the interactions between teachers and students in learning environments dependent on individual, social, political and historical prerequisites. For physics education, the theoretical ideas and practical references should enable teachers to plan and organise learning environments for successful performance in teaching their lessons, including interactions with their students. To be successful in their teaching, teachers need good knowledge of the related sciences about the effects of educational activities on their students, as well as about the theoretical attempts that may be helpful for understanding the process of teaching and learning physics.

1.2.1   The Content is Physics

Physics belongs to the natural sciences, which have been developing during the last 2000 years as part of the human history of culture. During the last three centuries, as foundation for technical development, the findings of research in biology (as life science), chemistry and physics (as natural sciences) have dramatically changed our world. Inventions and innovations include, for example, steam engines, airplanes, nuclear power plants, electrical energy, energy conversion, the atomic bomb, individual mobility, the Internet, electronic control, gene technology, fertilizers, herbicides, pesticides, advances in medicine and many others. All these inventions and innovations have changed and are still changing not only political relationships among nations and different aspects of our climate but also, for example, all kinds of arts and music and health care. The science domains affect aspects of our personal everyday life including our life expectancy. Therefore, individuals should know about physics, one of the science domains, and about how physicists work, which are necessary for understanding nature. This is also necessary for individuals to take an orientation for assessing their own role in society and in valuing the effects of human activities—manifested as technological progress or political statements and reports—on our planet's future ecological developments. Nations all over the world have recently valued sciences, especially physics, as an important resource of knowledge for their own economic growth and welfare. In the *Paris Agreement* of the United Nations (2015), 195 nations decided to cooperate in reducing global warming on the basis of joint research of physicists, biologists, chemists and meteorologists from all over the world.

Physics and Mathematics

The history of the development of physics is closely connected to the development of some fields of mathematics, which is what makes physics even harder to understand and teach than some other sciences. The close connection between physics and mathematics expressly started in the very beginning of modern physics when Galileo Galilei (1564–1642) and René Descartes (1596–1650) formulated the first descriptions and laws of modern physics. Galileo referred to the use of mathematics in describing the universe and he explicitly distinguished *theoretical physics* and *experimental physics* by emphasising the role of mathematics. In Florence (Toscana, Italy), Galileo was employed as a mathematician, whose role in those times included working also as an astronomer, an engineer and a musician.

> La filosofia è scritta in questo grandissimo libro che continuamente ci sta aperto innanzi a gli occhi (io dico l'universo), ma non si può intendere se prima non s'impara a intender la lingua, e conoscer i caratteri, né quali è scritto. Egli è scritto in lingua matematica, e i caratteri son triangoli, cerchi, ed altre figure geometriche, senza i quali mezi è impossibile a intenderne umanamente parola; senza questi è un aggirarsi vanamente per un oscuro laberinto. (Galileo 1623, p. 6, Capitolo VI)

> Philosophy is written in that great book which ever lies before our eyes. I mean the universe, but we cannot understand it if we do not first learn the language and grasp the symbols in which it is written. This book is written in the mathematical language, and the symbols are

triangles, circles, and other geometrical figures, without whose help it is humanly impossible to comprehend a single word of it, and without which one wanders in vain through a dark labyrinth. (translation by Ross 1951, p. 10)

Galileo's application of mathematical theories which guided experimentation in physics was innovative in those times; what he used was the standard mathematics already created by Archimedes of Syracuse (ca. 287–212 BC) and Leonardo of Pisa (1170–1250, named Fibonacci). Those mathematical theories all relied on the theory of proportion of Eudoxus of Cnidus (ca. 390–337 BC). But already during the last period of Galileo's lifetime, Descartes was one of the creators of analytical geometry (Linton 2004), which, by connecting algebra and geometry, opened opportunities for inventing many revolutionary descriptions of nature. Descartes lived in France and the Netherlands and was one of the precursors of the idea that humans accepted nature as given by God, but they also created and developed physics and mathematics through which they wanted to express generally valid laws of nature. Since then, every physics teacher and physics student teacher in the world has to take up the challenge of studying physics, including mathematics and physics concepts and modelling mostly in chronological order of content areas from the history of the development of physics.

Experimental physics describes and predicts physical events in time and space as well as processes quantitatively and qualitatively. Experimental physicists try to find empirical evidence by developing research questions from theoretical models and accordingly design and perform related experiments. Therefore, experimental physicists need theoretical frameworks, which are often developed in cooperation with theoretical physicists. Experiments and observations are mostly performed under defined conditions by avoiding any disturbances, often in a laboratory. Most of the physics faculties in universities all over the world investigate our physical world in laboratories. One of the biggest and most prestigious physics centres in the world is the Conseil Européen pour la Recherche Nucléaire (CERN; https://home.cern/) in Bern (Switzerland). Physicists from all over the world investigate fundamental laws of elementary particles using the biggest and most complex physics devices on earth at CERN. One basic demand of research in natural sciences is its replicability. Therefore, each laboratory activity has to be described as clearly as possible using basic procedures. The laboratory reports should at least include the related theoretical foundation, the design of respective experiments, the preparation and construction of measuring devices, the control of variables, the registration of measuring data and the production of results. Theories and hypotheses are needed for finding and formulating the research questions, which are guiding the design of the research and the production of results, which often leads to developing new experimental devices and measuring routines as result of experimental research. For example, the goal of astrophysics to look deeper into space leads to the development of the *Hubble Space Telescope* and its successor *James Webb Space Telescope* to avoid influences of the earth atmosphere for more precise measurements.

Theoretical physics deals with describing, explaining and predicting space and time changes of physical objects. For this purpose, theoretical physicists use concepts, connections and interdependencies of concepts such as theories, laws, rules, axioms, variables and constants in a physics terminology. This subject-dependent language is a first problem for learning physics at school because many words which are used to describe physics concepts, such as light, heat, work or force, originate from everyday language, but they are used in everyday life with many different meanings. In physics, the meanings of these words are clear and precise. One main goal of theoretical physics is to mathematically design and develop the *system of physics concepts* and validate whether or not the system fits the empirical findings of experimental physics and vice versa (see Chaps. 7, 8, 10 and 14).

Research findings and concepts of experimental and theoretical physics must be open for critical scrutiny. Therefore, the findings of both experimental and theoretical physicists always have been, and are today, published for discussion and for drawing conclusions, as well as for being available for replication in the community of physicists.

Physics teaching in schools has different aims depending on the type and level of schools or universities. Physics education at lower secondary levels mostly offers a systematic and experience-based empirical connection to the world. For higher education, physics teaching adds the orientation to more scientific and mathematically modelled content areas of modern physics. Therefore, students of different age, levels of cognitive development and in different types of schools are confronted with physics concepts and their mathematical descriptions of different complexities. To describe this process of physics teaching and learning, physics teachers, therefore, need knowledge of developmental psychology and psychology of learning and teaching, as well as knowledge of physics content, the related mathematics, and also the epistemology of physics (see Fig. 1.2).

1.2.2 Epistemology and Physics

Physics teaching and learning in schools and universities, and research and development in physics education at university, is not only related to physics content but also to the development of concepts, models and knowledge of all natural sciences. Epistemology is one of the main areas of philosophy which deals with the nature of knowledge and knowing, the justification of empirical experience and rationality of reality (Turri et al. 2019). The nature of knowledge and knowing is philosophically analysed in relation to concepts such as reality and truth, but also to justification and beliefs. The objects of analysis are sources and criteria for knowledge and knowing in a certain historical period. The criteria must be justified as being in line with the main stream and the doubts of the respective society for guiding the activities of research and human activities in general. Therefore, to complement justification, criteria are also needed for doubts which are developed as the general principle of *falsifiability* (Popper 1983 [reprinted 2000]) and as *sceptics* to develop critical arguments for further research and to challenge contentions of dogmatic scientists. Since more than 3000 years ago, epistemologists have asked questions such as "What is reality?" "When is the result of an observation valid?" or "What does it mean to know something?" Consequently, we need to understand the role of physics in our society, what we can accept as results of experiments and how the results should be interpreted. Physics teachers also need to have in mind the procedures of laboratory work and acceptance of experimental results in the historical past in order to understand the influence of social conditions on scientific experimentation.

Sir Francis Bacon (1561–1626) was one of the first who addressed the problem of *objectivity*, *self-critique* and *self-control* of scientists. He was a philosopher and lawyer in London, and in 1620, he published rules for scientific work. Bacon described them as "The idols and false notions that now possess the human intellect and have taken deep root in it don't just occupy men's minds so that truth can hardly get in, but also when a truth is allowed in they will push back against it, stopping it from contributing to a fresh start in the sciences" (Bacon 2017, pp. 7–8). The idols of Bacon are listed as follows:

1. Idola Specus (Idols of the Cave) point out that every (human) being lives in an individual cave and scientists should go beyond their cave and be self-critical in their thinking.
2. Idola Theatri (Idols of the Theatre or Tradition) are fallacies from convincing theorems (dogmata) presented by an authority. They should be scrutinised.
3. Idola Fori (Idols of the Market Place) are caused by everyday language and our custom to take the words as real.
4. Idola Tribus (Idols of the Tribe) are the most difficult to avoid. They are caused by our own mind, which tends to form opinions guided by biological properties of human beings.

In other words, the idols described problems for objective scientific work and, since the middle of the last century, they have been described, in a modern language, as part of the rules of scientific societies all over the world (Funari 2011).

For instance, the knowledge of how physicists investigate and how they explain their experimental results is strongly dependent on what their scientific community accepts as correct results and how the idols of Bacon are worked out in the respective society. For example, Charles Augustin de Coulomb (1736–1806), a French physicist, described the forces between stationary and punctiform electrical charges. Coulomb's law was first published in 1785. It was then possible to describe quantities of electrical charge, which was the basis for the later development of electro-magnetism. In the international system of units (CGS-system), in scalar form and in a vacuum, the force is defined as:

$$F = k_e \frac{q_1 q_2}{r^2}$$

k_e is Coulomb's constant, q_1 and q_2 are the magnitudes of the charges and r is the distance between the charges.

The main problem for all physicists who worked on this force in those times was to find the exponent of r experimentally. The way to find the results was different from what we accept today. Coulomb was strongly guided by Newton's law of universal gravitation; and therefore, he preferred the exponent of r to be 2. Coulomb published only the results of six experiments, with the first three experiments on repulsion. Later, because the related experiment was more difficult, he conducted another three on attraction and none of the values he found was exactly 2. Nevertheless, he was convinced that the value of exactly 2 was correct for describing the phenomenon and many of his colleagues accepted his results and were not sceptical of them. Bacon's idols were not discussed in those times; today such an experimental method would not be accepted. Coulomb's law has been accepted by physicists and later investigations with more precise experimental devices confirmed his theoretical attempt.

The awareness that physics is a shared and discussed cognitive construction—which depends on the respective society, the community of physicists and epistemologists—is fundamental for teaching physics. Physics is not dogma but a continuously discussed and changing cognitive construction of individuals. In Chap. 5, the nature of scientific knowledge is explained in detail.

1.2.3 Educational Psychology of Learning and Teaching

As a subarea of educational psychology, psychology of learning and teaching investigates processes, conditions and effects of teaching and learning inside and outside educational institutions (Duchesne and McMaugh 2018; Glynn et al. 2012). One focus of investigating teaching and learning, among others, is on characteristics of the learner such as memory and cognitive performance, motivation and social activities such as communication and interactions with others. Measuring performance and motivation of teachers and learners in controlled teaching–learning situations is an essential part of the related research. Results of these investigations provide criteria for the promoting and inhibiting conditions of learning situations and environments.

The aims of respective research studies always refer to psychological theories. Therefore, relevant theoretical models are outlined in the following sections.

Learning as Stimulus–Response Behaviour

At the beginning of the last century, psychologists started to describe and explain behaviour. A first and basic theory of Ivan Petrovich Pavlov (1849–1936) assumed that human behaviour could be caused and directed by stimuli from the environment. According to this theory of *classical conditioning*, simple learning processes are initiated by coupling external stimuli (Pavlov 1927). The reaction of a dog's salivation on a chime (stimulus 1) combined with the taste of an apple on the tongue (stimulus 2) could be triggered by the chime only (conditioned response). To be able to explain also more complex behaviour of human beings, Burrhus Frederic Skinner (1904–1990) established the concept of *operant conditioning* (Skinner 1938). In this model, behaviour is also initiated by the stimulus–response mechanism. However, Skinner's investigated learning processes that referred, in terms of modern concepts, to cognitive and affective processes as basis of behaviour. These processes, however, are not directly observable, and because there were not yet any existing research methods, these processes were not investigated. Accordingly, a human being was regarded as a black box with unknown contents, and mental processes were not a subject of his research. The theory was further developed as behavioural therapy to include rule-guided behaviour caused by positive or negative reinforcement, which should lead the learner to accept the consequences of his or her intended behaviour.

Until recently, some activities of teachers in the classroom were strongly guided and could be described by this more complex kind of Skinner's behaviourism. Parts of classroom management or direct instruction, which includes straightforward, explicit teaching techniques, were usually applied to prevent disturbances in the classroom or to teach a specific, well-defined skill (Kim and Axelrod 2005). In those cases, it is the aim of the teacher to teach students an expected behaviour or a skill through a single and simple intervention. In addition, more complex learning processes and interventions can be divided into simple steps to control students' learning activities through positive or negative reinforcement.

Cybernetic Theory of Learning and Transfer of Information

Cybernetics is the science of controlling and regulating machines, organisms and organisations. Accordingly, cybernetic theory describes learning as a process of reorganisation of sensory feedback within a closed loop. The aim of the respective learning processes is to increase the learner's level of control of both his or her own behaviour and the influences from the environment (Smith and Foltz Smith 1966).

According to cybernetic theory of von Cube (1968), the learning goals are implemented into the control loop of teaching and learning as the desired values. In a new situation, learning will be regulated by ordering the quantity of all available information, including the information from the learner's own memory. At the end of the learning situation, only a small amount of information is necessary for understanding and transferring to new learning in similar situations.

Some elements of the cybernetic learning approach still being applied to teaching are mainly part of teacher-centred and teacher-oriented direct education. In those lessons, the teachers' aims are developed in a way like a Socratic dialogue (Knezic et al. 2010). The answers of the students allow the teacher to control his or her next question and the next step in the learning process in order to follow his or her lesson plan. Necessary experiments are mainly integrated in the lessons as demonstrations. According to Hattie (2012), teacher-centred and teacher-oriented teaching leads to good results if it adequately fits the planned learning environment and the teacher's aims of the lesson. Mostly, these goals are not compatible with problem solving, conceptual thinking and laboratory activities planned and performed by the learner that are necessary for understanding physics.

Learning as Cognitive Process

Albert Bandura (born 1925), Jean Piaget (1896–1980) and Max Wertheimer (1880–1943) were the first scientists to open a small part of the black box of Skinner.

The main thesis of the cybernetic theory of learning implies that successful learning processes can be organised by *varying the offers* (or stimuli) and *controlling the reactions* (or behaviour) of the learner in a teacher-oriented way. But c*ognitivism* also adds to the theory of learning aspects of cognitive development (Piaget and Inhelder 1969), learning on social models (Bandura 1986) and learning by insight, understanding and reflecting the object of learning (Wertheimer 1945/1971). Increasingly, the learner is understood as an individual and not as an externally controllable machine where the action of learning is autonomous and goal driven. The teacher initiates dialogues between the participants of a learning process in a student-oriented setting. According to Anderson (1982), the general teaching aim at school is the *acquisition of cognitive skills* and, among others, Beckmann and Guthke (1995) identified *solving problems* instead of solving tasks as one of the main skills to be taught.

Processing perceptions in the brain developed as one central topic of biological psychology (Baars and Gage 2010a, b) in parallel with neurobiology (Schumann et al. 2004). Today, it is understood, for example, that the perception of different media (e.g., textual or pictorial), as well as its processing in different areas of the brain, can be optimised following certain rules as proposed by *cognitive load theory* (CLT, Sweller 2019). The research of human cognition suggests that knowledge is usually domain specific and that the brain has only a limited capacity and a limited duration of the working memory (see Chap. 7). In line with CLT, Mayer (2001) developed a cognitive theory of multimedia learning (CTML) which was further developed by Schnotz and Bannert (2003) for explaining the combination of textual and pictorial perceptions (see Chap. 7). An overview and guide is given by Jenlink (2019) for physics teachers and students, especially for the use of digital media in physics teaching (see Chap. 11).

Learning as Construction of Knowledge

According to Plato (428/427–348/347 BC), as father of (objective) idealism, we need reality individually for our physical and social life, but it does only exist as a general idea of all objects. His philosophical method is the *dialectic*, which, today, is what we would describe as the mental process of analysis and synthesis of concepts. Accordingly, logic induction leads the human thinking from concrete to general and from conditional to unconditional and back to conditional conceptualisation. The goal of thinking is to understand an *idea* through the synopsis of the plurality of thoughts to form one unique and completely separate concept (Ross 1951). Idealism is one of the main foundations of constructivism.

In addition, constructivism in education in general and science education in particular is strongly connected with constructivist epistemology, which developed during the last two centuries. During the last century, neurobiology and psychology added to epistemology some arguments for an individual development of knowledge by cognitive construction. Biologists such as Maturana and Varela (1980) identified full-grown brains as autopoietic systems which means that the brain is informationally closed and it forms perceptions by using cognitive constructions already stored during development of cognition. Allopoietic systems are in contrary controlled from outside the system. An important finding of neurobiology that the interface between inside and outside the brain does not allow transferring information supports this idea. Accordingly, human sensory receptors are the means by which humans react to changes in external and internal environments. For example, light or sound stimuli cannot be processed directly inside the brain. Photons or pressure differences of the air only produce -90 to $+30$ mV pulses of neuronal direct current voltage, but the action potentials of the neuronal receptors—in this case the retina in the eye and the cochlea in the inner ear—can vary in amplitude and frequency. This is the case for all possible external stimuli such as sight, hearing, smell, taste and touch but also for internal receptors in the muscles, tendons and joints, which give rise to the kinaesthetic sense. Receptors in the inner ear for the sense of balance or carbon dioxide receptors in the blood and hunger and thirst receptors in the hypothalamus of the brain produce stimuli of the internal environment. The frequency of the neuronal pulses correlates with the intensity of these stimuli but only when the pulses produced by neurones of the interface (e.g., of the retina, the skin or the inner ear) are higher than the threshold voltages of following neurones. This is the reason for the unbelievable ability of our brains when complex events are constructed. The neurones activated from all the above-mentioned interfaces and brain areas are able to get in contact because they *all speak the same language* and react accordingly.

As a consequence, it is not possible to input information or knowledge into the brain. Therefore, meaning is a product of the brain itself. For example, colours and tone pitches are produced in distinct brain areas, which are directly connected with the retina of the eye or the cochlea of the inner ear. A green colour always produces reactions of the same neurones, and other neurones are able to interpret the reactions always in the same way. The name *green* or other names and our assignment must be learned in communication. According to Magga (2006), the Sámi people of Norway, Sweden and Finland have about 180 words related to snow and ice and about 300 different words for types of snow, tracks in snow and conditions of how to use snow. It follows that the existence of the same reality valid for everybody, as Plato suggested, must be negated. As a consequence, von Glasersfeld (2001) suggested the existence of two realities. One can be described as *Ding an sich* (Kant 1968/1787, p. 306 ff), which is a fiction comparable with Plato's word of ideas, whereas the other reality is experienced, useful and proved in everyday life. Von Glasersfeld described this kind of reality as being *viable* and not transferable but responsible for the activities of a respective individual.

However, until now, most teachers have not used constructivist epistemology and neuronal functions to describe or control complex learning processes of individuals in school situations. Nevertheless, the idea of autopoiesis is used to substantiate individualised approaches to learning under classroom conditions. It is deduced hermeneutically that every learner with an individual access to the world needs an individual learning environment for constructing individual meaning (von Glasersfeld 2001). Therefore, the teacher appears as a moderator who supports individual meaning construction by evaluating each student's learning process and by varying the learning environment accordingly. For each individual student, cooperative, interactive and reflexive behaviours should be enabled to realise and recognise complex connections for action in complex situations. This attempt accepts a general and shared reality because the final consequence of negating reality would also mean the negation of the general usability of (real) learning environments and productive communication in a classroom.

To solve this dilemma, psychologists assume today that partners of a teaching–learning process operate relatively independently from each other and that their cognitive development appears to be also independent. Considering this development, Merrill (1991) discussed a *second generation instructional design* where the learning process is described as interaction between construction and instruction. The process of teaching emphasises a balance between self-determination and heteronomy in the sense of an adaptive design of learning environments (Ryan and Deci 2017; Wehmeyer et al. 2017). The corresponding model called *utilisation-of-learning-opportunities model* (see Fig. 1.1; Helmke 2009) suggests that the process of teaching and learning is the design of a learning invitation for the learner and the learner's utilisation of this invitation. The related *instructional design* is intended to optimise a certain result of the process by following the intended and predetermined learning goal. The learning opportunities are not seen as single and isolated activities of the teacher but as a system of multiple interdependencies (Fischer et al. 2005).

1.3 Instructional Design

1.3.1 General Principles of Instructional Design or General Didactics

General principles of instructional design or *general didactics*—as it is named in many regions of the world except in English-speaking countries—address theory, organisation and performance of teaching, including all prerequisites, procedures and theoretical and practical attempts of teaching and learning (Caillot 2007; Castiblanco Abril and Nardi 2018; Gundem 2000).

General didactics investigates history, structure and aims of organised teaching and learning (mainly in institutions), namely independent of individual learning processes or learning in groups and independent from subject-dependent teaching aims, especially independent from subjects taught. Nevertheless, anthropogenic and sociocultural preconditions of all participants in the learning process and the content of the subjects are considered by subject-dependent didactics/education (in this case, it is physics education). General didactics describes techniques and organisational and practical connections of organised learning and teaching in general and provides some foundation for subject education in particular to develop content-related theories and models for learning. Three examples are explicated as follows:

- Experimental group work is mainly a matter of the natural sciences; and therefore, science educators need to develop specific applications and learning environments from general principles of classroom management (see Chap. 10).
- Linguistic structure of speech is thematised by general didactics such as the connections between the syntactic, semantic, pragmatic and aesthetic dimensions of signs used in lessons. Physics educators apply these structures when causal linguistic structures are applied to teaching or writing an experimental protocol or explaining physics concepts (see Chaps. 10, 12 and 13).
- The connection between knowledge acquisition and organisation of activities for solving problems are applied to organising respective experimental or theoretical learning environments (see Chaps. 9 and 10).

Organising interactions in classrooms is another aim of general didactics. It gives a first frame for planning physics lessons. The following example illustrates the first step of how physics lessons and the learning processes can be planned and organised to achieve the lesson goals (see Chap. 4).

Interactions in the classroom can be guided by the teacher (teacher-centred) or by the students (learner- or learning-centred) (Lasry et al. 2014). *Teacher-centred* environments may be useful to introduce and explain a new physics concept or to prepare for a test, whereas a *learner-centred* environment may be necessary to explore the concept and to develop understanding by discourses and experimental work. Nevertheless, according to the lesson aims, both learning environments can be organised as teacher- or learner-oriented. *Teacher-centred* and *teacher-oriented* approaches mean that the teacher tries to lead the students' answers towards the planned direction. On the other hand, *student-centred* and *student-oriented* approaches lead students to engage in open discussions where the teacher leads communication. Both frameworks require and support different abilities.

Student-centred activities such as inquiry and laboratory work can be *teacher oriented* when the teacher strongly guides the activities like a recipe book. If students are free to follow their own ideas, the activity can be described as *student and learning oriented*. Which approach might lead to a higher quality of learning in the physics lesson depends on its appropriateness for achieving the instructional aim but also depends on whether it is appropriate in the context of the educational system and the specific school and classroom. The same student-oriented and student-centred approach may be successful in a class with 15 students but not in another class with 25 or 50 students.

In addition, the effect of different approaches are dependent on the students' age. Instead of choosing between a student- or teacher-centred approach, we should think of it more as a continuum. Planning, organising and moderating teacher–student interaction are a matter of utilisation-of-learning-opportunities for constructing concepts, which are compatible with the particular concepts of academic physics. Which combinations may have productive learning outcomes depends on the subject taught, the conditions of the classroom and, of course, on the cognitive, motivational and volitional prerequisites of the teachers and students.

During the course of the development of general didactics or instructional design, there are several directions of development. The two most important and relevant directions for today's physics education are described in the following sections.

1.3.2 Didactics from the Theoretical Perspective of Bildung

The concept of Bildung was strongly connected with a period of European political, philosophical and cultural development during the Age of Enlightenment (also known as the Age of Reason or the Enlightenment) from about 1700–1800. This period started with the reign of Louis XV and ended with the French Revolution and its consequences at the very beginning of the nineteenth century. These changes spread all over Europe and later had some influences on the cultural development of the USA.

The Enlightenment created some ideas and principles, which still influence cultural values even today, for example, *reason* as a primary basis of knowledge, *beliefs* in progress and toleration, *liberty* as a leading component of politics and *social affairs*, *secularism*, and *constitutions* as bases of governments. The Age of Enlightenment saw scientific methods and reductionism as motors for the development of societies and was vehemently opposed to obscurantism and mysticism, which was seen as the heritage of the Middle Ages. Immanuel Kant (1724–1904), one of the most prominent philosophers in those times, described Enlightenment in 1799 as follows (see the original text in Fig. 1.3):

> Enlightenment is human's awakening from self-inflicted immaturity. Immaturity is the inability to use the own intellect without control from another person. This immaturity is self-inflicted when its cause lies not in lack of understanding, but in lack of initiative and courage to use it without control of another person. Sapere Aude! Dare to be wise. Have courage to use your own intellect!—that is the motto of enlightenment. (Kant 1799, p. 697; translated by authors)

The development of the theory of *Bildung* started in the middle of the eighteenth century in the German-speaking world. Referring, for example to Plato, Rousseau and Kant, Wilhelm von Humboldt (1767–1835, scholar, writer and statesman in Prussia) and Friedrich Schleiermacher (1768–1834, theologian, classical philologist and pedagogue) developed the idea of Bildung that focussed on Bacon's idols of the upper classes in Ancient Greece. In the beginning of the development, they created the universally educated person whose knowledge and worldview corresponds with reality. Accordingly, to educate a person's ability to develop insight was the main goal of education at school. Later, in the 1970s, Humboldt developed this concept to *humanistic Bildung*, the optimistic view that humankind is in principle able to optimise the form of its own existence. Humboldt and others developed a picture of an ideal society and especially of ideal Bildung for which the educational system should enable each individual to have the best possible personality development. Also, as propagated by leading philologists of this time, Humboldt connected intellectual,

Beantwortung der Frage: Was ist Aufklärung?

„Aufklärung ist der Ausgang des Menschen aus seiner selbst verschuldeten Unmündigkeit. Unmündigkeit ist das Unvermö-
gen, sich seines Verstandes ohne Leitung eines andern
zu bedienen. Selbst verschuldet ist diese Unmündigkeit, wenn die Ursache derselben nicht am Mangel
des Verstandes, sondern der Entschließung und des
Muthes liegt, sich seiner ohne Leitung eines andern zu
bedienen. Sapere aude! Habe Muth, dich deines eige-
nen Verstandes zu bedienen! ist also der Wahlspruch
der Aufklärung.

Fig. 1.3 Kant's description of enlightenment (Kant 1799, p. 697)

aesthetical and ethical education for personality development with the necessity to study the philosophers of Ancient Greece according to Buck (1987):

> Greeks are a nation, under their felicitous hands all was matured to last perfection what, according to our profound inner feeling, contains the highest and richest human existence; we view them as a human tribe formed of a most precious and pure substance, … (p. 384; translated by authors)

The romantic idea of Greek philosophy as the basis for educating political and industrial managers led to an underrepresentation of mathematics, sciences and technology at institutions of higher education in German-speaking countries during the nineteenth century. However, this romantic idea was criticised by pedagogues and upper-class members of society who recognised the need for human resources for industrialisation. Only in 1900 was the newly established *Realgymnasium* allowed to prepare school students for studying biology, chemistry or physics at university (Fischler 2015). Since then, the concept of *Bildung for the upper classes* had developed in the twentieth century as the concept of *Bildung for all members of a society*. In the twenty-first century, we refer to a complex and differentiated concept of Bildung with several steps of actualisation and adaption to the needs of a society to be presented in the next sections.

1.3.3 Material Theory of Bildung

In the following, we distinguish between two categories of Bildung to express the different aspects of an educational situation and methods of teaching. There are *materials to learn, individual subjects* as ideal picture of responsible citizens and *respective methods* for teaching. Material theory of Bildung, formal theory of Bildung and methods are defined by the respective society, necessary for developing the social and economic system (see Fig. 1.4).

According to Terhart (2009), the content itself is the central object of the theory of Bildung. The choice of the content, its legitimation, its order in the process of teaching and explaining the concepts taught are of interest but teaching methods and media do not matter. Methodology of teaching is subordinate to Didactics. Klafki (2007) stated that the central content question of teaching is connected with social development. Human beings are understood, in the theory of Bildung, as holistic and the content taught as being the exemplary choice for gaining access to general principles of the mental order of the world (e.g., language, mathematical and scientific structures and concepts). Each respective society has to decide about the content that young citizens should be concerned with in order to become responsible members of the society. The access to these different areas is different in principle and, therefore, not disposable. Following this attempt, Baumert (2002, p. 113) described science subjects including physics as one of the *modes of exploring and encountering the world*; these modes are listed as follows:

1. Natural sciences as cognitive instrumental modelling of the world, with the main question: How can nature be described?
2. Policy and law as normative-evaluative analysis of economy and society, with the main question: How can we reach a binding agreement for our social world?
3. Arts as aesthetic-expressive encounter and composition with the main question: How can reality be cognitively constructed in an aesthetic-expressive mode as a consequence of individual experience and sensibility?

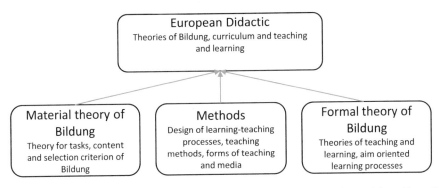

Fig. 1.4 Three categories of bildung and teaching methods related to the object and the subject of the learning process

4. Religion and philosophy as problems of constitutional rationality with the main question: How is it possible to verify different modes of constructing the world fundamentally and independently from individuals in philosophy? In contrast, how to explain individual meaning of individual decisions and activities as they appear in social reality from a religious point of view?

The modes of exploring and encountering the world become available through *basic cultural tools*:

1. A good command of the respective common language
2. The ability for modelling mathematically
3. The ability for being able to communicate at least in one foreign language
4. Basic IT-abilities
5. Self-regulation of the individual's own acquisition of knowledge.

Although the natural sciences are not listed as basic cultural tools, in nearly every country of the world today, knowledge and abilities to solve problems in sciences are seen as a key resource for economic and social development and scientific-technological progress.

1.3.4 Formal Theory of Bildung

Formal Theory of Bildung provides concepts and tools for analysing and planning lessons from a psychological point of view. It is assumed in general that there is a generalizable structure of instruction, focused on the intention (why), content (what), method (how) and media (with what). These questions are used and applied in a lesson. Intention, content and method are dependent on the respective socio-cultural and anthropologic-psychologic conditions of the learners and the school environments.

On the basis of the work of Piaget and Inhelder (1969) and Piaget (1978), Aebli (2006) integrated empirical research of teaching–learning psychology into planning and conducting lessons. Oser and Baeriswyl (2001) developed this idea and proposed prototypical learning sequences for lessons. They formulated basic models focused on learning processes such as the cognitive construction of concepts, learning through one's own experience or problem solving for structuring the lessons and many more (see Chap. 4). Meyer (2007) and Maier (2012, p. 171) outlined the following categories for planning lessons, based on learning-teaching psychology, neuroscience, learning and teaching research and pedagogical and psychological diagnostic. To control the effect of teaching, they suggested a professional analysis that takes into account the results of the intended learning process:

1. Identifying the curriculum and subject-dependent prerequisites and aims.
2. Identifying competence aims (competence is seen as the ability to use knowledge for task and problem solving)

3. Elaborating the learning requirements for a lesson schedule, which is based on learning theories.
4. Knowing and considering the methodical dimensions of designing learning-teaching processes.
5. Applying organisational aspects of holding classes.
6. Reflecting and evaluating of the applied learning and teaching processes.

Today, the complete picture of European Didactics covers most of the pedagogical areas relevant to teaching and learning in institutions. For teachers and student teachers, European Didactics provide an overview of which fields of knowledge are necessary to learn for high-quality teaching.

Horlacher (2015) gave an overview of the development of the theory of Bildung in German-speaking countries.

1.4 Summary

This chapter has emphasised the point that Physics Education is about research, teaching physics and learning physics, and designing lessons for all levels of schooling from preschool to university. The disciplines related to physics education as shown in Fig. 1.2 cover the most important skills and the professional knowledge that is necessary for high-quality physics teaching, but they are not sufficient. A thorough understanding of physics, mathematics and epistemology, and how they are related is essential. To utilise all resources of an educational system and additional international resources, it is, for example, necessary for physics teachers to be familiar with other languages, to know how to use digital media, to use related Internet sources and to decide if the information presented is trustable. Teacher education should therefore cover all these topics to orient students and in-service teachers towards a broad scholarly outlook for their work and to show them how the expertise of the related disciplines must be integrated into learning and teaching physics. The competence to teach physics needs all the above-mentioned skills and knowledge areas of these disciplines, especially knowledge about instructional design and the processes of teaching. All these skills, knowledge and procedures should be based on research on physics learning and teaching for a comprehensive physics teacher education. Therefore, teacher education at university should integrate the findings of research on teaching and learning physics, and on the development of new learning environments with considerations about relevant research from related disciplines. This integration can only be done by researchers in physics education and not by psychologists or pedagogues, because subject matter knowledge and orientation towards physics involve subject-dependent cognitive and motivational processes of students at school and in students' and in-service physics teachers' understanding of their discipline. On the other hand, physics educators must know enough about the psychology of learning and teaching and of instructional design or didactics as outlined above, and should cooperate with psychologists and pedagogues to apply the findings of empirical research to physics teaching and perform teaching accordingly.

Acknowledgements We would like to thank Prof. Richard Gunstone (Monash University, USA) and Prof. Detlev Leutner (Universität Duisburg-Essen, Germany), who carefully and critically reviewed this chapter.

Selected International Journals on Science Education

Eurasia Journal of Mathematics, Science and Technology Education (http://www.ejmste.com/)
European Journal of Science and Mathematics Education (http://www.scimath.net/)
German Journal of Science Education (https://link.springer.com/journal/40573)
International Journal of Environmental and Science Education (http://www.ijese.com/)
International Journal of Science and Mathematics Education
(https://www.springer.com/education+%26+language/mathematics+education/journal/10763)
International Journal of Science Education (https://www.tandfonline.com/toc/tsed20/current)
Journal of Astronomy and Earth Sciences Education (http://jaese.org/)
Journal of Science Education and Technology (https://www.springer.com/journal/10956)
Journal of Research in Science Teaching (https://onlinelibrary.wiley.com/journal/10982736)
Physics Education (http://iopscience.iop.org/journal/0031-9120)
Research in Science and Technological Education (https://www.tandfonline.com/loi/crst20)
Research in Science Education (https://www.springer.com/education+%26+language/science+education/journal/11165)
Science Education (https://onlinelibrary.wiley.com/journal/1098237x)
Science and Education (https://www.springer.com/education+&+language/science+education/journal/11191)
The Physics Teacher (https://www.aapt.org/Publications/tpt.cfm)
The Journal of Science Teacher Education (http://link.springer.com/journal/10972)

References

Aebli H (2006) Zwölf Grundformen des Lehrens: Medien und Inhalte didaktischer Kommunikation, der Lernzyklus [Twelve basic forms of teaching: media and content of didactic communication, the learning cycle] (13. Aufl. ed.). Klett-Cotta, Stuttgart
Anderson JR (1982) Acquisition of cognitive skill. Psychol Rev 89:369–406
Baars BJ, Gage NM (2010) Mind and brain. In: Baars BJ, Gage NM (eds) Cognition, brain, and consciousness, 2nd edn. Academic Press, London, pp 2–31
Baars BJ, Gage NM (2010) Thinking and problem solving. In: Baars BJ, Gage NM (eds) Cognition, brain, and consciousness, 2nd edn. Academic Press, London, pp 344–369
Bacon F (2017) The new organon: or true directions concerning the interpretation of nature (A. b. J. Bennett Ed.)
Bandura A (1986) Social foundations of thought and action: a social cognitive theory. Prentice-Hall, Englewood Cliffs
Baumert J (2002) Deutschland im internationalen Bildungsvergleich [Germany on an international comparison of educational systems]. In: Kilius JKN, Reisch L (eds) Die Zukunft der Bildung. Suhrkamp, Frankfurt, Main, pp 100–150
Beckmann JF, Guthke J (1995) Complex problem solving, intelligence, and learning ability. In: Frensch PA, Funke J (eds) Complex problem solving: the European perspective. Lawrence Erlbaum Associates, Hillsdale NJ, pp 177–200

Buck A (1987) Humanismus. Seine europäische Entwicklung in Dokumenten und Darstellungen [Humanism. Its European development in documents and presentations]. Freiburg Alber

Caillot M (2007) The building of a new academic field: the case of French Didactiques. Eur Educ Res J 6(2):125–130. https://doi.org/10.2304/eerj.2007.6.2.125

Castiblanco Abril O, Nardi R (2018) What and how to teach didactics of physics? An approach from disciplinary, sociocultural, and interactiona dimensions. J Sci Educ 19:100–117

Crease RP, Pesic P (2014) Physics and music. Phys Perspect 16(4):415–416. https://doi.org/10. 1007/s00016-014-0147-3

Duchesne S, McMaugh A (2018) Educational psychology for learning and teaching. Cengage AU, Australia

Fischer HE, Klemm K, Leutner D, Sumfleth E, Tiemann R, Wirth J (2005) Framework for empirical research on science teaching and learning. J Sci Teach Educ 16(4):309–349. Retrieved from http:// www.jstor.org/stable/43156373

Fischler H (2015) Bildung. In: Gunstone R (ed) Encyclopedia of science education. Springer Science+Business Media, Dordrecht

Funari AJ (2011) Francis Bacon and the seventeenth-century intellectual discourse, 1st edn. Palgrave Macmillan, New York

Galileo G (1623) Il Saggiatore

Glynn SM, Britton BK, Yeany RH (2012) The psychology of learning science, 1st edn. Routledge, New York

Gundem BB (2000) Understanding European didactics. In: Ben-Peretz M, Brown S, Moon B (eds) Routledge international companion to education. Routledge, London, pp 235–262

Hattie J (2012) Visible learning for teachers: maximizing impact on learning. Routledge, New York

Helmke A (2009) Unterrichtsqualität und Lehrerprofessionalität. Diagnose, Evaluation und Verbesserung des Unterrichts [Instruction quality and teacher profesionality. Diagnosis, evaluation and improvement]. Kallmeyer, Seelze

Horlacher R (2015) The educated subject and the German concept of bildung: a comparative cultural history. Routledge, New York

Jenlink PM (2019) Multimedia learning theory: preparing for the new generation of students. Rowman and Littlefield, Lanham

Kant I (1799) Beantwortung der Frage: Was ist Aufklärung [Answering the question: What is enlightenment?]. In: Tieftrunk JH (ed) Imanuel Kant's vermischte Schriften vol 2. In der Rengerschen Buchhandlung, Halle

Kant I (1968/1787) Kritik der reinen Vernunft [Critique of pure reason], vol 2. Walter de Gruyter, Berlin

Karlqvist A (1999) Going beyond disciplines. Policy Sci 32(4):379–383. https://doi.org/10.1023/ a:1004736204322

Kim T, Axelrod S (2005) Direct instruction: an educators' guide and a plea for action. Behav Analyst Today 6:111–123

Klafki W (2007) Neue Studien zur Bildungstheorie und Didaktik: Zeitgemäße Allgemeinbildung und kritisch-konstruktive Didaktik [New studies on theory of bildung and didactics: contemporary general bildung and critical-constructive didactics, 6 edn. Beltz, Weinheim and Basel

Knezic D, Wubbels T, Elbers E, Hajer M (2010) The socratic dialogue and teacher education. Teach Teach Educ 26(4):1104–1111. https://doi.org/10.1016/j.tate.2009.11.006

Lasry N, Charles E, Whittaker C (2014) When teacher-centered instructors are assigned to student-centered classrooms. Phys Rev Special Top Phys Educ Res 10(1):010116. https://doi.org/10. 1103/PhysRevSTPER.10.010116

Linton CM (2004) From Eudoxus to Einstein: a history of mathematical astronomy. Cambridge University Press, Cambridge

Magga OH (2006) Diversity in Saami terminology for reindeer, snow, and ice. Int Soc Sci J 58(187):25–34

Maier U (2012) Lehr-Lernprozesse in der Schule: Studium [Teaching and learning processes at school: Studies at university]. Klinkhardt, Bad Heilbrunn

Maturana HR, Varela FJ (1980) Autopoiesis and cognition: the realization of the living, vol 42. Reidel, Boston

Mayer RE (2001) Multimedia learning. University Press, Cambridge

Merrill MD (1991) Constructivism and instructional design. Educ Technol 31(5):45–53. Retrieved from Retrieved from http://www.jstor.org/stable/44427520

Meyer MA (2007) Didactics, sense making, and educational experience. Eur Educ Res J 6(2):161–173. https://doi.org/10.2304/eerj.2007.6.2.161

Oser F, Baeriswyl FJ (2001) Choreographies of teaching: bridging instruction to learning. In: Richardson V (ed) Handbook on research on teaching, 4th edn. American Educational Research Association (AERA), Washington, pp 1031–1065

Pavlov IP (1927) Conditioned reflexes. An investigation of the physiological activity of the cerebral cortex (G.V. Anrep, Trans.). Oxford University Press, London

Piaget J (1978) Das Weltbild des Kindes [The world view of children]. Klett-Cotta, Stuttgart

Piaget J, Inhelder B (1969) The psychology of the child. Basic Books, New York

Popper K (1983 [reprinted 2000]) Realism and the aim of science: from the postscript to the logic of scientific discovery, 1st edn. Routledge, London and New York

Reusser K (ed) (2008) Empirisch fundierte Didaktik – didaktisch fundierte Unterrichtsforschung. Eine Perspektive zur Neuorientierung der Allgemeinen Didaktik [Empirically based didactics - didactically based teaching research. A perspective on the reorientation of general didactics], Sonderheft ed. vol 9. VS Verlag für Sozialwissenschaften, Wiesbaden

Ross WD (1951) Plato's theory of ideas. Greenwood Press, Westport

Ryan RM, Deci EL (2017) Self-determination theory: basic psychological needs in motivation, development, and wellness. Guilford Press, New York, NY, US

Schnotz W, Bannert M (2003) Construction and interference in learning from multiple representation. Learn Instr 13(2):141–156. https://doi.org/10.1016/S0959-4752(02)00017-8

Schumann BH, Crowell SE, Jones NE, Lee N, Schuchert SA (2004) The neurobiology of learning: perspectives from second language acquisition. Routledge, New York

Skinner BF (1938) The behavior of organisms: an experimental analysis. Appleton-Century, New York

Smith KU, Foltz Smith M (1966) Cybernetic principles of learning and educational design. Holt, Rinehart and Winston, New York

Sweller J (2019) Cognitive load theory and educational technology. Educ Tech Res Dev. https://doi.org/10.1007/s11423-019-09701-3

Terhart E (2009) Didaktik. Eine Einführung [Didactics. An introduction]. Reclam, Stuttgart

Turri J, Alfano M, Greco J (2019) Virtue epistemology. In: Zalta EN (ed) The stanford encyclopedia of philosophy (Fall 2019 edition). Retrieved from https://plato.stanford.edu/archives/fall2019/entries/epistemology-virtue/

United Nations (2015) Paris agreement. Retrieved from https://unfccc.int/resource/docs/2015/cop21/eng/l09r01.pdf

van der Veen J (2013) Symmetry as a thematic approach to physics education. Sym Cult Sci 24:463–484

von Cube F (1968) Kybernetische Grundlagen des Lernens und Lehrens [Cybernetic basics of teching and learning], 2nd edn. Ernst Klett Verlag, Stuttgart

von Glasersfeld E (2001) The radical constructivist view of science. Found Sci 6(1):31–43. https://doi.org/10.1023/a:1011345023932

Wehmeyer ML, Shogren KA, Little TD, Lopez SJ (2017) Introduction to the self-determination construct. In: Wehmeyer ML, Shogren KA, Little TD, Lopez SJ (eds) Development of self-determination through the life-course. Springer, Dordrecht, pp 3–16

Wertheimer M (1945/1971) Productive thinking. Harper and Row, Evanston, San Francisco, New York, London

Chapter 2
Professional Competencies for Teaching Physics

Hans E. Fischer and Alexander Kauertz

Abstract To identify necessary content for teacher education, we first have to model professional competences of teachers as their abilities for applying a body of knowledge and concepts of physics to create an optimal learning environment. In general, competence can be seen as a cluster of characteristics including knowledge, skills, beliefs and abilities, which enable and improve the performance of a job. Consequently, competences always refer to a context and a task or problem to solve. Professional competence of physics teachers includes many areas of physics knowledge but also social and educational knowledge and skills to apply successfully to their teaching in the classroom. In the case of teaching physics, teachers' professional competence requires their content knowledge on an adequate level but also their abilities, motivation and volition to teach physics in their respective classes. Beliefs and values of teachers often have their sources in everyday life as well as in the related scientific communities at university. Therefore, teachers should know the actual scientifically acquired knowledge about attitudes and values in order to identify their own attitudes and values and to assess their impact on teaching and learning. In addition, self-regulation competencies are necessary for teachers to organize their own working and their learning processes for keeping the development of their own resources.

2.1 Overview

To identify necessary content for teacher education, we first have to model *professional competences* of teachers as their abilities for applying a body of knowledge and concepts of physics to create an optimal learning environment. In general, competence can be seen as a cluster of characteristics including knowledge, skills, beliefs

H. E. Fischer (✉)
Universität Duisburg-Essen, Essen, Germany
e-mail: hans.fischer@uni-due.de

A. Kauertz
Universität Koblenz-Landau, Landau, Germany
e-mail: kauertz@uni-landau.de

and abilities, which enable and improve the performance of a job (Eraut 1994). Consequently, competences always refer to a context and a task or problem to solve (Smith 1996/2005).

Professional competence of physics teachers includes many areas of physics knowledge but also social and educational knowledge and skills to apply successfully to their teaching in the classroom. In the case of teaching physics, teachers' professional competence requires their content knowledge on an adequate level but also their abilities, motivation and volition to teach physics in their respective classes (Darling-Hammond 2006; Kauertz et al. 2012). Beliefs and values of teachers often have their sources in everyday life as well as in the related scientific communities at university (Collins et al. 2020). Therefore, teachers should know the actual scientifically acquired knowledge about attitudes and values in order to identify their own attitudes and values and to assess their impact on teaching and learning. In addition, self-regulation competencies are necessary for teachers to organize their own working and their learning processes for keeping the development of their own resources.

Terhart (2011) described professionalism of teachers with three theoretical approaches: the *structural theoretical*, the *professional biographical approach* and the *competence theoretical approach.*

The *structural theory approach* points to that, during teaching, teachers are faced with a complex bundle of inherently contradictory tasks. They include, among others, proximity versus distance to students, individual needs of students versus general claims of the learning matter, equal versus individual treatment of students, administrative rules versus social interaction and aim to educate autonomous students versus heterogeneous characters of the learning environment. "Professionally" in this context means to handle these tensions and contradictory demands in a self-critical and reflective view of teachers' own professional development as momentum for improvement.

The *professional biographical approach* focuses on more individualized processes of gradual competence building and development, including the adoption of a professional habitus by newcomers, the continuity and fragility of professional development, the link between private life and professional career and similar issues. Competence development is understood as a long-term process over the whole span of professional life. The long-term perspective involves considerations of not only carer development but also individual difficulties of the development such as critical life events and their professional consequences, as well as other areas, such as resilience, stress management, reflection of own beliefs and values (e.g. those regarding students, parents, the school system and teaching). They are needed to handle further determining factors and workloads of the profession, stress experience and ways for coping with stress or resilience. As also demanded in both the competence theory approach and the structural theory approach, in the course of the carer, each teacher individually should develop a kind of general professional expertise.

The *competence-theoretical approach* focuses on describing competences and knowledge areas necessary for teaching as precisely as possible. According to Klieme et al. (2008), "The concept of competence is central to empirical studies dealing with the development of human resources and the productivity of education" (p. 4). To solve problems and to develop modern societies, knowledge only in a certain subject does not seem to be an adequate resource. Consequently, international empirical studies for assessing and comparing the development of educational systems at the end of schooling such as PISA and TIMMS refer to models of mathematical and science competences instead of knowledge acquisition. Referring to the applied model of scientific literacy, Bybee and Koeppen et al. (2008) defined competencies "… as context-specific cognitive dispositions that are acquired and needed to success-fully cope with certain situations or tasks in specific domains" (p. 62). Professional competences and professional knowledge are seen as an interplay between theory (theoretical model) and empirical research. The theoretical model is developed on the basis of past theories and research and must be permanently developed and verified. The object of research is the relation between teacher and student competences and knowledge and also that between the behaviour of teachers and students in teaching and learning situations. Teachers' competences and knowledge refer to their beliefs attitudes, activities and so forth but also to content knowledge, knowledge about classroom management, diagnosing, assessment, counselling and students leaning processes.

To physics, as an old empirical science, we have to add also the epistemolog-ical reflection of individual knowledge acquisition, as well as the philosophical and historical development of physics content knowledge. The demand for profound self-reflection of teachers' own beliefs, values, psychic and physical health (Herman et al. 2017) touches essential parts of their identity. Teacher educators at university should consider the above outlined aspects of professionally but can only initiate the teachers' own responsibility for lifelong learning and for considering not only subject matter but also their own mental and health resources.

The overall challenge for empirical research on teacher education is to find an optimal configuration and balance between teacher competencies and knowledge, and student competences and knowledge and well-being. One main effect of teacher professionalism is the greatest possible increase in learning and experience for as many of his or her students as possible, which can be measured. According to the complexity of teaching–learning situations in classrooms and the huge variety of student learning processes, success of teaching is described as statistical process to increase the average student outcome of a certain sample and to minimize the variance of its distribution. The complex model of professional competences of physics teachers is shown in Fig. 2.1.

Models of professional knowledge are explained briefly in the following sections and in detail in Chap. 3. Motivational orientations, beliefs and values and self-regulation skills should be the content of the general pedagogical and psycholog-ical part of teacher education. Teachers are also socialized by their own scientific community, which is the one of physicists, but even more, by the community of physics teachers (Aikenhead 2003). Glutsch and König (2019) argued that students of different subjects belong to different subcultures and differ in their academic languages and their attitudes. This is a particular challenge in countries where

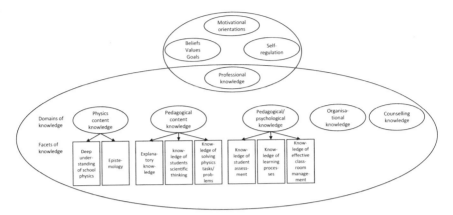

Fig. 2.1 Professional competence of teachers according to Baumert and Kunter (2013, p. 29)

teachers have to teach two or, for primary-school teachers, even more subjects. Additionally, they found that bachelor students' intrinsic, social-altruistic and pedagogical motivations depend on how they value their subject.

In the following sections, we briefly describe the general outlines of the related fields of physics teacher education to open the possibility to more intensive studies of the facets of these fields if necessary.

2.2 Professional Knowledge

According to Shulman (1987) and Baumert and Kunter (2013), physics teachers' professional knowledge contains in general content knowledge (CK), pedagogy/psychology knowledge (PK), pedagogical content knowledge (PCK) and knowledge for organizing and counselling/supervising. For physics teachers professional knowledge, as described by Fischer et al. (2012), is detailed in this section. Meta-knowledge—for example, knowledge on epistemology, learning processes and self-regulation (Schraw et al. 2006)—is also necessary for physics teachers to prepare learning environments adequately. Internationally, according to Shulman (1987), meta-knowledge is named and considered as one facet of professional knowledge. Consequently, in teacher education at university and high schools, teaching practice is already integrated; and teacher educators of the subjects, psychologists and pedagogues are responsible for teaching content and supervision.

Physics education is the discipline, which transfers research findings of other related disciplines to teaching and learning in physics lessons. Obviously, only physics educators can achieve this transfer. Physicists and pedagogues—including instructional psychologists and epistemologists—normally do not explicitly dispose towards knowledge about teaching and learning physics and the related epistemology.

To understand the integrative power of physics education, areas of the three most important related sciences, physics, teaching–learning psychology and pedagogy, are described in all chapters of this book, as far as they contribute to the necessary transfer of knowledge to physics education.

2.2.1 Content Knowledge

Content knowledge (CK) is a necessary precondition for successful teaching (Ball et al. 2001; Shulman 1987). Nevertheless, the effect of teachers' content knowledge on students' learning success was not measured very often with sufficient precision. According to Kunter et al. (2013), the quality of teachers' content knowledge is often indirectly measured by means of state certification, final grades of teacher education or numbers of attended specialized private or state courses. One difficulty for measuring content knowledge is to construct a valid test instrument. Validity in this case refers to several parameters like the grades and the characteristics of the school. High-school and elementary-school teaching needs different types of content knowledge. The projects Professional Knowledge in the Sciences (Professionswissen in den Naturwissenschaften [ProwiN]) defined content knowledge as deepened background knowledge related to the necessary school knowledge (Fischer et al. 2012). Another project *Professional Knowledge of Teachers, Science Education on the Crossover from Elementary to Secondary School* (Professionswissen von Lehrkräften, naturwissenschaftlicher Unterricht und Zielerreichung im Übergang von der Primar- zur Sekundarstufe [PLUS]) compared teachers' content knowledge with a test instrument which represents both school levels and the university level in addition (Kauertz et al. 2010). To know physics on the university level is necessary for planning lessons but also for getting a conceptual overview of teaching physics being flexible and oriented to the learning process in schools. A question on the university level, for example, asks for a phenomenon, which is based on the anomaly of water density (see Fig. 2.2). This anomaly is not explicitly taught at elementary school. Nevertheless, a teacher should be able to react in a sovereign and variable way to students' questions such as: "Why do lakes start to freeze at the surface?" or "Why does the ice in my water glass swim?".

Actually, there are many projects dealing with professional knowledge of teachers under different aspects. Among others, Wilson et al. (2019) reported on perspectives on the future of PCK research in science education, whereas Sorge et al. (2019) investigated the relationship between preservice physics teachers' professional knowledge. Liepertz and Borowski (2019) connected self-concept and interest relationships of physics teachers' and content structure and student achievement. Furthermore, Barenthien et al. (2020) investigated if preschool teachers' learning opportunities have an influence on in-service professional knowledge and motivation, whereas Gess-Newsome et al. (2019) correlated teachers' pedagogical content knowledge, their practice and the achievement of their students.

The picture shows the cross-section of a lake on a cold day in winter.
The temperature is far below the freezing point.

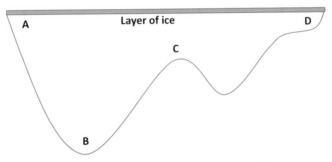

Where in the lake are the areas of highest temperature?

	true	false
In area A, because ice has an insulating effect		
In area B, because here accumulates warm water of high density		
In area C, because warm water accumulates between the cold water layers on the ground and on the surface		
In area D, because shallow water warms up more rapidly		

Fig. 2.2 A question about physics content on a scondary school level

Besides physics facts and concepts, teachers should also know why a certain physics content area should be taught. The justification for teaching certain content areas includes different systematizations of the subject such as topic specifications (e.g. mechanics, thermodynamics, etc.) but also their attribution to central physics concepts such as, for example, structural stability, matter conservation, energy flow, interaction and interdependency, relativity or causal and statistical physics.

Organizing physics content top down by using such kind of leading concepts as a network and following its historical development can give new perspectives on teaching and learning physics. Ausubel (1963) presented the procedure to initiate a learning process as an advance organizer:

> These organizers are introduced in advance of learning itself, and are also presented at a higher level of abstraction, generality, and inclusiveness; and since the substantive content of a given organizer or series of organizers is selected on the basis of its suitability for explaining, integrating, and interrelating the material they precede, this strategy simultaneously satisfies the substantive as well as the programming criteria for enhancing the organization strength of cognitive structure. (p. 81)

Studies showed that the combination of content knowledge and student learning processes using an advance organizer model could guide students' reorganization of physics content. Therefore, advance organizers are considered as a strong tool for teaching physics on all grade levels of the education system (e.g. Gidena and Gebeyehu 2017).

2.2.2 Pedagogical Content Knowledge

Pedagogical content knowledge (PCK) is supposed to help a teacher design learning environments as the basis for students' optimal learning processes. Shulman (1987) identified PCK as an amalgam of content knowledge and pedagogical knowledge. In a synopsis of different attempts to describe PCK, Lee and Luft (2008) identified knowledge about students' learning processes and their initial ideas and knowledge on teaching strategies and methods as central content areas of PCK.

PCK tests are not easy to construct. Test constructors have to operationalize processes and situations of teaching physics, which are different from content knowledge and pedagogy. The following example is an item of a test in an international study for comparing professional knowledge of teachers in Finland, Germany and Switzerland (Fischer et al. 2014). The item shows a lesson on the parameters of the electrical resistance starting with some demonstration experiments. The lesson shall be planned, respectively, by the teacher for testing the parameters and to continue the lesson adequately (see Fig. 2.3).

The answer is expected as a student-oriented summary of the results of the experiment or as a teacher-oriented explanation of the parameters. The example shows that, depending on the variety of possible answers, tasks may have different cognitive potentials, which were also found in Krauss et al. (2008a) study on PCK in teaching mathematics. This corresponds with the didactic triangle which includes

In grade 8, you want to teach the topic resistance of a conductor. Students shall understand on which parameter the resistance of a conductor depends.

Imagine the following situation in a lesson:
In the first part of the lesson, you showed your students different wires. Now you collect students' suggestions for the parameters of the resistance.

The suggestions include length, diameter, material and colour of the resistance wires and you test each parameter in a demonstration experiment. The parameter to be tested will be varied and the others have to be kept constant.

Question: Please describe how you would continue the started lesson. Please do NOT describe the lesson completely but only the task sequence of the lesson (the next task for the students or your next activity).

Fig. 2.3 A PCK task on planning and continuing a physics lesson

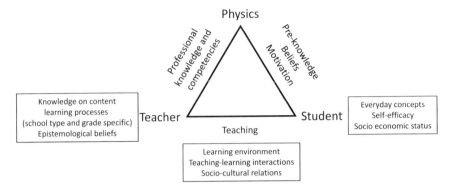

Fig. 2.4 Didactic triangle

the teacher with content-specific activities of depiction and intervention, the students with content and subject-specific ideas and the cognitive potential of tasks (see Fig. 2.4; Augustsson and Boström 2016).

The triangle represents the connections between physics, teaching and learning, and the actors in the classroom:

- the teacher and his individual cognitive abilities, methodical activities and capabilities for intervening,
- the students and their cognitive abilities and subject-oriented ideas and motivation to learn and
- the adequate content, especially the cognitive potential of the used tasks.

From a physics education perspective, knowledge about all three sides of the triangle is needed to organize learning opportunities by cognitively activating students for optimal learning.

Coherence Between Content Knowledge, Pedagogical Content Knowledge and Teaching

Among other studies, Neumann et al. (2019) gave an overview about different models and attempts to describe professional knowledge. However, until now, for science subjects, researchers have not investigated how the relations between CK and PCK of teachers affect their teaching. In contrary, Cauet et al. (2015) investigated the relations between physics teachers' professional knowledge and students' cognitive activation but could not find significant effects on teaching, neither for CK nor for PCK. But there might be light on the horizon as Kulgemeyer and Riese (2018) showed that, for student teachers teaching high-school physics, there was a mediation effect of their PCK on their CK which correlates with the quality of their explanation of a phenomenon in one particular teaching situation. This is in line with the findings in Krauss et al. (2008b) study on teaching mathematics that PCK and CK test scores of highly qualified mathematics teachers were significantly higher than those of teachers who are supposed to be not as qualified according to their teacher education track. The highly qualified teachers showed higher cognitive connectedness in their teaching between the two knowledge categories. Again, for physics teachers, Keller et al. (2017) showed in an international study comparing the quality of teaching in Germany and Switzerland that teachers' PCK positively predicted students' achievement.

Further research is needed to study the connection between teachers' CK and PCK and its influence on students' cognitive learning of physics, on the one hand, and on their effective learning on the other. This research is necessary to develop the contents and the methods of teacher education programmes in a more profound way (Kunter et al. 2020). One important finding of most of these studies shows that CK and physics content taught at school must be considered as strongly related to the learning processes of the students when their age, socioeconomic status, developmental stage and pre-knowledge are taken into account (Massolt and Borowski 2020; Ohle et al. 2015).

2.2.3 Pedagogical Knowledge

According to Shulman (1987), pedagogical knowledge (PK) refers to general principles of classroom organization and management. Today, the components of pedagogical knowledge are more clearly explicated and defined with details as declarative and procedural knowledge and part of professional knowledge responsible for teaching a trouble-free and effective course of lessons and for maintaining a social supportive learning climate (König et al. 2011; Sonmark et al. 2017). PK usually is characterized as *general* to point out that it is conceptualized as being subject independent and cross-disciplinary and as the basis for an optimal design and performance of lessons. Inclusion of not only declarative knowledge (knowing what) but also procedural knowledge (knowing how) in PK points to the conceptualization that it is necessary for teachers to know measures and strategies of teaching and the conditions of applying them effectively in their classroom performance. The general and cross-disciplinary measures and strategies refer to an effective classroom management and to creating a social climate, and designing and performing of learning environments, which support the learning processes of students. Because lessons are always addressing a certain subject, teachers' PK is seen as the necessary but not sufficient precondition for an optimal design and performance of teaching–learning processes in the classroom (Seidel and Shavelson 2007).

There are only a few studies investigating the effect of general PK on instructional quality. Most of these studies did not use any standardized or at least agreed measuring models for PK such as the validated instrument Pedagogical Knowledge and the Acquisition of Professional Competence in Teacher Education (Bildungswissenschaftliches Wissen und der Erwerb professioneller Kompetenz in der Lehramtsausbildung [BilWiss]) of Kunter et al. (2020).

In a study on a sample of 246 in-service teachers in Austria, König and Pflanzl (2016) correlated their model of PK with students' perceptions of instructional quality including effective classroom management, generic teaching methods, teacher clarity and teacher–student relationships. They found general PK—which is operationalized in a measuring model as knowledge on strategies and processes—to be a predictor for instructional quality controlling for the grades of the teachers' exams, their personality and their teaching experience but not students' performance. Their model of PK was supposed to be independent of the taught subject.

According to Schiefele (2017), six general principles and activities are important for establishing effective classroom management as one of the main facets of PK teachers should be able to apply:

- Establishing an efficient system of rules
- Preventing downtime
- Controlling disturbances
- Outsourcing of not subject-related activities
- Swift flow of the lesson
- Clarity of the explanations on an adequate level of requirements.

These principles cover strategies of disturbance prevention as well as handling actual disturbances in the classroom.

Seidel and Shavelson (2007) referred to knowledge on learning processes and personal characteristics of students and teachers—for example, self-efficacy and motivation, and knowledge on learning aims—as areas of PK (see also Shulman 1987). General principles of instruction cover knowing and being able to apply a number of teaching methods adequately in a framework of content, instructional aims and characteristics of the students. Because of the complexity of the social situation in a classroom and the individual cognitive architecture of learners, Kirschner et al. (2006) indicated that guidance of individual learning processes is an indispensable principle for organizing effective learning environments. Kirschner et al. (2006) strongly argued against minimal guidance as provided by theories of constructivism, discovery, problem-based teaching: "Recommendations advocating minimal guidance during instruction proceed as though working memory does not exist or, if it does exist, that it has no relevant limitations when dealing with novel information, the very information of interest to constructivist teaching procedures" (p. 77). Accordingly, knowing about how to formulate teaching goals, organize classroom management and how the cognitive structure directs learning processes is important. Therefore, designing and performing cognitively activating lessons demands teachers to apply theories of learning as well as principles of instruction to their teaching. Universities and high schools should be able to educate student teachers accordingly in this regard, and ideally, to cover theoretical a practical phases of teacher education.

2.3 Motivational Orientation and Self-Efficacy

According to Kunter et al. (2013), motivational orientations indicate teachers' maintenance of their intention to teach and their regulation of teaching and learning over a longer period. They are operationalized as *perceived self-efficacy* (being convinced to have the competence for acting successfully), *engagement* during lessons and the ability of *distancing* themselves from the emotional demands of profession related activities. According to Klusmann et al. (2008), successful teachers can be characterized through a combination of high engagement, good ability of distancing, positive perception and cognitive support of their students. Many empirical studies show that teachers with high self-efficacy are more open for using new teaching methods, learning new teaching concepts and designs and more satisfied with their activities in the classroom (e.g. Achurra and Villardón 2012). According to Ding et al. (2019), after a 17-week teaching practice at a German university, the self-efficacy of the participating preservice teachers increased significantly; and Toropova et al. (2019) also reported a positive influence of teachers' self-efficacy on teaching quality with respect to students' performance.

The effect of exposure to stress on motivation is an important factor for the quality of teaching in many countries of the world (Hattie 2012). Teaching seems to be one of the most demanding and stressful professions. Herman et al. (2017) found significant and complex effects of teacher stress, burnout, coping and self-efficacy and identified that teaching of teachers with high stress, high burnout and low coping skills in class correlated with low student outcomes. Following the definition of burnout as a psychological syndrome characterized by emotional exhaustion, depersonalization and reduced personal accomplishment (Maslach et al. 1997), Klusmann et al. (2016) found that German elementary-school mathematics teachers' emotional exhaustion was negatively related to students' mathematics achievement and pointed out that teachers' well-being has an important influence on students' success in mathematics. Accordingly, successful teachers can be characterized with combinations of *high engagement and high ability to dissociate*, and *positive perception and cognitive support of the students.*

The interdependence between personal (motivational) and environmental characteristics is known as *person-environment-fit (P-O)* (Chatman 1989). By generalization from findings on a sample of electrical workers, Mandalaki et al. (2019) found:

> …that the relationship between organizational identification and job performance is enabled through P-O fit. Additional analyses showed that personality and contextual characteristics differently guide identification and P-O fit work-related outcomes. Personality appears more closely related to the former and context to the latter. (p. 391)

The P-O has an effect on employees' satisfaction with their job, their performance, their engagement and their intention to quit the job. For prospective teachers, work environments positively directed to their future development increases their motivation to organize further successful teaching-related activities (Kaub et al. 2016). Accordingly, on the down side, school-related duties often cumulate to a high extent of subjective stress felt by the teachers. Harding et al. (2019) reported positive correlations between teacher and student well-being and psychological distress controlled by a negative correlation between teacher depression and student well-being. Zimmermann et al. (2012) and Kieschke and Schaarschmidt (2008) also investigated the mental health of 481 student teachers analysing how they perceived about their studies that prepared them to become teachers and how they were able to deal with stress caused by their work at university and teaching at school.

According to Kieschke and Schaarschmidt (2008), pressure on teachers teaching at school often cumulated to a considerable experienced burden. They found four central health patterns to describe in-service teachers' behaviour, motivation and management of resources (see Table 2.1). If appearing difficulties are coped with a passively shaded, resigned and consuming attitude, the risk of teachers falling psychically or mentally ill increases (see Table 2.1, Pattern of risk B).

Based on Kieschke and Schaarschmidt's (2008) study, Kassis et al. (2019) investigated teachers' and workers' individual stressful situations and pressures using the test called Work-related Patterns of Behaviour and Experience (Arbeitsbezogenes Verhaltens und Erlebensmuster [AVEM] Test). Stress experienced by teachers can be caused by external factors (environmental and school situations) and internal factors (experience and resilience patterns).

Figure 2.5 shows external and internal factors influencing teachers' psychic or mental health. The main individual-influencing factors (3) and medium-term and long-term consequences of stress (6) are open to other influences in their own configuration. Environmental factors such as conditions of the system or working conditions can only be affected directly to a small extent, and each teacher perceives the effect of burden or stress differently (4 and 7); and the reactions to stress comprises individual patterns influenced by different factors (5). Therefore, psychological and

Table 2.1 Patterns of behaviour and experience according to Kieschke and Schaarschmidt (2008, p. 430)

Pattern	Description
Health pattern	Engagement, adequate ability of distancing, low tendency of resignation, high level of abilities for problem solving. High resilience: Experience of success at school, high level of contentment, inner calm and social support
Pattern of conserving resources	Low importance of work and little professional ambition, willingness to extend oneself and striving for perfection and high ability of distancing. High resilience: high level of contentment, inner calm and social support
Pattern of risk A	Excessive engagement, high relevance of work, willingness to extend oneself and striving for perfection, low ability of distancing. Low resilience: low level of contentment, inner calm and social support
Pattern of risk B	Low importance of work and little professional ambition, low ability of distancing. Low resilience: high tendency of resignation, low level of contentment, inner calm and social support. But offensive coping with problems (proximity to burnout)

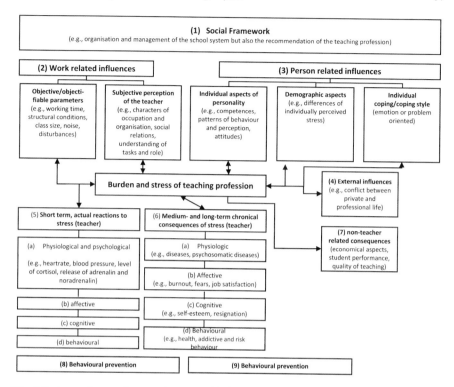

Fig. 2.5 Areas of research on teacher stress (Rothland 2019, p. 632, translated by the authors)

physiological reactions have direct influences on the experiences of stress and preventive psychic hygiene of teachers. They can be influenced by teachers' way of life and individual training (including physical health) (8 and 9).

2.4 Self-Regulation

Using tests such as AVEM test or other programmes allows researchers to classify teachers into stress types and offer them related courses. Among others, the prevention programme Work and Health in the Teacher Profession (Arbeit und Gesundheit im Lehrerberuf [AGIL]) by Lehr et al. (2009) for stress management is widely accepted in health support for European teachers (Scheuch et al. 2015). Storch (2004) offered a programme for self-management of personal resources, which is suitable also for teachers (Zurich Resource Model [ZRM]).

Compared to teachers of other subjects, physics teachers have to face additional time-consuming and management-intensive challenges. They have to prepare, try out and perform experiments for demonstrations, students' laboratory work and including the use of digital media. In addition, they have to agree in their team of physics teachers upon the use and maintenance of the equipment. In addition, many physics teachers have to convince their students that physics is interesting, joy creating and an opportunity for future careers. Students at lower and upper secondary levels regard physics as difficult to learn, abstract and little attractive. As a result, physics at school continuously loses importance and acceptance although a lot of work has been done to modernize and develop the related physics courses (Fischer and Horstendahl 1997; Kalender et al. 2019; Ornek et al. 2008; Salta and Koulougliotis 2020).

There is a consensus among many physics educators, pedagogues and psychologists that student teachers should learn how to preserve and further develop their motivation and self-efficacy by preventively taking care of their mental health. Educational sciences of some universities therefore provide teachers with compulsory courses dealing with stress and strain in the teaching profession and professorships in their own established fields.

2.4.1 Experience of Demands

Although the current-related empirical investigations in different countries are still not standardized and different models of stress experience, vulnerability and resilience are used, the discoveries and the statistical reports of health insurance companies are unambiguous. The conditions at school with increasing inclusiveness and cultural heterogeneity produce an increasing heterogeneity of students' learning conditions. Consequently, the school system, respectively, and the society permanently produce new demands, new aims and new content, for example, those on handling digitalization in physics teaching (see Chap. 11) or integrating disabled students and students with different cultures and languages into the school system. Lessons have to be responsive to manage the new demands by increasing diagnosis (e.g. Hjörne and Säljö 2019) and differentiation (e.g. Smale-Jacobse et al. 2019).

These demands cannot be avoided because societies always develop more or less rapidly. Therefore, depending on politics in each country, the related ministries of education and different public and private support systems develop more or less intensive and expensive programmes for teacher further education. The picture is very heterogeneous but mostly the density of further education is very low (OECD 2019). However, independent from offers of the educational system, teachers, often supported by their school management, should try to preserve and increase their professional competence individually or together with their team of colleagues, for example, in professional learning communities, in a usual way as do medical doctors or lawyers. This includes subject-related new developments in physics content, as well as further teacher education for learning processes of students, self-regulation of teachers in handling demands and their experience of stress.

2.4.2 System Conditions

Self-regulation is not only important to teachers for handling demands and stress but also for responsibly using personal resources and development, and using learning strategies related to the length of stay as teachers.

In addition, teachers need the motivational orientation for maintaining their working capacity and related self-regulation abilities, which include using personal resources responsibly and carefully handling of stress experience related to the expected period of the profession. As described above, the ability of maintaining and preventing psychic health (or mental hygiene) is a matter of teachers' professional competence (see Fig. 2.6).

According to Kieschke and Schaarschmidt (2008), research on demands and stress for teachers results in several options for their own activities. In the following sections, we describe three categories that are indications to identify parameters, which are system conditions but nevertheless partly accessible for self-regulation.

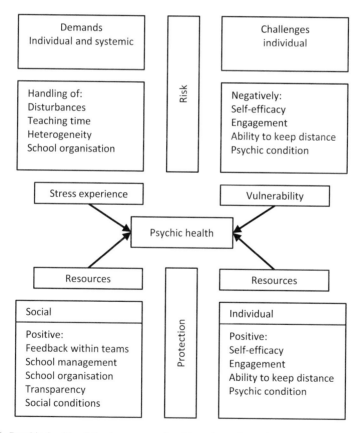

Fig. 2.6 Psychic health of teachers as result of interdependency between demand, stress and management of resources

Systemic Demands

Handling of disturbances, teaching time, heterogeneity and school organization (see Fig. 2.6) are part of the *systemic demands*. They are only to a limited extent influenced by individual characteristics. Handling the demands caused by these parameters is an important condition for successful teaching.

Working Conditions

The local working conditions are relevantly influenced by the social climate at school, and only to a limited extent, influenced by individual characteristics and school management. *Openness, interest for each other* and *mutual supporting* are better starting points for achieving teaching-related standards and aims, especially for teachers in small subject groups as physics, who may easily develop feelings of being a lone fighter left out in the cold. School management is very important for developing a cooperative school climate. They are able to release pressure from the teachers, for example, by flexible organizational procedures or the recreational value of breaks in timetabling of school lessons (Crisci et al. 2019; Suárez and Wright 2019).

Better Recruiting and Preparing of Young Teacher Trainees

Kieschke and Schaarschmidt (2008) indicate that the proof of qualification already before starting studies at university should be extended to work-related patterns of behaviour and experience (the AVME typology, p. 430). Teacher education students with the pattern of risk B (see Table 2.1 and p. 432) should be advised to take up a different profession, because it is not possible to process the expected deficits during their studies at university. In addition, according to Kieschke and Schaarschmidt (2008), teacher preparation should address the demands of teaching-dependent and general stress problems on teachers during their studies and to build up their "capability for effective self-management in stress situations" (p. 436).

Practical Consequences for Student Teachers and in Service Teachers

To handle the three working conditions above, individual initiation by caring leadership of the school management can provide further education to support in-service teachers, but final ways to handle these conditions can only be put into practice if the teachers actively take initiatives. Listed in the following are some examples of activities and behavioural patterns, which enable teachers to develop habits for regulating their individual management of stress:

Developing habits:

- Planning mandatory events (time) for cultivating social relationships (backing and recognition)
- Developing and increasing the physical fitness (resilience and resistance)
- Planning mandatory events for relaxation and recreation (together with others) like sport, relaxation exercises, meditation and music

- Understanding of relaxation and recreation as social obligation (in groups, teams, clubs, …)
- Limiting the amount of work time in a plan and adhering to it strictly
- Striving for regular and sufficient sleep
- Creating and receiving a private area.

Looking for help:

- Talking about work problems with a person of trust
- Reflecting on and explaining experiences appropriately and drawing conclusions for further activities
- Looking for professional help if necessary (from a coach or consultant independent of professional environment and family).

2.5 Self-Regulated Learning and Learning Strategies of Teachers as Learners

Self-regulated learning is based on knowing and applying learning strategies and an important ability for carefully handling personal resources (Winne 2015). Accordingly, learning strategies in general are activity sequences to reach learning goals and self-regulated learning refers to decisions learners make during their learning process. These decisions are dependent on the learners' knowledge about learning strategies and processes but also on their motivation, beliefs and values.

Under the assumption that teachers, who are also learners, are agents with inherent capability to make and act on decisions, self-regulated learning (SRL) is ubiquitous. Decisions teachers make about their own self-regulating learning, however can promote or impede achievement and other valued outcomes. Research on SRL and its roles in education faces multiple challenges. Evidence suggests that productive SRL can be fostered through evidenced-based designs for learning activities and environments, by promoting particular beliefs among teachers and helping them discover values in their own education (see Fig. 2.7).

The learning strategies in Fig. 2.7 correspond with Friedrich and Mandl's (2006) overview of all known and investigated aspects of learning strategies. Teachers as self-regulated learners should be able to apply them to their own learning processes in all areas of their professional competences. They should teach parts of them in their classes, and they should be able to diagnose their students-related abilities for organizing adequate learning processes in their subject. In the following section, the different strategies important for organizing teachers' own learning processes are indicated. These strategies are particularly important in the context of distance learning.

In the framework of learning strategies, *cognitive strategies* are necessary for organizing teachers' own use of knowledge, for applying *metacognition* to assess their already available knowledge and for adapting to new knowledge. These strategies are

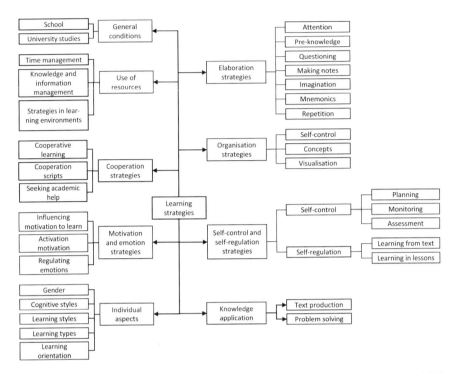

Fig. 2.7 Learning strategies (2020, March 02, retrieved from https://commons.wikimedia.org/wiki/File:Lernstrategien.png, translated by the authors)

important for self-control and self-regulation. Pintrich (1999) pointed out that *motivation and emotion strategies* are necessary for transferring new knowledge from working memory to long-term memory. *Cooperation strategies* and *strategies for using resources* are needed for unburdening the memory and necessary for applying individual habitual and gender-specific *learning styles* to the learning process. *Strategies of repetition* are needed to expand the stay of certain knowledge in the long-term memory, and *elaboration strategies* are used for activating already available knowledge and for transferring it to solve new tasks. *Organization strategies* are necessary for establishing a coherent picture of a certain content area, and *metacognitive strategies* are necessary for directly controlling, regulating and planning the learning process. Learning can be supported by *management strategies* such as time and information management. Some *strategies are domain-specific* and necessary to solve tasks and problems in a certain domain. The content of physics, for example, is organized in a concept-related and top-down way. For solving tasks, teachers need to know the underlying concepts (e.g. energy conservation) and connections to other concepts (interplay of kinetic and potential energy or thermal, kinetic and electric energy). Domain-specific strategies lead to *procedural knowledge* (production rules and procedures for problem solving) to solve tasks of a certain domain faster and better (Rheinberg et al. 2000).

It shall be assumed that learning strategies are mentally represented as individual plans of learning activities. In recent years, strategies for learning from text and pictures have been extensively researched and documented (Arndt et al. 2019; Leopold et al. 2015).

2.6 Beliefs and Values

Beliefs and values are based on epistemological competencies and competencies on theories about teaching and learning. Therefore, they are represented in many psychological and philosophical models.

2.6.1 Beliefs

Teachers' behaviour in a classroom is influenced by their *beliefs about the students' position in a society and teaching and learning.* At the beginning of the nineteenth century, the reputation of natural sciences, including physics, was low. It was not considered as important for educating the coming generation of elites. However, during the nineteenth century, physics became an important and separate discipline in universities all over the world and later also was taught in special schools at secondary levels (Morus 2005). In those times in the USA, pedagogy was Europe-centred and strongly influenced by a colonial view of education (De Lissovoy 2010). John Dewey was one of the first pedagogues in the USA who developed pedagogical attempts independent from the European development; he saw school and education as a democratic institution and concern (Dewey 1922). The natural sciences were described as an important social factor; and teaching sciences, especially physics, was a resource for economic progress (Moyer 1982). In many European countries, the change from estate-based to more democratic school systems started in the beginning of the twentieth century. In the face of an enormous economic development, this change in education systems was driven by the insight that mathematics and the science subjects were an important resource of social development.

Physics teacher education developed successively but dominated by physicists and the academic content. Teachers on higher secondary levels had to prepare students for their studies at university and special programmes for teaching and learning physics were not offered in teacher education. *A good physicist is always a good teacher* was the guiding idea, which is widely accepted even today (Seeley et al. 2019). Beliefs about teaching and learning changed slowly in the heads of teachers, still should change today and are obviously a factor of the quality of teaching and learning. Research in physics education shows influences of beliefs about learning and teaching physics on the instructional practices of teachers (e.g. Caleon et al. 2018).

Patrick and Pintrich (2001) characterized the interrelations between the characteristics of physics, taught content, beliefs on teaching and learning and teachers' behaviour in the classroom as relations between personal theories and general ideas about aims, perceptions, interpretations of classroom situations and anticipations of students' reactions. They pointed out that personal theories are always influencing the quality of teaching and student performance (see also Kunter et al. 2013). According to Schraw (2001), *epistemology* is also partly a matter of beliefs, which is known as epistemological beliefs. Epistemology is part of psychology as well as philosophy in the investigation of the individual generation and development of beliefs, knowledge and recognition and the conditions of the development of validated knowledge in a society (see Chap. 6).

Physics teachers' and their students' *epistemological beliefs* have direct effects on their teaching goals, their planning of learning environments and their activities in the classroom (Bernholt et al. 2019; Chen et al. 2019). A teacher who is a *realist* understands physics concepts as a real picture of nature, whereas another teacher, a *constructivist,* understands physics concepts as individual cognitive constructions. Their difference in understanding physics results directly in different contents, aims and methodological attempts of their respective lessons. In the first case, the teacher expects that students reproduce the truths of physics and in the second, the teacher expects that cognitively constructed physics concepts should be actively reconstructed. Mixing up of scientific and everyday concepts also occurs as issues because scientific concepts are often in contradiction with the experiences and religious and traditional beliefs (Yerdelen-Damar and Eryilmaz 2019).

A physics teacher should be aware that his teaching depends on his beliefs about the development of scientific knowledge (nature of knowledge) and about sources and justifications of scientific knowledge in society (nature of knowing). Personal theories and beliefs are developing through personal social and professional experiences during teachers' whole life. Even before the decision is made to become a teacher, everybody develops habitual ways of thinking, convictions and beliefs about teaching and rules in schools. They develop during an individual's own time as a student at school (Bigozzi et al. 2018). In addition, beliefs of physics teachers with direct effects on the content taught mostly refer to the historical and current development of physics and physics knowledge (see Chap. 5).

2.6.2   Values

Affective values mostly refer to connections between physics and society and students and their position in social systems. Veugelers and Vedder (2003), for example, discussed that values should be part of education. To establish the discussion about values for teaching physics, communicative skills to reflect on values should be part of teaching physics, and therefore, also of teacher education (Hansson and Leden 2016).

Integrating values in education refer to teaching social and aesthetic values, as well as political and cultural conditions (Oser 1994). Accordingly, individual value systems are modelled by an interplay of more general beliefs, for example, beliefs related to concepts of ethics, the obligation for assistance (e.g. promote and demand), fairness and trustfulness or teaching and learning in general. Regarding the *concept of value forming,* Veugelers and Vedder (2003) pointed out that values are being developed in society and that education can play an active role in their development. This is also one of the central functions of school theory (Fend 2001): integration into an existing norm and value structure (integration function), but also the further development of society within set limits (in the sense of an innovative function) (Hooghe and Marks 2019).

Initiated by humanism and enlightenment, discussions on physics and ethics already started in the nineteenth century with the European discussion on how physics should be integrated into the educational systems (Weber 1937). From a general position of education, Gore (1995) stated that "Educating is naming, communicating, and upholding norms—norms of behaviour, of attitudes, of knowledge" (p. 172). As an important science, physics is needed to develop resources and to build a basis of social development—including economy (all kinds of engineering), sciences (IT, physics theories and methods), politics (IT, military) and medicine (medical devices and physics methods), communication (instant publishing and social networks) and many more—which should be discussed in classrooms with all inherent ethical implications (see Fig. 2.8; Pope 2017).

Beliefs and values of physics teachers relevant for teaching refer to epistemology and teaching and learning. Unfortunately, in contrary to many school curricula, most of the curricula in teacher education do not explicitly thematize beliefs and values, which often remain as implicitly learned naive personal theories. These beliefs and

Fig. 2.8 Nothing would work without physics

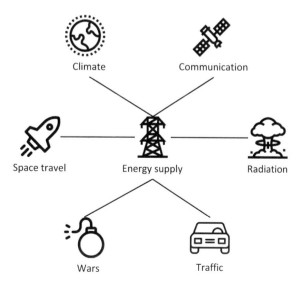

values are developed based on assumptions and individual experiences of students and teachers (in school situations and at university) as well as on unconsciously justified and characterized teachers' activities at school (Fischler 1989, 1994). Partly personal theories refer to the teachers' experience as a student at school and they inevitably do not contain validated speculations. Personal theories are very stable and difficult to change (van den Bogaart et al. 2018).

2.6.3 Teaching, Personal Theories and Reflection

Teaching at schools demands aim orientation and orientation on learning processes (see Chap. 4). In *planning processes*, teachers *operationalize* teaching aims based on curricula and often developed their plans as teaching sequences or units. The past interactions between teachers and students are explicitly or implicitly part of the planning process and their interactions during a lesson are influencing the *teaching–learning process* when the unit is taught. Structuring and performing a lesson is a very complex and difficult enterprise. Teachers are following the plans consciously but often need spontaneous reactions to students' activities in the lesson. Operationalizing means to perform teachers' own activities in the classroom based on theory and reflection and the best case would be to analyse a teaching situation before acting using academic and practice-approved professional knowledge (see Sect. 2.1). However, in addition to professional knowledge, classroom activities are also influenced by less reflective teaching activities and less mindful motivation orientations, self-efficacy and self-regulation abilities. This conglomerate appears to be successful in everyday (professional) life and it is very conservative (Thibaut et al. 2019). Therefore, professional competencies include teachers' ability to identify (not to validate) their personal theories and call them into question. To describe the process of how teachers perform their planned teaching and respond to students' activities in classroom situations, the *PID model* of Blömeke et al. (2015) can be a guideline for critical reflection. For example, in a video analysis of teachers' own teaching, the first step can be to control their teaching and to check if all particular events in an instructional setting have been *perceived* (P), if all perceived activities in the classroom have been adequately *interpreted* (I) and if the response to students' activities was adequately anticipated and led to an adequate *decision* (D). The PID model can help teachers to find alternative instructional strategies (Fig. 2.9).

Zlatkin-Troitschanskaia et al. (2019) pointed out that teachers' reactions in classroom situations not only depend on their knowledge, motivation and skills but also on the type of situation. Teacher-centred classroom situations are to be assessed differently compared to student-centred learning situations with more individualized parts, in which the teacher is more likely to be a learning assistant. Skills must be action related for situations in direct contact during instruction and reflective for planning lessons and their own activities and action-related decisions can be categorized as immediate and reflective. Immediate decisions are distinguished from reflective

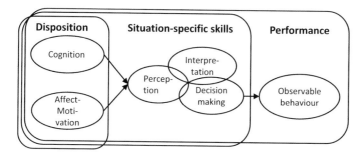

Fig. 2.9 Decision-making in classroom situations (Blömeke et al. 2015, p. 7)

ones as being consciously aware of the classroom situation. Dann and Haag (2017) described those decisions as:

- genuinely goal driven (planned, conscious, regulating and hierarchical),
- routine activities (high speed, not conscious, relief of cognitive load) and
- activities under pressure (fast decision required, mostly contradicting to teaching goals).

In the best case during teaching practice, decisions and the ensuing reactions become routines. All the other decisions are uncontrolled and sensitive to errors. For developing practicable and sound reactions in classroom teaching, teachers should be reflective, if possible, working together with experienced colleagues (Dann and Haag 2017).

Summary

The concept of competence is central to empirical studies dealing with the development of human resources and the productivity of education. Professional competences of physics teachers are needed when basic concepts and bodies of knowledge of physics and related sciences are applied for giving students' learning processes an optimal basis for reaching planned goals. As we argue in this chapter, physics teachers plan, conduct and reflect on their teaching, as well as, develop their professional competencies consciously. They are lifelong learners and actors in education as reflective practitioners (Schön 1987). The basis for their planning, conduction and reflection consists of their knowledge—about physics, physics education, psychology of teaching and learning and instructional design—and their self-reflective abilities to balance their own resources for maximum resilience in teaching. The basis for their lifelong learning and professional development consists of self-regulation and reflection on their beliefs and values regarding learning, the subject of physics and its role in society. For mastering the demands of their profession, physics teachers need a professional community of other teachers and awareness of their (mental and physical) health and needs, as well as their will, mental flexibility and cognitive ability for constant adaptation of their behaviour in communicative social situations. Only an academic education of physics teachers might guarantee that they can manage these complex problems successfully for a long time in their teaching professional career.

Acknowledgements We would like to thank Isabell van Ackeren (University Duisburg-Essen, Germany), who carefully and critically reviewed this chapter.

References

Achurra C, Villardón L (2012) Teacher' self-efficacy and student learning. Eur J Soc Behav Sci 2:366–383. https://doi.org/10.15405/FutureAcademy/ejsbs(2301-2218).2012.2.17

Aikenhead GS (2003) Chemistry and physics instruction: integration, ideologies, and choices. Chem Educ Res Pract 4(2):115–130.

Arndt J, Schüler A, Scheiter K (2019) Investigating the influence of simultaneous versus sequential–text-picture presentation on text-picture integration. J Exp Educ 87(1):116–127. https://doi.org/10.1080/00220973.2017.1363690

Augustsson G, Boström L (2016) Teachers' leadership in the didactic room: a systematic literature review of international research. Acda Didactica Norge 10. https://doi.org/10.5617/adno.2883

Ausubel DP (1963) The psychology of meaningful verbal learning. Grune and Stratton, Oxford, England

Ball DL, Lubienski ST, Mewborn DS (2001) Research on teaching mathematics: the unsolved problem of teachers' mathematical knowledge. In: Richardson V (ed) Handbook of research on teaching, 4th edn. Macmillan, New York, NY, pp 433–456

Barenthien J, Oppermann E, Anders Y, Steffensky M (2020) Preschool teachers' learning opportunities in their initial teacher education and in-service professional development—do they have an influence on preschool teachers' science-specific professional knowledge and motivation? Int J Sci Educ 42(5):744–763. https://doi.org/10.1080/09500693.2020.1727586

Baumert J, Kunter M (2013) The COACTIV model of teachers' professional competence. In: Kunter M, Baumert J, Blum W, Klusmann U, Krauss S, Neubrand M (eds) Cognitive activation in the mathematics classroom and professional competence of teachers: results from the COACTIV project (mathematics teacher education no. 8). Springer, New York, pp 25–48

Bernholt A, Lindfors M, Winberg M (2019) Students' epistemic beliefs in sweden and germany and their interrelations with classroom characteristics. Scand J Educ Res 1–17. https://doi.org/10.1080/00313831.2019.1651763

Bigozzi L, Tarchi C, Fiorentini C, Falsini P, Stefanelli F (2018) The influence of teaching approach on students' conceptual learning in physics. Front Psychol 9(2474). https://doi.org/10.3389/fpsyg.2018.02474

Blömeke S, Gustafsson J-E, Shavelson RJ (2015) Beyond dichotomies: competence viewed as a continuum. Zeitschrift Für Psychologie 223(1):3–13. https://doi.org/10.1027/2151-2604/a000194

Caleon IS, Tan YSM, Cho YH (2018) Does teaching experience matter? The beliefs and practices of beginning and experienced physics teachers. Res Sci Educ 48(1):117–149. https://doi.org/10.1007/s11165-016-9562-6

Cauet E, Liepertz S, Borowski A, Fischer HE (2015) Does it matter what we measure? Domain-specific professional knowledge of physics teachers. Schweizerische Zeitschrift Für Bildungswissenschaften 37(3):462–479

Chatman JA (1989) Improving interactional organizational research: a model of person–organization fit. Acad Manag Rev 14:333–349

Chen L, Xu S, Xiao H, Zhou S (2019) Variations in students' epistemological beliefs towards physics learning across majors, genders, and university tiers. Phys Rev Phys Educ Res 15(1):010106. https://doi.org/10.1103/PhysRevPhysEducRes.15.010106

Collins H, Evans R, Durant D, Weinel M (2020) How does science fit into society? The fractal model experts and the will of the people: society, populism and science. Springer International Publishing, Cham, pp 63–88

Crisci A, Sepe E, Malafronte P (2019) What influences teachers' job satisfaction and how to improve, develop and reorganize the school activities associated with them. Qual Quant 53(5):2403–2419. https://doi.org/10.1007/s11135-018-0749-y

Dann H-D, Haag L (2017) Lehrerkognitionen und Handlungsentscheidungen [Cognition of teachers and the infleuce on decisions for acting]. In: Schweer MKW (ed) Lehrer-Schüler-Interaktion: Inhaltsfelder, Forschungsperspektiven und methodische Zugänge [English book title?]. Springer Fachmedien, Wiesbaden, pp 89–120

Darling-Hammond L (2006) Constructing 21st-century teacher education. J Teach Educ 57(3):300–314. https://doi.org/10.1177/0022487105285962

Dewey J (1922) Democracy and education. An introduction to the philosophy of education. The Macmillan Company, New York

Ding K, Rohlfs C, Spinath B (2019) Selbstwirksamkeitserwartungen von Lehramtsstudierenden: Vorhersage von Veränderungen während des Semesterpraktikums durch Attributionsmuster und implizite Intelligenztheorien [Preservice teachers' self-efficacy: predicting changes over the internship period through attributional styles and implicit theories of intelligence]. Z Bild. https://doi.org/10.1007/s35834-019-00254-2

Eraut M (1994) Developing professional knowledge and competence. Routledge, London

Fend H (2001) Educational institutions and society. In: Smelser NJ, Baltes PB (eds) International encyclopedia of the social and behavioral sciences. Pergamon, Oxford, pp 4262–4266

Fischer HE, Horstendahl M (1997) Motivation and learning physics. Res Sci Educ 27(3):411–424. https://doi.org/10.1007/BF02461762

Fischer HE, Borowski A, Tepner O (2012) Professional knowledge of science teachers. Springer, New York

Fischer HE, Labudde P, Neumann K, Viiri J (eds) (2014) Quality of instruction in physics: comparing Finland, Germany and Switzerland. Münster Waxmann

Fischler H (1989) Orientations of the actions of physics teachers. Int J Sci Educ 11(2):185–193. https://doi.org/10.1080/0950069890110207

Fischler H (1994) Concerning the difference between intention and action: teachers' conceptions and actions in physics teaching. In: Carlgren I, Handal G, Vaage S (eds) Teachers' minds and actions: research on teachers' thinking and practice. London, Washington: Falmer, pp. 165–180

Friedrich H, Mandl H (2006) Lernstrategien: Zur Strukturierung des Forschungsfeldes [Learning strategies: to structure the research area]. In: Mandl H, Friedrich HF (eds) Handbuch Lernstrategien [Handbook learning strategies]. Hogrefe, Göttingen, pp 1–23

Gess-Newsome J, Taylor JA, Carlson J, Gardner AL, Wilson CD, Stuhlsatz MAM (2019) Teacher pedagogical content knowledge, practice, and student achievement. Int J Sci Educ 41(7):944–963. https://doi.org/10.1080/09500693.2016.1265158

Gidena A, Gebeyehu D (2017) The effectiveness of advance organiser model on students' academic achievement in learning work and energy. Int J Sci Educ 39(16):2226–2242. https://doi.org/10.1080/09500693.2017.1369600

Glutsch N, König J (2019) Pre-service teachers' motivations for choosing teaching as a career: does subject interest matter? J Educ Teach 45(5):494–510. https://doi.org/10.1080/02607476.2019.1674560

Gore JM (1995) On the continuity of power relations in pedagogy. Int Stud Sociol Educ 5(2):165–188. https://doi.org/10.1080/0962021950050203

Hansson L, Leden L (2016) Working with the nature of science in physics class: turning 'ordinary' classroom situations into nature of science learning situations. Phys Educ 51(5):055001. https://doi.org/10.1088/0031-9120/51/5/055001

Harding S, Morris R, Gunnell D, Ford T, Hollingworth W, Tilling K, Evans R, Bell S, Grey J, Brockman R, Campbell R, Araya R, Murphy S, Kidger J (2019) Is teachers' mental health and wellbeing associated with students' mental health and wellbeing? J Affect Disord 253:460–466. https://doi.org/10.1016/j.jad.2019.03.046

Hattie J (2012) Visible learning for teachers: maximizing impact on learning. Routledge/Taylor & Francis Group, New York and London

Herman KC, Hickmon-Rosa J, Reinke W (2017) Empirically derived profiles of teacher stress, burnout, self-efficacy, and coping and associated student outcomes. J Posit Behav Interv 20(2):90–100. https://doi.org/10.1177/1098300717732066

Hjörne E, Säljö R (2019) Diagnoses and their instructional implications-children's agency and participation in school activities. Emot Behav Diffic 24(3):219–223. https://doi.org/10.1080/136 32752.2019.1630999

Hooghe L, Marks G (2019) Grand theories of European integration in the twenty-first century. J Eur Publ Policy 26(8):1113–1133. https://doi.org/10.1080/13501763.2019.1569711

Kalender ZY, Marshman E, Schunn CD, Nokes-Malach TJ, Singh C (2019) Gendered patterns in the construction of physics identity from motivational factors. Phys Rev Phys Educ Res 15(2):020119. https://doi.org/10.1103/PhysRevPhysEducRes.15.020119

Kassis W, Graf U, Keller R, Ding K, Rohlfs C (2019) The role of received social support and self-efficacy for the satisfaction of basic psychological needs in teacher education. Eur J Teach Educ 42(3):391–409. https://doi.org/10.1080/02619768.2019.1576624

Kaub K, Karbach J, Spinath FM, Brünken R (2016) Person-job fit in the field of teacher education: an analysis of vocational interests and requirements among novice and professional science and language teachers. Teach Teach Educ 55:217–227. https://doi.org/10.1016/j.tate.2016.01.010

Kauertz A, Neumann K, Haertig H (2012) Competence in science education. In: Fraser BJ, Tobin K, McRobbie CJ (eds) Second international handbook of science education. Springer, Dordrecht, pp 711–721

Kauertz A, Ewerhardy A, Fischer HE, Kleickmann T, Lange K, Möller K, Ohle A (2010) Different perspectives on science teaching and learning in the transition from primary to secondary level. In: Taşar MF, Çakmakci G (eds) Contemporary science education research: teaching. Pegem Akademi, Ankara, Turkey, pp 419–436

Keller MM, Neumann K, Fischer HE (2017) The impact of physics teachers' pedagogical content knowledge and motivation on students' achievement and interest. J Res Sci Teach 54(5):586–614. https://doi.org/10.1002/tea.21378

Kieschke U, Schaarschmidt U (2008) Professional commitment and health among teachers in Germany: a typological approach. Learn Instr 18(5):429–437. https://doi.org/10.1016/j.learni nstruc.2008.06.005

Klieme E, Hartig J, Rauch D (2008) The concept of competence in educational contexts. In: Assessment of competencies in educational contexts

Kirschner PA, Sweller J, Clark RE (2006) Why minimal guidance during instruction does not work: an analysis of the failure of constructivist, discovery, problem-based, experiential, and inquiry-based teaching. Educ Psychol 41(2):75–86. https://doi.org/10.1207/s15326985ep4102_1

Klusmann U, Kunter M, Trautwein U, Lüdtke O, Baumert J (2008) Teachers' occupational well-being and quality of instruction: the important role of self-regulatory patterns. J Educ Psychol 100(3):702–715

Klusmann U, Richter D, Lüdtke O (2016) Teachers' emotional exhaustion is negatively related to students' achievement: evidence from a large-scale assessment study. J Educ Psychol 108. https://doi.org/10.1037/edu0000125

Koeppen K, Hartig J, Klieme E, Leutner D (2008) Current issues in competence modeling and assessment. Zeitschrift Fur Psychologie-J Psychol 216:61–73. https://doi.org/10.1027/0044-3409.216.2.61

König J, Pflanzl B (2016) Is teacher knowledge associated with performance? On the relationship between teachers' general pedagogical knowledge and instructional quality. Eur J Teach Educ 39(4):419–436. https://doi.org/10.1080/02619768.2016.1214128

König J, Blömeke S, Paine L, Schmidt WH, Hsieh F-J (2011) General pedagogical knowledge of future middle school teachers: on the complex ecology of teacher education in the United States, Germany, and Taiwan. J Teach Educ 62(2):188–201

Krauss S, Baumert J, Blum W (2008a) Secondary mathematics teachers' pedagogical content knowledge and content knowledge: validation of the COACTIV constructs. ZDM Math Educ 40(5):873–892. https://doi.org/10.1007/s11858-008-0141-9

Krauss S, Brunner M, Kunter M, Baumert J, Blum W, Neubrand M, Jordan A (2008b) Pedagogical content knowledge and content knowledge of secondary mathematics teachers. J Educ Psychol 100:716–725. https://doi.org/10.1037/0022-0663.100.3.716

Kulgemeyer C, Riese J (2018) From professional knowledge to professional performance: the impact of CK and PCK on teaching quality in explaining situations. J Res Sci Teach 55(10):1393–1418. https://doi.org/10.1002/tea.21457

Kunter M, Klusmann U, Baumert J, Richter D, Voss T, Hachfeld A (2013) Professional competence of teachers: effects on instructional quality and student development. J Educ Psychol 105(3):805–820

Kunter M, Kunina-Habenicht O, Holzberger D, Leutner D, Maurer C, Seidel T, Wolf K (2020) Putting educational knowledge of prospective teachers to the test: further development and validation of the bilwiss test, pp 9–28

Lee E, Luft J (2008) Experienced secondary science teachers' representation of pedagogical content knowledge. Int J Sci Educ 30:1343–1363. https://doi.org/10.1080/09500690802187058

Lehr D, Hillert A, Keller S (2009) What can balance the effort? Associations between effort-reward imbalance, overcommitment, and affective disorders in German teachers. Int J Occup Environ Health 15:374–384. https://doi.org/10.1179/107735209799160509

Leopold C, Doerner M, Leutner D, Dutke S (2015) Effects of strategy instructions on learning from text and pictures. Instr Sci 43(3):345–364. https://doi.org/10.1007/s11251-014-9336-3

Liepertz S, Borowski A (2019) Testing the consensus model: relationships among physics teachers' professional knowledge, interconnectedness of content structure and student achievement. Int J Sci Educ 41(7):890–910. https://doi.org/10.1080/09500693.2018.1478165

De Lissovoy N (2010) Decolonial pedagogy and the ethics of the global. Disc Stud Cult Polit Educ 31(3):279–293. https://doi.org/10.1080/01596301003786886

Mandalaki E, Islam G, Lagowskac U, Tobaced C (2019) Identifying with how we are, fitting with what we do: personality and dangerousness at work as moderators of identification and person–organization fit effects. Eur J Psychol 15(2):380–403

Maslach C, Jackson S, Leiter M (1997) The Maslach burnout inventory manual. In: Zalaquett CP, Wood RJ (eds) Evaluating stress: a book of resources, vol 3. The Scarecrow Press, Lanhm, MD, pp 191–218

Massolt J, Borowski A (2020) Perceived relevance of university physics problems by pre-service physics teachers: personal constructs. Int J Sci Educ 42(2):167–189. https://doi.org/10.1080/09500693.2019.1705424

Morus IR (2005) When physics became king. The University of Chicago Press, Chicago and London

Moyer AE (1982) John Dewey on physics teaching. Phys Teach 20(173)

Neumann K, Kind V, Harms U (2019) Probing the amalgam: the relationship between science teachers' content, pedagogical and pedagogical content knowledge. Int J Sci Educ 41(7):847–861. https://doi.org/10.1080/09500693.2018.1497217

OECD (2019) Education at a Glance 2019: OCED indicators. OECD Publishing, Paris. https://www.oecd-ilibrary.org/content/publication/f8d7880d-en

Ohle A, Boone WJ, Fischer HE (2015) Investigating the impact of teachers' physics CK on students outcomes. Int J Sci Math Educ 13(6):1211–1233. https://doi.org/10.1007/s10763-014-9547-8

Ornek F, Robinson WR, Haugan MP (2008) What makes physics difficult? Int J Environ Sci Educ 3(1):30–34

Oser FK (1994) Moral perspectives on teaching. Rev Res Educ 20:57–127

Patrick H, Pintrich PR (2001) Conceptual change in teachers' intuitive conceptions of learning, motivation, and instruction: the role of motivational and epistemological beliefs. In: Torff B,

Sternberg RJ (eds) Understanding and teaching the intuitive mind. Lawrence Erlbaum, Hillsdale, NJ, pp 117–143

Pintrich PR (1999) The role of motivation in promoting and sustaining self-regulated learning. Int J Educ Res 31(6):459–470. https://doi.org/10.1016/S0883-0355(99)00015-4

Pope TC (2017) Socioscientific issues: a framework for teaching ethics through controversial issues in science. TEACH J Christ Educ 11(2)

Rheinberg F, Vollmeyer R, Burns B (2000) Motivation and self-regulated learning. Pattern Recogn PR 131:81–108. https://doi.org/10.1016/S0166-4115(00)80007-2

Rothland M (2019) Belastung, Beanspruchung und Gesundheit im Lehrerberuf [burden, stress and health in the teacher profession]. In: Gläser-Zikuda M, Harring M, Rohlfs C (eds) Handbuch Schulpädagogik. Münster Waxmann, pp 631–641

Salta K, Koulougliotis D (2020) Domain specificity of motivation: chemistry and physics learning among undergraduate students of three academic majors. Int J Sci Educ 42(2):253–270. https://doi.org/10.1080/09500693.2019.1708511

Scheuch K, Haufe E, Seibt R (2015) Teachers' health. Dtsch Arztebl Int 112(20):347–356. https://doi.org/10.3238/arztebl.2015.0347

Schiefele U (2017) Classroom management and mastery-oriented instruction as mediators of the effects of teacher motivation on student motivation. Teach Teach Educ 64:115–126. https://doi.org/10.1016/j.tate.2017.02.004

Schön DA (1987) Educating the reflective practitioner: toward a new design for teaching and learning in the professions. Jossey-Bass, San Francisco

Schraw G (2001) Current themes and future directions in epistemological research: a commentary. Educ Psychol Rev 14:451–464

Schraw G, Crippen KJ, Hartley K (2006) Promoting self-regulation in science education: Metacognition as part of a broader perspective on learning. Res Sci Educ 36:111–139

Seeley L, Vokos S, Etkina E (2019) Examining physics teacher understanding of systems and the role it plays in supporting student energy reasoning. Am J Phys 87(7):510–519. https://doi.org/10.1119/1.5110663

Seidel T, Shavelson RJ (2007) Teaching effectiveness research in the past decade: the role of theory and research design in disentangling meta-analysis results. Rev Educ Res 77(4):454–499. https://doi.org/10.3102/0034654307310317

Shulman L (1987) Knowledge and teaching: foundations of the new reform. Harv Educ Rev 57:1–22

Smale-Jacobse AE, Meijer A, Helms-Lorenz M, Maulana R (2019) Differentiated instruction in secondary education: a systematic review of research evidence. Front Psychol 10:2366–2366. https://doi.org/10.3389/fpsyg.2019.02366

Smith MK (1996/2005) Competence and competencies. In: The encyclopedia of pedagogy and informal education. http://infed.org/mobi/what-iscompetence-and-competency/

Sonmark K, Révai N, Gottschalk F, Deligiannidi K, Burns T (2017) Understanding teachers' pedagogical knowledge: report on an international pilot study. OECD iLibrary, Paris. https://www.oecd-ilibrary.org/content/paper/43332ebd-en

Sorge S, Keller MM, Neumann K, Möller J (2019) Investigating the relationship between pre-service physics teachers' professional knowledge, self-concept, and interest. J Res Sci Teach 56(7):937–955. https://doi.org/10.1002/tea.21534

Storch M (2004) Resource-activating self-management with the Zurich Resource Model (ZRM®). Eur Psycho 1:27–64

Suárez MI, Wright KB (2019) Investigating school climate and school leadership factors that impact secondary STEM teacher retention. J STEM Educ Res 2(1):55–74. https://doi.org/10.1007/s41979-019-00012-z

Terhart E (2011) Lehrerberuf und Professionalität. Gewandeltes Begriffsverständnis - neue Herausforderungen Pädagogische Professionalität, vol 57. Beltz, Weinheim u.a., pp 202–224

Thibaut L, Knipprath H, Dehaene W, Depaepe F (2019) Teachers' attitudes toward teaching integrated STEM: the impact of personal background characteristics and school context. Int J Sci Math Educ 17(5):987–1007. https://doi.org/10.1007/s10763-018-9898-7

Toropova A, Johansson S, Myrberg E (2019) The role of teacher characteristics for student achievement in mathematics and student perceptions of instructional quality. Educ Inq 10(4):275–299. https://doi.org/10.1080/20004508.2019.1591844

van den Bogaart ACM, Hummel HGK, Kirschner PA (2018) Explicating development of personal professional theories from higher vocational education to beginning a professional career through computer-supported drawing of concept maps. Prof Dev Educ 44(2):287–301. https://doi.org/10.1080/19415257.2017.1288652

Veugelers W, Vedder P (2003) Values in teaching. Teach Teach Theory Pract 9(4):377–389. https://doi.org/10.1080/1354060032000097262

Weber LR (1937) Teaching physics with moral objectives. Christ Educ 20(5):350–354

Wilson CD, Borowski A, van Driel J (2019) Perspectives on the future of PCK research in science education and beyond. In: Hume A, Cooper R, Borowski A (eds) Repositioning pedagogical content knowledge in teachers' knowledge for teaching science. Springer, Singapore, pp 289–300

Winne PH (2015) Self-Regulated Learning. In: Wright JD (ed) International encyclopedia of the social and behavioral sciences, 2nd edn. Elsevier, Oxford, pp 535–540

Yerdelen-Damar S, Eryılmaz A (2019) Promoting conceptual understanding with explicit epistemic intervention in metacognitive instruction: interaction between the treatment and epistemic cognition. Res Sci Educ. https://doi.org/10.1007/s11165-018-9807-7

Zimmermann L, Unterbrink T, Pfeifer R, Wirsching M, Rose U, Stößel U, Nübling M, Buhl-Grießhaber V, Frommhold M, Schaarschmidt U, Bauer J (2012) Mental health and patterns of work-related coping behaviour in a German sample of student teachers: a cross-sectional study. Int Arch Occup Environ Health 85(8):865–876. https://doi.org/10.1007/s00420-011-0731-7

Zlatkin-Troitschanskaia O, Kuhn C, Brückner S, Leighton JP (2019) Evaluating a technology-based assessment (TBA) to measure teachers' action-related and reflective skills. Int J Test 19(2):148–171. https://doi.org/10.1080/15305058.2019.1586377

Chapter 3
How to Teach a Teacher: Challenges and Opportunities in Physics Teacher Education in Germany and the USA

Ben Van Dusen, Christoph Vogelsang, Joseph Taylor, and Eva Cauet

Abstract Preparing future physics teachers for the demanding nature of their profession is an important and complex endeavour. Teacher education systems must provide a structure for the coherent professional development of prospective teachers. Worldwide, physics teacher education is organised in different ways, but have to face similar challenges, like the relation between academic studies and practical preparation. In order to meet these challenges, it is worth taking look at different teacher education systems. In this chapter, we compare physics teacher education in two countries, representing two different educational traditions: Germany and the USA. Comparing different aspects of physics teacher education (standards, organisation and institutionalisation, content of teacher education, quality assurance), we describe both systems in their current state and why they are organised in the way they are. In doing so, we identify surprising commonalities but also different opportunities for both systems to learn from each other.

Introduction

This paper aims to inform physics teacher education by comparing two major traditions of physics teacher preparation. To compare different teacher education systems, it is necessary to understand how and why they were designed. Although every society has built unique educational systems grounded in a specific cultural context,

B. Van Dusen
School of Education, Iowa State University, Ames, IA, USA

C. Vogelsang
Paderborn Centre for Educational Research and Teacher Education—PLAZ-Professional School, University of Paderborn, Paderborn, Germany

J. Taylor
Department of Leadership, Research, and Foundations, University of Colorado, Colorado Springs, USA

E. Cauet (✉)
Institute for Natural Sciences Education, University of Koblenz-Landau, Landau, Germany
e-mail: cauet@uni-landau.de

© Springer Nature Switzerland AG 2021
H. E. Fischer and R. Girwidz (eds.), *Physics Education*, Challenges in Physics Education,
https://doi.org/10.1007/978-3-030-87391-2_3

55

scholars have identified two major educational traditions in Western education, distinguishing between two systems on a general level: an Anglo-American tradition and a Continental-European tradition. They represent specific sets of paradigms and philosophies of teaching and learning (Sjöström et al. 2017), influencing teachers' expected role and professional status and what and how teachers should learn during their preparation (Kansanen 2009).

The Anglo-American educational tradition is assumed to be a significant influence in the USA, UK, Australia and other primarily English-speaking countries. The US educational system is ostensibly designed to be a liberal education that provides students a "broad knowledge and transferable skills, and a strong sense of values, ethics, and civic engagement" (Association of American Colleges and Universities 2020). Examinations of US curricula, however have found its primary goal to be preparing K-12 students for the needs of society (Westbury 2000). In our understanding, a curriculum refers to a sequence of content and skills to be met in each year of instruction, often referred to as "standards" in the USA. For science teachers, this corresponds to promoting scientific literacy, which focuses on learning science concepts for later application and its usefulness in life and society (Roberts 2011). Exemplifying how the needs of society drive physics education in the USA is the fact that the last significant federal investment in physics education from the federal government was the Physical Science Study Committee in the late 1950s, during and in order to fight within the Cold War with Russia (Rudolph 2006). Similar calls for science education to meet the US workforce's needs can be seen in contemporary reports from the National Academies of Science (e.g. National Academies of Sciences, Engineering, and Medicine 2019; Committee on Prospering in the Global Economy of the 21st Century 2007; National Research Council 2010b; Committee on Underrepresented Groups and the Expansion of the Science and Engineering Workforce Pipeline 2011). Fensham (2009) describes the role of teachers in this tradition as "agents of the system" (p. 1082), who are responsible for meeting a set of standards given by an external authority. This perspective also influences how teacher education is designed. In a simplified way, Tahirsylaj et al. (2015) state that teacher education in this tradition emphasizes practical training. This emphasis is reflected in the curricular divide visible between foundations courses that cover topics such as learning theories and methods courses which cover the practical parts of teaching (Grossman et al. 2009).

The Continental-European educational tradition (Buchberger et al. 2000) is assumed to be a significant influence in Central and Northern Europe, especially in the German-speaking nations and Scandinavia. Central to this tradition is the concept of Bildung, a German word, that cannot be translated in English in one single term (Sjöström et al. 2017). Westbury (2000) provided one often-cited description:

"Bildung is a noun meaning something like, being educated, educatedness'. […] Bildung is thus best translated as 'formation,' implying both the forming of the personality into unity as well as the product of this formation and the particular 'formedness' that the person represents" (p. 24). Bildung frames the emancipation of an individual as the overall purpose of education. Following an influential concept, three abilities should be fostered by Bildung: self-determination (being able to determine one's own life and interpretations of meaning in interpersonal, professional and ethical areas), co-determination (being able to take part in the development of society) and solidarity (with other members of society, especially when whose opportunities for self- and co-determination are limited) (Fischler 2011).

Bildung refers to the overarching goal of education from childhood to adulthood. In this tradition, teachers at all levels are given a significant amount of autonomy and are expected to transform knowledge/content to contribute to this goal (Fensham 2009). Therefore, the task of teachers is to build a curriculum, roughly guided by rather brief standards, prescribed by an external authority and define the content and competencies students should learn in a subject. Within this tradition, teacher education emphasizes theoretical studies of education, structured in specific subdisciplines. Particularly, concepts dealing with the task of transforming subject matter content for learning are reflected in the subdiscipline of Fachdidaktik, in the case of physics called Physikdidaktik (see Fischler 2011 for a detailed description of these disciplines).

We must be aware that these traditions are products of complex historical and philosophical developments. They cannot fully represent the whole complexity and richness of one single teacher education system, as both traditions contain a considerable amount of simplification for sharp contrasting. In order to take a deeper look into the case of educating physics teachers in particular, we provide a comparative analysis of the teacher education systems for physics teachers in the USA and Germany. Both countries exemplify vital aspects of these traditions. Our analysis aims to identify the strengths and potentials of both systems—leaving the reader with a broader view of the different ways that successful physics teacher education can be established. Furthermore, we are interested in examining to what extent these reconstructions still apply to the practice in the systems in their current state.

Based on the work of Blömeke (2006), Darling-Hammond (2017) and Tahirsylaj et al. (2015), we developed a framework for comparing teacher education in both countries. We start with comparing the *standards for teacher education* in Germany (KMK 2019a, b) and the USA (NGSS Lead States 2013). Afterwards, we describe how the identified differences are manifested in the *organization of teacher education* for entry into the profession and ongoing professional development. At the heart of this chapter is the comparison of the *contents of physics teacher education* which reflects country-specific emphasis regarding different knowledge areas. We also discuss the role of theoretical education and practical preparation in the different systems and how these elements are linked to each other. Finally, we compare the *quality assurance and control* of physics teacher education in Germany and the USA.

3.1 Standards for Physics Teacher Education

What is the goal of physics teacher education? Teaching is a complex profession, and thus, teacher standards are an attempt to specify the competencies teachers need to acquire to be able to make sophisticated decisions multiple times a day about teaching and learning.

3.1.1 Standards for Physics Teacher Education in Germany

Germany's 16 states have a high degree of autonomy in educational politics. Each state independently defines the objectives of schooling and consequently determines the goals of teacher education. Despite the differences, there are many similarities between the states, and a lot is being done to harmonize systems while maintaining regional strengths.

The school system across the states can be characterized as follows. After completing a four to six-year elementary school beginning at the age of six, students attend a secondary school in one of three different tracks: One eight to nine-year academic track, the Gymnasium, leading to the highest possible school degree (*Abitur*), which allows students to attend university afterwards; one five-year track, the Realschule, for students seeking extended education, but do not wish to undertake an academic education; and one four-year track, the Hauptschule, focusing on preparing students for vocational training afterwards (i.e. learning a craft). However, many students switch tracks during their school career (e.g. to the upper classes of the Gymnasium after completing the Realschule). A number of states have begun to integrate Hauptschule and Realschule into one track and implement different forms of comprehensive schools, leading effectively to two-track school systems in some states. Future physics teachers are usually educated to teach in the Gymnasium and/or the combined tracks. Physics is taught as a mandatory subject (as well as chemistry and biology) at all secondary schools (cf. DPG 2016). In the first four to five years of some comprehensive schools, though science as a comprehensive subject is taught instead.

To enable a certain degree of comparability, teacher education programmes across the states are based on common standards formulated by the Standing Conference of the Ministers of Education and Cultural Affairs (KMK). They were first developed in 2004, partly due to international assessments such as TIMSS (Trends in International Mathematics and Science Study; https://timssandpirls.bc.edu) and PISA (Programme for International Student Assessment; https://www.oecd.org/pisa/). These standards exist for preparation in general educational sciences and also for each science subject. The standards for educational sciences (KMK 2019a) specify the professional competence teachers of all subjects and school types should achieve. They differentiate between four dimensions of competence: instruction, *Erziehen* (social and moral development and civil education), assessment and innovative development of schools. These dimensions formulate tasks that future teachers are expected to fulfil and cover a wide range of different topics (e.g. assessment approaches, teaching methods, socio-cultural influences on learning, etc.). Another set of standards (KMK 2019b) specifies the professional competence teachers should achieve for a specific subject. They also cover several aspects regarding physics education but are formulated relatively brief. For example, future physics teachers should have comprehensive content knowledge in physics, enabling them to design physics-related learning environments. This implicitly reflects the expectation that teachers have the task of selecting content for teaching. Also, the standards formulate a list of content of physics and physics education to be included in teacher education programmes.

Based on these brief standards, every institution for teacher education is responsible for designing its curriculum and specifying goals in detail, leading to various programmes differing between states and even between teacher education institutions within states. To ensure compliance with these standards, its programmes must be accredited by institutes, which, in turn, are also accredited by a statewide accreditation council (Neumann et al. 2017). In this process, teacher education programmes are also examined, whether they are in line with the Bologna agreement of the European Union, which aims at harmonizing the systems of higher education (Bauer et al. 2012).

3.1.2 Standards for Physics Teacher Education in the USA

As in Germany, each of the 50 states and the District of Columbia has its own independent status and local school systems that create what has been described as a "sprawling landscape" (National Academies for Science, Engineering, and Medicine 2020; Cochran-Smith et al. 2016). US elementary schools are often referred to as "K-5" schools, providing six years of schooling; kindergarten with students approximately at the age of five through 5th grade with students around ten years old. Students in grades six through eight (age 11–13) are typically schooled together in what is called "middle school". Finally, students usually attend a "high school" where they are educated with other 9th to 12th-grade students (age 14–17). Typical public US high schools are based on a liberal education model that eschews the orientation of having students locked into specialized career tracks in favour of requiring students to take a breadth of course work that prepares them for college or careers (Department of Education 2021). High school student graduation requirements are set by states but typically require 2–3 years of science classes that include at least one biological and one physics science course. A majority of high school students choose to take chemistry to fulfil their physics science courses. A 2013 survey found only 39% of students took any high school physics (White and Tesfaye 2014). Of those 39% of students, 65% took a non-college credit-earning physics course and 35% took physics courses that can lead to earning college credit (e.g. Advanced Placement, honours, or International Baccalaureate). While the goal of this system is to broadly educate students, the creation of remedial and advanced course tracks hinders these goals.

Each state has a teacher-certification department that sets the state's standards for kindergarten through 12th grade (K-12) education, teacher licensure and grants accreditation to pre-service teacher programmes. Most teacher credential programmes also receive external accreditation through a national organization called the Council for the Accreditation of Educator Preparation (CAEP 2020). While there is a national set of K-12 science education standards that states can choose to adapt (the Next Generation Science Standards; NGSS Lead States 2013), the effort to create a shared set of teacher licensure standards is not as well developed. The majority of states have content knowledge requirements measured through *Praxis* exams (Educational Testing Service 2020), although the specific scores and tests required vary by state. The education Teacher Performance Assessment (edTPA; Sato 2014) has also emerged as a more holistic assessment and support system, but it is still only used in a minority ($n = 18$) of states. The edTPA for secondary science teachers (middle and high school) assesses a host of teacher characteristics, including the following: the ability to plan effective instruction that is responsive to diverse student needs; knowledge of students prior conceptions and language demands; the ability to monitor learning and provide formative assessment feedback; the ability to engage students and design effective learning environments for deepening knowledge of concepts and processes of science; and the ability to analyse one's teaching effectiveness through the examination of student learning artefacts.

Each pre-service teacher programme determines its own set of objectives for its graduates based on the state teacher performance expectations and licensure requirements. While the state teacher performance standards vary by state, they share many attributes. California (the state that prepares the most teachers), for example, has six standards domains, each with a set of more specific substandards: (1) engaging and supporting all students in learning; (2) creating and maintaining effective environments for student learning; (3) understanding and organizing subject matter for student learning; (4) planning instruction and designing learning experiences for all students; (5) assessing student learning; and (6) developing as a professional educator (California Commission on Teacher Credentialing 2016). Even within states, however, teacher preparation programmes vary in the size, duration, curriculum and nature of field experiences (National Research Council 2010a).

3.2 Organization and Institutionalization of Teacher Education

Each teacher education system has specific pathways that prospective teachers typically have to follow to work in this profession. In this section, we describe these pathways for Germany and the USA.

3.2.1 How to Become a Physics Teacher in Germany?

The typical pathway leading to the teaching profession in Germany has a relatively stable structure, which is similar for all states (Cortina and Thames 2013). Teacher education is organized in two consecutive phases. The initial phase of preparation involves studies at a university followed by a structured induction to the field at a particular school in the second phase. Further professional development is considered as the third phase of teacher education, although this phase is not structured to the same extent as the first two phases.

Initial Preparation

In the first step, one must enrol in a teacher education study programme at a university. This requires a university entrance qualification, usually the *Abitur* (exceptions exist, e.g. for students from non-academic school tracks who completed vocational training and have some work experience). Teacher education programmes are aligned to the school tracks (e.g. you can study for teaching at Gymnasiums). Pre-service teachers have to study at least two subjects they later want to teach at schools, such as physics or chemistry. Each university is responsible for designing its curricula autonomously.

Since the Bologna agreement, pre-service teacher programmes in most states are organized in the Bachelor-Master-system. Students first have to obtain a Bachelor of Education (e.g. Bachelor of Arts or Science) before earning a Master of Education. Most students acquire both degrees at the same university since switching between universities is difficult and usually has disadvantages for the students because of the different curricula. Some states still organize their teacher education in the traditional study structure. Students obtain a *first state examination* at the end of their studies without a degree in between. However, the length and scope of studies are comparable in both systems. A master's degree or a first state examination is required to apply for the induction phase. Although the focus of the first phase is on the acquisition of theoretical professional knowledge, typically, several field experiences are integrated. The extent of these field experiences is usually defined by the legal requirements of the states, so the scope and location during a programme vary among universities (Gröschner et al. 2015). Many programmes include an initial orientation internship—typically a four-week school placement—in the first two semesters. Student teachers are supposed to reflect on their choice of teaching as a career, followed by one to two short school placements in the bachelor's programme. Eleven states have also implemented a one-semester internship at a school in their master's degree programmes (practical semester), accompanied by a shortening of the induction phase (Ulrich et al. 2020).

Induction Phase

After finishing their studies, future physics teachers are entitled to apply for in-service training in the second phase, lasting from one and a half to two years. Formally, they undergo their training at a seminar for teacher preparation. The states directly organize these seminars. Still, each seminar is responsible for designing their preparation programmes alongside the system of school tracks, which have to comply with the overall standards and state regulations.

Preparation takes place at two institutions. Most of the time, trainee teachers regularly teach a certain number of classes at a school. Experienced teachers from the same school usually mentor them, but they gradually teach more classes on their own during their traineeship in the majority of states. In most states, every teacher at a school is expected to be able to serve as a mentor. Very few states require them to undergo specific mentoring training. There is an ongoing discussion of whether and how mentors need to be trained on a mandatory basis (e.g. Weyland 2012). Mentors are expected to provide feedback on the trainee teacher's instruction and support them in lesson planning and reflection. Parallel to teacher training at a school, the trainee teachers attend courses at the seminar. Courses are taught by experienced teachers who have passed an examination to serve as teacher educators. Like their university studies, the trainee teachers take courses on subject matter education and general educational studies, focusing more on practical training than the theoretical focus of university studies. Courses are usually held one day a week.

During traineeship, teacher educators regularly observe lessons of their trainees to evaluate and provide feedback on their work. In many states, mentor teachers (and sometimes the principals of the schools) have to provide short written reports on the professional development of the trainee teachers. At the end of the induction phase, the trainees must undergo the second state examination. Elements of the test differ in detail from state to state, but typically the trainees are required to present one examination lesson in each subject and take an oral exam on the course contents. Some states also require a written thesis. Since much is at stake in this examination and the grades depend primarily on the examiners' subjective judgments, there is a constant criticism of too opaque grading criteria and unreliable assessment instruments (e.g. Strietholt and Terhart 2009). After completing a second state exam, the trainees are fully licensed teachers and can apply to ministries or private schools for an appointment.

Alternative Pathways

Like the USA, Germany has a significant shortage of qualified physics teachers, so many states provide alternative pathways for entry into the profession. The requirements for these pathways vary greatly and change from year to year, depending on the size of the shortage. Typically, two paths can be distinguished. In the first pathway, candidates with a master's degree of science related to physics (e.g. physics, engineering and architecture) can enter the induction phase. Some states require them to take a few courses in physics education and general educational studies at a university before or parallel to the induction phase. In the second pathway, teacher candidates with a master's degree are directly employed and work as teachers. They undergo on-the-job training to pass a second state examination. Sometimes, this training is parallel to the regular induction phase. Private schools are an exception, as they can decide on their staff independently. However, as private schools leading to secondary degrees are highly regulated and have to follow the same standards as public schools, they often hire teachers with state licensure.

In Germany, from 2002 to 2008, an average of 45% of new physics teachers entered the teaching profession following one of these alternative pathways, with an increasing proportion in the later years (DPG 2010; KMK 2020). As researchers are only granted access to this data upon request, more recent results are not available. Evaluations indicate that teachers entering the induction phase without a Master of Education achieve similar content knowledge and pedagogical content knowledge for teaching, but less pedagogical knowledge, at the end of the induction phase (Oetting-haus 2015). To mitigate potential negative impacts from increasing proportions of science teachers entering the profession on alternative pathways, several universities and organizations proposed providing additional support for these teacher candidates (DPG 2010).

Despite all these efforts, many teachers still have to teach out of field in many states to ensure there are enough physics teachers. Representative surveys reveal that roughly 6.5% of physics teachers teach out of their field (Stanat et al. 2019), with high variance between the states (between 1.7 and 17.9%) due to different and constantly changing entry requirements. This undermines the strategy of quality assurance through high entrance qualification.

3.2.2 How to Become a Physics Teacher in the USA?

There is no unified system for preparing physics teachers (Meltzer et al. 2012). The majority of physics teachers have neither a major nor a minor in physics (Banilower 2019; Meltzer et al. 2012). Further, the majority of physics teachers graduate from programmes in general education or science education programmes that do not offer any specialized instruction to prepare them for teaching physics. While 36% of physics departments report having a physics teacher education programme, barely half of them report graduating any majors (Meltzer et al. 2012). If physics teachers are primarily not coming from physics or physics teacher education programmes, where are they coming from?

Each year, about 3100 new high-school physics teachers enter the job market (White and Langer Tesfaye 2011). These 3100 physics teachers come from two sources: (1) in-service teachers who are transitioning to teaching physics ($n \sim 1700$) and (2) first-year teachers ($n \sim 1400$). The large number of in-service teachers transitioning to teaching physics reflects the severe shortage of physics teachers nationally. Many of these teachers transition from other science disciplines, while others are transitioning from unrelated disciplines. Both groups of teachers, however, are required to qualify for state teaching licensure.

There is a range of paths to earning state teaching licensure, and they vary by state. The paths are often described as either "traditional" or "alternative", but there is no commonly held agreement about how either of these categories is defined (National Academies of Sciences, Engineering, and Medicine 2020).

Traditional Preparation

The traditional pathway for licensure is typified by the requirement of completing a teacher preparation programme run by a university. These teacher preparation programmes are usually either a particular track within their undergraduate physics programme or a 1–2 year-long post-baccalaureate programme and offer students the opportunity to earn a master's degree in education while earning their licensure. Many states provide physics-specific teaching certifications, while others offer natural science (i.e. physics and chemistry) or general science certifications that allow teachers to teach any science discipline. Both types of certifications are likely to require some physics coursework to have been completed. Still, the requirements range from completing the introductory non-major sequence to completing several upper-division physics courses. While it is common for the teacher preparation programmes to offer science teaching methods courses, it is very uncommon for them to provide any courses specific to physics teaching preparation. This is likely due to the small number of pre-service physics teachers in any given degree programme, making it impractical to offer coursework for them.

A central feature of most traditional licensure programmes is an apprenticeship-based student teaching experience. Student teaching pairs pre-service teachers with one or more in-service teachers and provides an immersive teaching experience that ranges from weeks to months. Instead of offering traditional courses during student teaching terms, teacher educators from the pre-service teacher programmes typically observe their pre-service teachers in the classroom and provide them with feedback and support. These student teaching experiences are often the basis for capstone projects. Capstone projects are usually completed at the end of a licensure programme and are meant to develop and demonstrate the breadth and depth of student knowledge in the field. The projects often include an in-depth reflection on and assessment of their capstone teaching. There have been attempts to create multistate shared capstone expectations, such as edTPA (Sato 2014).

The final component of most traditional teacher preparation programmes is the completion of content-specific *Praxis* exams. The specific requirements and exams vary by state. States that offer physics-specific endorsements will typically require physics-specific examinations to be completed, while states that offer general science endorsements will typically require multidisciplinary examinations. Some programmes, however, have been accredited in ways that allow them to provide examination waivers if students complete a specific set of courses.

Each state offers a teaching license, but once a person has received licensure from one state, other states provide forms of reciprocity that facilitate the acquisition of state-specific licensures. State reciprocity programmes often require the passing of additional content assessments (Teacher certification degrees 2020). The only way to receive licensure in all 50 states is to earn a national board certification (Goldhaber and Anthony 2007). National board certification is only available to experienced teachers and employs a rigorous process of evaluating teacher quality that includes sharing and analysis of teaching videos.

Alternative Preparation

What qualifies as "alternative" preparation varies by state, and while there are exceptions, a common trait is that the programmes are not run by 2- or 4-year colleges. The acute lack of teachers in some disciplines (e.g. physics) and specific geographic regions has led many states to offer emergency credentials in high-need areas (Meltzer et al. 2012). Emergency credential standards vary, but they typically drop any post-baccalaureate programme requirements and focus on passing content assessments. Emergency credential programmes often lack any formal training to develop pedagogical knowledge or pedagogical content knowledge.

Some teachers skip the licensure and credential processes all together by working in private or charter schools. Private and charter schools are not bound by many state standards and often employ teachers who are not licensed. While most students attend public schools, 16% of the K-12 student population is enrolled in private or charter schools (Citylab 2014; In perspective 2018).

3.3 Ongoing Professional Development

Teachers, teacher educators, policymakers and researchers all agree that ongoing professional development (P.D.) is an essential and necessary part of being a teacher. In order to be able to orientate themselves within the different P.D. programmes, future physics teachers need to know the main characteristics of high-quality P.D. Darling-Hammond et al. (2017, p. 4) have identified seven criteria for effective professional development: "[Effective P.D.]

1. Is content focused
2. Incorporates active learning utilizing adult learning theory
3. Supports collaboration, typically in job-embedded contexts
4. Uses models and modelling of effective practice
5. Provides coaching and expert support
6. Offers opportunities for feedback and reflection
7. Is of sustained duration".

3.3.1 Ongoing Professional Development in Germany

In Germany, there are two kinds of professional development programmes: in-service training programmes aiming to preserve and improve teachers' professional competencies during their career and training programmes that are required to apply for specific positions (e.g. headmaster positions, teacher educators in the induction phase) (Eurydice 2003). All states in Germany require their teachers to engage in professional development. While some expect their teachers to fulfil their obligations within their course free time, others count at least a portion of the invested time as part of teachers' workload.

Terhart (2000) differentiates between supply-led and demand-led P.D. as well school-intern and school-extern P.D. Supply-led PD programmes are typically offered by school-extern institutions, mainly by the Landesinstitute (institutions responsible for quality assurance in a state-run school) but also by universities or organizations such as the German Physical Society. Teachers can individually decide whether to participate in these programmes if they are interested in the offered topics. School-extern P.D. is the most common type of such programmes in Germany. The purpose of these programmes is to engage teachers in content-specific learning processes absent of daily business. However, school-extern P.D. programmes are often responsible for implementing administrative reforms, which are increasingly based on plausibility rather than scientific evidence (Pasternack et al. 2017). More teachers prefer demand-led programmes, often realized within school-intern P.D. programmes, in which the staff of a school participates in so-called pedagogical days, conclaves or supervisions, independent of the question where these take place (e.g. school vs. extern venue) and who organizes and implements the events (e.g. collegium vs. extern referents) (Wenzel and Wesemann 1990). School-intern P.D. is mandatory in all states, but the specific obligations differ from state to state. These P.D. offerings usually concentrate on the particular school's needs (e.g. organizational development or teachers' professionalization; Deutscher Verein zur Förderung der Lehrerinnen und Lehrerfortbildung 2018).

Empirical data on German PD programmes is scarce—and for physics teachers in particular. In a nationwide survey of mathematics teachers and teachers of all science subjects in Germany (biology, chemistry and physics) (Richter et al. 2013), 85% of physics teachers reported to have participated in at least one P.D. within the last two years, while 15% of the teachers did not attend any P.D. programme (see also Stanat et al. 2019). The P.D. programmes which physics teachers (25%) most frequently attended focused on how to impart physics topics in a classroom setting (pedagogical content knowledge) followed by programmes on unspecific forms of teaching and methods (pedagogical knowledge) (20%). Teachers at the Gymnasium participate significantly more often in P.D., focusing on content knowledge or pedagogical content knowledge (see Sect. 3.5) than teachers of other school tracks. In contrast, the picture looks the opposite for pedagogical knowledge-related P.D. programmes. Teachers who did not participate in P.D. during the last two years reported organizational barriers (time conflicts 72%, difficulties in finding substitutes for their classes 53%), but 40–50% also reported little practical benefit or disappointing experiences from former P.D. participation (Krille 2020).

3.3.2 Ongoing Professional Development in the USA

In the USA, K-12 teachers have a vast array of professional development opportunities. While the extensive library of options is an inherently positive characteristic, such volume is often associated with a lack of systematicity and coherence (National Academies of Sciences, Engineering, and Medicine 2020). However, it is helpful to map the landscape of P.D. in the USA by describing themes in delivery formats, teachers' time participating, content foci and alignment of activities with principles of effective P.D. With regard to describing themes in professional development for science teachers and physics teachers, in particular, the most current and comprehensive source of empirical data is the nationally representative 2018 National Survey of Science and Mathematics Education (2018 NSSME+; Banilower et al. 2018). In the following section, we provide selected findings from this survey.

Among the many available delivery options, science teachers in the USA most often participate in P.D. via a workshop format. Subject-specific P.D. is often a part of what science teachers participate in, with approximately 80% of teachers in the sample participating in science-specific P.D. in the last three years. However, the quantity of P.D. was typically modest, with only about one-third of high school science teachers participating in more than 35 h of subject-specific professional development across those three years.

Science teacher respondents indicated that the alignment of their P.D. experiences with the elements of effective P.D. was moderate (average score of about 50 on a 100-point alignment scale) where the elements of effective P.D. included having teachers work with colleagues who face similar challenges, engaging teachers in investigations and examining student work/classroom artefacts. The results also indicate that just 63% of physics teachers participated in science-specific P.D. in the last year, and 85% had participated in such in the previous three years.

In terms of topical coverage, the survey collected teachers' ratings of the extent to which their P.D. offerings emphasized selected topics. The combined percentage of teachers rating each topic as a four or five on a five-point scale was used to rank the topics on perceived emphasis from those data. In terms of the most emphasized topics, the combined percentage of teachers giving a topic an emphasis score of four or greater was 54% for deepening understanding of how science is done, 43% for monitoring student understanding, 42% for developing science content knowledge, 38% for differentiating instruction and 38% for integrating STEM content.

The survey also provided data on how teachers were engaged in P.D. and what kind of learning opportunities teachers had during P.D. From those data, the combined percentage of teachers who gave their P.D. experience *an extent of opportunity* score of four or greater on a five-point scale was 51% for working with other teachers of the same subject or grade level, 49% for working with other teachers from the same school and 47% for engaging in scientific investigations or engineering design challenges.

3.4   Content of Teacher Education

The main goal of teacher education systems is to foster future teachers' development of professional knowledge and skills. There are many models of the professional knowledge base for teaching physics. We use the Refined Consensus Model of Pedagogical Content Knowledge (PCK) (Carlson and Daehler 2019) to give a comparative overview of the contents of teacher education in the USA and Germany. The model describes the interplay of different kinds of teacher knowledge. First, it distinguishes several professional knowledge bases: content knowledge, pedagogical knowledge, knowledge of students, curricular knowledge and assessment knowledge. These knowledge bases provide a foundation for teachers to develop their PCK, which, in short, describes the knowledge of how to teach a physics topic or concept so that their students develop an understanding of those concepts. A teacher's individual knowledge is referred to as personal PCK, whereas enacted PCK describes the "specific knowledge and skills utilised by an individual teacher in a particular setting" (Carlson and Daehler 2019, p. 84).

3.4.1   Content of Teacher Education in Germany

The goals of German teacher education programmes are often formulated alongside models of professional competence, which integrate models of professional knowledge (Baumert and Kunter 2013). From an overarching perspective, most programmes structure their curriculum into three knowledge areas: content knowledge of physics, knowledge in physics education (Fischler 2011) and general educational concepts. These areas can be found in all phases of German teacher education, but their scope and proportion change during the path of preparation.

In their university studies, future physics teachers mainly have to take courses focusing on physics content knowledge. In terms of the European Credit Transfer and Accumulation System (ECTS), a study programme for pre-service teachers has to cover 300 credit points. One point represents a study workload of 25–30 h. A typical study programme for physics teachers includes 90 ECTS-points for subject matter studies and 30 ECTS-points for studies in physics education (40 credit points accounts for general educational studies, 120 points for the second subject) (Deutsche Physikalische Gesellschaft 2016). Proportions differ between programmes focusing on different school tracks. Content courses typically reflect broad studies in various areas of physics as defined in the standards (KMK 2019b).

Regarding experimental physics, this includes lessons in mechanics, thermodynamics, electricity, optics, atoms and quantum physics (Neumann et al. 2017). The level and depth of studies also vary between the study programmes. Students studying for the tracks of Haupt- and Realschule also take basic classes on solid state, nuclear and particle physics. Students for the Gymnasium are expected to gain deeper knowledge in these areas. Regarding theoretical physics, students for Gymnasium have to participate in courses on theoretical mechanics, thermodynamics, electrodynamics and quantum mechanics. Students for Haupt- and Realschule are expected only to obtain a basic overview of the structure and main concepts of theoretical physics. All students have to take introductory laboratory work courses and courses on school-oriented experimentation; students for Gymnasium also take advanced laboratory work courses. Students for all tracks have courses on applied physics, leading to an overview of relevant topics for schooling (e.g. climate and weather, physics and sport). In addition, students are expected to learn aspects of the nature of physics. At most universities, pre-service teachers usually take courses together with students studying physics full-time. Still, most lecturers do not see themselves as teacher educators and thus do not prepare their courses for pre-service teachers primarily. Another critical factor is that problems concerning coping with content studies are among the main reasons for drop-outs (cf. Heublein and Schmelzer 2018).

Regarding physics education, courses cover physics education theories and conceptions, students' motivation and interest, learning processes, learning difficulties and students' conceptions of physics concepts, use of experiments, lesson planning and reflection on physics instruction, use of digital media in instruction, heterogeneity of students and topics of recent physics education research. The number of courses also varies depending on the focused school track. These courses contribute to the knowledge about students and provide collective PCK. Curricular knowledge and assessment knowledge are blind spots in German teacher study programmes, as they differ significantly in this respect. In addition, how courses are structured is highly variable between universities. In summary, the first phase focuses on learning theoretical professional knowledge and looks at physics instruction from the perspective of theory. In recent years, a lot of research was conducted to evaluate whether students acquire the knowledge as expected. Most studies found evidence for the positive development of content knowledge and PCK in general. In detail, differences in knowledge gains were identified regarding specific aspects, like significant differences in content knowledge and PCK between students studying for different tracks (e.g. Riese and Reinhold 2012).

The following induction phase focuses more on practical teacher training. The overall approach is similar to the concept of the reflective practitioner by Schön (1984). The trainee teachers are expected to apply their theoretical knowledge in actual classroom instruction and use it to reflect on their teaching (and the teaching of others). In terms of the model, future teachers develop mostly personal PCK and reflect on their enacted PCK in the induction phase. The content covered in the complementary courses during the traineeship contributes to this by focusing on practical issues on dealing with concrete, specific tasks and the trainees have to cope with at their schools. Many programmes also include curricular knowledge and knowledge of assessment, but usually with a strong focus on practical demands. In addition, course content on regulatory and school-law issues is part of the curriculum in most programmes. Similar to the first phase, content teaching varies significantly between different teacher preparation seminars. Compared to the first phase, there are fewer studies evaluating the effectiveness of the induction phase (e.g. Plöger et al. 2019), especially with a focus on physics.

In German teacher education, the theory–practice gap is a significant challenge. In physics education, research on the relationship between teachers' professional knowledge and teachers' performance shows inconclusive results (e.g. Vogelsang and Cauet 2017). Some studies in the field found little to no correlation between physics teachers' CK, PCK and the quality of their instruction or student achievement (e.g. Liepertz and Borowski 2019; Cauet et al. 2015). In instructional settings with reduced complexity, larger correlations were found, but only for prospective physics teachers in study programmes for teaching at Haupt- or Realschule (Korneck et al. 2017). In terms of the Consensus Model, only a few direct relationships between the knowledge bases or personal PCK and enacted PCK were observed. Implementing a one-semester internship (called the Praxis Semester) into master study programmes is a reaction to this and attempts to make more connections between theory and practice possible while students are still at university (Ulrich et al. 2020). The practical semester contains elements from the induction phase but also includes courses at university. This enables teacher educators from the first and the second phase to support their students in explicitly linking theory and practice when reflecting on their teaching together.

3.4.2 Content of Teacher Education in the USA

The content of teacher education ranges from traditional education programmes with comprehensive curricula to alternative education pathways with no pedagogical training. For this section, we will focus on traditional education programmes. Traditional teacher education programmes are post-baccalaureate programmes, which assume that their pre-service teachers have learned their content knowledge as part of their undergraduate coursework (National Academies of Sciences, Engineering, and Medicine 2019; Meltzer et al. 2012; Banilower 2019). That physics coursework might range from completing a pair of introductory physics courses to a traditional bachelor's degree in physics. While 36% of physics departments have a physics teacher education programme, barely half of those departments are actively graduating students (Meltzer et al. 2012). Nationally, about 270 students graduate from a physics teacher education undergraduate programme in either a physics department or a school of education each year (Meltzer et al. 2012). This means that only 8.7% of the 3100 first-year physics teachers earned a bachelor's degree from programmes that explicitly develop physics PCK. The lack of opportunities for pre-service physics teachers to develop PCK is a shortcoming of US teacher preparation programmes.

The post-baccalaureate teaching licensure programmes assume that students have sufficient content knowledge and focus on developing pedagogical knowledge and PCK. The coursework is a mix of graduate-level general education and science teaching courses. Typical course credit requirements total around 35 semester credits, with about one-third of those being science teaching-specific credits. Common general education course topics include educational theory (e.g. cognitive learning theory, behaviourism and constructivism), US educational history (e.g.

normal schools, school integration and the accountability movement), education law (Brown v. Board, compulsory education and teacher/student rights), educational technology (e.g. assistive technology, remote learning and asynchronous learning), educating diverse student populations and social justice (e.g. critical self-reflection and equitable pedagogical practices).

Science teaching courses are designed to develop general science PCK and are taken by a blend of pre-service teachers across the science disciplines. Despite being required to be taken for multiple terms in a programme, it is rare for a course to teach physics-specific PCK. The lack of physics PCK is likely due to the minimal number of pre-service physics teachers in a programme in any given year. Standard science teaching course topics include the nature of science, research on effective science teaching, creating equitable science learning outcomes, formative and summative assessment, fostering productive science talk and creating lesson plans that meet state science standards. These courses often use a book as a central organizing artefact (e.g. Ambitious Science Teaching; Windschitl et al. 2020).

Students' physics-specific PCK is primarily developed through apprenticeship as student teachers. Student teaching pairs pre-service teachers with in-service teachers, where they spend several months apprenticing in secondary school science courses. While student teaching, the pre-service teacher leads several units of instruction with the oversight and support of the in-service teacher and a university supervisor. It is common for pre-service teachers to lack coherence between their highly theoretical coursework and their real-world student teaching experiences (Zeichner 2010).

3.5 Quality Assurance and Control

Governments try to ensure the quality of their future teachers and their work using various strategies. Regarding the selection and recruitment of future teachers, governments can, for example, manage the total number of places available for teacher education students, try to influence the attractiveness and status of teaching as a profession and a career and specify the requirements and qualifications needed to enter the profession (Ingvarson and Rowley 2017).

Some of these quality assurance arrangements relate to teachers' working conditions, while others refer more to regulations of teacher education programmes, standards and the requirements for licensure. Many countries have also implemented quality control measures to improve the work of in-service teachers.

3.5.1   Quality Assurance and Control in German Teacher Education

The attractiveness of teaching as a career depends on various aspects such as status, working conditions and cost–benefit evaluations (study fees vs. expected salary scales). In Germany, teachers can generally have one of two different employment statuses. Most teachers are civil servants with lifetime tenure. They are working under a different regulatory framework (Eurydice 2003) and must follow a specific code of conduct/ethics. This status includes some privileges to have special health-care support and state-backed pension plans. However, civil servants are not allowed to go on strike and do not choose at which schools they are working. The minority of teachers have the status of employees, meaning they are employed on a contractual basis following general employment law. Most of them work under permanent contracts; temporary contracts are exceptions offered mainly to substitute teachers on sick or parental leave. Regardless of status, most teachers are employed directly by an individual state. For private schools, the respective school board is the employer. Overall, the job security of employed teachers is exceptionally high, resulting in a robust professional identity of teachers as officers of the state (Eurydice 2003). Salaries are based on collective bargaining, and teachers' salaries as employees are often a little lower than that of teachers as civil servants. However, the work requirements are the same.

Regardless of their status, teachers must teach 23–27 lessons (45 min) a week, depending on the school track and the state (KMK 2017).

How expensive is it to become a teacher in Germany? Students do not have to pay tuition fees at any public university in Germany—they simply have to cover their living expenses. In the induction phase, they even earn a reduced salary. Against the background of a general shortage of teachers, 15 out of 16 states employ new teachers as civil servants. The salary scales of secondary teachers, on average, are comparable to other professions requiring a master's degree in Germany (Ingvarson and Rowley 2017). However, once employed, there are few opportunities for promotion or different career paths; teachers can become principals (requiring further training), teacher educators, or take additional duties at school, for example, maintaining the computer laboratory to improve their pay grade. While teaching as a profession is quite attractive in Germany, there are still concerns that entrants in teacher education programmes are less qualified and, for example, have lower GPAs (*Abiturnote*) than entrants in other programmes. Still, studies show no evidence for this assumption in general (Henoch et al. 2015). However, regarding future physics teachers, students for the Gymnasium track have higher GPAs and begin their studies with a higher level of prior mathematics and physics knowledge than students for the other tracks (Riese and Reinhold 2012).

The governments of each state mainly regulate the number of places available for teacher education students. Because of the low number of enrolments, study programmes for physics teachers do not have to use any selection procedure. Dropout rates in study programmes in physics at German universities are relatively high, between 30 and 40% (Heublein and Schmelzer 2018).

As described in the previous section, following the ideal pathway, one must complete a long and highly structured qualification phase to be entitled to apply for an appointment as a teacher. However, once in-service little further professional development or certification is required. Physics teachers must obtain additional certificates for a few activities, such as being allowed to support students in conducting experiments with radioactivity in the classroom. Although teachers are obliged to engage in professional development, only 3 out of 16 states formulate verifiable criteria for professional development by quantifying the amount of training time they expect teachers to complete (12×5 h within four years, i.e. 15 h/year in Bavaria, 30 h/year in Hamburg and Bremen). Only nine states insist on explicit documentation (e.g. in a portfolio) of how teachers fulfil their P.D. obligations, and even fewer expect headmasters to use it as a base for individual career development during annual performance reviews (DVLfB 2018). For Germany as a whole, neither there are nationwide standards for assuring the quality of professional development programmes nor do national monitoring, evaluations or reporting exist to gather data for quality controls (DVLfB 2018). Even worse, only some states require governmental approval of P.D. offers—and in most cases, those are based on self-declarations of the P.D. suppliers and only require the adhesion of formal minimum standards (e.g. information on content and didactic and methodical design as well as on the acquirable competencies has to be provided, and programmes have to fit the school law and it aims) (Pasternack et al. 2017).

The responsibility for school regulation lies with the individual states. Following an evidence-based approach of school governance, all states have implemented some kind of school inspection as part of quality control measurements in recent years (Altrichter and Kemethofer 2016). The typical process is similar in all states. Each school is inspected once every several years (roughly five years). External inspectors visit the school for several days, observe classes, interview several stakeholders (principal, staff, parents, students), collect information on some school aspects provided by the school (e.g. management plans) and write an inspection report based on statewide standards. This report is given back to the school, and it is expected that the respective school will take it as a starting point for quality developments. On a national level, the Institute for Educational Quality Improvement (IQB) is responsible for providing the states with information for school development and monitoring the extent to which Germany's students are achieving educational standards. Therefore, the IQB carries out nationwide assessments based on representative samples of schools and students every one or two years. The IQB reports the results of these assessments to the states' governments, to the participating schools, and in some cases, to teachers and students. However, assessments regarding physics are seldom carried out (last in 2012 and 2018, e.g. Stanat et al. 2019), and these assessments have no direct consequences for individual teachers, neither positive nor negative.

3.5.2 Quality Assurance and Control in US Teacher Education

In the USA, states typically require teacher candidates to pass an assessment of relevant content knowledge, pedagogical knowledge and basic skills (e.g. reading, writing and mathematics). The *Praxis* tests (ETS 2020) are required for licensure in more than 40 US states, although the requirements for test content vary. For example, in some states, prospective physics teachers must pass a *Praxis General Science Content Knowledge* test, while in other states, passing the *Praxis Physics Content Knowledge* test is required. This variation is due in part to the fact that licensing specificity differs across states, some states offering only a secondary school science certification. In contrast, others certify prospective teachers specifically in physics or natural science.

Once a prospective teacher is hired for their first appointment in a public school, they usually become employees of a geographically defined school district. Still, they may draw retirement or other benefits through state-based programmes. In many states, teacher salaries are based on a collective bargaining agreement between the local teacher union and the school district board of directors. Collective bargaining agreements usually include salary schedules where one's salary is jointly determined by experience, degrees obtained and graduate or continuing education credits earned.

On the professional development front, newly hired physics teachers may have access to a teacher induction programme that includes being assigned a local mentor. According to Banilower et al. (2018), over two-thirds of US schools provide formal teacher induction programmes, with most lasting two or fewer years. When induction programmes exist, they are most often developed locally by the school or school district. This is somewhat consistent with previous work by Goldrick et al. (2012), who found that 27 states require some form of induction support for new teachers, with 11 requiring two or more years of support.

Twenty-two states were found to require completion of an induction programme for an advanced teaching certification. In the USA, some states offer two levels of certification, one is provisional/probationary that is concurrent with enrolment in an induction programme, and the second is received upon completion of an induction programme and an initial demonstration of teaching effectiveness. The second certification can also coincide with professional tenure. All individual states have ongoing professional development and license renewal requirements, which vary significantly across states.

Evaluation of teacher effectiveness for certification and other purposes is determined primarily by individual districts and states with some national input via the Every Student Succeeds Act (ESSA 2015), which provides high-level guidance that teacher effectiveness ratings be at least in part derived using evidence of student growth. The degree to which student growth on state tests is a factor in teacher evaluation and the choice of other evaluative factors is left entirely up to individual states and districts.

In 2019, only about one-half of US states required annual teacher evaluations (Ross and Walsh 2019). When teachers are evaluated, a primary source of evidence about their effectiveness, beyond student outcomes, is teaching observations. Citing research on the unreliability of a single observation for capturing a teacher's overall effectiveness, most US states require multiple observations of teachers within a given evaluation period. Any combination makes those observations of colleagues, school administrators, third-party evaluators and the teachers themselves.

Concerning the larger school context, the Every Student Succeeds Act (ESSA 2015) holds schools accountable for growth on several performance indicators: student achievement in mathematics and English/language arts, English proficiency for English language learners, graduation rates and school quality. While these five indicators are federally mandated, individual states may develop customized plans for using the indicators to identify schools in need of support and for correcting the course of low-performing schools.

3.6 Discussion

In this section, we provide summaries of the teacher education systems of Germany and the USA. We will compare these systems to identify each system's strengths and potential, offer insight into the different ways that physics teacher education can be designed, and how approaches borrowed from one country or tradition may help address the challenges of the other. Also, country comparisons can lead to a deeper knowledge of fundamental cultural concepts behind educational features (Blömeke and Paine 2008). Therefore, we chose to describe and compare these two countries, as they were identified by several scholars, using information available at the time, as representatives of two leading educational traditions (cf. Westbury 2000). In our comparison, we also try to identify how large the influence of these traditions is on both teacher education systems in their current state.

Looking into the school system in general, one significant difference between Germany and the USA is that physics is a mandatory subject for all students in secondary schools in Germany, compared to a system with more options for course choices in the USA. This can be traced back to the underlying concept of Bildung in the German educational tradition, in which it is assumed that every student should have a fundamental level of physics knowledge to become a self-determined citizen. This difference is also reflected in the teacher education system. Future teachers in Germany are prepared as teachers for physics. It is important to note that the German standards were formulated only recently (considering the long history of the German education system). They are rather influenced by results of large-scale international assessments, like PISA, than by the roots of Germany's educational tradition. This is an example of the adjustment of German teacher education to ideas based on developments from other educational systems.

Multidisciplinary science teachers are an exception in Germany. It is more common in the USA than in Germany for physics teachers to hold multidisciplinary science licensures rather than a physics-specific one. Both countries are similar in that the states have a significant amount of autonomy regarding educational policies. Therefore, both countries have a diversity of teacher education programmes throughout the country. However, in Germany, the states have agreed upon a set of standards for teacher education in general, particularly for physics. Hence, all teacher education programmes have to be designed to reflect those standards. In the USA, a shared set of standards has not been implemented to the same extent and is unlikely ever to be implemented because education is the purview of the states, and a unified approach is not promoted.

There are differences between Germany and the USA regarding the preparation programmes of future physics teachers. In the traditional pathway in Germany, future physics teachers are enrolling in teacher education programmes right from the beginning, developing PCK and PK. in their undergraduate studies. This corresponds to the underlying educational tradition, emphasizing theoretical studies in specialized subdisciplines like Physikdidaktik (Fischler 2011). Although US universities also offer undergraduate physics teacher preparation programmes, most pf them are post-baccalaureate and are not focused on physics. However, this makes it easier for students with science-related degrees to switch to the teaching profession. Also, CK-specific courses are less common in US post-baccalaureate programmes than in German teacher education programmes. The teaching experiences in US programmes are similar to the practical semester implemented in most states in Germany. They have elements similar to the German induction phase (like the capstone projects).

On the other hand, a structured induction phase organized by the states is a compulsory component in German preparation. In the USA, systemic induction phases are part of teacher preparation but are more locally based and mandatory in about half of the states. The overall length of teacher education in both countries, however, is relatively similar. In addition, the content of teacher education is reasonably similar in both, despite the differences between the educational traditions of the two systems. Differences lie in the emphasis on educational theories in the German academic part of teacher education, whereas US programmes often emphasize practical aspects of teacher preparation.

Both countries suffer from a shortage of physics teachers and offer alternative preparation. In all German states, the minimum requirement is to complete the induction phase for an alternative entry into the profession. Most US states offer alternative credential pathways that do not require any formal teacher training.

On the contrary, professional development could be regarded as the blind spot of the German teacher education system (DVLfB 2018). Compared to the USA, there are fewer options for professional development, and they are shorter in duration in Germany. In both countries, teachers prefer to attend subject-specific P.D. and requirements for teachers to attend P.D. are varying between the states. From an overall perspective, the German strategy is to ensure the quality of pre-service teacher education with high requirements for entrance into the profession. The US strategy is more focused on in-service professionalization, which is somewhat forgotten in Germany.

Although both countries operate under different educational traditions, there are many commonalities in practice. In recent decades, the requirements for adequate preparation of future physics teachers seem to have had more influence on teacher education than the educational tradition of each country. Most notable is that since the early 2000s, both countries are influenced by large-scale international assessments for student achievement and developed similar approaches to include the development of new standards. Both countries face similar challenges in physics teacher preparation. Germany and the USA each have to attend to a shortage of physics teachers and the small number of students enrolling in teacher preparation programmes. In both countries, a significant number of teachers teach physics out of field.

Despite all these commonalities, neither country seems to have found a comprehensive solution to these challenges. Therefore, a strategy for further research could be comparative analyses, examining physics teacher preparation in the high-achieving countries on assessments like PISA and TIMSS. It might also be promising to examine the differences between physics teacher education in Germany and the USA more closely than it was possible in this short chapter at the level of concrete programme designs. By looking at these details, other ideas and possibilities for improving one's own teacher preparation programmes could be gained.

Acknowledgements We would like to thank Janet Carlson (Stanford University) and Friederike Korneck (Goethe Universität Frankfurt) for helpful comments on the earlier draft of this chapter.

References

Altrichter H, Kemethofer D (2016) Stichwort: Schulinspektion [Keyword: School Inspection]. Z Erzieh 19(3):487–508

Association of American Colleges and Universities (2020) What is liberal education? https://www.aacu.org/resources/liberal-education

Banilower ER, Smith PS, Malzahn KA, Plumley CL, Gordon EM, Hayes ML (2018) Report of the 2018 NSSME+. Horizon Research, Inc

Banilower ER (2019) Understanding the big picture for science teacher education: The 2018 NSSME+

Bauer J, Diercks U, Rösler L, Möller J, Prenzel M (2012) Lehramtsstudium in Deutschland: Wie groß ist die strukturelle Vielfalt [Teacher education in Germany: how great is the structural variety?]. Unterrichtswissenschaft 40(2):101–120

Baumert J, Kunter M (2013) The COACTIV model of teachers' professional competence. In: Kunter M, Baumert J, Blum W, Klusmann U, Krauss S, Neubrand M (eds) Cognitive activation in the mathematics classroom and professional competence of teachers: results from the COACTIV project. Springer, Berlin, pp 25–48

Blömeke S, Paine L (2008) Getting the fish out of the water: considering benefits and problems of doing research on teacher education at an international level. Teach Teach Educ 24(8):2027–2037

Blömeke S (2006) Struktur der Lehrerausbildung im internationalen Vergleich. Ergebnisse einer Untersuchung zu acht Ländern [International comparison of the structure of teacher education. Results of a study on eight countries]. Zeitschrift für Pädagogik 52(3):393–416

Buchberger F, Campos B, Kallos D, Stephenson J (eds) (2000) Green paper on teacher education in Europe. Thematic Network of Teacher Education in Europe

California Commission on Teacher Credentialing (CCTC) (2016) California teaching performance expectations

Carlson J, Daehler KR (2019) The refined consensus model of pedagogical content knowledge in science education. In: Hume A, Cooper R, Borowski A (eds) Repositioning pedagogical content knowledge in teachers' knowledge for teaching science. Springer, Berlin, pp 77–92

Cauet E, Liepertz S, Borowski A, Fischer HE (2015) Does it matter what we measure? Domain-specific professional knowledge of physics teachers. Schweizerische Zeitschrift Für Bildungswissenschaften 37(3):462–479

Citylab (2014) Where private school enrollment is highest and lowest across the U.S. https://www.citylab.com/equity/2014/08/where-private-school-enrollment-is-highest-and-lowest-across-the-us/375993/

Cochran-Smith M, Ell F, Grudnoff L, Haigh M, Hill M, Ludlow L (2016) Initial teacher education: what does it take to put equity at the center? Teach Teach Educ 57:67–78

Committee on Prospering in the Global Economy of the 21st Century (2007) Rising above the gathering storm: energizing and employing America for a brighter economic future. National Academies Press

Committee on Underrepresented Groups and the Expansion of the Science and Engineering Workforce Pipeline (2011) Expanding underrepresented minority participation: America's science and technology talent at the crossroads. National Academies Press

Cortina KS, Thames MH (2013) Teacher education in Germany. In: Kunter M, Baumert J, Blum W, Klusmann U, Krauss S, Neubrand M (eds) Cognitive activation in the mathematics classroom and professional competence of teachers. Springer, Berlin, pp 49–62

Council for the Accreditation of Educator Preparation (CAEP) (2020) http://www.ncate.org/

Darling-Hammond L (2017) Teacher education around the world: what can we learn from international practice? Eur J Teach Educ 40(3):291–309

Darling-Hammond L, Hyler ME, Gardner M (2017) Effective teacher professional development. Learning Policy Institute

Department of Education (2021) College- and career-ready standards. https://www.ed.gov/k-12reforms/standards

Deutsche Physikalische Gesellschaft (DPG) [German Physical Society] (2010) Quereinsteiger in das Lehramt Physik - Lage und Perspektiven der Physiklehrerausbildung in Deutschland [Entering physics teaching from the side—situation and perspectives of physics teacher education in Germany]. DPG

DPG (2016) Physik in der Schule [physics in school]. DPG

Deutscher Verein zur Förderung der Lehrerinnen und Lehrerfortbildung e.V (DVLfB) [German Association for the Promotion of professional development in Teacher Education] (eds) (2018) Recherchen für eine Bestandsaufnahme der Lehrkräftefortbildung in Deutschland - Ergebnisse des Projektes Qualitätsentwicklung in der Lehrkräftefortbildung Teil 1 [Research for an evaluation of in-service teacher training in Germany—results of the project quality development in in-service teacher training Part 1]. Forum Lehrerfortbildung

Educational Testing Service (ETS) (2020) The Praxis tests. https://www.ets.org/praxis

ESSA, 20 U.S.C. § 6301 (2015) https://www.congress.gov/114/plaws/publ95/PLAW-114publ95.pdf

European Commission/Eurydice (2003) The teaching profession in Europe: profile, trends, and concerns. Eurydice Report. Publication Office of the European Union

Fensham PJ (2009) The link between policy and practice in science education: the role of research. Sci Educ 93(6):1076–1095

Fischler H (2011) Didaktik—an appropriate framework for the professional work of science teachers? In: Corrigan D, Dillon J, Gunstone R (eds) The professional knowledge base of science teaching. Springer, Berlin, pp 31–50

Floden R, Stephens A, Scherer L (2019) Changing expectations for the K-12 teacher workforce: policies, preservice education, professional development, and the workplace. Consensus Study Report. National Academies Press

Goldhaber D, Anthony E (2007) Can teacher quality be effectively assessed? National board certification as a signal of effective teaching. Rev Econ Stat 89(1):134–150

Goldrick L, Osta D, Barlin D, Burn J (2012) Review of state policies on teacher induction. New Teacher Center

Gröschner A, Müller K, Bauer J, Seidel T, Prenzel M, Kauper T, Möller J (2015) Praxisphasen in der Lehrerausbildung–eine Strukturanalyse am Beispiel des gymnasialen Lehramtsstudiums in Deutschland [Field experiences in teacher education–a structural analysis of secondary teacher education programmes in Germany]. Z Erzieh 18(4):639–665

Grossman P, Hammerness K, McDonald M (2009) Redefining teaching, re-imagining teacher education. Teach Teach Theory Pract 15(2):273–289

Henoch JR, Klusmann U, Lüdtke O, Trautwein U (2015) Who becomes a teacher? Challenging the "negative selection" hypothesis. Learn Instr 36:46–56

Heublein U, Schmelzer R (2018) Die Entwicklung der Studienabbruchquoten an den deutschen Hochschulen - Berechnungen auf Basis des Absolventenjahrgangs 2016 [Development of dropout-rates at German universities—analyses based on the 2016 graduate cohort]. HIS-Projektbericht

In perspective (2018) Key facts about charter schools. http://www.in-perspective.org/pages/introduction

Ingvarson L, Rowley G (2017) Quality assurance in teacher education and outcomes: a study of 17 countries. Educ Res 46(4):177–193

Kansanen P (2009) Subject-matter didactics as a central knowledge base for teachers, or should it be called pedagogical content knowledge? Pedagog Cult Soc 17(1):29–39

KMK (2017) Übersicht über die Pflichtstunden der Lehrkräfte an allgemeinbildenden und beruflichen Schulen [Overview of compulsory hours for teachers at general and vocational schools]. https://www.kmk.org/fileadmin/Dateien/pdf/Statistik/Dokumentationen/2019-09-16_Pflichtstunden_der_Lehrer_2019.pdf

KMK (2019a) Standards für die Lehrerbildung: Bildungswissenschaften - Beschluss der Kultusministerkonferenz vom 16.12.2004 i. d. F. vom 16.05.2019 [Standards for teacher education: educational sciences—resolution of the KMK of 16.12.2004 as amended on 16.05.2019]. https://www.kmk.org/fileadmin/veroeffentlichungen_beschluesse/2004/2004_12_16-Standards-Lehrerbildung-Bildungswissenschaften.pdf

KMK (2019b) Ländergemeinsame inhaltliche Anforderungen für die Fachwissenschaften und Fachdidaktiken in der Lehrerbildung. Beschluss der Kultusministerkonferenz vom 16.10.2008 i. d. F. vom 16.05.2019 [Common requirements for the content of studies of subjects and subject matter education in teacher education programmes - Resolution of the KMK of 16.10.2008 as amended on 16.05.2019]. https://www.kmk.org/fileadmin/Dateien/veroeffentlichungen_beschluesse/2008/2008_10_16-Fachprofile-Lehrerbildung.pdf

KMK (2020) Einstellung von Lehrkräften 2019 – Tabellenauszug [Recruitment of teachers 2019—table extract]. https://www.kmk.org/fileadmin/Dateien/pdf/Statistik/Dokumentationen/Tabellenauszug_EvL_2019.pdf

Korneck F, Krüger M, Szogs M (2017) Professionswissen, Lehrerüberzeugungen und Unterrichtsqualität angehender Physiklehrkräfte unterschiedlicher Schulformen [Professional knowledge, beliefs and instructional quality of prospective physics teachers in different school tracks]. In: Sumfleth E, Fischler H (eds) Professionelle Kompetenzen von Lehrkräften der Chemie und Physik [Professional competence of chemistry and physics teachers]. Logos, pp 113–134

Krille C (2020) Teachers' participation in professional development: a systematic review. Springer, Berlin

Liepertz S, Borowski A (2019) Testing the consensus model: relationships among physics teachers' professional knowledge, interconnectedness of content structure and student achievement. Int J Sci Educ 41(7):890–910

Meltzer DE, Plisch M, Vokos S (2012) Transforming the preparation of physics teachers: a call to action. Report: task force on teacher education in physics (T-TEP). American Physical Society

National Academies of Sciences, Engineering, and Medicine (2019) Science and engineering for grades 6–12: investigation and design at the center. National Academies Press

National Academies of Sciences, Engineering, and Medicine (2020) Changing expectations for the K–12 teachers: policies, preservice education, professional development, and the workplace. National Academies Press. https://doi.org/10.17226/25603

National Research Council (2010a) Preparing teachers: building evidence for sound policy. National Academies Press

National Research Council (2010b) Rising above the gathering storm, revisited: Rapidly approaching category 5. National Academies Press

Neumann K, Härtig H, Harms U, Parchmann I (2017) Science teacher preparation in Germany. In: Pedersen J, Isozaki T, Hirano T (eds) Model science teacher preparation programmes: an international comparison of what works. IAP, pp 29–52

NGSS Lead States (2013) Next generation science standards: for states, by states. National Academies Press

Oettinghaus L (2015) Lehrerüberzeugungen und physikbezogenes Professionswissen. Vergleich von Absolventinnen und Absolventen verschiedener Ausbildungswege im Physikreferendariat [Teacher beliefs and physics-related professional knowledge. Comparison of graduates of different academic pathways in the induction phase of physics teacher education]. Logos

Pasternack P, Baumgarth B, Burkhardt A, Paschke S, Thielemann N (2017) Drei Phasen: Die Debatte zur Qualitätsentwicklung in der Lehrer_innenbildung [Three phases: the debate on quality improvement in teacher education]. WBV

Plöger W, Scholl D, Schüle C, Seifert A (2019) Development of trainee teachers' analytical competence in their induction phase. A longitudinal study comparing science and non-science teachers. Teach Teach Educ 85:215–225

Richter D, Kuhl P, Haag N, Pant H-A (2013) Aspekte der Aus- und Fortbildung von Mathematik- und Naturwissenschaftslehrkräften im Ländervergleich [Aspects of initial and in-service training of mathematics and science teachers in cross-state comparison]. In: Pant H-A, Stanat P, Schroeders U, Roppelt A, Siegle T, Pöhlmann C (eds) IQB-Ländervergleich 2012 - Mathematische und naturwissenschaftliche Kompetenzen am Ende der Sekundarstufe I [IQB State Comparison 2012—mathematical and scientific competencies at the end of lower secondary school]. Waxmann, pp 367–390

Riese J, Reinhold P (2012) Die professionelle Kompetenz angehender Physiklehrkräfte in verschiedenen Ausbildungsformen [The professional competencies of trainee teachers in physics in different educational programmes]. Z Erzieh 15(1):111–143

Roberts DA (2011) Competing visions of scientific literacy: the influence of a science curriculum policy image. In: Linder C, Östman L, Roberts DA, Wickman P-O, Erickson G, MacKinnon A (eds) Exploring the landscape of scientific literacy. Routledge, pp 11–27

Ross E, Walsh K (2019) State of the States 2019: teacher and principal evaluation policy. National Council on Teacher Quality

Rudolph JL (2006) PSSC in historical context: science, national security, and American culture during the Cold War. American Association of Physics Teachers

Sato M (2014) What is the underlying conception of teaching of the edTPA? J Teach Educ 65(5):421–434

Schön DA (1984) Basic books. In: The reflective practitioner: how professionals think in action

Sjöström J, Frerichs N, Zuin VG, Eilks I (2017) Use of the concept of Bildung in the international science education literature, its potential, and implications for teaching and learning. Stud Sci Educ 53(2):165–192

Stanat P, Schipolowski S, Mahler N, Weirich S, Henschel S (eds) (2019) IQB-Bildungstrend 2018. Mathematische und naturwissenschaftliche Kompetenzen am Ende der Sekundarstufe I im zweiten Ländervergleich [IQB trends in student achievement 2018: mathematical and scientific literacy at the end of lower secondary education in the second country comparison]. Waxmann

Strietholt R, Terhart E (2009) Referendare beurteilen. Eine explorative Analyse von Beurteilungsinstrumenten in der Zweiten Phase der Lehrerbildung [Assessment of trainee teachers. An explorative analysis of assessment tools in the second phase of teacher education]. Zeitschrift für Pädagogik 55(4):622–645

Tahirsylaj A, Brezicha K, Ikoma S (2015) Unpacking teacher differences in Didaktik and curriculum traditions: trends from TIMSS 2003, 2007, and 2011. In: LeTendre GK, Wiseman A (eds) Promoting and sustaining a quality teacher workforce, vol 27. Emerald Group Publishing, pp 145–175

Teacher certification degrees (2020) The teacher certification reciprocity guide. https://www.teache rcertificationdegrees.com/reciprocity/

Terhart E (2000) Perspektiven der Lehrerbildung. Abschlussbericht der von der Kultusministerkonferenz eingesetzten Kommission [Perspectives on teacher education. Final report of the commission appointed by the conference of ministers of education and cultural affairs]. Beltz

Ulrich I, Klingebiel F, Bartels A, Staab R, Scherer S, Gröschner A (2020) Wie wirkt das Praxissemester im Lehramtsstudium auf Studierende? Ein systematischer Review [How does the practical semester in teacher education affect students? A systematic review]. In: Ulrich I, Gröschner A (eds) Praxissemester im Lehramtsstudium in Deutschland: Wirkungen auf Studierende [Practical semester in teacher training in Germany: impact on students]. Springer VS, pp 1–66

Vogelsang C, Cauet E (2017) Wie valide sind Professionswissenstests?: Zum Zusammenhang von erfasstem Wissen, Unterrichtshandeln und Unterrichtserfolg [How valid are tests for professional knowledge?: On the relationship between acquired knowledge, teaching practice and students achievement]. In: Fischler H, Sumfleth E (eds) Professionelle Kompetenz von Lehrkräften der Chemie und Physik [Professional competence of chemistry and physics teachers]. Logos, Berlin, pp 77–96

Wenzel H, Wesemann M (1990) Schulinterne Lehrerfortbildung. Begriffliche Klärungen, Abgrenzungen und Probleme [School internal professional development. Conceptual clarifications, delimitations and problems]. In: Wenzel H, Wesemann M, Bohnsack F (eds) Schulinterne Lehrerfortbildung. Ihr Beitrag zu schulischer Selbstentwicklung [Contribution of school internal professional development to school development]. Beltz, pp 14–40

Westbury I (2000) Teaching as a reflective practice: what might Didaktik teach curriculum. In: Westbury I, Hopmann S, Riquarts K (eds) Teaching as a reflective practice: the German Didaktik tradition. Routledge, pp 15–39

Weyland U (2012) Expertise zu den Praxisphasen in der Lehrerbildung in den Bundesländern [Expertise on the practical phases of teacher education in the federal states]. Hamburg

White S, Langer Tesfaye C (2011) Turnover among high school physics teachers, American Institute of Physics. http://www.aip.org/statistics/trends/reports/hsturnover.pdf

White S, Tesfaye CL (2014) High school physics courses and enrollments: results from the 2012–13 Nationwide Survey of High School Physics Teachers (American Institute of Physics, College Park, MD, 2014). https://www.aip.org/statistics/reports/high-school-physics-courses-enrollmen ts-0

Windschitl M, Thompson J, Braaten M (2020) Ambitious science teaching. Harvard Education Press

Zeichner K (2010) Rethinking the connections between campus courses and field experiences in college- and university-based teacher education. J Teach Educ 61(1–2):89–99

Chapter 4
Instructional Design

Heiko Krabbe and Hans E. Fischer

Abstract We present in this chapter an approach in which the design of physics lessons orients towards a framework of the surface and deep structure of physics teaching. In planning lessons and evaluating the quality teaching, we consider the learning goals as a decisive criterion. These goals are understood not only in terms of content in the sense of physics, but also in terms of research methodology in physics, epistemology and learning theory. According to empirically proven quality criteria for (physics) lessons, lesson design should be oriented towards learning processes (including learning theory and physics content) and cognitive activation, as well as being student-oriented and characterised by clear classroom management (clarity of rules and avoidance of disturbances). The practice of high-quality lessons should follow the overall goal of a conducive learning environment. These characteristics are part of the deep structure of a quality lesson, and they are described through theory-based analysis and reference to quality teaching. Teachers should follow the deep structure as closely as possible, but they are free to construct the surface structure according to the available equipment of the respective school. By surface structure, we mean the choice and organisation of teaching methods, media and social forms. The basic theoretical requirements are illustrated with lesson examples.

4.1 Design of Lessons

All teachers, preservice teachers and student teachers have already observed lessons with the aim of describing, commenting on and evaluating the interactions. However, lesson design is a difficult task that can only be solved very superficially without a scientific–theoretical analysis. There are many theoretical approaches and models with which observation and assessment can be purposefully practised and structured.

H. Krabbe (✉)
Ruhr-Universität Bochum, Bochum, Germany
e-mail: heiko.krabbe@rub.de

H. E. Fischer
Universität Duisburg-Essen, Essen, Germany
e-mail: hans.fischer@uni-due.de

© Springer Nature Switzerland AG 2021
H. E. Fischer and R. Girwidz (eds.), *Physics Education*, Challenges in Physics Education,
https://doi.org/10.1007/978-3-030-87391-2_4

In this chapter, we choose as a starting point the distinction between surface and deep structures of lessons (Aebli 1961; Lipowisky et al. 2018; Seidel 2003). Models based on this assumption have already been demonstrated in some studies when assessing quality teaching (Geller et al. 2014; Krabbe et al. 2015; Meyer 2016; Oser and Baeriswyl 2001).

The feature that all observers of a lesson can directly perceive is identified as the visual or surface structure of the lesson. The deep structure not only refers to *goals* that a teacher wants to achieve in teaching the students and eliciting their responses, but it also refers to *intentions* such as a constructive learning climate that is not always explicitly discernible. As already described in Chap. 1, the scientific view of learning has changed in recent decades from an information-theoretical perspective to a more cognitive model. Learning at school is now perceived on the level of students, teachers, school, family and society as an offer that learners can use with their opportunities for self-determined cognitive development (Fischer et al. 2005; Pauli and Reusser 2003). On the offer side, the professional knowledge of the teachers—on which the structuring of learning opportunities depends—plays an important role (see Chap. 1, Utilisation-of-learning-opportunities model by Helmke 2012). On the user side, that is, with the students, the most important parameters are cognitive abilities, prior knowledge, willingness and motivation to deal cognitively with the subject matter. For successful lessons, teachers must be able to use their professional knowledge in such a way that learners can use the content provided in the classroom to organise their learning process in the given time. In order to describe the design of lessons, characteristics on four different levels are defined in the following (cf. Kunter and Trautwein 2013, p. 62f):

1. **Forms of organisation**: Structural framework of teaching, such as classroom teaching, performance-based learning groups, supportive teaching and learning times (time on task).
2. **Instructional models**: Methodical large-scale forms of instructional planning and organisation such as direct instruction, research-discovery instruction, scientific inquiry, project work, classroom discussion and so on.
3. **Social forms**: Design of social interactions in the classroom, such as classroom teaching, group, partner and individual work.
4. **Teaching–learning processes**: Design of content-related interactions and learning climate by pre-structuring learning situations, cognitive activation, classroom management, constructive support, diagnostics and feedback.

Organisational forms of lessons, teaching methods and social forms of student activities belong to the surface structure, which is the subject of pedagogical and subject-specific education research. The teaching–learning processes—that are not directly observable, and therefore, the way in which learners deal with the learning content and how the people involved interact with each other—form the deep structure of the lesson. The deep structure is the subject of research in the field of psychology of learning and physics education (Meyer 2016).

Empirical research of classroom teaching showed that the surface structure and the deep structure of a lesson can be varied relatively independently of each other, but the surface structure supports the students' achievement of the intended learning outcomes of the deep structure of the lesson (Oser and Baeriswyl 2001). Empirical classroom research also showed that student learning success depends less on a lesson's surface structure but can be explained primarily by the quality of its deep structure, for example, the teaching methods and strategies that can be organised by the teacher as needed.

Treagust and Tsui (2014) gave an overview of six teaching methods, ordered from their being more to less teacher-centred: demonstrations, classroom explanations, questioning, representational learning, group and cooperative learning, and scientific reasoning and argumentation. Since Treagust and Tsui acknowledged to have a limited overview of the status of science education in non-English-speaking countries, our complementary insight into science education in Germany may be worthwhile here.

The IPN video study analysed physics lessons in 50 randomly selected ninth-grade school classes from four German states between 2002 and 2004 (Seidel et al. 2006). German physics lessons, like those in many other countries, proved to be predominantly teacher-centred with an inductive approach to science. The learning activities of the students consisted mainly of the receptive processing of physics concepts which were worked out in class discussion and only rarely made accessible through student experiments for deeper understanding. As a rule, the students followed the instructions of the teachers in the processing of tasks without the learning goals being made transparent to them and without a common thread being recognisable in the course of the lesson. It was also noticeable in the IPN video study that mistakes and misconceptions were not addressed in class, as student contributions merely served to provide keywords for the course of the discussion and feedback from the teacher was factual, constructive and conducive to learning in less than 13% of the cases. The characteristics of either teacher-centred or student-centred lessons proved in this study to be not significant for students' development of competences and interests in physics. In contrast, active participation of the students as equal partners in open classroom discussions had positive effects on their physics competences, especially among those with lower prior knowledge. Students with a high level of prior knowledge, on the other hand, felt that too close learning support hindered rather than encouraged their development. In a comparison of the lessons of six physics teachers, Fischer et al. (2002) described two different teaching patterns on the basis of the surface structure: teacher-centred instruction with demonstration experiments and student-centred work with experimental tasks. These two basic patterns of direct instruction and scientific inquiry learning can also be found in the German results of science teaching in PISA 2006 (Prenzel et al. 2007) and PISA 2015 (Schiepe-Tiska et al. 2016). German students were asked in these two PISA studies to rate the quality of their teachers' teaching on a four-level scale about how often they: (1) express their own ideas, (2) draw their own conclusions, (3) develop experiments themselves, (4) are allowed to carry out practical experiments and (5) relate to the everyday world.

In the German PISA 2015 study, four types of instruction were identified:

Type 1: Cognitively stimulating lessons, with autonomous experimenting by students, including explanations and conclusions (19%)
Type 2: Cognitively stimulating teaching with student experiments (13%)
Type 3: Average cognitively stimulating lessons with few experiments, explanations and conclusions (especially at upper secondary school, 54%)
Type 4: Less cognitively stimulating lessons without experiments (14%).

The German patterns of science teaching in PISA 2015 were found to be similar to those identified in PISA 2006 for all OECD countries (Prenzel et al. 2012), and a similar orchestration of science instruction was examined in an international comparison for PISA 2015 (Forbes et al. 2020). It seems that in many countries a more teacher-centred (Type 3) pattern of teaching still predominates, with a focus on conclusions and explanations of ideas and few students' experiments, even though scientific investigation and scientific inquiry is a central focus of science education reforms around the world. Looking at the secondary level in terms of scientific competence and interest, successful teaching is characterised by the combination of high cognitive activation through independent explanations and conclusions (minds-on) with regular experimental hands-on activities. In contrast, highly student-driven inquiry is least correlated with high levels of student science achievement. Latter finding indicates the important role of teacher guidance in science learning (Forbes et al. 2020).

Through a summary of many studies (meta-analysis), Hattie (2008) compiled a comprehensive collection of effective mechanisms (including methods, social forms, media, student/teacher behaviour, attitudes, teaching strategies and techniques) for successful learning, and ranked them hierarchically according to their effects on learning success. Hattie pointed out that teacher-oriented and teacher-centred teaching (direct instruction) is an effective teaching method which supports learning processes for the development of complex theoretical concepts better than group teaching with discovery learning. This is particularly plausible for physics concepts, which have usually been developed in difficult theoretical and experimental work by outstanding physicists. The independent development of physics concepts, such as Newton's concept of force or Maxwell's equations of electrodynamics, might be too demanding for the students. Therefore, concept development requires other social forms, methods and media for the organisation of the learning process than making one's own experiences with a physics phenomenon. Social forms and methods should be oriented towards the initiated learning process (deep structure) and of course be appropriately designed (surface structure). Poor direct teaching does not help in principle to achieve the planned learning goals, nor does inadequately designed experimentation—that needs a lot of teaching time with little effect—help to do so (Börlin 2012).

4.2 Social Forms, Methods and Media

According to Meyer (2002), social forms and methods used in class are " … the forms and methods by which teachers and students acquire the natural and social reality surrounding them, while considering the institutional framework of the school" (p. 109; translated by the authors). They are part of the surface structure, which must be adapted to the learning processes and to the respective conditions of the system (the school). This means that they must be planned with specific goals and times and their effects in the classroom must be controllable in an appropriate way.

In general education, one finds a wide variety of enumerations and systematisations of superficial teaching elements (e.g. Banner and Cannon 2008; Marzano 2007) of subject-related learning goals and contexts, and learning processes (see Chap. 1).

In addition, a group of methods and media are used in physics lessons that cannot be covered by general education and methodology, because they are determined by the subject itself. For the application of demonstration experiments and students' experiments in physics, it is most important that the students have actively learned to carry out the experiment in physics lessons and to assess the goals of experiments and the meaning of empirical investigations in physics before or while they are required to solve experimental tasks (see Chap. 10). Therefore, in some teaching phases of physics lessons, scientific work itself is a learning goal. There are also specific learning goals in physics teaching in relation to certain media. Many media can be used in any classroom, such as blackboards, whiteboards or tablets. However, some applications are only available in physics classes, such as tablets or PCs with applications for computer-based data collection, physics simulations or physics-related modelling. For lesson planning and design, this means that methods and media must partly be addressed at the surface structure level or they themselves become the learning goal, and therefore, must be planned and taught at the deep structure level, for example, in the design of a computer simulation to illustrate a functional relationship between physics meanings. In such cases, the application of simulation software or software for computer-based data acquisition becomes a learning goal (see Chap. 11).

4.3 Characteristics of Quality Teaching

Seidel and Shavelson (2007) found that the quality of teaching in terms of learning outcomes depends mainly on provision that is directly related to learning processes. This includes, among other things, the learning-process-oriented structuring of the lessons, the clearly presented content structure and the cognitive activation of the students.

Klieme and Rakoczy (2008) found efficient classroom management, teaching climate, supportive teacher behaviour and cognitive activation as categories of instructional quality by combining many other such characteristics. As a result, these categories are more clearly differentiated from one another than features that were derived individually and found to be unrelated (Praetorius et al. 2018).

Efficient classroom management is a necessary (but not sufficient) condition for good teaching. It is closely connected with the requirement to allow maximum learning time in the classroom (time on task). In this regard, Helmke and Helmke (2014) summarised classroom management as consisting of five components:

1. Obligatory agreements as rules and procedures.
2. Presence of the teacher through parallel control of several actions and situations (multitasking).
3. Effective use of teaching time (time on task), with an on-time start and end to a sequence of lessons and a rapid transition between teaching phases (e.g. from classroom teaching to group teaching).
4. Cognitive activation of the whole class (see below) in order to minimise the time that is left unused by inactive subgroups.
5. Sanctions for non-compliant behaviours and positive confirmations of compliant behaviours.

According to Wubbels et al. (2014) and Meyer and Bülter (2004), a positive and learning-supporting teaching climate consists of the following teacher behaviours and attitudes:

1. Mutual consideration and tolerance.
2. Responsible handling of persons and objects.
3. Satisfied and cheerful basic attitude.
4. Clearly structured guidance and direction by the teacher.
5. Politeness and mutual respect.
6. Support of self-esteem of each individual.
7. Willingness to cooperate of teacher and students.

Kunter and Trautwein (2013) characterised cognitively activating classroom management by the following recommendations for the teacher:

1. The teacher introduces the topic with exciting, demanding and challenging questions.
2. Opinions and answers to questions are explained and justified.
3. Different answers to questions and solutions are expected.
4. Inconsistencies in the argumentation are identified and discussed.
5. Different opinions are identified and discussed.
6. Asking questions and explaining among students is encouraged.
7. Feedback is given to encourage further work on the problem.

Meyer (2008) summarised ten characteristics as necessary boundary conditions for the planning and implementation of good instruction, which were modified by Helmke (2012) with results from his own and other empirical research. Accordingly, instruction should be conducted according to the following criteria:

1. Structuredness, clarity, comprehensibility.

2. Efficient classroom management and use of time.
3. Learner-friendly classroom climate.
4. Goal, effect and competence orientation.
5. Student-oriented support.
6. Appropriate variation of methods and social forms.
7. Encouraging active and independent learning.
8. Consolidation, backup, intelligent practice.
9. Various motivations.
10. Dealing with heterogeneous learning conditions.

According to Helmke (2012), the list of quality characteristics can be empirically confirmed, but must be constantly modified depending on new research results. It can therefore be summarised that classroom learning is mainly promoted by the following characteristics of the learning opportunities designed by the teacher:

1. A transparent lesson design allowing as much time on task as possible.
2. A friendly and open classroom atmosphere.
3. Cognitive activation of the students.
4. Communication in the classroom is oriented towards the students.

4.3.1 Articulation Schemes and Learning Process Orientation

In general terms, a lesson can be divided into stages (also phases or steps). In the German traditions of general education (see Arnold and Koch-Priewe 2011), an articulation scheme that was introduced by Roth (1983) comprises six learning steps, according to which, in his view, instruction can be organised. Each step includes a specific learning goal when students are asked to solve a task: (1) motivating to learn, (2) overcoming learning difficulties, (3) finding solutions to the task, (4) applying the solutions, (5) memorising and practising, and (6) transferring and integrating what has been learned. These steps are still used today in teacher education in order to offer a less complex way of structuring instruction (sometimes in modified formulations). They were normative in origin and the starting point for many very complex models, as described below.

However, the problem of structuring instruction cannot be answered independently of the subject-related goals of the instruction. The starting point for any instructional planning should therefore be a description based on learning theory and empirical evidence, of what the learners are supposed to have learned by the end of the instruction. It is not enough to state the subject-related goal, for example, that learners in an eighth-grade class should know and be able to apply Newton's laws after the lesson. It must also be planned whether at the end of the lesson the first experimental experiences with forces should be summarised (experiential learning) or whether the physics concept should be understood and problems or tasks should be solved with the concept (concept development). The goal can also be that the learners organise group work in physics exercises as social interactions (design of dynamic social relationships), through which they can look at the expected examination topic from a new, self-chosen perspective (hypertext learning) or that they practise applying a certain concept to solving corresponding tasks (routine formation by training of skills).

Furthermore, goals and teaching success cannot be assessed independently of empirical research on teaching (see Chaps. 1, 2, 16 and 17), which had not yet played a significant role in lesson planning and implementation during the time of Roth's (1983) study. The formulation of suitable teaching objectives on the basis of empirical research and professional criteria are a prerequisite for teaching quality. When determining the quality of teaching, many other variables like the students' performance, and their motivation, the professional knowledge of the teachers and their commitment must be taken into account. Measuring these variables, however, only permits statements of probability and not absolute statements about the quality of instruction. The more clearly the teaching objectives can be defined and expressed in the classroom for the students, the greater is the probability of achieving the average student's high performance, reducing differences in performance and creating motivation through a sense of success.

4.3.2   Instructional Design According to Gagné and Briggs

According to Reigeluth (1983), a theory of instructional design is the attempt to describe methods that best create the conditions under which the learning goal is most likely to be achieved. Gagné and Briggs (1979), starting from behaviourism (see Chap. 1), developed their first comprehensive draft of an instructional design theory in the 1960s over a period of more than 20 years, and they also integrated into this theory the subsequent cognitivist perspective of information processing and empirical findings on successful teaching practice (Gagné 1985; see Chap. 1). The theory, which has been widely used, especially in the USA, distinguishes five types of learning goals (see Table 4.1), different internal and external conditions for successful learning, and different instructional methods, each of which is structured according to the same nine instructional steps as outlined in the following section.

The core idea of Gagné and Briggs's theory is cumulative learning by selecting and structuring content as a hierarchical sequence for learners with simple to complex abilities. To this end, they proposed a top-down analysis of learning goals, which is then translated into a bottom-up strategy of instruction. The final objectives of the learning sequence or series of lessons are determined and necessary intermediate objectives are defined as performance targets (competencies) to be achieved for task solving. The hierarchical task analysis then specifies individual learning steps with their respective prerequisites, which are finally brought into a meaningful progression from the simple to the complex, in order to form a whole from the individual parts. The planning and structuring of lessons are usually based on the respective level of the learners' cognitive abilities. Linguistic information, cognitive strategies,

Table 4.1 Five types of learning goals according to Gagné and Briggs (1979)

Learning outcome	Definition	Example
1. Verbal information	Reproduction of previously learned information (facts, concepts, principles or procedures)	Identifying the characteristics of a convex lens. Reproducing the definition of the focal point
2. Intellectual skills		
a. Discriminations	Differentiation of objects, properties or symbols	Recognising that lenses magnify differently
b. Descriptive (concrete) concepts	Identification of classes of concrete objects, properties or events	Sorting lenses according to their external shape (curvature)
c. Abstract (defined) concepts	Classifying new examples of events or ideas based on their definition	Identifying a round filled water bottle as a convex lens
d. Rules/relationships	Applying a simple context to solve a particular set of tasks	Calculation of the image distance with the lens equation (focal and object distance are given)
e. High order rules/problem-solving	Application of a new combination of rules to solve a complex problem	Planning of a telescope that produces an image of a certain magnification at the correct height and on the correct side (Galilean telescope)
3. Cognitive strategies (self-regulation)	Use of individual possibilities to guide learning, thinking, acting and feeling	Choice of the form of representation to determine optical images (graphic design, mathematical calculation)
4. Motor skills (routines)	Performing certain sequences of physical movements	Adjusting the lens or screen for a sharp image
5. Attitudes	Choosing to behave in a particular way	Enjoy to investigate the effect of lenses

motor skills and attitudes are added at the points where the best connection can be made. Accordingly, Gagné and Briggs formulated nine steps of instruction that should be organised by the teacher: (1) gaining attention, (2) informing the learner of the lesson objective, (3) stimulating recall of prior learning, (4) presenting the stimulus material with distinctive features, (5) providing learning guidance, (6) eliciting performance (diagnosis; see Chap. 15), (7) providing informative feedback, (8) assessing performance, and (9) enhancing retention and learning transfer.

Petry et al. (1987) planned a teaching unit on geometrical ray optics using the instructional design theory of Gagné and Briggs (1979) (see Table 4.2). Even if this lesson plan, which is close to teacher-centred classroom instruction, does not correspond to today's view of instruction, it does well to illustrate the idea of a systematic, cumulative learning environment.

Table 4.2 Petry et al.'s (1987, pp. 11–44) lesson plan using the instructional design theory of Gagné and Briggs (1979)

Gaining attention	The teacher distributes magnifying glasses and encourages the students to look at many different objects. He or she asks what is happening and why it is happening
Informing the learner of the lesson objective	T: *In this lesson, you can learn about magnifying glasses and how they work. You will learn:* – to identify convex lenses, – graphically represent the trajectory of light beams through different parts of a convex lens, – to define and describe the focal distance, – to predict the focal distance and magnification based on the curvature of a convex lens Then, the students receive worksheets with tasks, which they are to work through independently. Sample solutions for self-control are provided
1st learning goal: identify convex lenses Presenting the stimulus material with distinctive features	Task 1: All convex lenses have at least one surface that is convex, so that the lens is thicker in the middle than at the edges. Here you can see different lenses in cross section: These are convex lenses. These are not convex lenses
Eliciting performance	The lenses below are shown in cross section. Circle the lenses that are convex:
Providing informative feedback	*Self-control based on sample solution. The teacher walks around and supports*
2nd learning goal: beam refraction at different parts of convex lenses Stimulating recall of prior learning	Task 2: Remember what happens to light when it passes through a medium like glass. Complement the sentence: *The light becomes (refracted)*
Presenting the stimulus material with distinctive features	What happens when light passes through a convex lens? When light beams pass through a convex lens at the edge, they are refracted towards the axis When light beams pass through a convex lens on the axis, they are not refracted
Providing learning guidance	Complete the following sentences: *At the edge, a convex lens refracts the light beams (towards the axis)* *Light beams on the axis of a convex lens are (not) refracted* *Students compare their solutions. The teacher goes around and supports student learning*

(continued)

Table 4.2 (continued)

Eliciting performance	Task 3: Draw the path of the rays parallel to the axis behind the convex lens:
Providing informative feedback	*Students compare their solutions. The teacher goes around and supports student learning by asking questions*
3rd learning goal: identify focal point and distance Stimulating recall of prior learning	Task 4: All beams parallel to the axis refracted inwards and the not refracted beam on the axis intersect at one point
Presenting the stimulus material with distinctive features	This point of intersection is called focal point The distance of the focal point from the centre of the lens is called focal length
Eliciting performance	Insert the focal point and the focal length into the illustration of Task 3
Providing informative feedback	*Self-check using a sample solution* *The teacher goes around and supports*
4th learning goal (a): relation of focal distance and curvature Presenting the stimulus material with distinctive features	Task 5: Here you can see how light rays pass differently shaped convex lenses. What is the relationship between the focal length and the radius of curvature of the lenses? Different convex lenses
Providing learning guidance	Complete the sentence *The larger the curvature of the lens, the (shorter) the focal length* *The smaller the curvature of the lens, the (longer) the focal length*
Eliciting performance	Task 6: Now draw the path of the light rays through the following convex lenses. Note the relationship between curvature and focal length. Mark the assumed focal point and focal length in the comparison of the lenses Path of the light rays behind the lenses
Providing informative feedback	*Students compare their solutions. The teacher goes around and supports*

Table 4.2 (continued)

4th learning goal (b): relation of focal distance and magnification Presenting the stimulus material with distinctive features	Convex lenses are used to magnify images The longer the focal length of a convex lens, the lower the magnification The smaller the focal length of a convex lens, the higher the magnification For the highest magnification, you have to use the convex lens with the smallest focal length
Providing learning guidance	What is the relationship between magnification and curvature of convex lenses? Fill in the blanks. Please note the results from Task 5 *The <u>greater</u> the curvature of the lens, the (<u>greater</u>) its magnification* *The <u>smaller</u> the curvature of the lens is, the (<u>smaller</u>) is its magnification* Order the lenses in Task 5 according to their magnification power. *Self-check using sample solution*
Eliciting performance	Task 6: You will get the following four lenses. Predict with which of the lenses you will achieve the highest magnification Lenses and their magnifications
Providing informative feedback	*To check their prediction, the students receive the lenses from the teacher. Result is discussed in plenum*
Assessing performance	As **homework** the students receive a worksheet with the following task: Draw the path of the light beams for the following convex lenses. Note how the shape of the convex lenses affects the focal length Mark the focal point and the focal length. Order the lenses according to their magnification power
Enhancing retention and learning transfer	In the further course of the series of lessons, concave lenses will be treated analogously. The students construct images using parallel beam, focal point beam and centre beam. In addition, it is investigated which lenses are used in different optical devices. The distinction between convex and concave lenses and the relationship between the curvature, focal length and magnification power of the lenses are always discussed

This lesson plan illustrates key elements of instructional design according to Gagné and Briggs, namely

1. clearly operationalised objectives which are accessible for assessment,
2. a small step-by-step, sequential decomposition of the learning content offered,
3. step-by-step practice of what has been learned,
4. opportunities to combine the newly learned with previous knowledge,
5. accompanying feedback on the individual steps.

Both *how* the lenses are presented by the teacher and *what* forms of representation are used (see Chap. 7) on the surface level change the content and complexity of the lesson. For example, if the topic is prepared as a students' experiment, the students' experimental skills must be taken into account. An experimental design by the students also needs a lot of time and is subject to errors, because, for example, incident light beams parallel to the optical axis and perpendicular to the lens axis are difficult to realise experimentally. These difficulties are avoided by a digital representation of the physics theme. With a suitable digital application, for example, the lens thickness can be changed and the focal point automatically adjusted. An additional difficulty arises because an application usually allows more variations than are needed for the concept being taught. The students therefore have to learn additionally how to use the digital instrument.

4.3.3 Basis Models According to Oser and Baeriswyl (2001)

In a way similar to the instructional design of Gagné and Briggs (1979), Oser and Baeriswyl (2001) also considered a variety of different types of learning goals that address different basic types of learning. They assumed that each type of learning objective has its own conditions and requires different preconditions, so that specific learning processes have to be followed. In the deep structure of the instruction, the respective learning processes provide a fixed sequence of mental operations for the learners, which can be choreographed in many different ways.

> Our hypothesis is that every sequence of (school) learning is based on a choreography that binds, on the one side, freedom of method, choice of social form, and situated improvisation with, on the other side, the relative rigor of the steps that are absolutely necessary in inner learning activity. (Oser and Baeriswyl 2001, p. 1043)

The sequences of steps derived from theories of psychology of teaching and learning are called basis models. According to Oser and Patry (1990), the sequence of every model usually has to be run through completely and in the correct order. These models form a general education framework for the deep structure of instruction, but do not make any binding regulations for concrete action at the surface structure. The relationship between surface and deep structure reflects the dichotomy of the pedagogical freedom of teaching and the strictness of the theories of learning psychology. This relationship enables teachers to creatively implement the basis models in accordance with their own teaching style and repertoire of methods as well as the requirements of instruction in each subject. With regard to the deep structure, Oser and Patry (1990) considered it important "that the view of the structured course of instruction should actually always be that of the student" (p. 2). Instruction according to the basis models is therefore student-oriented and learning-process-oriented, because the course of instruction is primarily oriented towards the needs of the learners in the acquisition process.

Table 4.3 Overview of the basis models adopted from Oser and Baeriswyl (2001, p. 1046, Table 46.1)

Name of the basis model	Type of learning goal	Examples from physics lessons
1a Learning through personal experience	Appropriating experiential knowledge	First experimental experiences with a phenomenon
1b Discovering learning	Appropriation through reality search processes	Discovering physics relationships in free experimentation
2 Development as an aim of education	Deep structures transformation	Creating cognitive conflict in students' perceptions
3 Problem-solving	Trial-and-error learning; analytical problem-solving	Search for an explanation or technical solution by hypothesis formation and experimental or theoretical hypothesis examination
4a Knowledge building	Explaining an object, understanding the meaning of a word	Introduction of technical terms or physics units using examples
4b Concept building	Constructing a knowledge network	Physics concepts (force, energy, elementary particles …) are developed theoretically and experimentally
5 Contemplative, meditative learning	Reflective abstraction	
6 Use of learning strategies	Learning to learn (meta-learning)	Use of concept maps as advance organisers or summarising presentation of content structures
7 Development of routines and skills	Automatisation	Repeated performing of calculations or constructions, practising tasks on specific concepts (e.g. conservation of momentum)
8 Learning through motility	Transformation of affective states in creative production	
9a Social learning	Development of the ability to relate to someone through social behaviour, social exchange	Cooperative experimental group work with the distribution of tasks (role cards)
9b. Learning through realistic discourses	Conflict resolution, need balance	Elaboration of a consensus for the acquisition of a solar plant

(continued)

Table 4.3 (continued)

Name of the basis model	Type of learning goal	Examples from physics lessons
10 Construction of values and value identity	Value clarification, value development, critical value analysis	(Social) evaluation of physics research
11 Hypertext learning	Reordering and revaluing of information bits	Preparation for an exam by organising a physics content from a new, self-chosen perspective
12 Learning to negotiate	Producing consensus in various situations	Discussion of (learners') own (research) results and conclusions

According to Oser and Baeriswyl (2001), there are 12 basis models as described in Table 4.3 (see also Elsässer 2000, p. 13).

For physics teaching in secondary and primary schools in particular, the three most important basis models—namely *learning through personal experience, problem-solving* and *concept building*—have been adapted and tested (Ohle 2010; Trendel et al. 2008; Zander 2016). According to Zander (2016), physics lessons organised according to learning processes are particularly helpful for students who belong to the lower- to medium-performance groups; they learn more than average. In the following sections, these basis models are presented in more detail.

Learning Through Personal Experience

In the basis model of *learning through personal experience* (LtE), students are supposed to gain experience by independently handling a physics learning object and thereby, acquire individual, mostly unstructured knowledge. In physics education, the learning object is usually a physics phenomenon that is changed through manipulation, thus allowing students to gain empirical experience and finally learn concept construction. Such experiences are therefore related to the context, which is very personal, unstructured and unsystematic at the beginning of the learning process. The individual episodic experiences are compared, connected and generalised in small steps. The goal is the design of lessons for experiencing physics concepts in the form of rules and functional relations. For the design it is assumed that the pre-knowledge of the learners is low.

The acquired experiential knowledge is personally significant and usually deeply anchored in the episodic long-term memory. Personal experiences can be made purposefully, but also through random, spontaneous discoveries (trial and error). They are usually developed from an existing conceptual framework of the learner through assimilation by adapting the newly perceived content to the existing mental patterns. The sequence of steps for the basis model of learning through personal experience is shown in a lesson example in Table 4.4. In this lesson, it is assumed

Table 4.4 Step sequence of the basis model learning through personal experience (LtE)

Step sequence	Procedure of the lesson
LtE 1: Anticipation and planning of possible actions	The teacher holds up a convex lens. She actually wants to use the lens as a visual aid, but the lens surprisingly reduces the image. A student looking through the lens confirms this. Another student suspects that this has to do with the distance between the lens and the object The teacher asks students to form groups. Each group is equipped with two lenses and asked to vary the distance of the lens from the object and from the eye. Also, the orientation and size of the image should be considered
LtE 2: Performance of the actions	The students observe various objects in the classroom through their lenses. They approach the objects in different ways and, as discussed, change the distance from the eye
LtE 3: Construction of meaning for the activity	The students report their observations unsystematically. The experiences are made conscious through verbalisation. The teacher records the observations: • Short distance → magnification • At the edge the picture becomes fuzzy • At large distance: mirror-inverted and reduced • There is a position where the picture vanishes (I can see my eye)
LtE 1: Anticipation and planning of possible action	The experiment execution is refined. Now, for the thicker lens, the distance between the object and the lens is to be systematically increased in 2 cm steps. The respective orientation and magnification of the image should be recorded in a table
LtE 2: Performance of the actions	The students carry out the observations
LtE 3: Construction of meaning for the activity	The results are almost identical. However, when estimating the magnification, there are differences in the position of the lens at which the image is viewed as large as the observed object. The teacher asks the students to describe the results: • In a distance up to 10 cm, the picture is upright and becomes larger with increasing distance • At 10 cm, the image disappears • At distances over 10 cm, the picture stands upright and becomes increasingly smaller

(continued)

Table 4.4 (continued)

Step sequence	Procedure of the lesson
LtE 4: Generalisation of the experience through analysis of common elements	The teacher asks students to clarify at what distance the picture is as large as the object. The class agrees that this is the case at 20 cm The teacher now states that the focal length of the thicker lens is 10 cm and that of the thinner lens 50 cm. The students decide for the thinner lens when the image will turn around and when it will be about the same size as the object being observed. After the results have been compared, general rules are formulated: • If the distance of the object is less than the focal length, the image is upright and magnified • If the object is in the focal point, no image is perceived • For distances greater than the focal length, the picture is upside down • The image is enlarged if the distance is less than twice the focal length and reduced if the distance is greater than twice the focal length • If the distance is exactly twice the focal length, then the image has the same size as the object
LtE 5: Reflection of similar experiences	The teacher holds up a water bottle and wants to know what objects look like when you look through a water bottle. Since the students have never consciously noticed this before, bottles are distributed. From their observations, the students realise that the focal point is very close behind the bottle The students are asked to examine their reflection on a soup spoon as homework

that the students are already familiar with converging lenses and the concept of focal length.

The lesson example shows that going back in the sequence of steps is quite possible. In contrast, skipping in the forward direction is not useful because it can be assumed that the students cannot follow. This also happens when a step is not performed carefully enough. For example, it is usually not possible for students to generalise their experience after the first phase of construction of meaning (Step 3 of the lesson plan) before the actions are performed more systematically (Steps 4–6). In the case that the students cannot fulfil the requirement in a current step, it is recommended that they return to the previous steps and work through each step more thoroughly.

Concept Building

The goal of *concept building* (CB) is to expand cognitive structures and build up physics concepts that students usually cannot reach without guidance through learning tasks and instruction (see Chap. 9). Concepts can be both simple physics terms, such as the name of an energy form, and more complex constructions such as the principle of energy conservation. Even more complex relationships between already known concepts can be the aim of teaching. For example, the Gaussian law, Faraday's law and Ampère's law can be combined in Maxwell's equations and thus can receive a different function and meaning. All physics concepts are theoretical models and creative inventions of scientists that students cannot easily find on their own (see Chap. 8). According to Oser and Baeriswyl (2001), concepts should therefore be developed in the physics classroom using a suitable prototype from which the concept is derived. In the end, the goal should be for students to be able to apply the concepts to solve other tasks or problems in different situations. For meaning making, it is important that students have the opportunity to match these concepts with their personal experiences and to be able to construct their individual meaning through active engagement.

It may occur that the learners are not able to follow the intended learning process related to the basis model because their necessary knowledge and competences are missing. In this case, according to Oser and Baeriswyl (2001), it is possible to create these prerequisites by inserting another basis model. For example, if the learners are lacking in basic experiences for concept building, the current basis model may be interrupted by the model of *learning through personal experience and* carried out completely before continuing and completing the model of concept building. Table 4.5 shows a course of instruction based on the basis model *concept-building* for image construction.

When transferring a new concept to other contexts, a reference to mathematics could also be established. For example, regarding the concept that just two rays are sufficient to clearly determine a point, it is also possible to derive the law relating

Table 4.5 Step sequence of the basis model concept building (CB)

Step sequence	Procedure of the lesson
CB 1: Stimulation of what the learners already know	The teacher first asks the students to report on their experiences with convex lens images from the previous lesson (see Table 4.4). This includes comparing the observations on the concave and convex mirror (spoon) from the homework
CB 2: Introducing and working through a prototype as valid example of the new concept	Students are handed an information sheet that shows in the case of $f < g < 2f$ the construction of an image of an object. The teacher uses the example to introduce the terms centre ray, focal ray and parallel ray and explains why these are sufficient to represent the construction of the image. Finally, it is demonstrated from the prototype that the picture is inverted and magnified

(continued)

Table 4.5 (continued)

Step sequence	Procedure of the lesson
CB 3: Analysis of essential categories and principles that define the new concept (abstraction)	The students should write down in their own words how to construct the image in a convex lens using the centre, focal and parallel rays and how to read the properties of the image from this construction. The student's texts are compared and, if necessary, refined. The essential elements of the construction are highlighted
CB 4: Active application of the new concept	The students are handed a worksheet on which they are asked to individually construct the cases $g > 2f$ and $g < f$. The results are first compared in partner work and then presented in the plenum by one student for each case Afterwards, the students should explain without further construction why in the case of $g = 2f$ the picture has the same size as the object In addition, the extreme cases $g \to 0$ and $g \to \infty$ will be considered
CB 5: Application of the new context in different contexts	The students should now transfer the construction of the image at the lens for the case $f < g < 2f$ to the reflection at the concave mirror. The teacher walks around to give assistance As homework, an unknown setting for the concave mirror has to be analysed

image to distance. $\frac{h'}{h} = \frac{i}{o}$ ($h' =$ image size, $h =$ object size, $i =$ image distance and $o =$ object distance) using the theorem of intersection lines based on the centre ray.

The lesson example in Table 4.5 shows that the basis model of concept building is close to the direct instruction of Gagné and Briggs (1979). This can be seen in the small-step progressive procedure in steps CB 4 and CB 5. In contrast to problem-based approaches of self-directed inquiry learning (Barrows 1996; Hmelo-Silver 2004), concept building does not expect students to inductively generate concepts themselves. The procedure is example-based (Renkl 2014) and often deductive, that is, once the students have learned the concept on the prototype, it has to be reconstructed in variations of the prototype. This is the precondition for the physics description of new situations and for the prediction of results of physics processes that can be confirmed experimentally.

Problem-Solving

The basis model of problem-solving (PS) is intended to develop the students' problem-solving competence. Problem-solving is thus not seen as a method of learning as in problem-based learning, but as the goal of learning (van Merriënboer 2013). For Oser and Baeriswyl (2001), problem-solving is the process of achieving a given goal without knowing exactly the theoretical or experimental method to achieve it; a cognitive barrier must be overcome along the solution process. This kind of problem-solving refers to *knowledge-based methods* for problems with multiple acceptable solutions in contrast to *strong methods* which offer knowable, comprehensible solutions to well-structured problems in a specific domain (van Merriënboer 2013). In contrast to solving tasks, where the solution is found using known solution approaches, the solution in solving problems can be more complicated. In solving a problem, the goal is to develop a solution approach, while the result of the process may be known as in the approach of analytical problem-solving (Nieswandt et al. 2020). If, in addition, the result is also not given, one refers to this process as complex problem-solving (see Chap. 9; Schmidt-Weigand et al. 2009).

For this purpose, the goal state is first specified as the success criterion for the solution of the problem. Based on existing experience and conceptual knowledge, possible alternative solutions (hypotheses) can be generated and tried out. Finally, it is tested whether the target state can be achieved with the presumed solution process, and different solution processes are reflected, characterised and evaluated. To solve analytical or complex problems, procedural knowledge (knowing how) is needed in addition to declarative knowledge (knowing what) (see Chaps. 1 and 9). In the exemplarily lesson plan for problem-solving, outlined in Table 4.6, it is assumed that the imaging of a diverging lens has already been addressed in previous lessons.

The basis model of problem-solving shows some similarities with the steps of scientific inquiry (SI) such as stating questions, deriving predictions from conjectures, and carrying out experiments based on those predictions to determine whether the conjectures are correct. Most SI approaches aim to generate understanding of concepts, which is subsequently to be generalised and transferred to new situations. As mentioned above, for the basis model of problem-solving, it is assumed that the

Table 4.6 Step sequence of the basis model problem-solving (PS)

Step sequence	Procedure of the lesson
PS 1a: Students perceive the problem	The teacher explains ametropia, an ocular disorder in which parallel rays fail to come to a focus on the retina. The students realise that ametropia can be corrected, for example, with glasses. The students should find out what type of lens can be used to correct myopia
PS 1b: Students understand the problem (clarification)	The problem is then specified: With myopia, the image in the eye is created in front of the retina. How can it be achieved that the image is once again displayed exactly on the retina?
PS 2: Students develop hypotheses about possible ways to find a solution	The students suggest to • create a model setup for myopia experimentally and try out vision correction • show, with the help of constructions of the ray paths, which correction is required

(continued)

Table 4.6 (continued)

Step sequence	Procedure of the lesson
PS 3: Students test the hypotheses	The students work in groups. They decide on a solution and try to realise it with given material
PS 4: Evaluation of solutions and possible approaches to solutions	The solution approaches of the different groups are compared and evaluated. The advantages and disadvantages of the different solution approaches are also reflected upon (test of hypotheses) • In the experimental approach, the solution can be found by trial and error. However, it does not provide a clear explanation • By construction, it can be explained how the correction would have to be made. However, it is not certain whether the solution actually work. This must be additionally checked

necessary knowledge elements and concepts are already present, but must be strategically rearranged to solve the problem (Deboer 2006; Reusser 2005; Shakhman and Barak 2019).

4.3.4 The 5E Learning Cycle According to Bybee

The 5E learning cycle model is an instructional model for inquiry in order to promote scientific reasoning (Karplus and Butts 1977). Atkin and Karplus (1962) addressed only three phases that were based on Piaget's model of *equilibration* (Piaget 1976): (1) In the *exploration phase,* learners collect relevant data and direct experiences about objects and events and assimilate them into their cognitive schemes until a discrepancy arises that triggers a disequilibration; (2) in the *concept introduction and development phase,* the learners are guided or engaged in the interpretation of the data and experiences through questions, explanations and discussions which leads to an accommodation of the new science concept; and (3) in the *concept application and expansion phase,* students are given the opportunity to explore the usefulness of the science concept by relating it to everyday applications and other concepts in the cognitive process of organisation (summarised after Marek et al. 2003). In the late 1980s, the model was extended to the five phases as presented in Table 4.7 (Bybee 2009).

Similar to the basis models of Oser and Patry (1990), the 5E learning cycle defines a sequence of steps in the deep structure of a lesson that can contain different tools and methods of instruction on the surface level. From the perspecitve of the basis models, this organisation can be seen as a kind of concept building, where learning through personal experience is embedded to elaborate a prototype which is framed by motivating and evaluating the learners. The model has been used extensively in

Table 4.7 Sequence of the 5E instructional model taken from Bybee (2009)

Phase	Procedure of the phase
Engagement	The teacher or a curriculum task assesses the learners' prior knowledge and helps them become engaged in a new concept through the use of short activities that promote curiosity and elicit prior knowledge. The activity should make connections between past and present learning experiences, expose prior conceptions, and organise students' thinking towards the learning outcomes of current activities
Exploration	Exploration experiences provide students with a common base of activities within which current concepts (i.e. misconceptions), processes, and skills are identified, and conceptual change is facilitated. Learners may complete laboratory activities that help them use prior knowledge to generate new ideas, explore questions and possibilities, and design and conduct a preliminary investigation
Explanation	The explanation phase focuses students' attention on a particular aspect of their engagement and exploration experiences and provides opportunities to demonstrate their conceptual understanding, process skills, or behaviours. This phase also provides opportunities for teachers to directly introduce a concept, process, or skill. Learners explain their understanding of the concept. An explanation by the teacher or the curriculum may guide them towards a deeper understanding, which is a critical part of this phase
Elaboration	Teachers challenge and extend students' conceptual understanding and skills. Through new experiences, the students develop deeper and broader understanding, more information, and adequate skills. Students apply their understanding of the concept by conducting additional activities
Evaluation	Evaluation phase encourages students to assess their understanding and abilities and provides opportunities for teachers to evaluate student progress towards achieving the educational objectives

science instruction around the world and scientifically evaluated as highly effective (Bybee et al. 2006).

4.4   Competences and Twenty-First-Century Skills

Gilbert (2010, p. v) pointed to three challenges that require physics education to evolve:

- the increasing intellectual isolation of science from other subjects in the school curriculum;
- how to accommodate a "science education for citizenship", and one that is still relevant to the needs of all students, in a curriculum which has traditionally been focused on … a preparation to be a scientist/engineer; and
- a consequence of the exponentially increasing gap between phenomena in which science is currently interested and what science education seems able to address.

It is therefore necessary to change traditional forms of teaching to include new content and to rebalance the objectives of teaching. The demand to move away from purely *subject-specific* or *subject-immanent* instruction—and instead, to take more account of life practices, student interests, interdisciplinary supra-subject contexts, more diverse forms of instruction, and the equivalent in lesson design—has permeated research on subject-specific education for many years and has become part of curriculum development (Venville et al. 2012). This approach is evident in the shift from teaching science to developing science literacy as a desirable outcome of learning science (Roberts and Bybee 2014) and from science education to STEM education and teaching socio-scientific issues (Sadler and Dawson 2012; Zeidler 2014).

There are always many new findings in physics research that play a role in social discussion, such as the recent measurement of gravity waves, the development of a theory on dark matter or new results in surface physics. On the other hand, most physics concepts, for example, those for the description of energy, force or elementary particles, change only slowly and they can remain in the curricula for a long time. Furthermore, in physics in particular, cognitive concepts and competencies to be learned, as well as those abilities, attitudes and beliefs—of students to learn physics and of teachers to teach physics—are clearly identified by research in the field of psychology of teaching and learning. It is therefore appropriate to change the orientation for planning teaching and learning physics from a content-specific approach to one that orients towards cognitive concepts and learning processes. The approach by Gagné and Briggs outlined in Sect. 4.3.2 already contains an orientation towards learning processes, although the lesson planning still starts from physics content. In the following sections, we show how a rigid orientation towards learning processes helps the teachers to structure the learning processes appropriately for the students and offers the teacher maximum flexibility to adequately design the lesson content. This opens the classroom for both changes in physics theory and empirical methods and to discuss social issues that are generated by physics and/or can be described and explained by physics concepts, at least partially.

As Bybee (2009) pointed out, to understand and participate in this discussion, students need physics-oriented abilities for adaptability, communicating complex social issues, solving non-routine problems, self-management and system thinking. The first theoretical step towards such a skill orientation is the consideration of cognitive activities as part of the concept of competence. Already in 2001, the Organisation for Economic Cooperation and Development (OECD 2001) also recommended a shift from accumulating knowledge about science to developing competence in science.

According to Mulder (2014), competence is defined as "the generic, integrated and internalised capability to deliver sustainable effective (worthy) performance (including problem-solving, realising innovation and creating transformation) in a certain professional domain, job, role, organisational context and task situation" (p. 11). This definition is in line with the definition of competence by Klieme et al. (2008), which was based on Weinert (2001). Competence is first differentiated from intelligence and knowledge and is defined as a context-related ability to cope with specific demands and is learnable and determined by demands in specific situations.

In contrast, intelligence is generalisable, stable over time and determined by biological factors and basic cognitive processes. According to Weinert (2001), competences are a combination of knowledge (cognitive abilities), skills and abilities, and values and attitudes.

Bybee (2009) suggested that learning-process-oriented instructional models, that are consistent with the 5E learning cycle or the basis models of Oser and Patry (1990) explained above, may be used to establish twenty-first-century skills abilities as learning outcomes of science education. For instance, the ability and willingness of individuals to cope with uncertain, new and changing conditions may be supported by activities such as investigations and lab work. Self-organisation is required when students need to persist at solving given problems and tasks and acquire new information on their own. Abilities to communicate based on evidence and in physics language (see Chap. 13) are needed for presenting students' own findings or physics concepts, including the abilities to use graphs, charts or other representations (see Chap. 7). To achieve such goals according to Taylor et al. (2007), instruction should:

- be organised around meaningful problems and goals;
- provide scaffolds for solving meaningful problems and supporting learning for understanding;
- provide opportunities for practice with feedback, revision and reflection; and
- promote collaboration and distributed expertise as well as independent learning.

This kind of instruction directs students' attention to the problems, tasks and materials used to orchestrate the choreography of the deep structure. For competency-based instruction, researchers—McRobbie et al. (1997), Tiruneh et al. (2018), and Struyf et al. (2019)—proposed models of instruction that focus on designing learning environments that engage learners in intensive, active and self-directed cooperative learning with the subject matter. Central to this instruction is students' autonomous use of handling learning materials that offer new information, data, experiences or stimuli using multiple representations such as texts, worksheets, pictures or experimental materials. The learning material is evaluated and transformed into learning products in visible and audible representations such as mind-maps, any kind of texts, sketches, pictures, diagrams, experiments or movies. At the same time, such learning products can serve as a diagnostic tool for the learners' individual level of competence and help teachers to evaluate their own teaching as part of *visible learning* (Hattie 2008). The learning processes is moderated by the learning environment as well as through the moderation and consultation of the teacher in case of student learning difficulties, the discussion in plenary phases as well as individual diagnosis and feedback.

In the future, digital computer technology will also change the way teachers teach and students learn scientific reasoning but the basic psychological learning processes will remain the same (Duijzer et al. 2019). Therefore, an approach focusing on deep learning structures also seems to be beneficial as a basis for future multimedia design of learning environments.

4.5 Summary

Physics teaching should be oriented towards learning processes, the content structure and other design features of instruction found through empirical research (see Chaps. 1, 2, 6, 11 and some parts among other chapters). These characteristics belong to the deep structure of instruction and cannot be described without theory-based analysis and reference to quality teaching. Table 4.8 summarises some important steps in the exemplary teaching process outlined in this chapter that should be followed for lesson planning.

Learning process orientation and subject content vary fundamentally depending on the lesson (see Table 4.8). Instruction begins with the instructional goals (learning by experience …) the topic (optics …), and a lesson design that is appropriate to the classroom situation, pre-knowledge and students' perceptions. The students should have sufficient time to organise their experiences with the physics objects and to discuss their ideas in their group and with their teacher. In principle, it is irrelevant in which social form this learning activity takes place. The social form only has to

Table 4.8 Exemplary teaching approach based on characteristics of teaching quality

Characteristics of the quality of teaching	Teaching goal	Offer	Use and student activity	Results
Learning process orientation and subject content	First experiences with physics situations	Learning task Experimental material	Experience with lenses in students' experiments	Summary of experience
Cognitive activation	Active experimental and cognitive activity of the students	Guiding questions, Cognitive prompts, Individual support	Autonomous examination and individual experiences with physics material	Autonomous verbalisation, Generalisation in class discussion
Student Orientation	Discussion is controlled by the learner	Dialogic discourse	Learners act self-determined	Learning processes run efficiently without distraction
Classroom management: clarity of rules, prevention of disturbances	Maximum time on task	Structured material Mandatory behavioural rules for student experiments	Learners work with the prepared material, teachers supervise the process	Learning processes run efficiently without distraction
Classroom climate supports learning	Self-determination, experience of autonomy, competence and relatedness	Freedom for self-determined, autonomous action Support for experiencing competence	Self-determined behaviour	Motivation, conceptual understanding

be suitable to optimise the students' activities in terms of content and time on task. Since the students have their first experience with a phenomenon, their experience plays a central role in their learning process. This is an example of how the learning goal of *learning through experience* (deep structure) determines the teaching method (surface structure). Then, in experimental group work, based on hypotheses formulated by the students, connections are made with already taught concepts and related experiences of the respective class and applications are discussed. Subsequently, the relations between newly discussed ideas and ongoing concepts are discussed and generalised with the whole class and formulated as a conclusion and outlook for further lessons. During the progressive differentiation in the learning process, the next level should only be worked on when the teacher has the impression that the last step has been addressed and comprehensively understood according to the possibilities for student learning. In the course of further lessons, different types of learning goals should be taken into account. A professional teaching design should be characterised by the fact that the teacher is familiar with different design models and can use them flexibly to achieve the intended learning goals. There are other demands on teaching that could not be addressed in this chapter, such as classroom management in general, teachers' attitudes and perceptions of learning, or the conditions in which students live or interact socially (see Chaps. 1 and 2). Teachers should be able to take as many of these conditions as possible into account when formulating the teaching goal and organising the physics lesson accordingly.

Acknowledgements We would like to thank David Treagust (Curtin University) and Anita Stender (Universität Duisburg-Essen) for carefully and critically reviewing this chapter.

References

Aebli H (1961) Grundformen des Lehrens. Ein Beitrag zur psychologischen Grundlegung der Unterrichtsmethode [Basic Forms of Teaching. A contribution to the psychological foundation of the teaching method]. Klett, Stuttgart

Arnold KH, Koch-Priewe B (2011) The merging and the future of the classical German traditions in general didactics: a comprehensive framework for lesson planning. In: Hudson B, Meyer MA (eds) Beyond fragmentation: didactics, learning and teaching in Europe. Verlag Barbara Budrich, Opladen, pp 252–264

Atkin JM, Karplus R (1962) Discovery or invention? Sci Teach 29(5):45–51. Retrieved from http://www.jstor.org/stable/24146536

Banner J, Cannon H (2008) The elements of teaching. Yale University Press, New Haven

Barrows HS (1996) Problem-based learning in medicine and beyond: a brief overview. New Dir Teach Learn 1996(68):3–12. https://doi.org/10.1002/tl.37219966804

Börlin J (2012) Das Experiment als Lerngelegenheit. Vom interkulturellen Vergleich des Physikunterrichts zu Merkmalen seiner Qualität [The experiment as a learning opportunity. From intercultural comparison of physics teaching to characteristics of its quality], vol 132. Logos, Berlin

Bybee RW, Taylor JA, Gardner A, Van Scotter P, Carlson Powell J, Westbrook A, Landes N (2006) The BSCS 5E instructional model: origins and effectiveness. Retrieved from Colorado Springs, CO

Bybee RW (2009) The BSCS 5E instructional model and 21st century skills. Retrieved from http://sites.nationalacademies.org

Deboer GE (2006) Historical perspectives on inquiry teaching in schools. In: Flick LB, Lederman NG (eds) Scientific inquiry and nature of science: implications for teaching, learning, and teacher education. Springer, Dordrecht, pp 17–35

Duijzer C, Van den Heuvel-Panhuizen M, Veldhuis M, Doorman M, Leseman P (2019) Embodied learning environments for graphing motion: a systematic literature review. Educ Psychol Rev 31(3):597–629. https://doi.org/10.1007/s10648-019-09471-7

Elsässer T (2000) Choreografien unterrichtlichen Lernens als Konzeptionsansatz für eine Berufs-felddidaktik [Choreographies of instructional learning as a conceptual approach for vocational field didactics]. Schweizerisches Institut für Berufspädagogik, Zollikofen

Fischer HE, Reyer T, Wirz T, Bos W, Höllrich N (2002) Unterrichtsgestaltung und lernerfolg im physikunterricht [Lesson design and learning success in physics lessons]. Zeitschrift Für Pädagogik, Beiheft 45:124–138

Fischer HE, Klemm K, Leutner D, Sumfleth E, Tiemann R, Wirth J (2005) Framework for empirical research on science teaching and learning. J Sci Teach Educ 16(4):309–349. Retrieved from Retrieved from http://www.jstor.org/stable/43156373

Forbes CT, Neumann K, Schiepe-Tiska A (2020) Patterns of inquiry-based science instruction and student science achievement in PISA 2015. Int J Sci Educ 42(5):783–806. https://doi.org/10.1080/09500693.2020.1730017

Gagné RM (1985) Conditions of learning and theory of instruction. Holt, Rinchart & Winston, New York

Gagné RM, Briggs LJ (1979) Principles of instructional design, 2nd edn. Holt, Rinchart & Winston, New York

Geller C, Neumann K, Fischer HE (2014) A deeper look inside teaching skripts: learning process orientations in Finland, Germany and Switzerland. In: Fischer HE, Labudde P, Neumann K, Viiri J (eds) Quality of instruction in physics: comparing Finland, Germany and Switzerland. Waxmann, Münster, pp 81–92

Gilbert JK (2010) Preface. In: Philsips LM, Norris SP, Macnab JS (eds) Visualization in mathematics, reading and science education. Springer, Dordrecht, The Netherlands, pp v

Hattie JAC (2008) Visible learning: a synthesis of over 800 meta-analysis relating to achievement. Routledge, London and New York

Helmke A (2012) Unterrichtsqualität und Lehrerprofessionalität: Diagnose, Evaluation und Verbesserung des Unterrichts [Teaching quality and teacher professionalism: diagnosis, evaluation and improvement of teaching], 4th edn. Klett-Kallmeyer, Seelze

Helmke A, Helmke T (2014) Wie wirksam ist gute Klassenführung? [How effective is good classroom management?]. Lernende Schule 65:9–12

Hmelo-Silver CE (2004) Problem-based learning: what and how do students learn? Educ Psychol Rev 16(3):235–266. https://doi.org/10.1023/B:EDPR.0000034022.16470.f3

Karplus R, Butts DP (1977) Science teaching and the development of reasoning. J Res Sci Teach 14(2):169–175. https://doi.org/10.1002/tea.3660140212

Klieme E, Hartig J, Rauch D (2008) The concepts of competence in educational contexts. In: Leutner D, Klieme E, Hartig J (eds) Assessment of competencies in educational contexts. State of the art and future prospects. Hogrefe, Göttingen, pp 3–22

Klieme E, Rakoczy K (2008) Empirische Unterrichtsforschung und Fachdidaktik: outcome-orientierte Messung und Prozessqualität des Unterrichts [Empirical classroom research and subject didactics: outcome-oriented measurement and process quality of teaching]. Zeitschrift Für Pädagogik 54(2):222–227

Krabbe H, Zander S, Fischer HE (2015) Lernprozessorientierte Gestaltung von Physikunterricht. Materialien zur Lehrerfortbildung [Learning process-oriented design of physics lessons. Materials for teacher training]. Waxmann, Münster

Kunter M, Trautwein U (2013) Psychologie des Unterrichts [psychology of teaching]. Schöningh, Paderborn

Lipowsky F, Drollinger-Vetter B, Klieme E, Pauli C, Reusser K (2018) Generische und fachdidaktische Dimensionen von Unterrichtsqualität - Zwei Seiten einer Medaille? [Generic and subject didactic dimensions of quality teaching - Two sides of the same coin?]. In: Martens M, Rabenstein K, Bräu K, Fetzer M, Gresch H, Hardy I, Schelle C (eds) Konstruktionen von Fachlichkeit [Constructions of subject specificity]. Klinkhardt, Bad Heilbrunn , pp 183–202

Marek EA, Laubach TA, Pedersen J (2003) Preservice elementary school teachers' understandings of theory based science education. J Sci Teacher Educ 14(3):147–159. https://doi.org/10.1023/A:1025918216347

Marzano R (2007) The art and science of teaching. Association for Supervision and Curriculum Development, Alexandria, VA

McRobbie CJ, Roth W-M, Lucas KB (1997) Multiple learning environments in the physics classroom. Int J Educ Res 27(4):333–342. https://doi.org/10.1016/S0883-0355(97)90015-X

Meyer H (2002) Unterrichtsmethoden [teaching methods]. In: Kiper H, Meyer H, Topsch W (eds) Einführung in die Schulpädagogik [Introduction to school pedagogy]. Cornelsen Scriptor, Berlin, pp 109–121

Meyer H (2008) Was ist guter Unterricht? [what is quality teaching?]. Cornelsen Scriptor, Berlin

Meyer H, Bülter H (2004) Was ist ein lernförderliches Klima? [what is a climate that supports learning?]. Pädagogik 11:31–36

Meyer H (2016) Unterrichtsmethoden 1. Theorieband [Teaching methods 1. Theory volume], 17 edn, vol 1. Cornelsen Scriptor, Berlin

Mulder M (2014) Conceptions of professional competence. In: Billett S, Harteis C, Gruber H (eds) International handbook of research in professional and practice-based learning. Springer, Dordrecht, pp 107–137

Nieswandt M, McEneaney EH, Affolter R (2020) A framework for exploring small group learning in high school science classrooms: the triple problem solving space. Instr Sci 48(3):243–290. https://doi.org/10.1007/s11251-020-09510-9

OECD (2001) Knowledge and skills for life: First results from PISA 2000. OECD, Paris

Ohle A (2010) Primary school teachers' content knowledge in physics and its impact on teaching and stundents' achievement, vol 110. Logos, Berlin

Oser F, Patry JL (1990) Choreographien unterrichtlichen Lernens: Basismodelle des Unterrichts [Choreographies of instructional learning: basis models of teaching]. Pädagogisches Institut der Universität Freiburg, Freiburg

Oser F, Baeriswyl FJ (2001) Choreographies of teaching: bridging instruction to learning. In: Richardson V (ed) Handbook on research on teaching, 4th edn. American Educational Research Association (AERA), Washington, pp 1031–1065

Pauli C, Reusser K (2003) Unterrichtsskripts im schweizerischen und im deutschen Mathematikunterricht [Lesson scripts in Swiss and German mathematics lessons]. Unterrichtswissenschaft 31(3):238–272

Petry B, Mouton H, Reigeluth CM (1987) A lesson based on the Gagné-Briggs theory of instruction. In: Reigeluth CM (ed) Instructional theories in action: lessons illustrating selected theories and models. Erlbaum Associates, Hillsdale, pp 11–44

Piaget J (1976) Piaget's Theory. In: Inhelder B, Chipman HH, Zwingmann C (eds) Piaget and his school: a reader in developmental psychology. Springer, Berlin, Heidelberg, pp 11–23

Praetorius A-K, Klieme E, Herbert B, Pinger P (2018) Generic dimensions of teaching quality: the German framework of three basic dimensions. Int J Math Educ 50(3):407–426. https://doi.org/10.1007/s11858-018-0918-4

Prenzel M, Seidel T, Kobarg M (2012) Science teaching and learning: an international comparative perspective. In: Fraser BJ, Tobin K, McRobbie CJ (eds) Second international handbook of science education. Springer, Dordrecht, pp 667–678

Prenzel M, Artelt C, Baumert JWB, Hammann M, Klieme E, Pekrun R (2007) PISA 06 - Die Ergebnisse der dritten internationalen Vergleichsstudie. [The results of the third international comparative study]. Waxmann, Münster

Reigeluth CM (ed) (1983) Instructional-design theories and models: an overview of their current status. Erlbaum Associates, Hillsdale

Renkl A (2014) Toward an instructionally oriented theory of example-based learning. Cogn Sci 38(1):1–37. https://doi.org/10.1111/cogs.12086

Reusser K (2005) Problemorientiertes Lernen – Tiefenstruktur, Gestaltungsformen, Wirkung. [Problem-based learning-deep structure, forms of design, effect] Beiträge zur Lehrerbildung 23(2):159–182

Roberts DA, Bybee RW (2014) Scientific literacy, science literacy, and science education. In: Lederman NG, Abell SK (eds) Handbook of research on science education, vol II. Routledge, New York, NY, pp 545–558

Roth H (1983) Pädagogische Psychologie des Lehrens und Lernens [Educational psychology of teaching and learning], 16th edn. Schroedel Schulbuchverlag GmbH, Hannover

Sadler TD, Dawson V (2012) Socio-scientific issues in science education: contexts for the promotion of key learning outcomes. In: Fraser BJ, Tobin K, McRobbie CJ (eds) Second international handbook of science education. Springer, Dordrecht, pp 799–809

Schiepe-Tiska A, Rönnebeck S, Schöps K, Neumann K, Schmidtner S, Ilka Parchmann I, Prenzel M (2016) Naturwissenschaftliche Kompetenz in PISA 2015 – Ergebnisse des internationalen Vergleichs mit einem modifizierten Testansatz [Scientific literacy in PISA 2015 - Results of the international comparison with a modified testing approach]. In: Reiss K, Sälzer C, Schiepe-Tiska A, Klieme E, Köller O (eds) PISA 2015. Eine Studie zwischen Kontinuität und Innovation [A study between continuity and innovation]. Waxmann, Münster, New York, pp 45–98

Schmidt-Weigand F, Hänze M, Wodzinski R (2009) Complex problem solving and worked examples. Zeitschrift Für Pädagogische Psychologie 23(2):129–138. https://doi.org/10.1024/1010-0652.23.2.129

Seidel T, Shavelson RJ (2007) Teaching effectiveness research in the past decade: the role of theory and research design in disentangling meta-analysis results. Rev Educ Res 77(4):454–499. https://doi.org/10.3102/0034654307310317

Seidel T, Prenzel M, Rimmele R, Dalehefte IM, Herweg C, Kobarg M, Schwindt K (2006) Blicke auf den Physikunterricht. Ergebnisse der IPN Videostudie [Views on physics teaching: Results of the IPN video study]. Zeitschrift für Pädagogik 52(6):799–821

Seidel T (2003) Lehr-Lernskripts im Unterricht. Freiräume und Einschränkungen für kognitive und motivationale Lernprozesse - eine Videostudie im Physikunterricht. [Teaching-learning scripts in the classroom, free spaces and constraints for cognitive and motivational learning processes: A video study in physics lessons]. Waxmann, Münster

Shakhman L, Barak M (2019) The physics problem-solving taxonomy (PPST): development and application for evaluating student learning. Eurasia J Math Sci Technol Educ 15. Retrieved from https://doi.org/10.29333/ejmste/109266

Struyf A, De Loof H, Boeve-de Pauw J, Van Petegem P (2019) Students' engagement in different STEM learning environments: integrated STEM education as promising practice? Int J Sci Educ 41(10):1387–1407. https://doi.org/10.1080/09500693.2019.1607983

Taylor JA, Van Scotter P, Coulson D (2007) Bridging research on learning and student achievement: the role of instructional materials. Sci Educ 16(2):44–50. Retrieved from https://files.eric.ed.gov/fulltext/EJ783420.pdf

Tiruneh DT, De Cock M, Elen J (2018) Designing learning environments for critical thinking: examining effective instructional approaches. Int J Sci Math Educ 16(6):1065–1089. https://doi.org/10.1007/s10763-017-9829-z

Treagust DF, Tsui C-Y (2014) General instructional methods and strategies. In: Lederman NG, Abell SK (eds) Handbook of research on science education, vol II. Routledge, New York, pp 303–303

Trendel G, Wackermann R, Fischer HE (2008) Lernprozessorientierte Fortbildung von Physiklehrern [Learning process oriented further education of physics teachers]. Zeitschrift Für Pädagogik 54:322–340 van Merriënboer JJG (2013) Perspectives on problem solving and instruction. Comput Educ 64:153–160. https://doi.org/10.1016/j.compedu.2012.11.025

Venville G, Rennie LJ, Wallace J (2012) Curriculum integration: challenging the assumption of school science as powerful knowledge. In: Fraser BJ, Tobin K, McRobbie CJ (eds) Second international handbook of science education. Springer, Dordrecht, pp 737–749

Weinert FE (2001) Concept of competence: a conceptual clarification. In: Rychen DS, Salganik LH (eds) Defining and selecting key competencies. Hogrefe & Huber Publishers, Cambridge, MA, pp 45–65

Wubbels T, Brekelmans M, den Brok P, Wijsman L, Mainhard T, van Tartwijk J (2014) Teacher–student relationships and classroom management. In: Emmer E, Sabornie E (eds) Handbook of classroom management, 2nd edn. Routledge, New York, pp 373–396

Zander S (2016) Lehrerfortbildung zu Basismodellen und Zusammenhänge zum Fachwissen [In-service teacher training on basis models and connections to content knowledge]. Logos, Berlin

Zeidler DL (2014) Socioscientific issues as a curriculum emphasis: theory, research and Practice. In: Lederman NG, Abell SK (eds) Handbook of research on science education, vol II. Routledge, New York, pp 697–726

Chapter 5
Nature of Scientific Knowledge and Nature of Scientific Inquiry in Physics Lessons

Burkhard Priemer and Norman G. Lederman

5.1 Introduction

This chapter is an introduction to concepts of how to understand and describe the nature of physics or more specifically nature of scientific knowledge (NOSK) in physics and nature of scientific inquiry (NOSI) in physics. Current science education reforms and curriculum put a strong emphasis in supporting students to become scientifically literate. NOSK and NOSI are seen as critical components of this educational aim, so teaching physics means not only a focus on physics content but on NOSK and NOSI as well. Besides describing why it is important to discuss NOSK and NOSI in the classroom, this chapter also discusses inadequate and adequate views. Based on this background, we outline concepts how NOSK and NOSI can be implemented in physics lessons without compromising the learning of foundational physics knowledge. Finally, we answer the question how NOSK and NOSI can be assessed. Throughout the text we will refer to examples in physics.

5.2 Definitions of Fundamental Terms

5.2.1 What Is Science?

Before carefully considering the importance of nature of scientific knowledge (NOSK) and nature of scientific inquiry (NOSI) to science education, it is critical to

---

Sadly Norman Lederman died a few weeks after we finished writing this chapter. His ideas will stay alive.

B. Priemer (✉)
Humboldt-Universität zu Berlin, Berlin, Germany
e-mail: priemer@physik.hu-berlin.de

N. G. Lederman
Illinois Institute of Technology, Chicago, IL, USA

© Springer Nature Switzerland AG 2021
H. E. Fischer and R. Girwidz (eds.), *Physics Education*, Challenges in Physics Education,
https://doi.org/10.1007/978-3-030-87391-2_5

"define" what is meant by "science." There are many conceptualizations of science. The rotunda in the National Academy of Science in the USA (in 1936) contains the following inscription: Science is "pilot of industry, conqueror of disease, multiplier of the harvest, explorer of the universe, revealer of nature's laws, eternal guide to truth." Nobel Prize winning physicist Richard Feynman defined science in the 1970s as "the belief in the ignorance of experts" (Feynman and Cashman 2013). Most recently, Arthur Boucot, a famous paleobiologist, in a personal conversation characterized science as "an internally consistent set of lies designed to explain away the universe." These statements are quite varied, and as provocative as Boucot's and Feynman's definitions, they may be close to how science is characterized in recent reform documents such as the Next Generation Science Standards (NGSS Lead States 2013) and the National Science Education Standards (National Research Council 1996) in the USA, the Bildungsstandards in physics in Germany (KMK 2004), or the national science curriculum in England (UK department for education 2015). The question "What is science?" still remains and will always have multiple answers. But what conceptualization would be most appropriate for primary and secondary education in physics? Commonly, the answer to this question has three parts.

First, science can be seen as a body of knowledge. This refers to concepts, laws, and theories that are taught for instance in classical domains like biology, chemistry, physics and which are presented in textbooks, e.g., the law of gravitation, the theories of relativity, or the phenomenon of the optical refraction.

5.2.2 What Is Nature of Scientific Inquiry?

The second part refers to how the knowledge is developed. That is scientific inquiry. As a student outcome, it usually includes the doing of inquiry (e.g., asking questions, developing a design, collecting and analyzing data, and drawing conclusions). Additionally, it also includes knowledge about inquiry (e.g., knowing that all investigations begin with a question, there is no single scientific method, research questions guide the procedures, etc.). For the purposes of this chapter, knowledge about inquiry is referred to as nature of scientific inquiry, NOSI. Even though there are multiple scientific methods to generate knowledge, one prominent way is an interplay between predictions, e.g., the Higgs boson or gravitational waves, and experimental confirmations, e.g., by the ATLAS and CMS collaboration or the LIGO consortium.

5.2.3 What Is Nature of Scientific Knowledge?

Third, because of the way the knowledge is developed, scientific knowledge has certain characteristics. These characteristics of scientific knowledge are often referred to as nature of scientific knowledge, NOSK (Lederman et al. 2013). They usually include, but are not limited to the idea that science is empirically based, involves human creativity, is unavoidably subjective, and is subject to change (Lederman et al. 1998). For example, the understanding of the concepts of mass and energy changed over time. Before the early years of the twentieth-century, mass and energy were to *different* concepts. Energy, for example, had *unprecise meanings*. Scientists like Kelvin, Clausius, Mayer, Joule, Helmholtz, and others *refined* the concept of energy by distinguishing it from and relating it to entropy, work, and heat. Further, the properties "conservation" and "transformation" were discovered and *added* to the concept of energy. The concept of mass was developed by scientists like Kepler, Galilei, Newton, and others. Here is the origin of differing between mass as a resistance to acceleration and mass as quantity that influences the strength of gravitational attraction. Both concepts, energy and mass, were seen as *different* but *related*, e.g., through the dependence of the kinetic energy on the mass of a moving object. Research in electrodynamics led Thomson, Poincaré, Hasenöhrl, and others at the turn of the century to introduce an "electromagnetic mass" that is directly related to energy by a constant factor, the speed of light. By doing this, energy and "electromagnetic mass" became *equivalent* entities. However, it was Einstein in 1905 who explained the *general meaning* and *relevance* behind the equivalence of mass and rest energy (independent of electromagnetism).

Often individuals conflate NOSK with scientific inquiry. Lederman (2007) also notes that the conflation of NOSK and scientific inquiry has plagued research on NOSK from the beginning and, perhaps, could have been avoided by using the phrase "nature of scientific knowledge" as opposed to the more commonly used nature of science (NOS). In this chapter, we will use the term NOSK instead of NOS. As it more accurately represents its intended meaning (Lederman and Lederman 2004). Now the critical point is what is the appropriate balance among the three components of science in the science curriculum and science instruction? Current reforms have appropriately recognized that the amount of emphasis has traditionally emphasized the body of knowledge to the detriment of any emphasis on inquiry or NOSK.

5.2.4 What Is Scientific Literacy?

Current visions of science education are returning to the perennial goal of scientific literacy. In general, the goal of scientific literacy is to help students use their scientific knowledge to make informed decisions about scientifically based global, societal, or personal decisions. The literate individual cannot make such decisions based on scientific knowledge alone. They must also understand the source of the knowledge (NOSI) and the ontological characteristics of the knowledge (i.e., NOSK). In this chapter, we will clarify how NOSK and NOSI are related to scientific literacy.

5.2.5 What Differs Science from Engineering and Technology?

Throughout this chapter, we refer to science. However, some of the issues we raise hold true for the fields of engineering and technology as well. When this is not the case, we point out differences. To relate science to engineering, we can state that the latter uses scientific knowledge to develop devices that solve problems or addresses certain needs and, thus, can be seen as applied science and mathematics. Technology encompasses methods, procedures, or tools to serve certain purposes in scientific or engineering approaches like helping humans communicate with each other, manufacturing goods, or supporting the development of knowledge. However, these "definitions" frame the three fields in a pretty simplistic way (for the sake of giving the reader a first idea). We like to stress that science, engineering, and technology cannot be easily defined or described because these fields are complex by nature, influence each other, and are not disjunct but interwoven.

Besides describing the fields of technology and engineering, questions like "What is engineering?", "What do engineers do?", or "How is knowledge generated in engineering?" relate to *nature of engineering* (Pleasants and Olson 2019), or, more precise, the *nature of engineering knowledge* and *nature of engineering processes*. Because science is closely linked to engineering, we will mention aspects of nature of engineering when relevant.

5.3 The Relevance of NOSK and NOSI for Teaching Physics

Why should our students learn science and to what extent? Are we teaching our students to make them scientists? What happens to those students who do not continue studying science? Do not they need to learn a minimum amount of science? These questions are critical to portray the goal of science education.

5.3.1 Legitimation to Teach NOSK and NOSI

To begin with, we raise the question why it is relevant for students to learn about NOSK and NOSI in physics lessons. The following six arguments (Driver et al. 1996, 11) are often put forward for legitimacy:

- *The economic argument*: An understanding of NOSK and NOSI is important for recruiting future scientists and technicians on which our societies depend. For example, science laboratory work activities can prepare students to conduct scientific investigations.
- *The utilitarian argument*: An understanding of NOSK and NOSI is important for understanding and handling science and scientific products in everyday life.

For example, when purchasing technical equipment such as mobile phones or refrigerators, it is important to understand manufacturers arguments about the quality of products and how they know that the products work.

- *The democratic argument*: An understanding of NOSK and NOSI is important to make informed decisions and to be able to participate in societal scientific issues. For example, in elections, citizens may have to decide whether to support or oppose the closure of power stations of certain energy sources such as nuclear energy or coal.
- *The cultural argument*: An understanding of NOSK and NOSI is important to understand our culture which is strongly influenced by science and technology. For example, the implementation of automated processes by computers in many areas of life have a cultural impact.
- *The moral argument*: An understanding of NOSK and NOSI is important to understand the general importance of scientific standards and values in our societies. For example, it is important to understand that a detailed disclosure and sharing of scientific knowledge in journals and on conferences is a key concept in science.
- *The psychological argument*: An understanding of NOSK and NOSI helps to better understand scientific content. For example, knowledge about the historical development of nuclear models can promote an understanding of the Bohr model.

It is also assumed that an understanding of NOSK and NOSI can increase interest in the natural sciences, students' self-concept, the use of learning strategies, the willingness to change concepts, and the ability to solve problems (Lederman and Lederman 2014).

The six arguments for addressing NOSK and NOSI in the physics classroom seem plausible. However, it is important to note that these arguments must be subject to an empirical examination. Studies have already been conducted on the psychological argument which show a positive influence of adequate views about NOSK and NOSI on learning success (Lederman 2007). However, the other arguments have hardly been empirically examined.

5.3.2   Reform-Based Rationale for Teaching NOSK and NOSI

Science educators believe that the goal of science education is to develop scientific literacy. Since the first use of "scientific literacy" in the late 1950s, science educators and policy makers have gradually reconceptualized the term to such an extent that one author remarked relatively recently that "scientific literacy is an ill-defined and diffuse concept" (Laugksch 2000, p. 71). Policy makers and educators often get confused between "science literacy" and "scientific literacy." Often, they are considered synonymous, although the two have very different meanings. Science literacy focuses on how much science you know, so it is solely based on content knowledge. It is not about applying knowledge and making decisions.

On the other hand, scientific literacy deals with the aim of helping people use scientific knowledge to make informed decisions. This is a goal that science educators have been striving to achieve, but unfortunately many of us have not truly realized the importance of scientific literacy or might have misrepresented the goal in various platforms. DeBoer (2000) states that the term scientific literacy, since it was introduced in the late 1950s has defied precise definition. Although it is widely claimed to be a desired outcome of science education, not everyone agrees with what it means.

The national review of Australian science teaching and learning (Goodrum et al. 2001) defined the attributes of a scientifically literate person. In particular, it stated that a scientifically literate person is (1) interested in and understands the world about him, (2) can identify and investigate questions and draw evidence-based conclusions, (3) is able to engage in discussions of and about science matters, (4) is skeptical and questioning of claims made by others, and (5) can make informed decisions about the environment and their own health and well-being. PISA defines scientific literacy as the ability to engage with science-related issues, and with the ideas of science, as a reflective citizen. PISA's definition includes being able to explain phenomena scientifically, evaluate and design scientific enquiry, and interpret data and evidence scientifically. It emphasizes the importance of being able to apply scientific knowledge in the context of real-life situations (OECD 2017).

Current science standards (e.g., the NGSS in the USA or the Bildungsstandards in physics in Germany) stress science practices, but there is very little emphasis on understanding NOSI and NOSK. Doing science is necessary as a means, but it should not be the end goal. The end goal should be scientific literacy, which unfortunately is often not explicitly mentioned in the standards.

5.4   Inadequate Views About NOSK and NOSI

"In physics, researchers usually use experiments to prove knowledge about the reality." This quote of a student reflects ideas about scientific practice that indicate a largely inadequate understanding of NOSI. Experiments do not constitute the only standard method and source of knowledge in physics—there are multiple ways how knowledge is generated in physics -, nor can any measurement result be considered a proof. Numerous studies have shown that students express a variety of such inadequate views (Lederman and Lederman 2014). A summary of "typical" inadequate views is given in the following list of so-called myths (McComas 1998):

1.   Hypotheses become theories and that in turn become laws.
2.   Scientific laws and other such ideas are absolute.
3.   A hypothesis is an educated guess.
4.   A general and universal scientific method exists.
5.   Evidence accumulated carefully will result in sure knowledge.
6.   Science and its methods provide absolute proof.
7.   Science is procedural more than creative.

8. Science and its methods can answer all questions.
9. Scientists are particularly objective.
10. Experiments are the principle route to scientific knowledge.
11. Scientific conclusions are reviewed for accuracy.
12. Acceptance of new scientific knowledge is straightforward.
13. Science models represent reality.
14. Science and technology are identical.
15. Science is a solitary pursuit.
16. Scientific knowledge is independent from cultural and sociological backgrounds of the researcher.
17. Scientific knowledge—after being discovered and accepted by the community—is conserved in books in its original form.
18. Research in science is only driven by the curiosity of scientists.

One source of these inadequate views—which are shared by some teachers as well—can be found in the representation of physics in public (e.g., media) and in schools. For example, students could transfer the simple and unproblematic way in which laws are derived in school lessons and textbooks to research in science. Research in science education has shown that adequate views about NOSK and NOSI do not automatically develop by attending physics lessons (Lederman and Lederman 2014). With no explicit discussion of NOSK and NOSI, students will refer to intuitive ideas and beliefs. This is why the inadequate views of students can be both persistent and unstable, not consistent in themselves, or logically constructed, but are often formed spontaneously.

5.5 Adequate Views About NOSK and NOSI

Before presenting views about NOSK and NOSI that are often referred to as adequate, we would like to give a brief historical outline of the discovery of the dwarf planet Pluto (Messeri 2010). We would like to use this scenario to illustrate generally formulated views.

5.5.1 A Historical Example

Percival Lowell was convinced that disturbances in the orbit of Uranus must originate from a so far undiscovered planet. However, after more than 10 years of unsuccessful searching, Lowell died in 1916, and the search for the ninth planet X was continued in 1929 by Clyde Tombaugh in the same location in Flagstaff (Arizona, USA). In 1930, based on the documented preliminary work of Lowell, in which the presumed position of Pluto was estimated, Tombaugh found a planet after only a few months. However, following investigations showed that the mass of the ninth planet named

Pluto was too small to explain disturbances in the orbit of Uranus and Neptune. Finally, a reanalysis of the orbit and mass of Neptune showed that additional masses are not necessary to describe the observed orbit of Uranus. So, the discovery of Pluto was more of a coincidence. Years later, other trans-Neptunian objects (bodies farther from the sun than Neptune in the so-called Kuiper belt) have been discovered, in particular those with a mass comparable to the mass of Pluto. In 2006, the International Astronomical Union decided to redefine planets and classify Pluto as a dwarf planet. This decision was preceded by controversial discussions, as Pluto had acquired a cultural significance as a planet, and his planetary status was a general knowledge. The controversy over Pluto has shown that changing a planetary definition does not only create differences between cultural and scientific perspectives, but also between different scientific disciplines and schools that use different cosmologies. Thus, it becomes clear that science is not as immutable and objectively correct as often taught and accepted.

5.5.2   Adequate Views About NOSK

The relationship and differences between nature of scientific knowledge (NOSK) and nature of scientific inquiry (NOSI) is often discussed and confused within existing literature (Lederman and Lederman 2014). NOSK, as opposed to the more popular nature of science (NOS), is used here to be more consistent with the original meaning of the construct (Lederman 2007). Given the manner in which scientists develop scientific knowledge (i.e., SI), the knowledge is engendered with certain characteristics. These characteristics are what typically constitute NOSK (Lederman 2007). As mentioned before, there is a lack of consensus among scientists, historians of science, philosophers of science, and science educators about the particular aspects of NOSK. This lack of consensus, however, should neither be disconcerting nor be surprising given the multifaceted nature and complexity of the scientific endeavor. Conceptions of NOSK have changed throughout the development of science and systematic thinking about science and are reflected in the ways the scientific and science education communities have defined the phrase "nature of science" during the past 100 years (e.g., AAAS 1990, 1993; Central Association for Science and Mathematics Teachers 1907; Klopfer and Cooley 1963; NSTA 1982).

However, many of the disagreements about the definition or meaning of NOSK that continue to exist among philosophers, historians, and science educators are irrelevant to K-12 instruction. The issue of the existence of an objective reality as compared to phenomenal realities is a case in point. There is an acceptable level of generality regarding NOSK that is accessible to K-12 students and relevant to their daily lives. Moreover, at this level, little disagreement exists among philosophers, historians, and science educators. Among the characteristics of the scientific enterprise corresponding to this level of generality are that scientific knowledge is tentative (subject to change), empirically based (based on and/or derived from observations of the natural world), subjective (theory-laden), necessarily involves human inference, imagination, and creativity (involves the invention of explanations), and is socially and culturally embedded. Two additional important aspects are the distinction between observations and inferences, and the functions of, and relationships between scientific theories and laws. What follows is a brief consideration of these characteristics of science and scientific knowledge.

As with the myths, there are views about NOSK and NOSI (McComas et al. 1998, p. 6), which are often described as adequate, and on which there is a certain consensus in science. We will explain why we speak of a *certain* consensus here after we have presented the views.

1. Science has the aim to explain natural phenomena and is strongly related to technology.

As outlined above, science in general is difficult to define. Men and woman have been trying to make sense of the world around for as long as men exist. Understanding the world, predicting that certain things will happen under certain circumstances—with or without own interferences—and to make use of gained knowledge are cognitive desires that seem to drive peoples' curiosity for a long time. Technology—seen as an application and a source of science knowledge—has been around since men have found out that it may make life easier.

In their search for planet X, Lowell and Tombaugh tried to explain the orbits of the planets. Tombaugh used a completely new telescope, which was clearly superior to Lowell's. This new technology has allowed him to carry out a systematic research that brought new scientific knowledge about the number of planets in our solar system.

2. Scientific knowledge is continuously developed; thus, it has both a durable and a tentative character.

Scientific knowledge is never absolute or certain. This knowledge, including "facts," theories, and laws, is tentative and subject to change. Scientific claims change as new evidence, made possible through advances in theory and technology, is brought to bear on existing theories or laws, or as old evidence is reinterpreted in light of new theoretical advances or shifts in the directions of established research programs. It should be emphasized that tentativeness in science does not only arise from the fact that scientific knowledge is inferential, creative, and socially and culturally embedded. There are also compelling logical arguments that lend credence to the notion of tentativeness in science. Indeed, contrary to common belief, scientific hypotheses, theories, and laws can never be absolutely "proven." This holds irrespective of the amount of empirical evidence gathered in the support of one of these ideas or the other (Popper 1968, 1988). For example, to be "proven," a certain scientific law should account for every single instance of the phenomenon it purports to describe at all times. It can logically be argued that one such future instance, of which we have no knowledge whatsoever, may behave in a manner contrary to what the law states. As such, the law can never acquire an absolutely "proven" status. This equally holds in the case of hypotheses and theories.

The search for Pluto was started because of observed disturbances in the orbit of Uranus. So, there was a need to develop a more coherent understanding of the solar system. However, the assumption that the orbit disturbance of Uranus was produced by Pluto had to be rejected, as the mass of Pluto was too small. The phenomenon was later explained by a more precise determination of the Neptune's mass.

3.   In science, the terms *theory* and *law* denote different concepts; therefore, theories do not become laws.

Closely related to the distinction between observations and inferences is the distinction between scientific laws and theories. Individuals often hold a simplistic, hierarchical view of the relationship between theories and laws, whereby theories become laws depending on the availability of supporting evidence. It follows from this notion that scientific laws have a higher status than scientific theories. Both notions, however, are inappropriate because, among other things, theories and laws are different kinds of knowledge and one cannot develop or be transformed into the other. Laws are statements or descriptions of the relationships among observable phenomena. Boyle's law, which relates the pressure of a gas to its volume at a constant temperature, is a case in point (Lederman and Adb-El-Khalick 1998). Theories, by contrast, are inferred explanations for observable phenomena. The kinetic molecular theory, which explains Boyle's law, is one example. Moreover, theories are as legitimate a product of science as laws. Scientists do not usually formulate theories in the hope that one day they will acquire the status of "law." Scientific theories, in their own right, serve important roles, such as guiding investigations and generating new research problems in addition to explaining relatively huge sets of seemingly unrelated observations in more than one field of investigation. For example, the kinetic molecular theory serves to explain phenomena that relate to changes in the physical states of matter, others that relate to the rates of chemical reactions, and still other phenomena that relate to heat and its transfer, to mention just a few.

The discovery of Pluto was based on a comprehensive theory of gravitation that includes, for example, Newton's and Kepler's laws. So, the theory of gravitation is not becoming Newton's laws with more evidence at hand.

4.   Science is imbedded in historical, social, and cultural settings.

Science as a human enterprise is practiced in the context of a larger culture, and its practitioners (scientists) are the product of that culture. Before 1909, when women were not allowed to study in Prussia, Marie Curie had to enter the building of the chemical institute in Berlin via the back entrance and could not use the lecture and laboratory rooms. Together with Otto Hahn, she worked in Max Planck's working room, a former wood workshop. Science, it follows, affects and is affected by the various elements and intellectual spheres of the culture in which it is embedded. These elements include, but are not limited to, social fabric, power structures, politics, socioeconomic factors, philosophy, and religion. An example may help to illustrate how social and cultural factors impact scientific knowledge. Telling the story of the evolution of humans (*Homo sapiens*) over the course of the past seven million years is central to the biosocial sciences. Scientists have formulated several elaborate and differing story lines about this evolution. Until recently, the dominant story was centered about "the man-hunter" and his crucial role in the evolution of humans to the form we now know (Lovejoy 1981). This scenario was consistent with the white-male culture that dominated scientific circles up to the 1960s and early 1970s. As the feminist movement grew stronger and women were able to claim recognition in the various scientific disciplines, the story about hominid evolution started to change. One story that is more consistent with a feminist approach is centered about "the female-gatherer" and her central role in the evolution of humans (Hrdy 1986). It is noteworthy that both story lines are consistent with the available evidence.

The reclassification of Pluto as a dwarf planet was controversial at the International Astronomical Union. The reasons for this were not only cultural views—as a small planet, Pluto had his own "reputation" among the population—but also, to a certain extent, the fact that Pluto was the only planet that was discovered by an American.

5. Science is made by humans that are creative, inquisitive, rational, and—to a certain extend—intuitive and subjective.

Even though scientific knowledge is, at least partially, based on and/or derived from observations of the natural world (i.e., empirical), it nevertheless involves human imagination and creativity. Science, contrary to common belief, is not a totally life-less, rational, and orderly activity. Science involves the invention of explanations, and this requires a great deal of creativity by scientists. The "leaps" from atomic spectral lines described by Balmer to Bohr's model of the atom with its elaborate orbits and energy levels to the modern description of quantum mechanics is a case in point. This aspect of science, coupled with its inferential nature, entails that scientific concepts, such as atoms, black holes, and species, are functional models that relate phenomena to theories rather than faithful copies of reality.

Additionally, scientific knowledge is to a certain extend subjective and theory-laden. Scientists' theoretical commitments, beliefs, previous knowledge, training, experiences, and expectations actually influence their work. All these background factors form a mind-set that affects the problems scientists investigate and how they conduct their investigations, what they observe (and do not observe), and how they make sense of, or interpret their observations. It is this (sometimes collective) individuality or mind-set that accounts for the role of subjectivity in the production of scientific knowledge. It is noteworthy that, contrary to common belief, science never starts with neutral and objective observations (Chalmers 1999). Observations (and investigations) are always motivated and guided by previous theories or concepts and acquire meaning in reference to questions or problems. These questions or problems, in turn, are derived from certain theoretical perspectives.

The discussion of the International Astronomical Union and the struggle for a consensual definition of planets and dwarf planets show creativity and subjectivity in scientific work. Further, the story of the discovery of Pluto shows that intuition and good or bad luck can also play a role in knowledge acquisition.

It is clear from the attributes of a scientifically literate individual espoused by Showalter (1974) and NSTA (1982) that NOSK is considered a critical component of scientific literacy. If precollege and postsecondary students are expected to make informed decisions about scientifically based personal and societal issues, they must have an understanding of the sources and limits of scientific knowledge. For example, it is becoming increasingly common for the public to hear alternative viewpoints presented by scientists on the same topic. Does living close to a nuclear power plant increase your risk of dying from cancer? Do mobile phones harm your health? Are meteor showers threatening life on earth? Is using electric motors instead of combustion engines more environmentally friendly? Sometimes, the claims are based on pseudoscience, like current claims that there really is no global warming or the claim that biological evolution never occurred. Alternatively, these differences in perspectives and knowledge are the result of science in action. It is the results of the nature of scientific knowledge. Science is done by humans, and it is limited, or strengthened by the foibles that all humans have. Scientific knowledge is tentative or subject to change. We never have all of the data, and if we did we would not know it. If you look up in the sky on a clear night, you will see a white, circular object. We would all agree that the object is the moon. Three hundred years ago if we looked at the same object, we would call it a planet. This is because the current view of our solar system is guided by heliocentric theory. This theory places the sun at the center of the solar system and certain objects orbiting the sun are planets (e.g., the earth) and certain objects orbiting a planet are moons or satellites. Three hundred years ago, our view was guided by the geocentric theory which places the earth at the center and anything orbiting the earth was considered a planet (e.g., our current moon). The objects and observations have not changed, but our interpretation has because of a change in the theories we adopt. You could say that our theories "bias" our interpretations of data. Scientists make observations, but then eventually make inferences because all the data are not accessible through our senses. This is why scientific knowledge is tentative and partly a function of human subjectivity and creativity.

5.5.3 Adequate Views About NOSI

At a general level, scientific inquiry can be seen to take several forms (i.e., descriptive, correlational, and experimental). Descriptive research is the form of research that often characterizes the beginning of a line of research. This is the type of research that derives the variables and factors important to a particular situation of interest. Classifying stars by the elements they are built of or by the radiation they emit or classifying planets by their orbit and additional mass they have collected are examples. Whether descriptive research gives rise to correlational approaches depends upon the field and topic. For example, defining the solar system with its planets or looking for exoplanets is descriptive in nature. The purpose of research in these areas is sometimes simply to describe. However, in physics research very often goes beyond descriptions, e.g., investigating how stars develop or the origin of planets.

To briefly distinguish correlational from experimental research, the former explicates relationships among variables identified in descriptive research (e.g., that objects which are immersed in water appear to be lifted upwards when seen from outside of the water depending the angle of view and their depth) and experimental research involves a planned intervention and manipulation of the variables studied in correlational research in an attempt to derive causal relationships (e.g., Snell's law). In some cases, lines of research can be seen to progress from descriptive to correlational to experimental, while in other cases (e.g., astronomy) such a progression is not possible (e.g., astronomical objects cannot be manipulated). This is not to suggest, however, that the experimental design is more scientific than descriptive or correlational designs but instead to clarify that there is not a single method applicable to every scientific question.

Inquiry is perceived in science education in different ways. The National Research Council developed an addendum of sorts titled Inquiry and the National Science Education Standards (NRC 2000). First, scientific inquiry was conceptualized as a teaching approach. That is, the science teacher would engage students in situations (mostly open-ended) in which they could ask questions, collect data, and draw conclusions. In short, the purpose of the teaching approach was to enable students to learn science subject matter in a manner similar to how scientists do their work. Second, although closely related to science processes, scientific inquiry extends beyond the mere development of process skills such as observing, inferring, classifying, predicting, measuring, questioning, interpreting and analyzing data. Scientific inquiry includes the traditional science processes, but also refers to the combining of these processes with scientific knowledge, scientific reasoning, and critical thinking to develop scientific knowledge. From the perspective of the science standards (e.g., in the USA the National Science Education Standards NRC 1996; the Bildungsstandards for physics in Germany, KMK 2004; the national science curriculum in England; UK department for education 2015), students are expected to be able to develop scientific questions and then design and conduct investigations that will yield the data necessary for arriving at answers for the stated questions.

Third, inquiry can also be viewed from a perspective of meta-cognitive skills that students are to achieve. In particular, the current visions of reform (e.g., NGSS Lead States 2013; NRC 1996) are very clear (at least in written words) in distinguishing between the performance of inquiry (i.e., what students will be able to do) and what students know about inquiry (NOSI). For example, it is one thing to have students set up a control group for an experiment, while it is another to expect students to understand the logical necessity for a control within an experimental design. So, understanding how science "works" is a learning goal of its own, e.g., because the way knowledge is developed in science distinguishes it from other ways of knowing. This has been elaborated extensively by prominent scientists and philosophers such as Popper, Kuhn, and Feyerabend. In summary, the knowledge about inquiry included in current science education reform efforts includes the following (NGSS Lead States 2013; NRC 1996):

1. There are multiple ways to do science.

The contemporary view of scientific inquiry advocated is that the research questions guide the approach and the approaches vary widely within and across scientific disciplines and fields (Lederman and Adb-El-Khalick 1998). Pre-college students, and the general public for that matter, believe in a distorted view of scientific inquiry that has resulted from schooling, the media, and the format of most scientific reports. This distorted view is called *the scientific method*. That is, a fixed set and sequence of steps that all scientists follow when attempting to answer scientific questions. A more critical description would characterize the scientific method as an algorithm that students are expected to memorize, recite, and follow as a recipe for success. The perception that a single scientific method exists owes much to the status of classical experimental design. Usually, scientific investigations presented in textbooks are experiments and no examples of descriptive and correlational research are provided or identified. The problem, of course, is not that investigations consistent with *the scientific method* do not exist. The problem is that experimental research is not representative of scientific investigations as a whole.

Approaches to explain disturbances in planet orbits can be made in different ways: new influencing masses can theoretically be predicted and empirically searched, the known masses of the existing planets can be checked for correctness, the observed disturbances can be verified and described in more detail, new approaches—like the theories of relativity—can be introduced and applied, etc. Therefore, there are different approaches and methods to explain detected irregularities.

2.   Scientific investigations are driven by curiosity and questions that may or may not be based on hypotheses.

Scientific investigations and inquiry procedures begin with a question or a certain interest, but do not necessarily test a hypothesis. For example, exploratory investigations—like Faraday's initial work in electrodynamics—may thrive to derive hypotheses.

Clearly, Lowell's and Tombaugh's driving question was to explain orbit disturbances by searching for a new planet. For this, they had a hypothesis. However, they had no specific idea where the planet X exactly was. So, they scanned the night sky exploratively in order to find the missing mass.

3.   Doing science can be a complex task depending on certain conditions like content knowledge, theoretical assumptions, available experimental material, different methods to analyze data, etc.

Because of this complexity, different scientists can come to different results and conclusions even though they use the same data and methods. The researchers reasoning is influenced by the experimental devices available, theory referred to, and results of former investigations.

In principle, Lowell and Tombaugh had the same aim and used the same methods, e.g., telescopes to search the night sky. However, Tombaugh's telescope had a better quality and he succeeded in finding Pluto. Later, he looked at Lowell's data and found Pluto on Lowell's photo plates as well. So, Tombaugh and Lowell came to different conclusions (identifying Pluto and not) based on the same data.

4.   Standards in science require that new knowledge must be reported verifiable and openly so that in principle it could be reviewed and replicated.

Research in science has developed certain standards over the years. These agreed roles of conduct include for example that research conclusions must be consistent with the data collected, that other researchers—at least in principle—must be able to verify the results, that results are reviewed by critical experts, and that results are presented openly. Lately, the science community puts additional emphasis in making results more "objective." The term "open science" describes efforts to make the process and results of doing science accessible to the public, for example by providing raw data or pre-registering hypothesis and research methods.

The discovery of Pluto based on Tombaugh's photo plates can be reconstructed years after the documentation. This and additional documentations and publications of Pluto has enabled astronomers worldwide to find Pluto and to make own measurements.

5. In the history of science, there are developments based on accepted and agreed conditions ("evolution") and developments that reject accepted and agreed conditions in favor of a completely new approach ("revolution").

In his well-known work "The Structure of Scientific Revolutions" Kuhn (1962) describes that science—generally seen—begins with forming a paradigm in which concepts, theories, and methods are developed and agreed upon, followed by a period of "normal sciences" when scientist refine and expand their knowledge based on the underlying paradigm. However, major and fundamental anomalies may arise that challenge or even question basic assumptions or concepts of the current paradigm. If strong and convincing enough and incompatible with the current paradigm, they may—over the time—lead to a revolution in which a new paradigm is developed that "explains" the observed anomalies and that is able to explain the long know phenomena as well.

The continuous development of Newtonian mechanics has explained many phenomena that can be described as "evolutionary," such as the prediction of the existence of Pluto. On the other hand, the observed rotation of Mercury's perihelion (point of the orbit nearest to the sun) could not be explained on the basis of Newtonian mechanics, but with the help of the general theory of relativity. Thus, the explanation is based on completely different basic conditions, which are "revolutionary" in light of classical mechanics.

6. In science, the terms data and evidence denote different concepts; therefore, data are different from evidence.

Data are qualitative or quantitative observations, for example a set of measurements of the environment around us like air temperature, air humidity, air pressure, air density, solar radiation, sound pressure level, amount of CO_2 or fine dust in the air. However, not all of these data are relevant to answer a certain question. For example, to determine the speed of sound in air in the laboratory, the measured air temperature, distance, and time it took the sound travel that distance are data that count as evidence. However, data of the experimenter's mass and body temperature or the temperature and air pressure outside of the laboratory are irrelevant for the experiment and therefor no evidence.

Tombaugh collected a fast amount of data (e.g., photograph plates with numerous stars), but not all these data were relevant and, thus, are evidence.

The six aspects of the development of scientific knowledge show that science follows multiple ways. In summary, Kind and Osborne (2016, p. 11) identified six styles of scientific reasoning: *mathematical deduction* (e.g., deriving the ideal gas law from a set of conditions and assumptions), *experimental evaluation* (e.g., empirically checking if the ideal gas law describes real gases), *hypothetical modeling* (e.g., constructing and describing an ideal gas), *categorization and classification* (e.g., the standard model of particle physics), *probabilistic reasoning* (e.g., the probability that a measurement in a system produces a certain value), and *historical-based evolutionary reasoning* (e.g., the derivation of the development of the solar system). These styles give a broad view of how new knowledge in science is constructed. A more fine-grained description of detailed paths of knowledge acquisition in science, technology, engineering, mathematics (STEM) and computer science are described in Priemer et al. (2019). The framework shows the multiple options a researcher or learner has at hand—e.g., exploring, prototyping, proofing, experimenting, and deducing—to solve domain-specific and interdisciplinary problems.

In addition to producing pieces of new knowledge science develops new methods and new fields as well. An example of the former is *blinded analysis.* Saul Perlmutter observed a confirmation bias—the tendency to search for, analyze, interpret, and choose information that confirms own beliefs and hypotheses—in astrophysics research. Two different groups of researchers kept on confirming their own results for the Hubble constant while criticizing the other groups' results. The Hubble constant $H = v/r$ (today we know that H is not really constant …) describes the expansion of the universe by determining the ratio of the apparent velocity v of an astronomical object and its distance to the earth r. McCoun and Perlmutter (2015) show that the confirmation bias—that is based on subjective influences of the researchers—can be avoided to a large extend by *blinded analysis.* Blinded analysis is a method that has the following steps:

1. All methods to analyze data are agreed on and then fixed *before* any data are used.
2. A trustworthy person manipulates the data temporarily and in a reversible way and conceals the changes so that other researchers who analyze the data cannot control if results fit their expectations.
3. After the analysis is completed the manipulation is undone and the final results emerge.
4. These results will be published no matter what outcome they bring.

An example that illustrates how new fields in science are developed is the World Weather Attribution project. The research calculates the human influence on extreme weather events immediate after they occur. New in this field is the approach that weather events can be described almost simultaneously by models based on climate data. For example, in July 2019, there was a heatwave in Western Europe with temperatures above 40 °C in the Netherlands. Researchers of the World Weather

Attribution (WWA Website) project estimated that in the Netherlands "such temperatures would have had extremely little chance to occur without human influence on climate—return periods higher than ~ 1000 years."

5.5.4 Dimensions of Nature of Engineering

As is was done in science, Pleasants and Olson (2019) reviewed texts that described the engineering discipline to frame *nature of engineering*. The analysis brought forth nine aspects. In the following, we present these aspects accompanied with an illustrating question (taken from Pleasants and Olson 2019, 161) and leave it to the reader to relate these aspects to the views about NOSK and NOSI stated above:

1. Design in engineering ("To what extent is engineering design problem-solving?"),
2. Specifications, constraints, and goals ("How are the specifications and constrains of an engineering project determined?"),
3. Sources of engineering knowledge ("How do engineers use knowledge from other disciplines, such as science?"),
4. Knowledge production in engineering ("How do engineers produce the knowledge needed to engage in design?"),
5. The scope of engineering ("What kinds of technological activities do engineers *not* generally do?"),
6. Models of design processes ("How well do models of the design process capture the real work of designers?"),
7. Cultural embeddedness of engineering ("In what ways is engineering design influenced by the culture in which it is practiced?"),
8. The internal culture of engineering ("To what extend does there exist an 'engineering culture' ?"),
9. Engineering and science ("How do engineering and science influence one another?").

5.6 Questioning a Consensus View About NOSK and NOSI and a Critical Understanding of NOSK and NOSI

5.6.1 Questioning the Consensus View About NOSK and NOSI

Recently, there have been direct criticisms of the popular consensus lists used to describe important understandings of NOSK and NOSI (Allchin 2017; Erduran and Dagher 2014). Alchin's work primarily conflated NOSI and NOSK and was not aligned with current curriculum reforms in science education. Most importantly, the assessments he developed did not have established validity or reliability. A more developed "challenge" to the consensus view that has gained some traction comes from Erduran and Dagher (2014). Their "new vision" (as they would call it) of NOSK is derived from Wittgenstein's concept of "family resemblance." Although presented as an alternative to consensus views, the two views are not at odds. The family resemblance conception is answering a different question than the consensus view. The family resemblance viewpoint is actually attempting to answer the question "What is science?" It attempts to demarcate science from non-science. And, as previously discussed in this chapter, the answer to that question generally has three components: subject matter, development of subject matter knowledge (i.e., inquiry), and characteristics of knowledge (i.e., NOSK). The consensus lists do not attempt to demarcate science from non-science. Indeed, many of the aspects of NOSK identified overlap with the knowledge in a variety of disciplines. For example, the idea that knowledge is subject to change is not unique to scientific knowledge. It is also true of knowledge in the social sciences, mathematics, and history. The family resemblance view is much broader than consensus views, but it includes the aspects identified in the consensus lists. The two views are not at odds or contradictory.

The primary weaknesses of the family resemblance view is that, because of its conceptual breadth, it is not well aligned with existing curricula and empirical support for its efficacy with precollege students is yet to be demonstrated.

Those who present alternatives to the consensus view are also critical of "lists." For some reason, they (e.g., Matthews 2000) are convinced that instruction that follows this view is reduced to students simply being asked to memorize the list of aspects related to NOSK. This is not true and is easily debunked by referring to the empirical literature. The NOSK aspects identified in consensus views are analogous to chapters in a book. Each aspect identified is simply the label for a wealth of information students are engaged with during instruction using a variety of pedagogical approaches. This will be demonstrated later in this chapter.

5.6.2 A Critical Understanding of NOSK and NOSI

Although the cited adequate views of NOSK and NOSI appear convincing at first glance, they are nevertheless not unproblematic (Hodson and Wong 2017). This is partly due to the fact that the views are formulated in a very short and general way and thus reflect incompletely or even incorrectly the complex knowledge-gathering processes of the natural sciences. For example, a statement like "Scientist require accurate record keeping, peer review and replicability" (McComas et al. 1998, p. 6) has only limited validity in light of research at the Large Hadron Collider. Since there is no second such accelerator, a completely independent execution of the experiments cannot be done in any another location. Further, there are examples where the statement "New knowledge must be reported clearly and openly" (McComas et al. 1998, p. 6) is not true. It is known, for example, that Robert Andrews Millikan was very "picky" when selecting which of his measurements to include to determine the elementary charge. There are also examples where the statement "Scientific findings are being examined by experts" (McComas et al. 1998, p. 6) has not led to sufficient quality testing. One example is the Sokal-Hoax in 1996, where the physicist Alan D. Sokal got a non-sense paper published in a cultural science journal to illustrate the low academic standards in the humanities. The "reverse" effect took place in 2002 when the two Bogdanov brothers published papers of questionable quality in physics journals. In this way, we can find counter examples to adequate views stated above that cast doubt on the general validity which they imply: A view that is prominently and very well described by Feyerabend (1975) in his book "Against Method." So, there are reasons to agree to this lists, and good reasons to be careful with it, too!

So, how can a consensus look like? In answering this question, we argue that an adequate view of NOSK and NOSI cannot, in principle, be formulated as a general, brief statement (as we did above). Rather, there are different perspectives from which a statement can be seen. These various "right" perspectives make views about NOSK and NOSI a discursive area of content without a clear "true" or false."

The statement "Scientific knowledge while durable, has a tentative character" (McComas et al. 1998, p. 6) can illustrate the problem of such a statement and show that it is difficult to describe such short statements as fundamentally adequate. A naive understanding interprets this statement in the simple sense that all knowledge in science is stable, but changes when its time has come.

A more profound understanding interprets this statement as follows:

- Scientific findings are subject to developmental processes. This is not a shortcoming, but an advantage without which innovation is not possible. This process of development fascinates many people (e.g., in quantum cryptology).
- There is knowledge that seems very reliable and stable from the current perspective. For example, the theory of relativity has worked extremely well in its existing form.
- However, a knowledge base that is considered complete at a certain time can be expanded, questioned, or changed in detail or in its foundation. The story of the interpretation of light with the Young's and Fresnel's wave theory, the corpuscle theory of Newton, and the wave–particle–dualism is an example for this.
- New concepts appear from a new perspective superior to the old, but a final comparative assessment is not possible. The interpretation of light by photons to explain the photoelectric effect appeared superior to the wave theory at that time. However, the wave interpretation of light was not discarded.
- Correcting, supplementing or rejecting theories and concepts does not mean that they become worthless. Even after the development of the theories of relativity, there are numerous applications of classical mechanics—in particularly in daily life—which are valuable and work fine.
- The adoption of a new concept is not necessarily based only on purely scientific reasoning. For example, it is assumed that Galilei used propaganda methods to convince opponents of the use of telescopes (Chalmers 2007, p. 125).

The adequate and inadequate views are therefore neither catalogues of rules to be memorized on NOSK and NOSI, nor a definition of learning objectives, because they do not do justice to the diversity of science in general and physics in particular. However, they can be an excellent starting point—but not the result—to offer the opportunity to discuss views about NOSK and NOSI critically and discursively.

5.7 Teaching NOSK and NOSI in Physics and Science Classrooms

5.7.1 Teaching NOSK and NOSI to Achieve Scientific Literacy

We have outlined above how scientific literacy—a main goal of science education—is related to NOSK and NOSI. To better understand the suggestions in the following paragraphs of how to teach NOSK and NOSI, we briefly summarize dimension of scientific literacy. The attributes of a scientifically literate individual were for example described by the National Science Teachers Association [NSTA] (1982) which were based on work from Showalter (1974). A scientific literate person …

- uses science concepts, process skills, and values making responsibly everyday decisions;
- understands how society influences science and technology as well as how science and technology influence society;
- understands that society controls science and technology through the allocation of resources;
- recognizes the limitations as well as the usefulness of science and technology in advancing human welfare;
- knows the major concepts, hypotheses, and theories of science and is able to use them;
- appreciates science and technology for the intellectual stimulus they provide;
- understands that the generation of scientific knowledge depends on inquiry process and conceptual theories;
- distinguishes between scientific evidence and personal opinion;
- recognizes the origin of science and understands that scientific knowledge is tentative, and subject to change as evidence accumulates;
- understands the application of technology and the decisions entailed in the use of technology;
- has sufficient knowledge and experience to appreciate the worthiness of research and technological developments;
- has a richer and more exciting view of the world as a result of science education; and
- knows reliable sources of scientific and technological information and uses these sources in the process of decision making.

5.7.2 General Aspects to Teach NOSK and NOSI

There are several ways to address NOSK and NOSI in the physics classroom. Before we outline these, we want to draw the attention to some general recommendations concerning teaching NOSK and NOSI. A prerequisite to all of these recommendations is, that NOSK and NOSI are addressed explicitly. Research shows that adequate views about NOSK and NOSI do not develop "automatically" through learning physics content.

- *Reflexive.* Addressing NOSK and NOSI involves critical and discursive discussions with a multi-perspective view. Thus, we recommend to teach NOSK and NOSI by implementing repeated phases of reflections.
- *Integrated into physics content.* NOSK and NOSI can directly be linked to physics content, so we argue that NOSK and NOSI and physics content should be taught and learned together.
- *Authentic.* We recommend to teach NOSK and NOSI by means of "real" problems in schools, by societal, or historical contexts, so that the relevance of NOSK and NOSI is recognized by the students.
- *Exemplarily.* NOSK and NOSI cover a wide range of content, so we suggest to choose suitable examples that highlight certain aspects of NOSK and NOSI.

5.7.3 General Approaches to Teach NOSK and NOSI

NOSK and NOSI can be taught in the physics classroom whenever knowledge-generating processes are addressed and reflected. For example, if a general principle—e.g., force and acceleration are proportional—is inferred from experimental results. In many physics classrooms, only few measurements are carried out and these often have comparatively high uncertainties. However, on the basis of this "low" evidence, laws are often deduced in school practice. Without addressing problems of this inductive conclusion (general laws are concluded from single measurements), it is easy to give students an inadequate impression of how scientific knowledge is generated. However, it is possible to compare, contrast, and evaluate knowledge generation in the classroom to science research. This "honest" approach to use data, with its potential to draw conclusions from it, is important in order to acquire adequate views about NOSK and NOSI. We argue that this does not limit the acceptance of experiments in the classroom, as pupils understand that schools do not have the same possibilities and aims as research institutions.

In physics and more generally in science education various teaching approaches have been developed which are particularly suitable for teaching NOSK and NOSI (Höttecke 2001, p. 85). Before we outline these in the following we like to remind the reader that they all require an explicit discussion of NOSK and NOSI:

- *Historical developments of physics and case studies.* The history of science offers numerous examples how knowledge was generated and what characterizes science knowledge. Examples are the history of the Pluto discovery, Konrad Röntgens discovery of the X-rays, Lise Meitner's scientific work and the political–social circumstances of women in science at the time, the exploratory experiments on the electric charge of Charles Dufay, the story of the N-rays and fallacy and fraud in science, Gustav Robert Kirchhoff's difficulties with the recognition and publication of his scientific works, the nuclear bomb research in the USA around the Chicago Pile and the Manhattan project regarding political–social influences on research (see Höttecke 2012, for further historical lessons).

- *Science-oriented or research-oriented teaching*: NOSK and NOSI can be taught when students work in a research-oriented learning environment, that is to say when they try to "find something out." This may involve purely domain-specific questions (examining the factors on which the oscillation time of a pendulum depends), everyday life objects ("How does a pocket heater work?"), or natural phenomena ("How does a thunderstorm develop?"). In addressing these questions, problems must be identified, objectives defined, strategies developed, solutions followed and solutions reflected. In all these "phases," a reflection of the own approach is also an insight into NOSK and NOSI. For example, it can be made clear that creativity plays a role, that there may be several solutions, that results obtained must be revoked or modified by new findings, that observations should be based on prior knowledge, that a phenomenon must be explored first, etc. In connection with this approach the term "research-based learning" is often used. However, in most cases (an exception may be physics tournaments) students are not in a research situation, as results, for example, are already known, the methods available are very limited, the time frame is different, the prior knowledge is smaller, etc. We point to this not to diminish the value of research-based learning in principle, but to argue that it is not the same as scientific research. There is a learning situation constructed on the basis of educational considerations, and it should also be addressed to students that it is different from scientific research.

- *Physical-technical questions in social contexts*. The interplay between science and society—which is reflected in various views about NOSK and NOSI—can also be addressed in physics classes. Addressing these socio-scientific issues can highlight that politics play a key role in determining the budgets and contents of civil and military research (e.g., renewable energies or arms development), that the public has ethical, cultural or religious reasons to support or reject certain areas of research (such as genetic engineering), that precise scientific knowledge on current important issues may not yet be sufficiently available (such as climate change or long-term damage to the health of humans caused by mobile telephones), that technological achievements and developments have a significant impact on our lives (aircraft, communications, the Internet), that risks and benefits of technical developments must be weighed up for society (such as civil use of nuclear energy, civil supersonic aircrafts), and that the handling of data (reception, production, dissemination, publication) calls for a critical reflection.

- *Philosophy of science*: Although it has been said that NOSK and NOSI should be taught together with physics content, addressing philosophical theories about knowing and scientific knowledge is an option, too. This approach focusses on the content of NOSK and NOSI and physics is "used" as an example. Topics are, for example, the induction problem in physics ("Can a general law be inferred from experimental data?"), the construction of scientific arguments (structure and logic), the role of experiments in the scientific knowledge-generating process, the role of axioms in the natural sciences (such as the "indemonstrable" assumption that periodic processes exist), and the question of what distinguishes science from pseudoscience. Lederman and Adb-El-Khalick (1998) have compiled a series of activities for students which focus on different aspects of NOSK and NOSI. For example, the "Tricky Tracks" show the image of regularly occurring black spots on a white background, which look like the traces of two birds in the snow. The task of describing the observation of these patterns shows how difficult it is to make objective statements without prior knowledge or assumptions: most descriptions of students assume that these patterns are in fact bird tracks, although this context was never given.

These teaching approaches can be implemented in different learning settings. Besides discussing NOSK and NOSI in physics lessons or in school science projects, out-of-school activities like visits to museums or research institutes as well as using different educational media channels can trigger discussions of "how science works." These settings can as well serve to distinguish science from technology and engineering. Developing airplanes—for example—calls for a fundamental understanding of aerodynamics, a field of science. Modeling lifting forces in physics can predict in principle if a new type of plane will fly. Here lies a strength of scientific argumentation. However, many engineering problems and technological questions remain when building an airplane that can be solved for example by optimizing effects that do not necessarily have to be based on a detailed theory but rely on empirical investigations, e.g., improving fuel efficiency, using lighter material, or optimizing engines.

5.7.4 Classroom and Empirically Based Sample Activities

In order to illustrate how activities to teach NOSK and NOSI may look like, we present two examples: the *air pressure activity* and the *Twirlie activity*. References to more examples can be found at the end of this chapter.

The Air Pressure Activity

Purpose: Pressure is a difficult concept for students to understand. It is defined as (the perpendicular component of a) force per area and, thus, includes force as another difficult concept. What makes it even more difficult to understand is that pressure as a physics quantity cannot be seen. We do observe effects of pressure differences when, for example, a resulting force accelerates an object.

This activity can be used in middle- and high-school classes after pressure (in or on liquids or gases) was introduced to teach about air pressure and its interaction with other forms of matter.

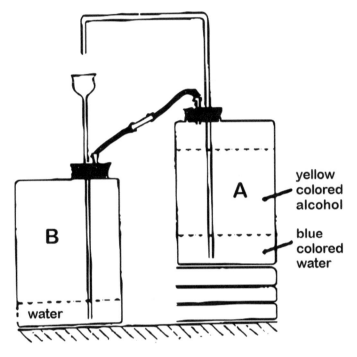

Fig. 5.1 Construction of the experiment

Material: Two translucent (like "frosted glass") one-gallon or four-liter gasoline containers (made of plastic), 2 rubber stoppers, rubber tubing, one thistle glass tube funnel, glass tubing, ethyl alcohol, and food coloring (three different colors).

Preparation: Set up the experiment as indicated in Fig. 5.1. Container B contains plain tap water; container A contains tap water colored with blue food coloring at the bottom (about 1/3 of the container), and the rest of the container is filled with ethyl alcohol colored with yellow food coloring. When adding the yellow colored alcohol, do this slowly to prevent any unnecessary mixing or agitation. A small segment of class tubing should be inserted within the rubber tubing that connects containers A and B. The thistle glass tube allows water to be poured into container B; another gas tube allows blue water to flow from container A via the thistle tube to container B.

Understandings about NOSK:

- Science is inherently tentative, yet robust.
- Science is based on evidence (empirical).
- Science is a human endeavor, and so is influenced by the perspective or theoretical framework of the investigator (subjective/theory-laden).
- Science is based on both observation and inference.

Understandings about NOSI:

- Different kinds of questions suggest different kinds of investigations

- Current scientific knowledge and understanding guide scientific investigations
- Scientific explanations emphasize evidence, have logically consistent arguments, and use scientific principles, models, and theories.

Advices for teaching:

- Begin by having the students make any kind of observations of the experimental setup as illustrated in Fig. 5.1.
- Begin the experiment by pouring a red colored liquid from a reagent bottle (labeled with a "bogus" chemical formula) into the thistle tube. The red liquid is actually tap water colored red with food coloring. Pour enough liquid into the thistle tube until blue liquid starts flowing out of container A via the funnel inserted into container B. The water will now keep running for approximately 20 min. If the flow stops you can just pour more red liquid into the thistle tube to restart the flow.
- Initially, blue colored water will flow from container A to container B. Blue will eventually turn into green (which is actually the interface of the blue water and yellow alcohol), and finally into yellow.
- As the experiment proceeds, ask students, in groups, to explain what they see and offer inferences about the original contents of the containers and how far each glass tube extends into each container.
- If you want, students can be invited to the front of the room to look at the containers more closely.
- After students have speculated about the contents of the containers at the start of the demonstration, ask the class if there is anything they want you to do to the apparatus. As students make suggestions discuss that what they are really doing is testing their hypotheses, just as scientists would do. Typically, students will ask you to squeeze some of the rubber tubing, lower container A to a level lower than container B, or squeeze either container.
- Have a representative from each group provide their conclusions, which should include the initial contents of the cans, how far each glass tube extends into each container, and an explanation for the continuous flow of water and noted color changes.
- Typical answers will state that the red liquid created a chemical reaction in container B, which then liberated a gas. This gas traveled to container A and then chemically interacted with the liquid in container A to change its color.
- Continued discussion will end with an understanding that the flow of liquid was created by changes in air pressure within the container to cause the movement of liquid. Students usually never arrive at the idea that the color change was not related to any chemical reaction.
- As the demonstration proceeds the teacher can emphasize how the pressure of air in the container change and how air pressure, or any pressure in a gas, can provide enough force to move liquids or solids if the liquids or solids if there is a pressure gradient. That is, air moves from high pressure to low pressure, and if there is a solid or liquid between the two pressures, enough force can be created to move the liquid or solid.

In addition to teaching students about air pressure and its potential effects, this demonstration also allows the teacher to ask students to reflect on NOSK and NOSI:

- The nature of the scientific knowledge created based on their prior knowledge (i.e., the model about the contents of the containers).
- In particular, students can easily realize that scientific knowledge is tentative, their models were a function of their own creativity and subjectivity,
- Students can also be directed to reflect on the observations and inferences that were a focal point of the lesson
- Teachers should discuss the occurrence of different methods and conclusions and point out that one conclusion may not necessarily be better than another. The approach is guided by the question asked. Multiple viable conclusions are possible, as long as they are consistent with the available information. Different groups may have gone about studying the variable in a different way from another group. Methods of investigation influence the results. Interpretation of data may differ from group to group. Scientists have a framework through which to examine data just like the students do. This framework (prior knowledge or expectations) influences how that data are analyzed and interpreted. This subjectivity and creativity are a part of nature of science.
- Our understanding keeps changing as we get more and more information or as we look at something in a different way. Science is dynamic, not static.
- Regardless of your designs, the question being asked guided scientific investigations; also, knowledge from past experiences and science content influenced the work of scientists. Scientists use their prior knowledge and experiences to make predictions and decisions about their investigations and conclusions.
- Scientists cannot always control all variables in real systems. Models are created that control some of the variables and help enable the scientists to offer explanations and make predictions. Sometimes, none of the variables can be controlled. For example, investigations in astronomy and ecology often cannot be controlled. Manipulation is undesired. The goal might be to describe what is observed or to study relationships. These types of investigations are still science because they rely on observations of the natural world, but they do not require identification and manipulation of variables to do an investigation. There are multiple methods in science.

The Twirlie Activity

Purpose: This activity is useful for introducing students to doing scientific inquiry as well as learning knowledge about scientific inquiry and nature of scientific knowledge. The paper helicopter *twirlie* is simple to manipulate so that students can ask many questions and conduct a variety of investigations while learning about forces and motion, speed, velocity, and acceleration. Secondary-level physics teachers can also use them to investigate circular motion.

Material: Twirlie templates in various sizes (see Fig. 5.2), scissors.

Preparation: Built the twirlies as follows (see result in Fig. 5.3). (1) Trace the pattern of Fig. 5.2 on a piece of paper. (2) Cut out the pattern, cutting along all solid

Fig. 5.2 Twirly template

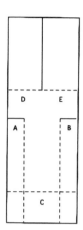

Fig. 5.3 Ttwirlie

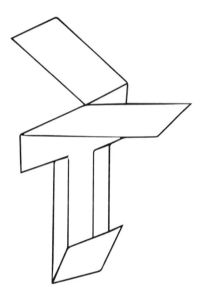

lines. (3) Fold "A" inward. (4) Fold "B" inward. (5) Fold "C" upward. (6) Fold "D" backward. (7) Fold "E" forward. (8) Hold with flaps up and drop from a high place. (9) Watch it twirl and flitter like a falling leaf.

The following aspects about NOSI and NOSK that can be explicitly addressed during this lesson include:

Doing Scientific Inquiry:

- Identify questions that can be answered through scientific investigations.
- Design and conduct a scientific investigation
- Develop descriptions, explanations, predictions, and models using evidence

- Think critically and logically to make the relationships between evidence and explanation
- Communicate scientific procedures and explanations.

Understandings about NOSK:

- Science is inherently tentative, yet robust.
- Science is based on evidence (empirical).
- Science is a human endeavor, and so is influenced by the perspective or theoretical framework of the investigator (subjective/theory-laden).
- Science is based on both observation and inference.

Understandings about NOSI:

- Different kinds of questions suggest different kinds of investigations
- Current scientific knowledge and understanding guide scientific investigations
- Scientific explanations emphasize evidence, have logically consistent arguments, and use scientific principles, models, and theories.

Advices for Teaching

- Science involves observations and inferences. This activity has students making observations and then design investigations to infer understandings of what they observed.
- Students receive a copy of a flat twirlie template (see Figs. 5.2 and 5.3). They are asked to observe what they see. Then, following the teacher's directions, they fold the twirlie. Then, they are all instructed to hold them up and drop them at the same time.
- Students are prompted to share their observations of the spinning twirlie. Drop twirlie several times so that class can see. It should be stressed that what they have done to this point is not an investigation because their work was not guided by trying to answer a question.
- Students will usually begin to make inferences also. Clarify the difference between observation and inference by asking them what observation they make that leads them to their inference (such as "the air is pushing up on the wings as the twirlie falls"). We do not directly see the air pushing on the wings, but we see the effect (observe that the wings are extended from the stem as it falls...we infer that the air molecules push against the wings, causing them to extend during the fall).
- Lead the discussion to explore what could be investigated about the twirlie to get a better understanding of why it behaves this way. A list of class questions can be generated including: "What do you think will happen to the rate of fall if we change the size of the twirlie?" Other questions may suggest the investigation of the effect on fall rate by changing the mass, shape, or materials the twirlies are made of.
- For this activity the students are asked to investigate the effect of size on the fall rate of the twirlie.

- Depending on how familiar the class is with experimental design, this is a good opportunity to either teach them or review ideas of variables, controls, and what makes a fair test.
- Students are divided into working groups and each group is given a copy of the page of different size twirlies. They are instructed to plan, design and carry on an investigation to answer the question: What happens to the fall rate of the twirlie when the size is changed? Each group is free to use whatever they decide to help them answer the question and develop an evidence-based conclusion. Let students be creative.
- At the end of the activity, each group is asked to share what their procedures were to answer the question. It will be clear that each group had slightly different procedures but nonetheless all conclusions must be consistent with the collected evidence.

The teacher then leads the discussion to bring out NOSK and NOSI aspects.

Discussion

- Teachers should discuss the occurrence of different methods and conclusions and point out that one conclusion may not necessarily be better than another. The approach is guided by the question asked. Multiple viable conclusions are possible, as long as they are consistent with the available information. Different groups may have gone about studying the variable in a different way from another group. Methods of investigation influence the results. Interpretation of data may differ from group to group. Scientists have a framework through which to examine data just like the students do. This framework (prior knowledge or expectations) influences how that data are analyzed and interpreted. This subjectivity and creativity are a part of nature of science.
- Our understanding keeps changing as we get more and more information or as we look at something in a different way. Science is dynamic, not static.
- Regardless of your designs, the question being asked guided scientific investigations; also, knowledge from past experiences and science content influenced the work of scientists. Scientists use their prior knowledge and experiences to make predictions and decisions about their investigations and conclusions.
- Scientists cannot always control all variables in real systems. Models are created that control some of the variables and help enable the scientists to offer explanations and make predictions. Sometimes, none of the variables can be controlled. For example, investigations in astronomy and ecology often cannot be controlled. Manipulation is undesired. The goal might be to describe what is observed or to study relationships. These types of investigations are still science because they rely on observations of the natural world, but they do not require identification and manipulation of variables to do an investigation. There are multiple methods in science.

5.8 Assessing Students' and Teachers' Understandings of NOSK and NOSI in Classroom Practice and Research

Teachers need indicators of students' understandings of NOSK and NOSI to plan, conduct, and evaluate their lessons. So, it is important to have appropriate tools at hand. Below, we present two research instruments that were designed for summative assessment. However, both tools can as well be used by teachers in practice for formative assessment. That is, teachers can include similar questions during instruction to make NOSK and NOSI explicit. Some of the items can also be used on periodic exams during the year and within specific activities students perform in laboratory work. Of course, there are scoring schemes for both tests that can be obtained (see references given). However, since teachers in practice do not necessarily need to compare their class to a bigger sample, students answers can as well be evaluated by the teacher individually. Thus, the presented tests can be seen as a set of guiding questions that teachers can use in class. A subset of the questions that align with what the teacher is stressing can also be used. Students' answers will give valuable information how students in the own class understand NOSK and NOSI.

5.8.1 Assessing NOSK

Assessment of students' and teachers' understandings of NOSK began in 1957, but became more formalized in 1963. A detailed compendium of assessment for NOSK can be found in Lederman (2007). For over a decade, it has become clear that the most accurate assessments of NOSK was with the use of open-ended questions where respondents have a chance to elaborate on their views and provide examples to support their views. The two most widely used assessments are the VNOS-C and the VNOS-D+. Each of these instruments has been used with middle-school and high-school students, as well as undergraduate students. Although space does not permit a detailed scoring guide for these two instruments, such information can be found in Lederman et al. (2002). One of the two instruments mentioned above is presented here: the *Views About Nature of Science—D+ (VNOS D+)* questionnaire.

Instructions

- Please answer each of the following questions. You can use all the space provided and the backs of the pages to answer a question.
- Some questions have more than one part. Please make sure you write answers for each part.
- This is not a test and will not be graded. There are no "right" or "wrong" answers to the following questions. I am only interested in your ideas relating to the following questions.

1. What is science?
2. What makes science (or a scientific discipline such as physics, biology, etc.) different from other subject/disciplines (art, history, philosophy, etc.)?
3. Scientists produce scientific knowledge. Do you think this knowledge may change in the future? Explain your answer and give an example.
4. (a) How do scientists know that dinosaurs really existed? Explain your answer.
    (b) How certain are scientists about the way dinosaurs looked? Explain your answer.
    (c) Scientists agree that about 65 millions of years ago the dinosaurs became extinct (all died away). However, scientists disagree about what had caused this to happen. Why do you think they disagree even though they all have the same information?
    (d) If a scientist wants to persuade other scientists of their theory of dinosaur extinction, what do they have to do to convince them? Explain your answer.
5. In order to predict the weather, weather persons collect different types of information. Often, they produce computer models of different weather patterns.

    (a) Do you think weather persons are certain (sure) about the computer models of the weather patterns?
    (b) Why or why not?

6. The model of the inside of the Earth shows that the Earth is made up of layers called the crust, upper mantle, mantle, outer core and the inner core. Does the model of the layers of the Earth *exactly* represent how the inside of the Earth looks? Explain your answer.
7. Scientists try to find answers to their questions by doing investigations/experiments. Do you think that scientists use their imaginations and creativity when they do these investigations/experiments?

    (a) If *no*, explain why.
    (b) If *yes*, in what part(s) of their investigations (planning, experimenting, making observations, analysis of data, interpretation, reporting results, etc.) do you think they use their imagination and creativity? Give examples if you can.

8. Is there a difference between a scientific theory and a scientific law? Illustrate your answer with an example.
9. After scientists have developed a scientific theory (e.g., atomic theory, evolution theory), does the theory ever change? Explain and give an example.
10. Is there a relationship between science, society, and cultural values? If so, how? If not, why not? Explain and provide examples.

5.8.2 Assessing NOSI

As with NOSK, the most useful assessment of students' and teachers' understandings of NOSI are open-ended questions. The Views About Scientific Inquiry (VASI) is a valid and reliable assessment of NOSI. Details of its development and scoring can be found in Lederman et al. (2014). The assessment presented below has been successfully used with middle-school and high-school students, as well as college-level students: the *Views About Scientific Inquiry (VASI)* questionnaire.

Instructions

- Please answer each of the following questions. You can use all the space provided and the backs of the pages to answer a question.
- Some questions have more than one part. Please make sure you write answers for each part.
- This is not a test and will not be graded. There are no "right" or "wrong" answers to the following questions. I am only interested in your ideas relating to the following questions.

1. A person interested in birds looked at hundreds of different types of birds who eat different types of food. He noticed that birds who eat similar types of food, tended to have similar shaped beaks. For example, birds that eat hard-shelled nuts have short, strong beaks, and birds that eat insects have long, slim beaks. He wondered if the shape of a bird's beak was related to the type of food the bird eats and he began to collect data to answer that question. He concluded that there is a relationship between beak shape and the type of food birds eat.

    (a) Do you consider this person's investigation to be scientific? Please explain why or why not.
    (b) Do you consider this person's investigation to be an experiment? Please explain why or why not.
    (c) Do you think that scientific investigations can follow more than one method?

         If *no*, please explain why there is only one way to conduct a scientific investigation.
         If *yes*, please describe two investigations that follow different methods, and explain how the methods differ and how they can still be considered scientific.

2. Two students are asked if scientific investigations must always begin with a scientific question. One of the students says "yes" while the other says "no." Whom do you agree with and why?

3.

    (a) If several scientists ask the *same question* and follow the *same procedures* to collect data, will they necessarily come to the same conclusions? Explain why or why not.

(b) If several scientists ask the *same question* and follow *different procedures* to collect data, will they necessarily come to the same conclusions? Explain why or why not.

4. Please explain if "data" and "evidence" are different from one another.
5. Two teams of scientists were walking to their laboratory one day and they saw a car pulled over with a flat tire. They all wondered, "Are certain brands of tires more likely to get a flat?"

Team A went back to the laboratory and tested various tires' performance on one type of road surface.
Team B went back to the laboratory and tested one tire brand on three types of road surfaces.
Explain why one team's procedure is better than the other one.

6. The data table below show the relationship between plant growth in a week and the number of minutes of light received each day. Given this data, explain which one of the following conclusions you agree with and why.

Minutes of light each day	Plant growth-height (cm per week)
0	25
5	20
10	15
15	5
20	10
25	0

Please circle one:

(a) Plants grow taller with *more* sunlight.
(b) Plants grow taller with *less* sunlight.
(c) The growth of plants is *unrelated* to sunlight.

Please explain your choice of a, b, or c below:

7. The fossilized bones of a dinosaur have been found by a group of scientists. Two different arrangements for the skeleton are developed as shown (Figs. 5.4 and 5.5).
   (a) Describe at least two reasons why you think most of the scientists agree that the animal in Fig. 5.4 had the best sorting and positioning of the bones?
   (b) Thinking about your answer to the question above, what types of information do scientists use to explain their conclusions?

Fig. 5.4 Arrangement 1 of
skeleton

Fig. 5.5 Arrangement 2 of
the skeleton

5.9 Summary

Adequate views of the development and of characteristics of knowledge in physics
and science are part of the physics curriculum. That means, physics teaching should
not only deal with physics contents, but also with content relating to ways of
generating knowledge (NOSI) and characteristics of that knowledge (NOSK). This
is supported by arguments, which, for example, confirm that content learning is
supported by an adequate view of physics. Since many students have rather inadequate views of the natural sciences, the aim of teaching physics is to counteract this
image about physics. This is best achieved within a discursive and critical process
with different views, so that the complexity of NOSK and NOSI becomes clear. There
are no simple short statements or phrases about what are adequate views about NOSK
and NOSI. On the contrary, the appropriateness of the views lies in students' ability
to analyze and judge different perspectives. This teaching goal can be achieved by
an instruction that is not only reflective, but also explicit, integrated into the physics
content, authentic, and exemplary. In addition to address NOSK and NOSI in "usual"

lessons, NOSK and NOSI can be dealt with, in particular, in historical, research-orientated, physical–social, and philosophy-based contexts. Thus, it is essential that teachers have *content knowledge in physics, knowledge about NOSK and NOSI*, and *knowledge about teaching NOSK and NOSI* (such as student-centered, exploratory experimental learning). The latter is often termed as *pedagogical content knowledge of NOSK and NOSI* (Abd-El-Khalick 2013).

Acknowledgements We would like to thank Andreas Borowski (Universität Potsdam, Germany) and Jari Lavonen (University of Helsinki, Finland), who carefully and critically reviewed this chapter.

References

Teaching Resources

Bell RL (2008) Teaching the nature of science through process skills: activities for grades 3–8. Pearson, New York

Clough MP (2011) Teaching and assessing the nature of science: how to effectively incorporate the nature of science in your classroom. Sci Teach 78(6):56–60

Clough MP, Olson JK (2004) The nature of science: always part of the science story. Sci Teach 71(9):28–31

Höttecke D (2012) History and philosophy in science teaching: a European project. Sci Educ 21(9) (special issue)

Lederman NG (Website, n.d.) Resources for teaching and assessment of nature of scientific knowledge and scientific inquiry. Retrieved from https://ledermanscience.app.box.com/s/1l6hopa5k2wi nok8fzrcl4qgvg2khv32

Lederman NG, Abd-El-Khalick F (1998) Avoiding de-natured science: activities that promote understandings of the nature of science. In: McComas WF (ed) The nature of science in science education: rationales and strategies. Kluwer Academic Publishers, The Netherlands, pp 83–126

McComas WF (2019) Nature of science in science instruction: rationales and strategies. Springer Publishing, Dordrecht, The Netherlands

National Academy of Sciences (1998) Teaching about evolution and the nature of science. National Academies Press, Washington, DC

References

Abd-El-Khalick F (2013) Teaching with and about nature of science, and science teacher knowledge domains. Sci Educ 22:2087–2107

Abd-El-Khalick F, Lederman NG (2000) Improving science teachers' conceptions of the nature of science: a critical review of the literature. Int J Sci Educ 22(7):665–701

Allchin D (2017) Beyond the consensus view: whole science. Can J Sci Math Tech Educ 17(1):18–26

American Association for the Advancement of Science (1990) Science for all Americans. Oxford University Press, New York, NY

American Association for the Advancement of Science (1993) Benchmarks for science literacy. Oxford University Press, New York. https://doi.org/10.1080/14926156.2016.1271921

American Association for the Advancement of Science (1994) Benchmarks for science literacy. Oxford University Press, New York

Angier N (2010) STEM education has little to do with flowers. The New York Times, D2

Ayala FJ (2008) Science, evolution, and creationism. Proc Natl Acad Sci USA 105(1):3–4. https:// doi.org/10.1073/pnas.0711608105

Bell RL (2008) Teaching the nature of science through process skills: activities for grades. Pearson, New York, pp 3–8

Bronowski J (1956) Science and human values. Harper & Row Publishers Inc., New York

Bybee RW (2013) The case for STEM education: challenges and opportunities. NSTA Press, Washington, DC

Bybee RW (2013) Translating the NGSS for classroom instruction. NSTA Press, Arlington, VA

Central Association for Science and Mathematics Teachers (1907) A consideration of the principles that should determine the courses in biology in secondary schools. Sch Sci Math 7:241–247

Chalmers AF (1999) What is this thing called science? University of Queensland Press, Queensland, AU

Chalmers AF (2007) Wege der Wissenschaft [What is this thing called science?]. Springer, Berlin

Clough MP (2011) Teaching and assessing the nature of science: how to effectively incorporate the nature of science in your classroom. Sci Teach 78(6):56–60

Clough MP, Olson JK (2004) The nature of science: always part of the science story. Sci Teach 71(9):28–31

Czerniak CM (2007) Interdisciplinary science teaching. In: Abell SK, Lederman NG (eds) Handbook of research on science education. Lawrence Erlbaum Publishing, Mahwah, NJ, pp 537–560

Czerniak CM, Johnson CC (2014) Interdisciplinary science teaching. In: Lederman NG, Abell SK (eds) Handbook of research on science education. Routledge, New York, pp 395–411

DeBoer GE (2000) Scientific literacy: another look at its historical and contemporary meanings and its relationship to science education reform. J Res Sci Teach 37(6):582–601

Dewey J (1916) Democracy and education. The Free Press, New York

Driver R, Leach J, Millar R, Scott P (1996) Young people's images of science. Open University Press, Buckingham

Erduran S, Dagher ZR (2014) Reconceptualizing the nature of science for science education. Springer, Dordrecht, The Netherlands

Feyerabend PF (1975) Against method: outline of an anarchistic theory of knowledge. Redwood, Burn Limited, Great Britain

Feynman RP (1965) The character of physical law. MIT Press, Cambridge, MA

Feynman RP, Cashman D (2013) The pleasure of finding things out. Blackstone Audio, Incorporated

Fund RB (1958) The pursuit of excellence: education and the future of America. Panel report V of the Special Studies Project. Doubleday

Goodrum D, Rennie LJ, Hackling MW (2001) The status and quality of teaching and learning of science in Australian schools: a research report. Department of Education, Training and Youth Affairs, Canberra

Gould SJ (1981) The mismeasure of man. W.W. Norton and Company, New York

Hodson D, Wong SL (2017) Going beyond the consensus view: broadening and enriching the scope of NOS-oriented curricula. Can J Sci Math Technol Educ 17(1):3–17

Hoffman R, Torrence V (1993) Chemistry imagined: reflections on science. Smithsonian Institution Press, Washington, DC

Höttecke D (2001) Die Natur der Naturwissenschaften historisch verstehen [understanding the nature of science historically]. Logos, Berlin

Höttecke D (2012) History and philosophy in science teaching: a European project. Sci Educ 21(9) (Special issue)

Hrdy SB (1986) Empathy, polyandry, and the myth of the coy female. In: Bleier R (ed) Feminist approaches to science. Perganon Publishers, pp 119–146

Hurd PD (1958) Science literacy: its meaning for American schools. Educ Leadersh 16(1):13–16

Kind P, Osborne J (2016) Styles of scientific reasoning: a cultural rationale for science education? Sci Educ 101(1):8–31. https://doi.org/10.1002/sce.21251

Klopfer LE, Cooley WW (1963) The history of science cases for high schools in the development of student understanding of science and scientists. J Res Sci Teach 1(1):33–47

KMK—Sekretariat der Ständigen Konferenz der Kultusminister der Länder in der Bundesrepublik Deutschland (2004) Bildungsstandards im Fach Physik für den Mittleren Schulabschluss [Science standards for middle school graduation for the school subject physics]. Wolters Kluwer, München

Kuhn TS (1962) The structure of scientific revolutions. The University of Chicago Press, Chicago

Lakatos I (1983) Mathematics, science, and epistemology. Cambridge University Press, Cambridge, UK

Laudan L (1977) Progress and its problems: towards a theory of scientific growth. University of California Press, Berkeley, CA

Laugksch RC (2000) Scientific literacy: a conceptual overview. Sci Educ 84(1):71–94

Lederman JS, Lederman NG, Bartos SA, Bartels SL, Antink Meyer A, Schwartz RS (2014) Meaningful assessment of learners' understandings about scientific inquiry—the views about scientific inquiry (VASI) questionnaire. J Res Sci Teach 51(1):65–83

Lederman NG, Wade P, Bell RL (1998) Assessing understanding of the nature of science: a historical perspective. In: McComas W (ed) The nature of science in science education. Springer, Dordrecht, pp 331–350

Lederman NG (2007) Nature of science: past, present, and future. In: Abell SK, Lederman NG (eds) Handbook of research on science education. Lawrence Erlbaum Associates, Mahwah, NJ, pp 831–880

Lederman NG (Website, n.d.) Resources for teaching and assessment of nature of scientific knowledge and scientific inquiry. Retrieved from https://ledermanscience.app.box.com/s/1l6hopa5k2wi nok8fzrcl4qgvg2khv32

Lederman NG, Abd-El-Khalick F (1998) Avoiding de-natured science: activities that promote understandings of the nature of science. In: McComas WF (ed) The nature of science in science education: rationales and strategies. Kluwer Academic Publishers, The Netherlands, pp 83–126

Lederman NG, Lederman JS (2014) Research on teaching and learning of nature of science. In: Lederman NG, Abell SK (eds) Handbook of research on science education, vol II. Routledge, New York, pp 600–620

Lederman NG, Lederman JS (2004) The nature of science and scientific inquiry. Art Teach Sci 2–17

Lederman NG, Abd-El-Khalick F, Bell RL, Schwartz RS (2002) Views of nature of science questionnaire: toward valid and meaningful assessment of learners' conceptions of nature of science. J Res Sci Teach 39(6):497–521

Lederman NG, Lederman JS, Antink A (2013) Nature of science and scientific inquiry as contexts for the learning of science and achievement of scientific literacy. Int J Educ Math Sci Technol 1(3)

Lederman N, Adb-El-Khalick F (1998) Avoiding de-natured science: activities that promote understandings of the nature of science. In: McComas W (ed) The nature of science in science education. Springer, Dordrecht, pp 83–126

Lovejoy CO (1981) The origin of man. Science 211:341–350

MacCoun R, Perlmutter S (2015) Hide results to seek the truth. Nature 526:187–189

Matthews M (2000) Time for science education. Springer, Dordrecht, The Netherlands

Mayr E (1988) Toward a new philosophy in biology. Harvard University Press, Cambridge, MA

McComas W (1998) The principle elements of the nature of science: dispelling the myths. In: McComas W (ed) The nature of science in science education. Springer, Dordrecht, pp 53–70

McComas WF (2019) Nature of science in science instruction: rationales and strategies. Springer Publishing, Dordrecht, The Netherlands

McComas W, Almazroa H, Clough M (1998) The nature of science in science education: an introduction. Sci Educ 7(6):511–532

Messeri L (2010) The problem with pluto: conflicting cosmologies and the classification of planets. Soc Stud Sci 40(2):187–214

Moore J (1993) Science as a way of knowing: the foundation of modern biology. Harvard University Press, Cambridge, MA

National Academy of Sciences (1998) Teaching about evolution and the nature of science. National Academies Press, Washington, DC

National Education Association (1918) Cardinal principles of secondary education: a report of the commission on the reorganization of secondary education. U.S. Bureau of Education Bulletin No. 35. U.S. Government Printing Office, Washington, DC

National Education Association (1920) Reorganization of science in secondary schools: a report of the commission on the reorganization of secondary education. U.S. Bureau of Education Bulletin No. 20. U.S. Government Printing Office, Washington, DC

National Research Council (1996) National science education standards. National Academy Press, Washington, DC

National Research Council (2000) Inquiry and the national science education standards. National Academy Press, Washington, DC

National Research Council (2008) Research on future skill demands: a workshop summary. National Academies Press, Washington, DC

National Research Council (2010) Exploring the intersection of science education and 21st century skills: a workshop summary. National Academies Press, Washington, DC

National Research Council (2012) A framework for K–12 science education: practices, crosscutting concepts, and core ideas. National Academies Press, Washington, DC

National Science Teachers Association (1982) Science-technology-society: science education for the 1980s. Author, Washington, DC

National Science Teachers Association (2000) The nature of science: NSTA position statement. Author, Arlington, VA

National Science Teachers Association (2018) Transitioning from scientific inquiry to three-dimensional teaching and learning. Author, Arlington, VA

National Society for the Study of Education (1960) Rethinking science education. In: Yearbook of the national society for the study of education, vol 59. University of Chicago Press, pp 113

NGSS Lead States (2013) Next generation science standards: for states, by states. National Academies Press, Washington, DC. Retrieved from www.nextgenscience.org/next-generation-science-standards

OECD (2017) PISA for development brief—2017/2 (February). Retrieved from https://www.oecd.org/pisa/pisa-for-development/10-How-PISA-D-measures-science-literacy.pdf

Pleasants J, Olson JK (2019) What is engineering? Elaborating the nature of engineering for K-12 education. Sci Educ 103(1):145–166. https://doi.org/10.1002/sce.21483

Popper KR (1968) The logic of scientific discovery. Harper & Row Publishers, New York

Popper KR (1988) The open universe: an argument for indeterminism. Routledge, London

Priemer B, Eilerts K, Filler A, Pinkwart N, Rösken-Winter B, Tiemann R, Upmeier zu Belzen A (2019) A framework to foster problem-solving in STEM and computing education. Res Sci Technol Educ. https://doi.org/10.1080/02635143.2019.1600490

Ruse M (ed) (1998) Philosophy of biology. Prometheus Books, New York

Showalter VM (1974) What is unified science education? Program objectives and scientific literacy. Prism 2(3–4):1–6

UK Department for Education (2015) National curriculum in England: science programmes of study. Retrieved from https://www.gov.uk/government/publications/national-curriculum-in-england-science-programmes-of-study/national-curriculum-in-england-science-programmes-of-study

World Weather Attribution Project (Website, n.d.). Retrieved from https://www.worldweatherattribution.org

Chapter 6
Instructional Coherence and the Development of Student Competence in Physics

Knut Neumann and Jeffrey Nordine

Abstract The world is rapidly changing, driven by ever-accelerating scientific progress (e.g. Organisation for Economic Cooperation and Development (OECD) in The future of education and skills: education 2030. OECD, 2018). The number of problems with conceptual or technological ties to science that societies as a whole as well as individuals in their everyday lives are facing is continuously increasing (e.g. Sadler in Journal of Research in Science Teaching: The Official Journal of the National Association for Research in Science Teaching 41(5):513–536, 2004). In order to meet these problems, individuals need more than knowledge of isolated science ideas or skill to engage in isolated scientific practices. Instead, individuals need to develop a competence in science that enables them to explain phenomena and solve problems in their everyday lives and prepares them for future learning and development (OECD in PISA 2015 assessment and analytical framework: science, reading, mathematic, financial literacy and collaborative problem solving. OECD, 2017, p. 20). Physics education should contribute to the development of such competence (Sekretariat der Ständigen Konferenz der Kultusminister der Bundesrepublik Deutschland [KMK] in Beschlüsse der Kultusministerkonferenz – Bildungsstandards im Fach Physik für den Mittleren Schulabschluss (Jahrgangsstufe 10). München Neuwied, 2005). The development of competence in physics requires engaging students in the scientific endeavour to learn about a small number of core physics ideas (e.g. energy). Developing such competence requires coherent effort over multiple years of schooling. Single lessons or single instructional units do not suffice in developing the envisioned competence. Instead, the development of competence requires instruction to cohere with educational aims, across multiple lessons (i.e. within a unit), across units (i.e. within and across school years). This chapter discusses the notion of competence in science as the vision for twenty-first-century science education. It further argues that such competence requires coherent efforts across multiple years of schooling and delineates this idea with several examples from coherent

K. Neumann (✉) · J. Nordine
IPN–Leibniz Institute for Science and Mathematics Education, Kiel, Germany
e-mail: neumann@leibniz-ipn.de

J. Nordine
University of Iowa, Iowa City, IA, USA

© Springer Nature Switzerland AG 2021
H. E. Fischer and R. Girwidz (eds.), *Physics Education*, Challenges in Physics Education,
https://doi.org/10.1007/978-3-030-87391-2_6

151

curricula as well as corresponding assessments to monitor student learning across grades and grade bands from K to 12. The chapter concludes by discussing the role of learning progressions as a basis for organizing coherent instruction and thus allowing for a smooth progression of students in developing competence in physics and science, respectively, over extended periods of schooling.

6.1 Introduction

The twenty-first century places new demands on students' education. Technological advancements have transformed the workplace. Information technologies provide access to almost any information almost anywhere; and production technologies have taken over many of the tasks originally performed by humans in manufacturing. As a result, the ability to competently operate and maintain such technologies has taken precedence over the knowledge of specific information and selected manufacturing skills as a prerequisite for finding a job. At the same time, the technological advancements have created a range of challenges. Such challenges include, for example, the need to meet the increased demand for energy or water, the changing climate, or the complex flows of pathogens, information and goods in a globalized world. Taking up these challenges requires exploring their origins and implications, as well as potentially developing and testing solutions. Education cannot possibly provide individuals with all the individual abilities and skills needed to meet these challenges, in part because some of these challenges are not even known yet. As a consequence, education can no longer be about providing students with specific knowledge or select skills. Instead, education must provide students with the competence that prepares them for future learning, enabling occupational, societal and cultural participation.

The competence required for occupational, societal and cultural participation differs substantially from previous goals of science or physics education, respectively. Instead of comprehensive knowledge about, for example, physics, students are now expected to know only about the most core ideas of the domain in question. In addition, instead of a broad range of additional skills, students are envisioned being able to integrate their knowledge about the core ideas with the core skills of the domain to make sense of authentic phenomena or problems, engage in further learning about them and develop explanations or solutions, respectively. For example, in order to make informed decisions about local or national policies regarding the production and storage of energy, it is essential to have knowledge of energy and be able to actively use this knowledge in seeking and evaluating information and making decisions. Such competence requires (deep) understanding (e.g. Bransford et al. 2000), or integrated knowledge (e.g. Linn 2006), but includes going beyond simply knowing in order to successfully activate knowledge relevant for making sense of meaningful phenomena and problems. Such successful activation is also known as knowledge-in-use (e.g. Harris et al. 2015).

Fostering students' development of the envisioned competence is a complex effort that requires engaging students in constructing explanations of phenomena or finding solutions to problems as contexts in which students can learn in conjunction about the core ideas and master the core skills in physics, or more broadly, science. Just providing students with opportunities to acquire select knowledge and skills does not suffice. Instruction needs to be organized such that students use their skills to extend their knowledge, or their knowledge to refine their skills. Accordingly, instructional activities must require students to combine both aspects of the envisioned competence in differing situations, while also systematically varying the combinations of knowledge and skills required to address phenomena and problems in order to support students in developing robust competence that spans a range of meaningful contexts (see Schwartz and Goldstone 2016). Science education has so far largely failed in developing the envisioned competence in students (OECD 2013, 2016). Reasons for this failure include a historical emphasis on the breadth instead of the depth of the addressed content (Schmidt et al. 1997), the engagement in inquiry activities in the absence of meaningful content (Kirschner et al. 2006), and a lack of coherence in instruction within and across grades (National Research Council [NRC] 2007).

In this chapter, we focus on the idea of instructional coherence as a means to support students in developing competence. We begin by reviewing the idea of competence, its origins, meaning and recent delineations, and theoretical frames for guiding the development of student competence—so-called learning progressions or models of competence. We then introduce the idea of instructional coherence, discuss different forms or types of instructional coherence and present promising strategies for establishing instructional coherence. We conclude by reviewing issues closely related to instructional coherence, such as assessment practices, supporting teachers, systemic requirements, all of which are affected by the assertions we are making about the role of instructional coherence in developing the competence in science that students needed for occupational, societal and cultural participation.

6.2 Student Competence

The idea of competence is not new. White (1959) used the term *competence* to describe an individual's ability to effectively interact with the environment, and McClelland (1973) did so to acknowledge that a single cognitive ability such as intelligence would not sufficiently explain successful the performance of more complex, authentic tasks such as driving a car.

6.2.1 Defining Student Competence

The idea of competence as the capacity to perform more complex and authentic tasks successfully has made it a particularly popular concept in the context of occupational education and training (Franke 2005). In contrast to general education, the aims of occupational education and training are typically defined according to the range of activities associated with a particular occupation. For example, a competent car mechanic should be able to diagnose a faulty car and explain his findings to a customer. The finding, that cognitive abilities such as intelligence were poor predictors of successful performance of such tasks (e.g. McClelland 1973; Pottinger and Goldsmith 1979), has led to an outcome-oriented approach, analysing the respective tasks for the combinations of knowledge and skills needed to perform them successfully (Winterton et al. 2006, p. 14).

The initiation of PISA, the international large-scale assessment programme, has led to a renewed popularity of the idea of competence with regard to science education. Prior to PISA, large-scale international assessments focused on knowledge and skills at the intersection of curricula across countries (Beaton et al. 1996). These large-scale assessments repeatedly demonstrated that, while students could successfully reproduce factual knowledge and schematically apply skills to routine tasks, they struggled to apply their knowledge to solve more complex problem-solving tasks. In addition, students largely failed at modelling, explanation or argumentation tasks (Baumert et al. 2000, 2002). The PISA Framework (OECD 1999), which represents the next generation of large-scale international assessments, therefore, instead of representing the curricular core of the participating countries, aimed to articulate the goals of twenty-first-century science education. More specifically, the Framework articulated the competence in science that students need to develop in order to be prepared for future, life-long learning and thus for occupational, societal and cultural participation. The Framework defined competence as 'the capacity to use scientific knowledge, to identify questions and to draw evidence-based conclusions in order to understand and help make decisions about the natural world and the changes made to it through human activity' (OECD 1999, p. 60).

Students' repeated subpar achievement in PISA led to a critical review of the aims of science education in many countries (for an overview see Waddington et al. 2007). In Germany, for example, the substantial number of students failing to meet the minimum level of competence required for occupational, societal and cultural participation led to the introduction of science education standards (for details see Neumann et al. 2010). These standards are based on a definition of competence given by Weinert (2001), according to which competence entails the cognitive abilities as well as the motivational and volitional attitudes needed to solve problems across a variety of different situations (p. 27ff). The standards list a range of performance statements that define the competence students are expected to develop in science in four areas:

1. **Content knowledge** (e.g. use analogies to solve tasks or problems),
2. **scientific inquiry** (e.g. plan, perform and analyse simple experiments),
3. **communication** (e.g. adequately present results of their work) and
4. **evaluation** (e.g. use their knowledge to evaluate risks and security measures in experiments, in their everyday lives and with respect to modern technologies).

Students who would be able to show most or all of the performances listed would be considered to have reached the level of competence envisioned as a result from schooling.

In the USA, the discussion led to a revision of the existing science education standards (NRC 2012). A National Research Council committee was charged with the task of developing a Framework for articulating new science education standards. These standards were expected to lay out the expectations for the competence students are expected to have developed at the end of formal schooling (i.e. 12th grade). The committee outlined that competent students have:

> some appreciation of the beauty and wonder of science; possess sufficient knowledge of science and engineering to engage in public discussions on related issues; are careful consumers of scientific and technological information related to their everyday lives; are able to continue to learn about science outside school; and have the skills to enter careers of their choice, including (but not limited to) careers in science, engineering, and technology. (NRC 2012, p. 1)

Based on this definition, the Framework identified three interdependent dimensions of student competence in science: disciplinary core ideas (e.g. energy, evolution), science and engineering practices (e.g. modelling, argumentation) and crosscutting concepts (e.g. mechanisms of cause and effect). The level of competence students should have reached by the end of high school is marked by the ability to integrate knowledge about one or more core idea(s) with knowledge about one more crosscutting concept, while engaging in one or more science/engineering practices to make sense of phenomena or solve real-world problems.

In summary, the idea of competence has a long-standing history as a construct to explain successful or unsuccessful performance within a domain and as a means to outline the aims of education in this domain. Not surprisingly, this has resulted in a multitude of differing definitions of the term competence (for an overview see Winterton et al. 2006). The fact that sometimes the ability to exhibit a particular performance (e.g. plan, perform and analyse experiments) or combination of traits (e.g. modelling all kinds of situations using knowledge about energy) is referred to as 'a competence' has added to the confusion. However, three features have emerged as common themes across the different definitions of competence. First, competence refers to the construct underlying students' performance across a range of different situations. Second, competence is specific to a domain in the sense that it explains an individual's performance only across situations from a specific domain. Third, competence represents a conglomerate of different knowledge, abilities and skills, including meta-cognitive skills such as self-regulation (Erpenbeck 1997). While some definitions also include non-cognitive dispositions such as beliefs (e.g. Baumert and Kunter 2013), most conceptualizations of competence include only cognitive abilities and skills—mostly for the sake of simplifying assessment (Koeppen et al. 2008, p. 62).

6.2.2 Delineating Student Competence

The nature of competence as a complex construct requires to delineate the construct into subconstructs (defined by certain groups of tasks) or to delineate the specific knowledge, skills and abilities that makeup competence in a domain and how they do so. That is, it is necessary to specify (a) the situations and demands that require the corresponding competence, as well as (b) criteria that specify what it means to cope with these demands successfully. This process is also referred to as *competence modelling* and result in a *model of competence* or *competence model* (Ufer and Neumann 2018; see Leutner et al. 2017).

Models of competence commonly delineate student competence into areas of competence and define levels of competence with respect to these areas (e.g. Neumann et al. 2010) or into the individual knowledge, skills and abilities constituting competence (e.g. Neumann et al. 2019). For example, in the case of the German science education standards, a model has been proposed that describes competence in each of the four areas of competence identified by the standards, as a combination of the complexity of students' knowledge (e.g. the knowledge of facts vs. conceptual understanding), and cognitive processes in using this knowledge (e.g. remembering vs. elaborating). Different levels of competence are then characterized by a hierarchy among the elements of each dimension (Neumann et al. 2010), so that a more competent student can engage in higher-order cognitive processes.

In case of the US Framework for *K*-12 Science Education, the model describes student competence in science in terms of three dimensions: For each of four major domains of science, the Framework identifies a small set of disciplinary core ideas (DCIs), the knowledge of which is needed to make sense of the wealth of scientific information. However, just knowledge of DCIs is not enough to make sense of phenomena or develop solutions to design problems. In addition to the DCIs, the Framework lists a set of eight science and engineering practices (SEPs), practices that scientists and engineers employed in their attempt to investigate and explain the world and design solutions to practical problems. Finally, in addition to the DCIs and the SEPs, the Framework names seven crosscutting concepts (CCCs). The CCCs are a set of conceptual tools that are critical to applying the DCIs and SEPs in practice and that are used across all science and engineering domains. The Framework emphasizes that these three dimensions (DCIs, SEPs and CCCs) do not exist in isolation from each other. Instead, the Framework calls for engaging students in the SEPs in conjunction with DCIs and CCCs to make sense of phenomena or find solution to problems and in so doing develop deeper knowledge of the dimensions.

In summary, the idea of competence in a domain (e.g. physics or science) as such is somewhat broad and requires further delineation through models of competence. Such models commonly specify the knowledge, abilities and skills that constitute competence in a domain. However, designing instruction that supports students in developing competence in a domain requires to not only awareness of the knowledge, abilities and skills that are specific to that domain, but also how students develop them; and, more importantly, knowledge of how to combine them into true competence in the respective domain.

6.3 Developing Student Competence

Models of competence underlie the formulation of standards or, in case of the German Science Education standards, delineate the knowledge, skills and abilities underlying a set of standards. However, to support students in developing the competence envisioned, instruction is needed that systematically works towards that goal. This is one of the most neglected yet core challenges of science education: systematically building on students' current level of competence and developing it further through instruction. Students are not taught by the same teacher throughout their whole school career. Teachers change when moving from elementary to middle school, and oftentimes within middle school. Nevertheless, how does a new teacher know where his students are in their endeavour to develop competence in science? What level of competence have they reached, and what should be the next steps to take? Obviously, the teacher needs to assess students' level of competence—but against what metric? What can the teacher reasonably expect from students? This is where models of competence development come in. Models of competence development describe how students' competence in a domain typically develops or would typically develop under ideal conditions. That is, models of competence are progressions of students' learning (or: learning progressions, see Duncan and Rivet 2018) that describe ideal trajectories of students' learning. Such models of competence development or learning progressions exist at different levels of grain size. Obviously, the most coarse-grained version would be a model describing how students' competence develops over grades or even grade bands, rather than within grades.

Models of competence development spanning multiple grades or even grade bands can guide the development of standards, as well as state- or school-level curricula. In case of the German standards, models of competence development have been developed and studied to guide the development of so-called core curricula. State-wide curricula break down the competence envisioned by the standards into smaller aspects called performance expectations that can reasonably be expected by students at the end individual grades or grade bands. In case of the USA, the Next Generation Science Standards (NGSS) already arrange the performance expectations across grades and grade bands from K to 12 so that they describe a progression in developing competence in science over extended periods of schooling—based on respective findings from science education research. The NGSS lists performance expectations for each grade from *K*-5 and for the middle and high school grade bands. Each performance expectation (PE) is written as a statement integrating elements of a SEP with elements of a DCI and a CCC and provides an example of the type of performance a student who has completed the respective grade or grade band should be able to demonstrate. By the end of high school, for example, students are expected to be able to '[c]reate a computational model to calculate the change in the energy of one component in a system when the change in the energy of the other component(s) and energy flows in and out of the system are known' (NGSS Lead States 2013, HS-PS3-1,). This performance expectations integrates the development of a scientific and engineering practice called computational modelling (i.e. the use of mathematics and computational thinking, SEP-5) with a disciplinary core idea about changes in energy and energy transfers (i.e. energy, DCI-PS3) into and out of a system (i.e. systems and system models, CCC-4). Since DCIs, CCCs and SEPs are large and complex concepts, the NGSS breaks each of them down into smaller elements at a particular grade level or grade band.[1] A PE at a particular grade level or band therefore represents a snapshot of what students should be able to do at this grade level or band on their way of developing competence in science throughout their K-12 experience. By the end of middle school, for example, students are expected to be able to '[d]evelop a model to describe that when the arrangement of objects interacting at a distance changes, different amounts of potential energy are stored in the system' (NGSS Lead States 2013, MS-PS3-2). At the end of 5th grade, students are supposed to be able to '[u]se models to describe that energy in animals' food (…) was once energy from the sun' (NGSS Lead States 2013, 5-PS3-1). Whereas the high school PE involves any kind of change in energy due to energy flows into or out of the system, the middle school PE focuses only on changes to potential energy. The system is supposed to involve fewer components, and the model can be a representation, diagram, picture or a written description. The models referred to in the 5th grade PE include only diagrams or flowcharts, and students are only expected to know that energy released from food came from the sun and was captured by plants. These example PEs show that the elements of the SEPs, DCIs and CCCs included in a PE are more complex and sophisticated at upper grades or grade bands. Across grades, PEs represent the learning that develops as students' competence at using their knowledge of the DCIs and the CCCs while engaging in the SEPs grows deeper and more integrated. That is, the PEs and thus the NGSS describe how students are expected to develop competence in science from K to 12 by building on previous knowledge.

In modelling the development of competence, it is important to note that students essentially must progress in each of the dimensions. In case of the NGSS, for example, students do not just progress in their knowledge about the DCIs but also in their knowledge about the CCCs and their abilities and skills to engage in the scientific and engineering practices. In fact, over the past decades, models of competence development (or learning progressions respectively) have been developed for a broad range of ideas, concepts and practices. This includes learning progressions for the concept of energy (e.g. Lee and Liu 2010; Neumann et al. 2013), matter (Hadenfeldt et al. 2016; Smith et al. 2006) or force (Alonzo and Steedle 2009). Learning progressions have also been proposed for scientific practices such as the developing and using models (Schwarz et al. 2009), constructing explanations (e.g. Berland and McNeill 2010), or argumentation (e.g. Osborne et al. 2016). These learning progressions vary in their grain size (i.e. the time they span and the level of detail identified in the progression). However, all of them articulate an upper anchor defined by the envisioned outcomes to be supported by the learning progression, a lower anchor describing students learning prerequisites upon entering the learning progression

---

[1] Appendix F—Scientific and Engineering practices of the NGSS, for instance, presents the elements associated with each practice at each grade band.

and intermediate steps. The energy learning progression proposed by Neumann et al. (2013), for example, identifies four aspects of the energy concept that students need to understand:

1. that energy manifests itself in different forms,
2. that energy can be converted from one form into another,
3. that whenever an energy conversion happens, some of the energy is converted into thermal energy that dissipates into the environment,
4. the total energy in an isolated system remains constant over time.

Neumann et al. (2013) then show that students' progress in their learning about energy by successively developing understanding of these ideas in the above order (i.e. forms, transformation, dissipation, conservation). Interestingly, this does not imply that students do not begin learning about conservation early in their schooling. They will however develop an understanding of energy that entails the idea of conservation last (if at all). This finding is consistent with much of the literature on students learning about energy (e.g. Duit 1984; Lee and Liu 2010; Liu and McKeough 2005; Trumper 1993). Later, Neumann and Nagy (2013) suggested that the mechanism underlying students progression may be best described by a knowledge integration approach, where the integration of higher-order aspects (e.g. conservation) requires an integrated knowledge about lower-order aspects. The *Framework* and the NGSS build on a synthesis of existing work in each of the three dimensions (see Appendix E, F and G). The NGSS then integrates the synthesis about students learning about each of the dimensions into PEs that reflect students' progression in each of the dimensions.

In summary, since competence is a complex, multifaceted construct, developing student competence is a long-term, multi-year effort. In fact, some aspects of student competence require a repeated, systematic and coherent effort from K to 12. In order to guide the design of instruction, models are needed that describe how students would ideally progress in their learning about certain aspects of student competence. For most ideas or concepts, as well as for practices, science education research has developed so-called learning progressions describing idealized trajectories of students learning for individual grades, across multiple grades and even across grade bands. It is important to note that although instruction often foregrounds ideas or concepts, it is important to attend to students' progress in the other dimensions of competence as well. A unit on energy may, for example, be based on a particular sequencing of ideas about the energy concept (e.g. introducing energy as an idea, followed by the learning about energy forms, learning about the transformation of energy from one of these forms into another, etc.). However, in order to support students in developing competence in science they will need to be engaged in the construction and use of models, the planning, performance and analysis of experiment, or the construction explanations; and the respective activities need to be organized in alignment with models of how students best develop these aspects of their competence as well. This is the reason, why supporting students in developing competence requires a coherent effort across multiple years of schooling.

6.4 Instructional Coherence and the Development of Competence

The idea of developing competence in a domain as an educational goal and the idea of instructional coherence are closely related (Fortus and Krajcik 2012). The idea of competence became popular because students had been found to possess broad factual knowledge in domains such as science, but not the ability to explain phenomena or solve problems in this domain (e.g. Baumert et al. 2002; Baumert et al. 2000). Research suggested that this was paralleled by curricula which were characterized by comprehensive coverage of content instead of any significant depth with which the content was covered (Schmidt et al. 1997). This led to calls for more coherent curricula, which focused on fewer, core ideas and sequencing these ideas in a way that supports cumulative learning (Schmidt et al. 2005).

The idea of curriculum coherence has gained traction since then (for an overview see Kali et al. 2008). Over the years, the idea has continued to be elaborated and has become broader and more comprehensive. Fortus and Krajcik (2012), for example, subsume under curriculum coherence the systems through which educational goals build on each other across grade bands; the congruence of instructional activities and learning goals, a sequencing of learning goals aimed at supporting students in developing a deeper understanding, as well as approaches for ensuring that instructional units build on each other. Research findings confirm the importance of educational goals systematically sequenced across grade bands (e.g. Schmidt et al. 2005) and the careful sequencing of learning goals within instructional units (e.g. Drollinger-Vetter 2011) as well as across units for student performance (e.g. Fortus et al. 2015). These findings highlight the importance of the coherence of the implemented curriculum, or the instruction. While curriculum may be classified as coherent or not (e.g. Kesidou and Roseman 2002), students do not directly experience curriculum, but rather instruction. Thus, it is critical to consider how students experience coherence within classroom instruction (Sikorski and Hammer 2017); we therefore focus on the importance of 'instructional coherence' for supporting the development of student competence.

Based on Fortus and Krajcik (2012), we differentiate between four types of instructional coherence: (1) coherence of educational goals, (2) coherence of instructional activities, (3) coherence of instructional units and (4) coherence between instructional units. The coherence of educational goals constitutes the degree to which educational goals (e.g. performance expectations for grades or grade bands) are arranged in a meaningful way. The coherence of instructional activities represents the fit between the learning goals for an instructional activity and the actual activity (i.e. does the activity support meeting the learning goal). The coherence of instructional units refers to the extent to which the sequence of activities within a unit is suitable to support students in meeting the goals of the unit. The coherence between units similarly refers to the extent to which different units build on each other such that they support meeting overarching educational aims over time. These educational aims are of course informed by coherence in educational goals. Thus, the four types of instructional coherence are mutually reinforcing.

6.5 Designing Coherent Instruction

6.5.1 Coherence of Educational Goals

Coherent instruction begins with clearly articulated instructional goals. In most places, instructional goals begin with curriculum standards or frameworks relevant to the context, such as the NGSS in the USA (NGSS Lead States 2013), the Bildungsstandards in Germany (KMK 2005) or the National Core Curriculum in Finland (Vahtivuori-Hänninen et al. 2014). These and other documents provide specific statements regarding outcomes that are central characteristics of competence, but it is important to note that these statements on their own are not sufficient for the design of instruction. Educational standards are written either broadly, to describe general student competencies, e.g. 'Die Schülerinnen und Schüler nutzen diese Kenntnisse zur Lösung von Aufgaben und Problemen [Students use their knowledge of physics to solve tasks and problems]' (KMK 2005, p. 11); or, they are written specifically to describe specific measurable performances, e.g. 'Use mathematical representations to support the claim that the total momentum of a system of objects is conserved when there is no net force on the system' (NGSS Lead States 2013, HS-PS2-2). Whether written broadly or specifically, standards statements alone simply do not convey either the breadth or depth of competence that students should demonstrate within a single instructional unit. For example, the NGSS standard HS-PS2-2 states that students should engage in the practice 'use mathematical representations', yet this practice is not the only practice that students should be able to use in the context of momentum conservation. They should also, for example, be able to design investigations, construct arguments and communicate conclusions. In this way, the standard alone does not describe the breadth of competence that students should achieve. Similarly, this standard does not specify the target depth of competence. The statement 'use mathematical representations' does not identify which types of mathematical representations should be used or their expected level of sophistication. The NGSS offers some guidance regarding the expected level of sophistication by providing an 'assessment boundary' for this standard that states 'Assessment is limited to systems of two macroscopic bodies moving in one dimension' (NGSS Lead States 2013, HS-PS2-2). However, the mathematical symbols and concepts to be used (e.g. describing momentum only before/after a collision or calculating rates of change during an interaction) are left to the teacher's discretion.

Whether they are written broadly or specifically, standards statements alone cannot fully specify the range of learning goals that students should achieve during instruction. To set the stage for the design of coherent instruction, educational standards must be bundled and elaborated.

Bundling standards involve identifying groups of related standards that can be reasonably investigated within a particular learning context. Standards are necessarily worded so that they are applicable to a range of problems and phenomena, e.g. HS-PS2-2 could be connected to the contexts of bowling, celestial motion, vehicle safety and many others. In setting the stage for coherent instruction, it is critical that these contexts, not standards, drive student learning. Learning is fundamentally situated within the contexts that motivate and sustain inquiry, generate a need-to-know about new ideas and provide a set of real-world problems and/or phenomena to be explained (Lave and Wenger 1992)—contexts, rather than standards, should be used to drive coherent instruction. Thus, in designing instruction in which HS-PS2-2 is addressed, it is critical to identify contexts that motivate a need to develop the competence that this standard describes. Meaningful contexts will involve multiple standards. For example, HS-PS2-2 could be addressed through the context of vehicle safety, and there are other standards that become relevant through this instructional context, such as:

- Apply science and engineering ideas to design, evaluate and refine a device that minimizes the force on a macroscopic object during a collision (HS-PS2-3)
- Analyse data to support the claim that Newton's second law of motion describes the mathematical relationship among the net force on a macroscopic object, its mass, and its acceleration. (HS-PS2-1)

Bundling standards involve identifying sets of related standards that can be naturally connected (e.g. avoid forced connections of tangentially related standards) within coherent instruction that focuses on a meaningful context.

Once a bundle of standards is identified, these standards must be elaborated. Elaborating standards is a process by which teachers make general standards statements specific for their students. This process is especially important when standards describe complex competencies that develop over extended time, such as learning to quantitatively model energy transfers between interacting systems. The process of elaborating standards involves several steps:

1. **clarifying the key ideas** contained within a standard (e.g. what students should know about 'net force' in HS-PS2-2),
2. **specifying relevant prior knowledge** and upper limits of expectations (similar to the lower and upper limits in learning progressions),
3. **cataloguing common student ideas** related to the standard, identifying real-world problems and phenomena, and
4. considering how students should be expected to **demonstrate their learning** during instruction.

There are several tools and heuristics available for supporting the standards elaboration process, such as the 'unpacking standards' procedure (e.g. Krajcik et al. 2008) and the Content Representations (CoRe) tool (e.g. Loughran et al. 2008). Regardless of the particular tool or heuristic used, the process of elaborating upon a set of bundled standards that can be naturally connected within a real-world context sets the stage for developing coherent instructional activities.

6.5.2   Coherence of Instructional Activities

The process of elaborating standards lays the conceptual foundation for developing instructional activities that are aligned to a set of coherent educational goals. Elaborating standards provides a map of the conceptual terrain that students will encounter during instruction, and instructional activities are designed to support students in navigating this terrain in order to develop target competencies. Critical to supporting students along this conceptual journey is providing landmarks along the way in the form of learning performances (LPs), which students demonstrate during the course of instructional activities, that are aligned with broader educational goals. These LPs are smaller in grain size relative to the standards statements that are addressed in the elaboration process, but most critically, they are specifically connected to phenomena/problems that are investigated during instructional activities. LPs therefore serve a critical function of ensuring that instructional activities are aligned with the broader educational goals, which describe target student competence at the end of instruction.

LPs serve a critical function of clarifying how and why students should engage within a particular activity. Too often, instruction consists of individual activities that are cobbled together based on connection to a physics topic without regard to how students should demonstrate competence as they engage in activities. For example, physics students commonly investigate collisions between carts on a track and are simply asked to collect data to verify that the momentum before the collision and the momentum after the collision is roughly equal. However, simply collecting data to verify a physics law misses valuable opportunities to more deeply engage with physics ideas to demonstrate competence in physics. Consider again HS-PS2-2 which specifies that students use mathematical representations to support a claim that momentum is conserved. In the context of a collision cart investigation, a learning performance may be 'students use data collected during an inelastic collision between moving carts to construct an argument about whether it is reasonable to assume that net external forces are zero'. Table 6.1 shows the standard and the LP side-by-side.

Note that this LP differs from the standard statement in several ways. First, the LP is situated within a particular phenomenon, namely a perfectly inelastic collision (i.e. when objects collide and stick together) between two moving carts. The standard, on the other hand, refers to no specific phenomenon. Second, the LP highlights a different science practice (engaging in argument from evidence) compared to the

Table 6.1 NGSS standard HS-PS2-2 and an associated learning performance

HS-PS2-2	Learning performance
Use mathematical representations to support the claim that the total momentum of a system of objects is conserved when there is no net force on the system	Students use data collected during a perfectly inelastic collision between moving carts to construct an argument about whether it is reasonable to assume that net external forces are zero

standard (using mathematical and computational thinking), while highlighting the same physics principles (momentum conservation and net force). Yet, the physics principles are narrower in the LP, since they deal with only one type of collision (perfectly inelastic). This restriction by its nature simplifies the analysis required (in a perfectly inelastic collision, the final velocity of both carts will be the same); likewise, the LP may only emphasize a specific part of a science practice (e.g. constructing a single argument without considering alternative explanations)

The LP shown in Table 6.1 builds towards the standard, but it is a smaller grain size and requires that students engage in different science practices. Thus, while building towards the standard, the LP is more contextualized and broadens the ways in which students engage in science. Designing instruction such that students only engage with the ideas represented specifically within the standards is unlikely to be coherent. Standards statements are designed to either communicate a broad vision of competence (e.g. the German Bildungsstandards) or guide the development of assessment tasks that sample elements of broader physics competence. In either case, standards must be translated into learning performances, based on an elaboration of the broader educational goals (which is built upon existing learning progressions research and/or models of competence development), in order to guide the development of instructional activities that are coherent and support students in making progress towards the broader educational goals that outline the contours of what it means for a student to become competent in physics.

LPs that align with standards help to ensure that instructional activities align with broader standards and focus on building student competence rather than the simple transmission of content knowledge, yet individual activities typically last only one or two instructional periods and by design target only portions of broader educational goals. It is therefore critical to consider how individual instructional activities connect to each other throughout an instructional unit.

6.5.3   Coherence of Instructional Units

Coherent instructional units, similar to authentic science investigations, are motivated by the investigation of meaningful phenomena or problems and require engagement in a sequence of activities that systematically contribute to a process of making sense of the phenomena or problems under investigation. Instructional unit coherence is perhaps the type of coherence most directly apparent to students, since individual units frame coherent longer-term investigations within classrooms. Also, students play—or should play—a critical agentic role in constructing coherence for themselves through carefully designed classroom learning activities (Sikorski and Hammer 2017). Thus, coherent instructional units must be designed such that instructional activities meaningfully connect to build competence over time, and they must include supports for students to actively consider and construct these connections through reflection and metacognition.

There are a variety of models for developing coherent instruction, project-based science (see Krajcik and Czerniak 2018), problem-based learning (see Hmelo-Silver 2004), design-based learning (see Fortus et al. 2004) and model-based inquiry (see Windschitl et al. 2012). No matter their form, each of these instructional models is based upon a shared set of learning principles identified by Krajcik and Shin (2014). First, students actively construct new knowledge based on their existing knowledge; it is not simply transmitted from teacher to student (see Bransford et al. 2000). Second, learning is inseparable from the context in which it occurs; learning is most effective when new ideas are situated within contexts that have relevance and meaning for learners (see Lave and Wenger 1992). Third, learning is fundamentally social; knowledge is constructed through active collaboration and debate with peers and teachers (see Vygotsky 1978). Fourth, conceptual tools (e.g. learning technologies, graphical representations) support learners in their ability to engage socially and construct new knowledge over time; these conceptual tools are often simplified versions of what is used within the community of professional scientists and engineers (see Salomon et al. 1991). In addition to drawing upon similar learning principles, models for designing coherent instructional units also share a key design feature: using meaningful problems/phenomena to drive learning.

Learning in coherent instructional units is organized around, and motivated by, the goal of figuring out interesting phenomena and solving meaningful problems (Blumenfeld et al. 1991). That is, rather than organizing instruction by topic (e.g. energy, electricity), instruction is organized around relevant context(s) that make learning about physics and engaging in science practices necessary. For example, a unit designed to support student understanding about energy might organize instructional activities through investigating the driving question (Krajcik and Czerniak 2018), 'How can I use trash to power my stereo?' (Nordine et al. 2011). Likewise, a design-based unit focused on electricity might present students with the challenge of designing an electric alarm (Engineering is Elementary 2011). Putting phenomena and problems at the centre of instruction supports students in answering two key questions as they engage in instructional activities: What am I doing? Why am I doing it? In a traditional instruction in which activities are structured by progressing through physics topics, students may struggle to go beyond answering these two questions with something like 'I'm working through the circuits' lab' and 'Because my teacher told us to'. Conversely, coherent instruction that motivates learning through meaningful phenomena/problems supports students in answering these questions with responses like, 'We are learning about different ways to arrange batteries in a circuit' and 'We want to figure out how to power an electric alarm'. Coherent instructional units support students in knowing where they have been (i.e. what they have learned), where they are going (i.e. the problem/phenomena under consideration), and—critically—to perceive a need-to-know about new ideas.

A perceived need-to-know about new ideas is critical for learning (Bransford et al. 2000; National Academies of Sciences, Engineering, and Medicine 2018), and coherent instruction should be designed not only by logically sequencing physics topics, but also through a careful consideration of how learners' curiosity can be piqued and maintained. A promising strategy for designing instructional sequences that generate a need-to-know about new ideas is the 'storyline' (Nordine et al. 2019). A storyline leverages the power of narrative to draw learners into learning activities by a thoughtful sequence of activities. Just like exciting movies often begin with a compelling event that engages the audience (e.g. an opening action sequence in a James Bond movie), coherent instructional units can compel student interest and attention through 'anchoring event' (The Cognition and Technology Group at Vanderbilt 1990). An anchoring event is designed to capture students' interest and set the stage for investigating the problem/phenomena under consideration. For example, in a unit about light and colour with the driving question 'Can I believe my eyes?' (Fortus et al. 2013), students view a message illuminated by different colours of light, revealing different letters. This experience sets up a discussion about what students have observed, when they might have seen something similar before, and how light influences human vision. An anchoring event draws students into the storyline of a unit, and subsequent instructional activities are arranged with special attention to what students figure out in each activity and—critically—what new questions that each activity will prompt. In this way, storylines help to generate a perceived need-to-know. Each activity builds on the ones before and sets the stage for the ones that follow both in terms of prerequisite content knowledge and student curiosity. Just like a key element of a good story is surprise (e.g. a character betraying a friend), coherent instructional units leverage may leverage surprise by using, for example, embedded discrepant events that are put in front of students at key moments. A discrepant event is one in which most students expect one result and observe another—such a surprising or unexpected result sets the stage for learning (González-Espada et al. 2010). While many focus on surprise as not knowing something, it takes particular prior knowledge to experience surprise. In order to experience the intended emotion when a character in a story betrays a friend, the audience must first know something about their relationship. Similarly, in order to experience the full power of a discrepant event, students must have the appropriate prior knowledge to experience surprise when a system behaves in ways that run counter to their expectations. Constructing coherent instructional units in terms of storylines can help to ensure that students are both conceptually and affectively prepared to experience a need-to-know about new ideas.

Storylines can help set the stage for learning and for supporting students in answering questions like 'What are you doing?' and 'Why are you doing it?', but storylines are fundamentally a tool for teachers rather than students (just like storyboards are a behind-the-scenes tool in filmmaking and not intended for audience consumption). Yet, it cannot be taken for granted that just because an instructional unit is coherently designed that students will notice this coherence. No matter how coherently designed instruction might be, students may miss this coherence without specific supports for reflection and metacognition. One potentially powerful support for making the instructional unit coherence visible to students is the driving question board (DQB) (Nordine and Torres 2013; Weizman et al. 2008). A DQB is a graphic organizer that is used throughout the course of an instructional unit; its purpose is to remind students of the main objective of the unit (e.g. a design challenge or driving question), post-artefacts of learning (e.g. record relevant science principles or post-photographs of key experimental results) and log relevant questions (especially student questions posed during class discussions that warrant further investigation). DQBs may be constructed using physics materials or maintained digitally; by referring to DQBs periodically throughout instruction, especially at the beginning or end of a new lesson, students have concrete representation of what they have learned, how they know it, and where they are going throughout an instructional unit. This helps make the coherence of instructional activities more apparent to students, and it helps to ensure that students' attention remains focused on the most central ideas during a unit.

Instructional unit coherence is critical for supporting students in building integrated knowledge; while individual instructional activities typically focus on one key idea, instructional units provide students with the opportunity to connect ideas across several contexts. Encountering and using ideas across multiple contexts help learners to recognize the deep structure of knowledge—ideas that are the same across many instances (Fortus and Krajcik 2020; Schwartz and Goldstone 2016). To develop well-integrated knowledge, students must be supported in using the same set of core ideas across a range of contexts and in reflecting on their use of those ideas (Linn et al. 2004). Thus, coherent instructional units are the most fundamental way to support students in developing well-integrated knowledge.

6.5.4 Coherence Between Instructional Units

Coherent instructional units are the most fundamental way in which learners are supported in connecting ideas across instructional activities and beginning to develop well-integrated knowledge; yet, instructional units unfold over the course of weeks while competence in physics is developed over the course of many months and years. Thus, coherence between instructional units is critical for supporting student competence over time. For example, an individual instructional unit for middle school students may focus on using the idea of energy transformation to interpret and explain a range of phenomena, but the idea of energy transformation is only aspect of the broader energy concept (Neumann et al. 2013). Building competence with the full energy concept requires integrating across multiple aspects of energy (e.g. forms, transformation, dissipation, conservation); learning progressions research has consistently demonstrated that developing such multifaceted competence takes many years (Alonzo and Gotwals 2012).

Coherence between instructional units can be manifest horizontally or vertically. We refer to horizontal coherence as coherence across units in which learners apply the same ideas across instructional units. For example, students should learn to use the idea that energy can be transferred between systems to investigate phenomena/problems in the instructional units focused on mechanics, thermodynamics and electromagnetism. Horizontal coherence is typically manifest across the course of a single academic year. Vertical coherence refers to how instructional units connect to deepen learners' understanding of concepts over multiple years; this type of coherence is most closely connected to the idea of learning progressions. For example, learners in early grades may focus on classifying macroscopic properties of materials (e.g. rigidity, density), while older students begin to use a particle model of matter to explain phenomena like evaporation and melting, and upper-level students begin to learn about the structure of atoms in order to learn about nuclear phenomena (see Hadenfeldt et al. 2016). Too often, learners are introduced to ideas they are not ready to understand. For example, it is common for young students to be told that 'energy is never created nor destroyed', but the evidence is overwhelming that young students simply are not equipped to go beyond simple repetition of this phrase and to use the idea of energy conservation to interpret and explain phenomena (see Herrmann-Abell and DeBoer 2018; Neumann et al. 2013). Inter-unit coherence refers to how ideas are used consistently across contexts (horizontal coherence) and how complex understanding is built systematically over longer periods of time (vertical coherence).

Support for horizontal and vertical coherence can be found in standards documents that outline the broader contours of competence in physics. The German Bildungsstandards (KMK 2005) identify four 'basic concepts' in physics: systems, energy, material and interaction. These ideas, which are core aspects of understanding and doing physics, should be used across instructional units and built systematically over the course of years. The US *Framework* (NRC 2012) identifies a set of seven 'crosscutting concepts' (e.g. patterns, mechanisms of cause and effect, systems modelling, etc.), which are ideas that are used across all disciplines of science and engineering. Basic concepts and crosscutting concepts describe conceptual tools in science that are both powerful and complex; by identifying these conceptual tools, standards documents can provide an important lens for evaluating inter-unit coherence both within and across academic years.

A powerful example of inter-unit coherence is a curriculum called *Investigating and Questioning our World through Science and Technology (IQWST)* (Krajcik et al. 2012), which is a comprehensive middle school science curriculum for use in grades 6–8. An important feature of the IQWST curriculum is the inclusion of explicit features for supporting inter-unit coherence (see Schwarz et al. 2009). IQWST designers carefully considered how science practices and ideas are built across units and how these connections are made explicit to students. For example, in a unit on light, students use a light metre to explore the light that is reflected and transmitted through a sample and learn that the total light hitting the sample must be the sum of light reflected, transmitted and absorbed. This investigation is then specifically referenced in a later unit on energy in the context of energy transfers and transformations within and between systems. Additionally, IQWST designers gave careful consideration to how learners should be supported in engaging in science practices (e.g. constructing explanations) and in how these supports should be faded over the course of multiple units (McNeill et al. 2006).

Inter-unit coherence supports students in building more complex ideas over time and assuming increasing agency over their participation in science. To design for inter-unit coherence, it is therefore critical to consider both the sophistication with which students should understand and use core ideas and also the ways in which students can begin to function independently to apply their knowledge and seek new understanding. In this way, inter-unit coherence supports students in developing competence over time.

6.6 Summary and Conclusion

To be successful in twenty-first-century society, students must do more than accumulate knowledge during schooling—they must develop competence for using this knowledge in the context of relevant social and cultural problems as they engage in sense-making, problem-solving and future learning. The development of such student competence in physics requires many years, and it depends on the instructional environment that student's experience. Recent research into learning progressions has provided empirically derived roadmaps that begin to identify how students might develop competence in understanding and using core ideas in science over the course of many years. This research has begun to influence the development of standards documents throughout the globe, which in turn form a foundation for designing classroom instruction that is coherent both within individual learning activities and across multiple units that span multiple years and a range of learning contexts.

Standards documents such as the US Next Generation Science Standards, German Bildungsstandards and Finnish National Core Curriculum provide a vision for how student competence may be developed over time, but the challenge of supporting learners in developing physics competence is too vast in scope to be addressed by any one teacher. Student competence development requires systemic solutions. In particular, systems of assessment (see Chap. 15, this volume) and professional learning (see Chap. 2, this volume) are critical for providing students and teachers the resources necessary to evaluating/reflecting upon student progress and to collaborating with other educators in service of supporting students coherently over time and across learning contexts. Furthermore, a robust exchange between practicing physics teachers and physics education researchers (see Chaps. 16 and 17, this volume) is essential to ensure that physics instruction and assessment are empirically grounded and that physics education research is useful for informing the work of physics teachers.

The development of student competence requires systemic solutions, so physics teachers should look for opportunities to leverage connections with other stakeholders (e.g. teacher colleagues, policymakers, researchers) and access to knowledge bases (e.g. research literature) in order to participate in systemic initiatives that support student competence. Such engagement with broader systems may include participation in professional conferences, collaborating in research–practice partnerships, or simply maintaining subscriptions to physics teaching journals (e.g. *The Physics Teacher, Unterricht Physik*).

The world that students will inherit is changing quickly, and students must be prepared to learn and adapt to new situations. By developing a robust understanding of the most central ideas in physics through active engagement and collaboration with others, physics instruction can play an important role in promoting the competencies that students will need to navigate the world they are set to inherit.

Acknowledgements We would like to thank David Fortus (Weizman Institute of Science, Israel) and Hans E. Fischer (Universität Duisburg-Essen) for their thorough reading of our manuscript and the many truly helpful comments and suggestions.

References

Alonzo AC, Gotwals AW (eds) (2012) Learning progressions in science: current challenges and future directions. Sense Publ.

Alonzo AC, Steedle JT (2009) Developing and assessing a force and motion learning progression. Sci Educ 93(3):389–421. https://doi.org/10.1002/sce.20303

Baumert J, Kunter M (2013) The COACTIV model of teachers' professional competence. In: Kunter M, Baumert J, Blum W, Klusmann U, Krauss S, Neubrand M (eds) Cognitive activation in the mathematics classroom and professional competence of teachers. Springer US, pp 25–48. https://doi.org/10.1007/978-1-4614-5149-5_2

Baumert J, Bos W, Watermann R (2000) Mathematisch-naturwissenschaftliche Grundbildung im internationalen Vergleich. In: Baumert J, Bos W, Lehmann R (eds) TIMSS/III Dritte Internationale Mathematik- und Naturwissenschaftsstudie—Mathematische und naturwissenschaftliche Bildung am Ende der Schullaufbahn. VS Verlag für Sozialwissenschaften, pp 135–197. https://doi.org/10.1007/978-3-322-83411-9_5

Baumert J, Artelt C, Klieme E, Neubrand M, Prenzel M, Schiefele U, Schneider W, Tillmann K-J, Weiß M (eds) (2002) PISA 2000—Die Länder der Bundesrepublik Deutschland im Vergleich. VS Verlag für Sozialwissenschaften. https://doi.org/10.1007/978-3-663-11042-2

Beaton AE, Michael OM, Ina VSM, Eugenio JG, Teresa AS, Dana LK (1996) Science achievement in the middle school years: IEA's third international mathematics and science study (TIMSS). Center for the Study of Testing, Evaluation, and Educational Policy, Boston College

Berland LK, McNeill KL (2010) A learning progression for scientific argumentation: understanding student work and designing supportive instructional contexts. Sci Educ 94(5):765–793. https://doi.org/10.1002/sce.20402

Blumenfeld P, Soloway E, Marx RW, Krajcik JS, Guzdial M, Palincsar AS (1991) Motivating project-based learning: sustaining the doing, supporting the learning. Educ Psychol 26:369–398

Bransford JD, Brown AL, Cocking RR (2000) How people learn: brain, mind, experience, and school: expanded edition. National Academies Press

Drollinger-Vetter B (2011) Verstehenselemente und strukturelle Klarheit. Waxmann Verlag

Duit R (1984) Learning the energy concept in school—empirical results from the Phillippines and West Germany. Phys Educ 19:59–66

Duncan RG, Rivet AE (2018) Learning progressions. In: International handbook of the learning sciences, pp 422–432

Engineering is Elementary (2011) An alarming idea: designing alarm circuits. Museum of Science, Boston

Erpenbeck J (1997) Selbstgesteuertes, selbstorganisiertes Lernen. Kompetenzentwicklung 97:309–316

Fortus D, Dershimer C, Krajcik JS, Marx RW, Mamlok-Naaman R (2004) Design-based science and student learning. J Res Sci Teach 41(10):1081–1110

Fortus D, Sutherland Adams LM, Krajcik J, Reiser B (2015) Assessing the role of curriculum coherence in student learning about energy. J Res Sci Teach 52(10):1408–1425. https://doi.org/10.1002/tea.21261

Fortus D, Krajcik J (2012) Curriculum coherence and learning progressions. In: Fraser BJ, Tobin K, McRobbie CJ (eds) Second international handbook of science education. Springer, Netherlands, pp 783–798. https://doi.org/10.1007/978-1-4020-9041-7

Fortus D, Krajcik J (2020) Supporting contextualization: lessons learned from throughout the globe. In: Sánchez Tapia I (ed) International perspectives on the contextualization of science education. Springer International Publishing, pp 175–183. https://doi.org/10.1007/978-3-030-27982-0_9

Fortus D, Grueber D, Nordine J, Rozelle J, Schwarz C, Vedder-Weiss D, Weizman A (2013) Can i believe my eyes? In: Krajcik JS, Reiser BJ, Fortus D, Sutherland L (eds) Investigating and questioning our world through science and technology. Activate Learning

Franke G (2005) Facetten der Kompetenzentwicklung [Aspects of Competence Development]. Bertelsmann, Bielefeld

González-Espada WJ, Birriel J, Birriel I (2010) Discrepant events: a challenge to students' intuition. Phys Teach 48(8):508–511. https://doi.org/10.1119/1.3502499

Hadenfeldt JC, Neumann K, Bernholt S, Liu X, Parchmann I (2016) Students' progression in understanding the matter concept: students' progression in understanding matter. J Res Sci Teach 53(5):683–708. https://doi.org/10.1002/tea.21312

Harris CJ, Penuel WR, D'Angelo CM, DeBarger AH, Gallagher LP, Kennedy CA, Cheng BH, Krajcik JS (2015) Impact of project-based curriculum materials on student learning in science: results of a randomized controlled trial. J Res Sci Teach 52(10):1362–1385. https://doi.org/10.1002/tea.21263

Herrmann-Abell CF, DeBoer GE (2018) Investigating a learning progression for energy ideas from upper elementary through high school. J Res Sci Teach 55(1):68–93. https://doi.org/10.1002/tea.21411

Hmelo-Silver CE (2004) Problem-based learning: what and how do students learn? Educ Psychol Rev 16(3):235–266. https://doi.org/10.1023/B:EDPR.0000034022.16470.f3

Kali Y, Linn MC, Roseman JE (eds) (2008) Designing coherent science education: implications for curriculum, instruction, and policy. Teachers College Columbia University

Kesidou S, Roseman JE (2002) How well do middle school science programs measure up? Findings from project 2061's curriculum review. J Res Sci Teach 39(6):522–549

Kirschner PA, Sweller J, Clark RE (2006) Why minimal guidance during instruction does not work: an analysis of the failure of constructivist, discovery, problem-based, experiential, and inquiry-based teaching. Educ Psychol 41(2):75–86. https://doi.org/10.1207/s15326985ep4102_1

Koeppen K, Hartig J, Klieme E, Leutner D (2008) Current issues in competence modeling and assessment. Zeitschrift Für Psychologie/J Psychol 216(2):61–73. https://doi.org/10.1027/0044-3409.216.2.61

Krajcik JS, McNeill KL, Reiser BJ (2008) Learning-goals-driven design model: developing curriculum materials that align with national standards and incorporate project-based pedagogy. Sci Educ 92:1–32

Krajcik JS, Czerniak CL (2018) Teaching science in elementary and middle school: a project-based learning approach, 5th edn. Routledge, Taylor & Francis Group

Krajcik JS, Shin N (2014) Project-based learning. In: KeithEditor Sawyer R (ed) The Cambridge handbook of the learning sciences, 2nd edn. Cambridge University Press, pp 275–297. https://doi.org/10.1017/CBO9781139519526.018

Krajcik JS, Reiser BJ, Sutherland L, Fortus D (2012) IQWST: investigating and questioning our world through science and technology. Activate Learning

Lave J, Wenger E (1992) Situated learning: Legitimate peripheral participation. Cambridge University Press

Lee H-S, Liu OL (2010) Assessing learning progression of energy concepts across middle school grades: the knowledge integration perspective. Sci Educ 94(4):665–688

Leutner D, Fleischer J, Grünkorn J, Klieme E (eds) (2017) Competence assessment in education. Springer International Publishing. https://doi.org/10.1007/978-3-319-50030-0

Linn MC (2006) The knowledge integration perspective on learning and instruction. In: Sawyer RK (ed) Cambridge handbook for the learning sciences. Cambridge University Press, pp 243–264

Linn MC, Eylon B-S, Davis EA, Linn MC, Davis EA, Bell P (2004) The knowledge integration perspective on learning. In: Internet environments for science education. Lawrence Erlbaum Associates, Inc.

Liu X, McKeough A (2005) Developmental growth in students' concept of energy: analysis of selected items from the TIMSS database. J Res Sci Teach 42(5):493–517

Loughran J, Mulhall P, Berry A (2008) Exploring pedagogical content knowledge in science teacher education. Int J Sci Educ 30(10):1301–1320. https://doi.org/10.1080/09500690802187009

McClelland DC (1973) Testing for competence rather than for 'intelligence.' Am Psychol 28(1):1–14. https://doi.org/10.1037/h0034092

McNeill KL, Lizotte D, Krajcik JS, Marx RW (2006) Supporting students' construction of scientific explanations by fading scaffolds in instructional materials. J Learn Sci 15(2):153–191

National Academies of Sciences, Engineering, and Medicine (2018) How people learn II: learners, contexts, and cultures. National Academies Press. https://doi.org/10.17226/24783

National Research Council (2007) Taking science to school: learning and teaching science in grades K-8. In: Duschl RA, Schweingruber HA, Shouse AW (eds). The National Academies Press. https://doi.org/10.17226/11625

National Research Council (2012) A framework for K-12 science education: practices, crosscutting concepts, and core ideas. The National Academies Press

Neumann K, Fischer HE, Kauertz A (2010) From PISA to educational standards: the impact of large-scale assessments on science education in Germany. Int J Sci Math Educ 8(3):545–563. https://doi.org/10.1007/s10763-010-9206-7

Neumann K, Viering T, Boone WJ, Fischer HE (2013) Towards a learning progression of energy. J Res Sci Teach 50(2):162–188. https://doi.org/10.1002/tea.21061

Neumann K, Schecker H, Theyßen H (2019) Assessing complex patterns of student resources and behavior in the large scale. Ann Am Acad Pol Soc Sci 683(1):233–249. https://doi.org/10.1177/0002716219844963

Neumann K, Nagy G (2013) Students' progression in understanding energy. In: Annual international conference of NARST, San Juan, Puerto Rico

NGSS Lead States (2013) Next generation science standards: for states, by states. National Academies Press

Nordine J, Torres R (2013) Enhancing science kits with the driving question board. Sci Child 50(8):57–61

Nordine J, Krajcik J, Fortus D (2011) Transforming energy instruction in middle school to support integrated understanding and future learning. Sci Educ 95(4):670–699. https://doi.org/10.1002/sce.20423

Nordine J, Krajcik J, Fortus D, Neumann K (2019) Using storylines to support three-dimensional learning in project-based science. Sci Scope 42(6):85–91

OECD (2016) PISA 2015 results (volume I): excellence and equity in education. OECD. https://doi.org/10.1787/9789264266490-en

OECD (1999) Measuring student knowledge and skills: a new framework for assessment. Organisation for Economic Co-operation and Development

OECD (2013) PISA 2012 results in focus: what 15-year-olds know and what they can do with what they know. OECD Publishing

OECD (2017) PISA 2015 assessment and analytical framework: science, reading, mathematic, financial literacy and collaborative problem solving. OECD. https://doi.org/10.1787/9789264281820-en

Organisation for Economic Cooperation and Development (OECD). (2018) The future of education and skills: education 2030. OECD

Osborne JF, Henderson JB, MacPherson A, Szu E, Wild A, Yao S-Y (2016) The development and validation of a learning progression for argumentation in science. J Res Sci Teach 53(6):821–846. https://doi.org/10.1002/tea.21316

Pottinger PS, Goldsmith J (1979) Defining and measuring competence. Jossey-Bass

Sadler TD (2004) Informal reasoning regarding socioscientific issues: a critical review of research. J Res Sci Teach: Official J Nat Assoc Res Sci Teach 41(5):513–536

Salomon G, Perkins DN, Globerson T (1991) Partners in cognition: extending human intelligent with intelligent technologies. Educ Res 20(3):2–9

Schmidt WH, Wang HC, McKnight CC (2005) Curriculum coherence: an examination of US mathematics and science content standards from an international perspective. J Curric Stud 37(5):525–559. https://doi.org/10.1080/0022027042000294682

Schmidt WH, McKnight CC, Raizen SA (1997) A splintered vision: an investigation of U.S. science and mathematics education. Kluwer Academic Publishers

Schwartz DL, Goldstone R (2016) Learning as coordination: cognitive psychology and education. In: Corno L, Anderman EM (eds) Handbook of educational psychology, 3rd edn. Routledge

Schwarz CV, Reiser BJ, Davis EA, Kenyon L, Achér A, Fortus D, Shwartz Y, Hug B, Krajcik J (2009) Developing a learning progression for scientific modeling: making scientific modeling accessible and meaningful for learners. J Res Sci Teach 46(6):632–654. https://doi.org/10.1002/tea.20311

Sekretariat der Ständigen Konferenz der Kultusminister der Bundesrepublik Deutschland (2005) Beschlüsse der Kultusministerkonferenz – Bildungsstandards im Fach Physik für den Mittleren Schulabschluss (Jahrgangsstufe 10). München Neuwied. https://www.kmk.org/fileadmin/Dateien/veroeffentlichungen_beschluesse/2004/2004_12_16-Bildungsstandards-Physik-Mittleren-SA.pdf

Sikorski T-R, Hammer D (2017) Looking for coherence in science curriculum. Sci Educ 101(6):929–943. https://doi.org/10.1002/sce.21299

Smith CL, Wiser M, Anderson CW, Krajcik J (2006) Implications of research on children's learning for standards and assessment: a proposed learning progression for matter and the atomic-molecular theory. Measure Interdisc Res Persp 4(1–2):1–98. https://doi.org/10.1080/15366367.2006.9678570

The Cognition and Technology Group at Vanderbilt (1990) Anchored instruction and its relationship to situated cognition. Educ Res 18:2–10

Trumper R (1993) Children's energy concepts: a cross-age study. Int J Sci Educ 15(2):139–148

Ufer S, Neumann K (2018) Measuring competencies. In: International handbook of the learning sciences. Routledge, pp 433–443

Vahtivuori-Hänninen S, Halinen I, Niemi H, Lavonen J, Lipponen L (2014) A new finnish national core curriculum for basic education (2014) and technology as an integrated tool for learning. In: Niemi H, Multisilta J, Lipponen L, Vivitsou M (eds) Finnish innovations and technologies in schools. SensePublishers, pp 21–32. https://doi.org/10.1007/978-94-6209-749-0_2

Vygotsky LS (1978) Mind in society: the development of higher psychological processes. Harvard University Press

Waddington D, Nentwig P, Schanze S (2007) Making it comparable: standards in science education. Waxmann Verlag GmbH. https://books.google.de/books?id=rleKAwAAQBAJ

Weinert FE (2001) Concept of competence: a conceptual clarification. In: Rychen DS, Salganik LH (eds) Defining and selecting key competencies. Hogrefe & Huber Publishers, pp 45–65

Weizman A, Shwartz Y, Fortus D (2008) The driving question board: a visual organizer for project-based science. Sci Teach 75(8):33–37

White RW (1959) Motivation reconsidered: the concept of competence. Psychol Rev 66(5):297–333. https://doi.org/10.1037/h0040934

Windschitl M, Thompson J, Braaten M, Stroupe D (2012) Proposing a core set of instructional practices and tools for teachers of science. Sci Educ 96(5):878–903. https://doi.org/10.1002/sce.21027

Winterton J, Delamare-Le Deist F, Stringfellow E (2006) Typology of knowledge, skills and competences: clarification of the concept and prototype. Office for Official Publications of the European Communities Luxembourg

Chapter 7
Multiple Representations and Learning Physics

Maria Opfermann, Annett Schmeck, and Hans E. Fischer

Abstract The following chapter will give an overview on learning with multiple representations and why they are so relevant for acquiring knowledge in physics. This will comprise the classical multimedia view of multiple representations in terms of text picture combinations and conceptualisations that broaden this view by taking into account any representations, including tables, graphs and more. To shed more light on these different views, the chapter will begin with definitions and an overview of what can be understood as single and multiple representations. This will be followed by a range of popular theories that explain why learning with multiple representations is beneficial, especially regarding information processing in working memory. We will then focus a little closer on specific characteristics of textual and pictorial representations and on individual learner characteristics that should be taken into account for successful learning. Finally, we will introduce views on internal mental representations and in this regard the so-called theory of choreographies of teaching.

Introduction

Guido B. is an 8th-grade physics and math teacher at a large German high school. He loves his job, and he loves teaching the two domains to his students, but at the same time, he knows that not all of them love physics and math as much as he does. On contrary, especially physics is often perceived as being difficult because of the many complex interrelations students have to understand at once to develop a schema of a physics concept, for instance when learning about the concept of block and tackle. At the same time, it is often hard for students to build up mental models, when they cannot imagine such concepts mentally, for instance the physics concept of force. In sum, this can lead to cognitive overload and decrease the motivation of students. Guido B. knows this from his classes and even remembers this very well from his time at university, when many students did not even finish their first year of physics studies (cf. Chen 2013; Heublein 2014).

M. Opfermann (✉) · A. Schmeck · H. E. Fischer
Essen, Germany

© Springer Nature Switzerland AG 2021

H. E. Fischer and R. Girwidz (eds.), *Physics Education*, Challenges in Physics Education,
https://doi.org/10.1007/978-3-030-87391-2_7

To increase motivation and accordingly the learning success of his students, Guido B. asks for their attention at the end of one lesson and announces "Listen, next week, we're gonna do something really cool and learn with multiple representations!" This leads to different reactions within the class. While Emilia, a 16-year-old girl who has pretty good grades in physics and math so far, looks rather blankly and asks "Isn't that just what we always did?" her classmate Leo replies "Cool, that's the same as multimedia, isn't it?".

Both, Emilia and Leo are right in a certain kind of way. The term *multiple representations* as a means to enhance learning has been used and understood in a widespread fashion in instructional research. The following chapter will explain the different conceptualisations and theories. This includes the very basic distinction between internal and external representations. With regard to external (multiple) representations, we will go into more detail with regard to their conceptualisation as multimedia learning materials, which in a very simple way refers to the use of (written or spoken) words and pictures in learning materials. The reason for doing so is obvious. Many concepts, processes or relations in physics can be comprehended much more quickly when pictures are provided in addition to the text, because pictures are able to show at once, what would be impossible or at least take much longer to be described with words only.

We will also go beyond the classical multimedia view and introduce theories on multiple external representations that conceptualise them as any kind of representation combination, that is, not only text and pictures, but also text plus tables, tables plus graphs and many more. One other advantage of using such multiple sources of information is that learners are able to choose the sources with which they prefer to learn.

Another reason for using multiple representations for physics teaching and learning is the structure of physics itself and refers to the above-mentioned view of internal representations and the construction of internal mental models. Physics uses mathematical modelling to describe phenomena and to explain relations between variables. Therefore, teaching and learning physics necessarily includes both the conversion of physics modelling into mathematical modelling (e.g. regarding functional relations) and the interpretation of mathematical models from a physics point of view (cf. Bing and Redish 2009; Nielsen et al. 2013). Newton's law of gravity, for example, can only be understood and applied to different problems when the functional relation is used in a mathematical form.

In sum, it seems advisable to use more than only one representational format to convey information and to support knowledge construction, and this is what Guido B. takes into account when making his announcement at the end of his lesson. Accordingly, and developed not only for teaching physics, a number of well-established theories claim that the use of multiple representations can enhance learning. These theories describe the basics of human cognitive architecture, in particular the processing limitations of working memory (e.g. Baddeley 1992; Paivio 1986; Sweller 2010), and consider how instructional materials in general should be designed to support learning (e.g. Ainsworth 2006; Mayer 2020; Schnotz 2005). In this chapter, we will discuss these theories and link them to the *choreographies of teaching* approach by

Oser and Baeriswyl (2001), who emphasise the need to distinguish between the sight structure of a learning scenario (e.g. instructional materials in a physics lesson) and the underlying deep structure, which refers to the way in which learners process and comprehend information. We will start with an attempt to clarify what multiple representations actually are.

7.1 Multiple Representations—One Term for Different Concepts?

The term *representation* is used in a wide fashion in previous and current educational research literature. As mentioned in the introduction, one should be aware of whether an *external representation* (such as a text, a graph, or a picture) or an *internal representation* (the mental model a learner builds with regard to a certain learning content) is being described. In this chapter, we will mainly focus on external representations.

In a fundamental classification approach for external representations in chemistry, Gilbert and Treagust (2009) distinguish three types: a phenomenological or macro type (representations of the empirical properties of bonding), a model or submicro-type such as atom or molecule models (e.g. visual models that depict the assumed arrangement of entities, see Fig. 7.1), and a symbolic type (the submicro-type is further simplified to symbols such as H or O, see Fig. 7.1).

While Gilbert and Treagust (2009) use the term *representation* for *external*, visible representations (cf. *visualisations*; Dickmann et al. 2019) as well as for *internal* representations (cf. *mental models*; Harrison and Treagust 1996), a remarkable amount of instructional design research that deals with multiple representations refer more or less explicitly to *external* representations only, that is, any kind of visualisations. For instance, theories such as the *Integrated Model of Text and Picture Comprehension* (ITPC; Schnotz 2005) or the *Cognitive Theory of Multimedia Learning* (CTML; Mayer 2005, 2020) focus on a multimedia concept of multiple representations— that is, a combination of textual and pictorial information. When a physics teacher would use this conceptualisation to teach the block and tackle concept to his class, his (multimedia) learning material would for instance include one or more pictures of a block and tackle that are accompanied by explanatory text (see Fig. 7.2).

A broader view of external multiple representations is given by Ainsworth (2006). According to her DeFT (Design, Functions, Tasks) taxonomy, learning with multiple representations includes two or more external representations that, in an optimal

Fig. 7.1 Example for submicro-type representation (**a**) and symbolic type representations (**b**) of a water molecule

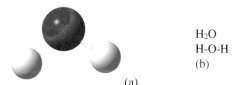

H_2O
H-O-H
(b)

(a)

The mechanical advantage of a block and tackle system is equal to the number of supporting ropes or cables. Notice how the pulling force advantage of a pulley varies depending on the number of strands that it has. If it has a single strand, then the pulling force advantage is 1, which is not an advantage at all. Two strands give a pulling force advantage of 2, three strands give a pulling force advantage of 3, and so forth.	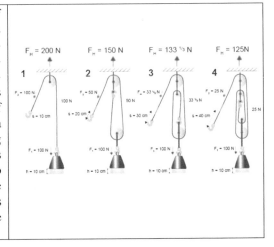

Fig. 7.2 Multimedia learning material consisting of text and accompanying pictures for the concept of block and tackle

case, are processed simultaneously. This can include the classical text-picture-combinations that are described in the ITPC and the CTML, but also goes beyond this view by considering any other kind of combinations of external representations as well. For instance, with regard to the block and tackle example, instead of showing the picture with accompanying text, one might also show the text accompanied by a table that systematically lists examples for weights, number of strands, length of ropes and resulting pulling power. Multiple external representations according to this view could also mean that all of these are used together—e.g. a teacher would show the above-mentioned text and picture combination and then the table, which could also be filled out actively with the students while they look at the figure and read the text.

The mathematical description of technical or physical events can be traced back to about 2500 years ago and has been the result of theory-based empirical research since Galileo (1564–1641) at the latest. According to Simonyi (2003, p. 89), Archimedes (287–212 B.C.E.) had already combined mathematics with physics and technology and described, among other things, static buoyancy as follows: "Any body lighter than water strives upwards during diving with a force resulting from the difference between the weight of the water displaced by the body and the weight of the body itself. However, if the body is heavier than the water, it is pulled downwards with a force resulting from the difference between the weight of the body and the weight of the water it displaces" Simonyi (2003, p. 89 ff., translated by the authors).

In physics textbooks, such text-based or mathematical descriptions are often combined with drawings as shown in Fig. 7.2.

In this regard, a specific form of representation is characteristic especially for physics education, namely mathematical expressions, that is, equations or functions (cf. Angell et al. 2008). Generally, mathematics is used to express models of the physical world to describe physics relations between space, time and matter by means of functions that establish functional relations between a set of variables (see Chap. 8). For instance, Newton's empirically found laws causally describe and predict space–time relations of matter in the meso-world. Such mathematical descriptions as physics models might be added to other formats of multiple representations such as written text or instructional pictures to build the multiple external representations that authors such as Ainsworth (2006) focus on in their models. In physics, those representations are not only used for teaching and learning but also as theoretical basis for empirical research.

In another popular strain of research, a theoretical rationale behind all these approaches to external multiple representations refers to the way in which they can be processed cognitively. In short, learning with multiple representations enables us to make use of all processing channels in our working memory instead of overloading only one (cf. Sweller et al. 2011). This can lead to dual coding and thus deeper cognitive processing, which is taken up in the theories described in the following paragraph.

7.2 Theories on Learning with Multiple Representations

The question of why using multiple representations in instructional materials fosters meaningful learning has been addressed in a remarkable number of studies and led to several well-established theories. Most of these theories are based on assumptions about cognitive processing of information and the structure of the human mind. That is, working memory is assumed to be limited with regard to the amount of information it can process at a certain time (Baddeley 1992). This information can consist of multiple forms of representations, which are either processed in a verbal/auditory or a visual/pictorial channel (cf., *dual channel assumption*; Paivio 1986), depending on the modality of the information. Similar to the overall capacity of working memory, both channels are assumed to be limited regarding the amount of information they can process at a time and in parallel. In this regard and as mentioned above, it is recommended to make optimal use of both channels instead of overloading only one of them.

7.2.1 The Cognitive Theory of Multimedia Learning (CTML)

Based on this view of information processing, the CTML (Mayer 2014, 2020) proposes to use multimedia instructional materials to support deep-level understanding and thus meaningful learning. The CTML mainly focuses on multiple representations in the form of text and picture combinations. In this regard, Mayer (2020) states in his *multimedia principle* that "Students learn better from words and pictures than from words alone" (p. 117). This principle is based on the assumption that words and pictures are qualitatively different with regard to the information they contain; and because of the different channels in which they are processed, different information contents are being learned and (when learning takes places optimally) integrated to one coherent mental model. It seems that humans intuitively apply this idea of picture-supported explanations for more than 4000 years already. For example when looking at the more than 4500-years-old Egyptian stone carving showing Nut, the queen of the sky, spanning the dome of the sky (Metropolitan Museum, New York) or drawings of Galileo (1610), who illustrated valleys and hills on the moon accompanied by verbal descriptions. The multimedia principle has been shown to work in several studies using paper-based as well as computer-based learning materials (e.g. Mayer 2020; Schwamborn et al. 2011; Schwan et al. 2018).

However, just combining words, pictures, mathematical expressions or other kinds of visualisations does not automatically guarantee meaningful learning. The CTML states several further principles that go into more detail with regard to *how* multimedia materials should be presented and combined. For instance, the *modality principle* states that when using text and pictures together, the text should be spoken rather than written, because in this case both the auditory and the visual channels are used instead of overloading the visual channel only. While this principle could be supported in a large number of studies (cf., Ginns 2005; Harskamp et al. 2007; Mayer 2020), others argue that written text can be as effective given that there is enough time to process both the text and the pictures (e.g. Kalyuga 2005; Tabbers et al. 2004). In this case, written text might even be superior to spoken text, because while the latter is transient in nature, written text can be re-read and scanned for relevant information selectively. This might be especially important for texts and learning materials that are perceived as being complex and that students have to re-read several times before comprehension takes place—with physics being one of the domains where this appears to happen regularly in learning scenarios.

Two less controversial principles that could be shown for auditory as well as for visual multiple representations are the *spatial contiguity principle* and the *temporal contiguity principle*. These principles state that when using multimedia learning materials, the different representations (e.g. the text and pictures shown in Fig. 7.3) should be presented closely together (Ginns 2006; Mayer and Fiorella 2014; Mayer and Moreno 1998). That is, in physics textbooks, paragraphs explaining a certain phenomenon should be placed right beside the respective picture. Optimally, text parts might even be integrated into the respective parts of the picture. For instance,

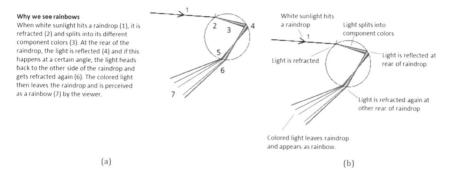

Why we see rainbows
When white sunlight hits a raindrop (1), it is refracted (2) and splits into its different component colors (3). At the rear of the raindrop, the light is reflected (4) and if this happens at a certain angle, the light heads back to the other side of the raindrop and gets refracted again (6). The colored light then leaves the raindrop and is perceived as a rainbow (7) by the viewer.

(a)

White sunlight hits a raindrop

Light splits into component colors

Light is refracted

Light is reflected at rear of raindrop

Light is refracted again at other rear of raindrop

Colored light leaves raindrop and appears as rainbow.

(b)

Fig. 7.3 Examples for learning materials that are spatially separated and might cause split attention (**a**) or that adhere to the spatial contiguity principle (**b**)

when explaining the refraction of light in raindrops when teaching about how rainbows develop, learning materials as shown in Fig. 7.3a might be less helpful than the more spatially contiguous presentation in Fig. 7.3b, because in the first case, associated parts of the learning materials are presented far from each other. In this case, *split attention effects* can occur; that is, working memory capacities are stressed with visual search processes that are actually unnecessary and do not contribute to comprehension and learning (Ayres and Sweller 2005; Kalyuga et al. 1999). The same assumption applies to instructional materials that are presented in temporal contiguity—very simply stated, when explaining the concept of block and tackle, the teacher should not talk first and then show the respective picture, but show the picture at the same time when describing the principles depicted there.

The redundancy principle states that when presenting text and pictures together, using identical written and spoken text at the same time is unnecessary and can even hinder learning, because in this case, the same kind and amount of information is presented and has to be processed twice at a time (Craig et al. 2002; Mayer 2020; Sweller 2005). This double attention to text and pictures stresses the respective working memory channels, but no additional knowledge gains can be expected. However, it should be noted that avoiding redundancy does not mean that the multiple representations used in instructional materials are completely different from each other with regard to the information they contain. In contrast, a certain amount of overlap is necessary so that the relations between the representations (that should all aim at conveying knowledge on one certain topic, model, etc.) become clear and support the integration of information and thus the construction of one coherent mental model (Scheiter et al. 2008).

The signalling principle (Mayer 2005, p. 183) states that "people learn better when cues that highlight the organisation of the essential material are added" (cf., Mayer and Fiorella 2014). That is, when using multiple representations such as text plus picture or a table with an accompanying graph, highlighting techniques such as colour coding or printing parts of the text in bold or cursively can off-load working memory and thus free capacities that can be used for meaningful learning. In the same way, when explaining physics concepts, the teacher might raise his voice or use gestures when important phrases or terms come into play. In these cases, learners would not have to use their cognitive capacities to search for the most relevant information in instructional materials, because it is made obvious to them already (Beege et al. 2020; Harp and Mayer 1998; Mautone and Mayer 2001).

Finally, the coherence principle states that despite the benefits of multimedia learning and multiple representations, all materials that do not directly contribute to the comprehension of the content to be learned and are thus extraneous materials should be excluded (Mayer and Fiorella 2014). For instance, according to Mayer (2020), text parts and pictures that are interesting but irrelevant for the actual information processing process should be removed from learning materials. In this case, working memory capacities are used for paying attention to these unnecessary details, while at the same time, they cannot be used for the construction of schemas, integration of information sources or meaningful learning. Such *seductive details* can even be detrimental for learning when learners are tempted to focus their attention around the wrong kind of information. For instance, when the teacher wants to explain how lightning works during thunderstorms, mere movies of airplanes that are struck by lightning and continue to fly are not a good way to convey knowledge about the Faraday cage concept according to the coherence principle. However, the principle has also led to some controversies in educational research, as it tends to ignore affective variables such as motivation and interest. In this regard, recent research (e.g. Lenzner et al. 2013; Lindner 2020; Park et al. 2015) has shown that seductive details such as decorative pictures are not necessarily harmful and can even foster learning, when they are able to induce and thus have an indirect positive impact on learning. That is, the teacher could show pictures of cars or airplanes struck by lightning to introduce the topic and raise the awareness of his class and then continue with explanatory instructional materials.

Taken together, according to the CTML, multimedia learning materials and thus the use of multiple representations are recommended because, compared to learning with single representations such as text only, they address different processing channels in working memory, contain information of different kinds and different qualities and support the construction of coherent and integrated mental models. In other words, using multiple representations can foster learning.

7.2.2 The Integrated Model of Text and Picture Comprehension (ITPC)

Closely related to the CTML, the ITPC (Schnotz 2005, 2014; Ullrich et al. 2012) also assumes that the processing of multiple representations takes places in two different channels, which are called the auditive and the visual channel. In a first step, all incoming information is processed on a perceptual level (e.g. text-surface representations or visual images). This perceptual level is followed by a cognitive level when information is being processed in working memory in a verbal and/or pictorial channel. Contrary to the CTML, which states that visual and verbal information first leads to the construction of visual and verbal mental models, which are later on integrated into one coherent mental model, the ITPC assumes that this integration and building of one mental model takes place right from the beginning of the processing of multiple representations. That is, information being processed in each of the two channels is aligned and matched from the start of information processing.

In the ITPC, the benefits of learning with multiple representations (in this case, again, primarily with text–picture combinations) are based on this assumption of an integrative processing of verbal and pictorial sources of information. However, an important condition for these benefits to take place is that the respective "verbal and pictorial information are simultaneously available in working memory" (Horz and Schnotz 2008, p. 50). Only in this case, learners are able to recognise that the different representations belong together and can map them to their respective counterparts to make use of the information contained in both of the sources. This view has also been taken up recently under the labels of visual model comprehension (Dickmann et al. 2019) or representational competence (Daniel et al. 2018) especially with regard to science education.

In line with the CTML and the ITPC, Ainsworth (2006, 2014) proposes that learning with multiple representations is not automatically effective, but that these representations should fulfil certain functions. In contrast to the CTML and the ITPC, however, Ainsworth's view of multiple representations comprises more than only text-picture combinations. Her DeFT framework will be described in detail in the following.

7.2.3   The DeFT (Design, Functions, Tasks) Framework for Learning with Multiple External Representations

According to Ainsworth (2014; see also Tsui and Treagust 2013), learning with multiple representations takes place when any two or more external representations are used in instructional materials. This means that in addition to (written or spoken) text and picture combinations, multiple external representations (MERs) can include photos, diagrams, tables, graphs, formulas, concept maps, or even notes taken during learning. In this regard, specific combinations of MERs are not effective in themselves, but they should fulfil certain functions for learning (Fig. 7.4).

First, learners can benefit from MERs if the different representations fulfil complementary functions; that is, each of the single representations should at least partly offer unique information or support different inferences (Ainsworth 2014). That is, multiple representations support comprehension, when they either contain qualitatively different aspects of the information to be learned, or when they convey the same information, but in different ways. For instance, when the concept of acceleration is taught for rectilinear motion, the teacher could just tell (or write on the

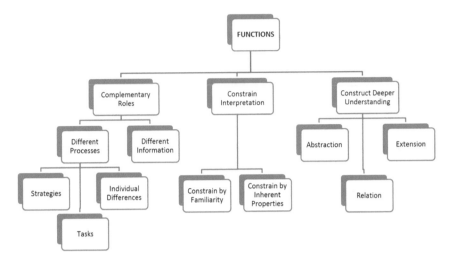

Fig. 7.4 Functions of multiple external representations according to the DeFT framework (Figure adapted from Ainsworth 2006)

blackboard) that the average acceleration ($\overline{a}(t)$) can be expressed as the differential quotient of (change in) velocity (Δv) divided by the time interval (Δt) and could show this quotient along with the formula for acceleration. In addition, he could present a table with exemplary values for the acceleration of a vehicle and depict these values with the help of the respective graph (see Fig. 7.5).

In Fig. 7.5, the table and the graph contain partly the same but, due to different types of representations, partly also complementary information, in addition generalised by the function $\overline{a}(t)$. In this case, if the teacher presented all three multiple representations to the class, he or she would support the steps of a learning process from changing velocity in time and for rectilinear motion to the notion of acceleration as a general description of a phenomenon related to space and time.

Time in seconds	Velocity Car 1	Velocity Car 2
0	0	0
1	4	1
2	8	4
3	12	5
4	16	7

$$\overline{a}(t) = \frac{\Delta v}{\Delta t}$$

Fig. 7.5 Example for learning materials on acceleration using multiple external representations

Another advantage of MERs is that they can support different cognitive processes, because individual differences can be taken into account. That is, learners could "choose to work with the representation that best suits their needs" (Ainsworth 2006, p. 188). Similarly, MERs can foster learning, when learners can choose the representation that best fits the requirements of a certain task—that is, performance is enhanced when the structure of the external representation is similar to the structure of information required to solve the respective problem (cf., Gilmore and Green 1984). Finally, with regard to different processes, providing learners with MERs might encourage them to use more than one strategy to solve a problem (Ainsworth and Loizou 2003; Won et al. 2014). This, in turn, might also be accompanied by higher motivation, which most teachers also want to increase in their physics classes.

In addition to fulfilling complementary functions by supporting different cognitive processes, MERs can also provide complementary information; that is, the single representations contain (partly) different but complementary aspects of the physics concept. It would be harder or even impossible to learn with one single representation in isolation, for instance, learning about how a block and tackle works would probably be possible with just a written text describing the mechanisms behind and the relations between the number of ropes, position of roles and pulling force. It would, however, be much easier (and learning might take place much quicker) with an accompanying picture (*cf.,* Fig. 7.2), because this picture contains visual/spatial information that can be seen at once, which is not possible to realise in a sequentially organised text. The picture in this case would complement the text by providing additional information.

Besides taking into account that different representations contain complementary information, MERs can also support learning if they constrain each other's interpretation possibilities when being presented together. As can be seen in Fig. 5.4, this can be done in two ways.

First, if one representation is significantly more familiar to the learner than the other, this familiar representation can constrain the interpretation of the other one. According to Ainsworth (2006), this is often the case when complex graphs are used in instructional materials. Interpreting these graphs can be challenging for less experienced learners. Consequently, providing a table or a picture or an explanatory text along with the graph would help learners make sense of the data depicted in the graph and thus foster learning.

Second, besides familiarity, also inherent properties of the representations can constrain each other's interpretation. For instance, imagine working in a high-class restaurant and having to learn how the cutlery has to be positioned around the plates. Just being told "Put the dessert spoon and the cheese fork above the plate; thereby the spoon should be above the fork" might give you some information, but not enough with regard to the directions in which the spoon and fork should point or the distances between plate, fork and spoon. Showing a picture at the same time that depicts a standard cutlery arrangement would immediately constrain the interpretation options for the above instruction. Similarly, the interpretation of descriptive representations such as text can be constrained by presenting them along with a depictive representation (Schnotz 2014)—for example and as mentioned in the introduction, the word "force" might lead to very different internal images. While in physics classes, one student might immediately think of centrifugal force and see her favourite merry-go-round, another might rather have his favourite wrestling star in mind. These mental pictures would take place, unless the word *force* is accompanied by a picture, such as a pulley with indicated absolute values of pulling and lifting forces and their directions (cf. Fig. 7.2).

Finally, the third function of multiple external representations according to the DeFT framework is that such combinations are able to promote a deeper level of understanding. This is the case when learners are able to integrate information from the different representation modes and thus gain knowledge that would be hard to infer from just one representation alone (Ainsworth 2006), a view that is very similar to the above-described multimedia principle (Mayer 2020). In order for MERs to construct such a deep conceptual understanding, three processes need to be considered. Learners should be able to *abstract* relevant information from the representations and by doing so, construct references across the multiple representations that represent the underlying structure of a content to be learned (cf. visual model comprehension; Dickmann et al. 2019). In addition, learners should be able to *extend* the knowledge they have with regard to one representation to learning with other representations without fundamentally reorganising the actual knowledge. For instance, when having learned about Ohm's law by means of the formula $I = \frac{V}{R}$ (with $R =$ constant and independent of current, voltage and temperature). In addition to a graph depicting the electric current as a function of the ratio between voltage and resistance, learners should be able to generalise this knowledge to the comprehension of respective tables or to a related solution of the equation. Third, learners should be able to *relate* representations to each other; that is, they should be able to translate between representations—for instance by being able to draw a graph when the acceleration formula is given—along with the table with exemplary values. According to Ainsworth (2006), "this goal of teaching relations between representations can sometimes be an end in itself" (p. 189).

To sum up, multiple external representations according to Ainsworth (2006, 2014) can support learning when they are designed in a way that they (a) support different cognitive processes or include complementary information, (b) constrain interpretation options, thereby preventing inaccurate interpretations and (c) promote deep level understanding by means of abstraction, extension and relation (cf. Rau 2017; Tsui and Treagust 2013). Especially with regard to the third proposed function of MERs, an overlap with the Cognitive Theory of Multimedia Learning and the Integrated Model of Text and Picture Comprehension can be seen. All three theories, multiple representations, especially multimedia learning materials, are assumed to be beneficial for learning only if learners are able to recognise that the different representations express the same cognitive concepts. In addition, the learners must be able to mentally relate the different types of representation to each other and integrate them with existing knowledge and concepts already stored in long-term memory. However, these benefits of multiple representations depend on the *kind of external representation* (text and/or pictorial representations) used as well as on the cognitive characteristics of individual learner. These aspects are shortly taken up next.

7.3 Types of External Representations and Their Benefits for Learning

In the previous section, we have discussed the coherence principle of the CTML (Mayer 2020), which states that interesting but irrelevant materials should be excluded from learning contents. This already points to whether external representations have any instructional value. Although there are newer strains of research that also support the assumption that seductive details such as decorative pictures can be (indirectly) beneficial for learning (Lindner 2020; Lenzner et al. 2013; Opfermann et al. 2014) because of their motivational potential. However, most research still focuses on external representations (and pictorial representations) that are, at least to some degree, instructional and have some kind of explanatory value. Such representations can be divided into verbal representations such as written or spoken text and pictorial representations such as pictures, graphs, photographs or drawings.

7.3.1 Characteristics of Text That Are Beneficial for Learning

When one or more of the multiple external representations used for learning contains text, an important aspect in this regard is that the text is comprehensible for the addressed individual (cf. Leutner et al. 2014). To ensure text comprehensibility, Langer et al. (2006) introduced the *"Hamburg Approach" for language comprehension*, which proposes four characteristics that written or spoken text should fulfil to foster learning. The first characteristic is *simplicity*, that is, sentences should be formulated concisely, and complicated words and phrases should be avoided whenever possible. Second, *organisation* means that text should be clearly arranged, and an internal as well as external structure should be visible. Third, *conciseness* is important in that sentences should be short and not long-winded. Fourth, text should be able to support some kind of *motivational–affective stimulation*; that is, it should be able to arouse the interest of learners (important for teachers, who do not only want to teach physics to their class, but are genuinely interested in arousing their enthusiasm about this great domain). Overall, it is recommended that the longer a text is and the more complex the topic to be learned, the better it is not to present the respective text as a whole (van Hout-Wolters and Schnotz 2020). It should be split up into smaller and meaningful units that can be processed consecutively—a suggestion that is also reflected in the *segmenting principle* of the CTML (Mayer and Pilegard 2014). Particularly important for teaching is the reference to the students' level of development and cognitive physics and linguistic abilities (see Chap. 13).

7.3.2 Characteristics of Pictorial Representations that Are Beneficial for Learning

In general, pictorial forms of multiple representations (in short, visualisations, cf. Dickmann et al. 2019) can be classified in different ways. For instance, a distinction can be made between static (e.g. pictures, photographs or drawings) and dynamic visualisations (e.g. videos, animations or interactive graphs). In this regard, Höffler and Leutner (2007) found that animations are on average superior to static pictures (their meta-analysis shows a small to medium effect of $d = 0.37$), but that this superiority mainly shows up when the visualisations are realistic (i.e. real videos) or when dynamic contents have to be learned, such as steps of a certain process. For instance, for a learner to *understand how* a block and tackle works, an animation might be preferable, while a static picture such as the one in Fig. 7.3 would be sufficient if the goal was to learn about the *relation* between ropes, roles and pulling power (see also Höffler et al. 2013).

Another distinction that has attracted a considerable amount of recent research refers to the question, whether a visualisation such as a picture is *presented* to the learners along with other representations such as the text, or whether learners are requested to generate external representations by themselves. In this regard, the *generative drawing principle* (Schwamborn et al. 2010; Schmeck et al. 2014) states that asking learners to draw pictures of the instructional contents themselves while reading a text can enhance learning, because it encourages learners to engage in deeper cognitive and metacognitive information processing and thus fosters generative processing. The finding that self-generated drawings can improve learning has been confirmed in several studies (e.g. Ainsworth 2014; Hellenbrand et al. (2019); Van Meter and Garner 2005). However, these benefits are also subject to several preconditions such as the quality of the drawing or the question whether instructional support for instance by means of drawing tools is available (cf. Leutner and Schmeck 2014).

Irrespectively of whether a visualisation is static or dynamic and whether it is presented or self-generated, a further distinction can be made following an approach by Schnotz (2005), who distinguishes between descriptive (or propositional) and depictive representations. *Descriptive representations* do not have any structural similarity with the content matter they are supposed to describe and are often used synonymously with symbols (see Fig. 7.6a). For instance, the letter *l* or the expression

Fig. 7.6 Examples for descriptive (**a**) and depictive (**b**) representations of a car according to Schnotz (2005)

Car

Automobile

Skoda

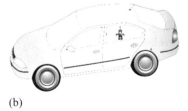

(a)   (b)

electric current do not look like electricity, they are just meant to describe the concept. Similarly, —⊗—does not look like a light bulb, but once a learner is used to this symbol in technical and physics domains, he or she is able to learn and work with it. *Depictive representations,* on the other hand, can be compared to icons that show similarities or structural commonalities with the respective object they are supposed to depict. For instance, the drawing of a car is depictive—although not being identical to a real car, it shows enough overlap to be recognised as a car (see Fig. 7.6).

While descriptive representations appear to be more suitable to convey abstract knowledge, depictive representations are informationally more complete (e.g. the drawing or photograph of a car contains more details at one sight than the word *car* or even the more concrete word *Skoda Octavia*). Concepts and the drawing of inferences can thus be better supported by providing depictive representations. It has to be noted, however, that this advantage of being informationally more complete can also cause opposite effects when there are too many details that are not needed for learning and that distract learners and stress cognitive capacities that could otherwise be used for meaningful learning (cf., *extraneous cognitive load;* Sweller et al. 1998).

Furthermore, pictorial representations can be classified according to an approach by Niegemann et al. (2008). The authors distinguish between *realistic pictures* (e.g. the drawing or photograph of a block and tackle such as in Fig. 7.3), *analog pictures* (e.g. depicting the limited capacity of working memory by means of a bottle that can only be filled to a certain extent until it overflows) and *logical pictures* (such as diagrams and graphs; see Fig. 7.6). With regard to realistic pictures, research has shown that the degree of realism that is beneficial for learning depends upon several factors such as the prior knowledge of learners (Klauer and Leutner 2012). For instance, a highly realistic picture might overburden learners because—as mentioned in the previous paragraph—there is extraneous load created through the attempt to process all the details that are actually not necessary for comprehension (see also Dwyer 1978; Rieber 2000). According to Niegemann et al. (2008), a medium level of realism should be beneficial for learning in most cases.

Analog pictures (in terms of the assumptions by Schnotz 2005, these would be descriptive rather than depictive visualisations) do not necessarily show structural similarities with the contents or object that they are supposed to depict on a visual or surface level. However, they relate to each other in some kind of analogy relationship (Leutner et al. 2014; Niegemann et al. 2008). Such representations are especially suitable when abstract concepts have to be illustrated—for instance, "electrical energy" is an abstract term and rather a mental model in itself, but it can partly be visualised by using water circuits as done by Paatz et al. (2004). Furthermore, analogy pictures support transfer abilities, given that learners understand that they are learning with analogies (cf. DiSessa et al. 1991; Glynn 1991; Leutner et al. 2014).

Finally, logical pictures also do not have structural or obvious similarities with the contents they represent, but they depict these contents schematically. For instance, the graph in Fig. 7.6 is a schematic comparison of two cars with different amounts of acceleration. In this regard, all kinds of diagrams can be classified as logical pictures. According to Niegemann et al. (2008; see also Leutner et al. 2014; Schnotz 2002), diagrams such as pie charts, bar charts or line charts are more effective for learning than other forms of diagrams, because they are more familiar to learners. Further-more, pie charts are especially suitable to convey information about the composition of a certain content to be learned and should be used when the learning content as a whole is of particular interest—for instance, when the distribution of the capacity of power generation for different sources in a country is demonstrated. When, on the other hand, quantitative differences between elements or information units need to be depicted, bar charts should be used (e.g. to show the development of alternative power generation over time). In short, when using logical pictures in multiple repre-sentations, one should make sure that learners are familiar with the conventions of how to process and interpret such representations (cf. Schnotz 2002; Weidenmann 1993).

Park et al. (2020) analysed the drawing processes of fifth and sixth graders learning physics concepts in mechanics. They identified three pictorial representational levels. A sensory level such as laboratory equipment, non-visible physics entities such as atoms or molecules and non-visible effects such as forces or energy. All students started drawing on the sensory level and added explanations using the non-visible levels to describe and explain the physics concepts.

The example of Qasim et al. (2019) shows the role of constructing pictorial repre-sentation in physics research in cooperation between CERN (Geneva, Switzerland) and the National University of Sciences and Technology (Islamabad, Pakistan). They described the data flow in a constructed neural network that learned to process spatial information to represent a certain space including different characteristics. It is used

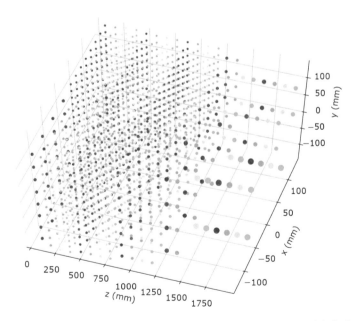

Fig. 7.7 Calorimeter geometry. The markers indicate the centre of the sensors and their size. Layers are colour-coded for better visualisation. Qasim et al. (2019, p. 608)

to describe the generation of the inside of a calorimeter and the places for sensors for measuring π-mesons that are generated at different places in the calorimeter. The instrument was built and optimised according to the simulations with pictorial representations shown in Fig. 7.7.

To sum up, multiple external representations can be presented to learners in many different forms with each being more or less suitable to convey certain kinds or aspects of knowledge. In physics learning, all of such aspects (e.g. the retention of facts, the comprehension of mechanisms, the application and generalisation of functional relations) have to be considered when an overall and complete range of conceptual and procedural knowledge and transfer are to be acquired. This includes students' production of pictorial representations during a learning process and their use in physics research as a complex method for modelling, for example space–time relations. Besides the inherent characteristics of each representation, a second important factor that needs to be considered is individual learner characteristics, which can serve as moderators between instructional design and learning outcomes. These characteristics will be focused on in the next section.

7.3.3   The Role of Individual Learner Characteristics for Learning with Multiple Representations

An engaged and careful physics teacher is always motivated to incorporate current research insights into his way of teaching and thus uses a broad range of multiple external representations to explain physics phenomena to his class. However, he or she also recognises that among other characteristics there is a distribution, for example of individual cognitive abilities, prior knowledge, motivation and self-efficacy expectation in a class. Therefore, one instructional approach alone still does not guarantee a good physics understanding for all students of a class, and the teacher should try to consider the needs and preferences of each individual student as well. Such student characteristics and individual prerequisites that learners bring into a certain learning scenario have been subject to a large amount of empirical studies, not only in physics and science teaching (e.g. Aufschnaiter et al. 1970; Duit 2008; Incantalupo et al. 2013), but also for learning with multiple representations in general. For instance, Mayer (2020) in his *individual differences principle* of the CTML states that multimedia design effects that are beneficial for low prior knowledge learners do not necessarily need to be as effective for high prior knowledge learners and can even be detrimental for them, for instance because of redundancy effects (Kalyuga and Sweller 2014). In line with this, Kalyuga (2005) assumes an expertise reversal effect for instructional materials. Experts in a certain domain get along much better with reduced multimedia materials and less guidance than beginners as, for example, students. For instance, university students at later stages of their studies who are learning about the relativity theory might need remarkably less information than high school students who are just introduced to the theory.

A second learner prerequisite that appears to be important when learning with any kind of visual information is the spatial ability of a learner. For instance, according to Mayer and Moreno (1998), students with high spatial ability can better retain multiple visual representations in their working memory, relate such visual/spatial elements to each other and thus better learn when words are presented together with pictures. Spatial ability has been investigated especially with regard to dynamic representations (e.g. animations or videos), where Höffler (2010) and Höffler and Leutner (2011) found evidence for the so-called *ability-as-compensator* hypothesis. They assume that learners with low spatial ability might benefit more external representation of a process or procedure (dynamic visualisation) that helps learners to build an adequate mental model of the information to be learned. Constructing such a mental model by using static pictures is expected to be more difficult for low spatial ability learners (Hays 1996). Going beyond the scope of spatial ability, the already mentioned concept of visual model comprehension has recently gained greater attention by Dickmann and colleagues. They state that the ability "to extract relevant information from visualisations, to 'translate' them and to relate them to each other and to their respective textual counterparts" (Dickmann et al. 2019; p. 805) impacts how successful learning with multiple representations can be. This might be especially important in a domain that is as abstract as physics.

In addition to prior knowledge, spatial ability and visual model comprehension, several other learner characteristics have been focused on by research, including the cognitive style of learners, their epistemological beliefs, metacognitive and self-regulatory abilities and motivational as well as other effective variables. While these variables do not explicitly relate to learning with multiple representations, such variables can have an impact on how learners approach a learning situation, how they structure and regulate their learning process, or how much attention and perseverance they show during learning (cf., Duit 1991; Höffler et al. 2013).

Taken together, individual learner characteristics should be taken into account when designing instructional materials that contain multiple representations, as such characteristics can serve both as moderators (Schraw et al. 1995) or mediators (Davis et al. 1989; Opfermann 2008) between instructional design, strategies and activities deployed during learning as well as cognitive load and learning outcomes. Such characteristics determine how multiple representations are processed individually and whether learners are able to translate the external representation into an internal and coherent mental model (see also Höffler et al. 2013; Mayer 2020). This view is closely related to the *choreographies of teaching* view introduced by Oser and Baeriswyl (2001), which we will take up in the last section of this chapter.

The Theory of Choreographies of Teaching

When a physics teacher takes all of the above-mentioned theories and research findings on learning with multiple representations into account, he would develop instructional materials with as many external representations (text, pictures, formulas, graphs, tables) as possible and try to ensure that the different students in his class benefit from these materials as equally as possible. In turn, knowing as many external representations of a concept as possible, the logical connections between them and differences in individual learner characteristics should be an important part of physics teachers' professional knowledge. This is in line with the *choreographies of teaching* approach introduced by Oser and Baeriswyl (2001; see Chap. 4 and Geller et al. 2014; Ohle 2010). Oser and colleagues emphasise the need to design teaching explicitly according to pre-defined learning goals. That is, not only the *sight structure* (everything that is visible, such as instructional materials including multiple external representations; see also *surface structure*; Reyer 2004) of a physics lessons is important, but also the *deep structure* as well, which comprises so-called basis models and underlying processes of learning that should be supported. Teachers can introduce the deep structure of a discipline by providing learners with different instructional designs to choose from, which is also called *offers or opportunities to learn* in this approach (Schmidt et al. 2011).

For example, to understand the effect of gravity, learners need to *develop a prototype* in the first step. In showing the legendary apple of Newton paradigm in a demonstration, a teacher might support this as a part of the sight or surface structure of the lesson. To build up a coherent mental model (as part of the deep structure), learners then need to interpret this visual external representation and to transfer the information into an internal text-based representation. This approach is very similar to the ITPC approach by Schnotz (2005) and its description of building (text-based) propositional mental representations and (visual) mental models and integrating them into one coherent schema of the learning content (cf. visual model comprehension; Dickmann et al. 2019).

To organise their offers to learn and to support the students' construction of mental models, teachers should know several of such prototypes for the same concept. In the next step, the text-based prototype must be described in detail by *analysing its essential categories and principles,* which include the reconstruction of measurements and of mathematical modelling of underlying concepts (see Chap. 8 and Müller et al. 2018). Third, to *deal actively with the concept* requires mental activities like the application of mathematical formalisms or of Newton's law of gravity or Newton's second law of motion, for instance by students conducting their own experiments. In this regard, a physics law is represented differently. In the first case, it is used as the *description* of a phenomenon, and in the second case, it must be interpreted as a *source* to design an experiment. As the last step, learners should be able to *apply the developed concept in different situations,* like, for example, to describe the sun–earth–moon system or the gravity conditions in the International Space Station (ISS).

Following the model of Oser and Baeriswyl (2001), multiple external representations as part of the sight structure of a lesson should be used as a means to support processes related to the deep structure of the lesson. From a pragmatic perspective, teachers should not only present multiple representations and consider their individual students' prerequisites; they should also make sure that their classes understand what is being taught on a deep cognitive level.

Summary

In this chapter, we have presented theories and approaches on the role of multiple external representations (MERs) for (physics) learning. Based on well-established views on information processing and working memory, multiple external representations are suitable to foster learning, because they address different sensory and working memory channels instead of overloading only one channel. By using MERs, several functions that are beneficial for learning can be fulfilled—they can support learning and deep level comprehension by providing complementary information or addressing different cognitive processes and/or by constraining each other's interpretation possibilities. We also emphasised that such beneficial effects do not only depend on the kind of representation combination (e.g. their spatial and temporal contiguity), but also on the inherent characteristics of the representations as well as on individual learner characteristics. Finally, the instructional design side and thus the sight structure of learning scenarios (multiple external representations, their characteristics, combination and presentation) should be distinguished from the underlying deep structure, and multiple external representations as part of instructional materials should be designed according to this deep structure, for instance by offering learners different opportunities to learn using respectively related representations.

To sum up, using multiple external representations can facilitate learners' comprehension of concepts and support them in task solving. In addition, using MERs is a necessary prerequisite for students' own construction and reconstruction of meaning not only from instructional materials provided but also to understand the internal structure of physics concepts expressed in different forms of representations. In turn, a broad knowledge of adequate external representations that can be used within instructional materials is a necessary constituent of Guido B.'s and other teachers' professional knowledge not only in physics education, but also in general.

Acknowledgements We would like to thank John Airey (Stockholm University) and Tim N. Höffler (Universität Kiel) for carefully and critically reviewing this chapter.

References

Ainsworth SE (2006) DeFT: A conceptual framework for considering learning with multiple representations. Learn Instr 16:183–198

Ainsworth SE (2014) The multiple representations principle in multimedia learning. In: Mayer RE (ed) The Cambridge handbook of multimedia learning, 2nd edn. Cambridge University Press, Cambridge, pp 464–486

Ainsworth SE, Loizou A (2003) The effects of self-explaining when learning with text or diagrams. Cogn Sci 27:669–681

Angell C, Kind PM, Henriksen EK, Guttersrud O (2008) An empirical mathematical modeling approach to upper secondary physics. Phys Educ 43(3):256–264

Aufschnaiter SV, Duit R, Fillbrandt H, Niedderer H (1970) Vorkenntnisse, Unterrichtserfolge und Begriffsstrukturen bei der Behandlung des einfachen elektrischen Stromkreises im 5. und 6. Schuljahr. [Prior knowledge, teaching success and conceptual structures for the simple electrical circuit in grades 5 and 6.] Naturwissenschaften im Unterricht 18:135–143, 182–188

Ayres P, Sweller J (2005) The split-attention principle in multimedia learning. In: Mayer RE (ed) The Cambridge handbook of multimedia learning. Cambridge University Press, Cambridge, pp 135–146

Baddeley A (1992) Working memory. Science 255:556–559

Beege M, Ninaus M, Schneider S, Nebel S, Schlemmel J, Weidenmüller J, Moeller K, Rey GD (2020) Investigating the effects of beat and deictic gestures of a lecturer in educational videos. Comput Educ 156

Bing TJ, Redish EF (2009) Analyzing problem solving using math in physics: epistemological framing via warrants. Phys Rev Special Top Phys Educ Res 5(2):020108

Chen X (2013) STEM attrition: college students' paths into and out of STEM fields. Statistical Analysis Report. National Center for Education Statistics, Institute of Education Sciences, U.S. Department of Education. Washington DC

Craig SD, Gholson B, Driscoll DM (2002) Animated pedagogical agents in multimedia educational environments: effects of agent properties, picture features, and redundancy. J Educ Psychol 94:428–434

Daniel KL, Buck CJ, Leone EA, Idema J (2018) Towards a definition of representational competence. In: Daniel KL (ed), Towards a framework for representational competence in science education (pp 3–11). Springer, Cham

Davis FD, Bagozzi RP, Warshaw PR (1989) User acceptance of computer technology: a comparison of two theoretical models. Manage Sci 35:982–1003

Dickmann T, Opfermann M, Dammann E, Lang M, Rumann S (2019) What you see is what you learn? The role of visual model comprehension for academic success in chemistry. Chem Educ Res Pract 20:804–820

DiSessa AA, Hammer D, Sherin B, Kolpakowski T (1991) Inventing graphing: Metarepresentational expertise in children. J Math Behav 10:117–160

Duit R (1991) Students' conceptual frameworks: consequences for learning science. In: Glynn SM, Yeany RH, Britton BK (eds) The psychology of learning science. Hillsdale: Lawrence Erlbaum, pp 65–85

Duit R (2008) Zur Rolle von Schülervorstellungen im Unterricht. [The role of student concepts for learning.] Geographie heute, 30:2–6

Dwyer FM (1978) Strategies for improving visual learning. State College: Learning Services

Frigg R, Hartmann S (2012) Models in science. In: The stanford encyclopedia of philosophy, last download 9 Aug 2020. http://plato.stanford.edu/archives/fall2012/entries/models-science/

Galilei G (1610) Sidereus nuncius. last download 9 Aug 2020. http://www.rarebookroom.org/Control/galsid/index.html

Geller C, Neumann K, Fischer HE (2014) A deeper look inside teaching scripts: learning process orientations in Finland, Germany and Switzerland. In: Fischer HE, Labudde P, Neumann K, Viiri J (eds) Quality of instruction in physics—results from a tri-national video study. Waxmann, Münster, pp 81–92

Gerjets P, Hesse FW (2004) When are powerful learning environments effective? The role of learning activities and of students' conceptions of educational technology. Int J Educ Res 41:445–465

Gilbert JK, Treagust D (2009) Introduction: macro, submicro and symbolic representations and the relationship between them: key models in chemical education. In: Gilbert JK, Treagust D (eds) Multiple representations in chemical education. Springer, The Netherlands, pp 1–8

Gilmore DJ, Green TRG (1984) Comprehension and recall of miniature programs. Int J Man Mach Stud 21:31–48

Ginns P (2005) Meta-analysis of the modality effect. Learn Instr 15:313–331

Ginns P (2006) Integrating information: a meta-analysis of the spatial contiguity and temporal contiguity effects, pp 511–525

Glynn SM (1991) Explaining science concepts: a teaching-with-analogies mode. In: Glynn SM, Yeany RH, Britton BK (eds) The psychology of learning science. Erlbaum, Hillsdane, NJ, pp 219–239

Harp SF, Mayer RE (1998) How seductive details do their damage: a theory of cognitive interest in science learning. J Educ Psychol 90:414–434

Harrison AG, Treagust DF (1996) Secondary students' mental models of atoms and molecules: implications for teaching chemistry. Sci Educ 80:509–534

Harskamp EG, Mayer RE, Suhre C (2007) Does the modality principle for multimedia learning apply to science classrooms? Learn Instr 17:465–477

Hays TA (1996) Spatial abilities and the effects of computer animation on short-term and long-term comprehension. J Educ Comput Res 14:139–155

Hellenbrand J, Mayer RE, Opfermann M, Schmeck A, Leutner D (2019) How generative drawing affects the learning process: an eye-tracking analysis. Appl Cogn Psychol 33(6):1147–1164. https://doi.org/10.1002/acp.3559

Heublein U (2014) Student drop-out from German Higher Education Institutions. Eur J Educ 49:497–513

Höffler TN (2010) Spatial ability: Its influence on learning with visualizations—a meta-analytic review. Educ Psychol Rev 22:245–269

Höffler TN, Leutner D (2007) Instructional animation versus static pictures: a meta-analysis. Learn Instr 17:722–738

Höffler TN, Leutner D (2011) The role of spatial ability in learning from instructional animations—evidence for an ability-as-compensator hypothesis. Comput Hum Behav 27:209–216

Höffler T, Schmeck A, Opfermann M (2013) Static and dynamic visual representations: individual differences in processing. In: Schraw G, McCrudden MT, Robinson D (eds) Learning thru visual displays: current perspectives on cognition, learning, and instruction. Information Age Publishing, Charlotte, NC, pp 133–163

Horz H, Schnotz W (2008) Multimedia: how to combine language and visuals. Language at Work 4:43–50

Incantalupo L, Treagust DF, Koul R (2013) Measuring student attitude and knowledge in technology-rich biology classrooms. J Sci Educ Technol 23(1):98–107

Kalyuga S (2005) Prior knowledge principle. In: Mayer R (ed) The Cambridge handbook of multimedia learning. Cambridge University Press, Cambridge, pp 325–337

Kalyuga S, Sweller J (2014) The redundancy principle in multimedia learning. In: Mayer RE (ed) Cambridge handbook of multimedia learning, 2nd edn. Cambridge University Press, Cambridge, pp 247–262

Kalyuga S, Chandler P, Sweller J (1999) Managing split-attention and redundancy in multimedia instruction. Appl Cogn Psychol 13:351–371

Klauer KJ, Leutner D (2012) Lehren und Lernen. Einführung in die Instruktionspsychologie. [Teaching and learning. Introduction into instructional psychology.], 2nd edn. Beltz, Weinheim

Langer I, Schulz von Thun F, Tausch R (2006) Sich verständlich ausdrücken [Expressing yourself clearly.], 8th edn. Ernst Reinhard, München

Lenzner A, Schnotz W, Müller A (2013) The role of decorative pictures in learning. Instr Sci 41:811–831

Leutner D, Opfermann M, Schmeck A (2014) Lernen mit Medien [Learning with media.] In: Seidel T, Krapp A (eds) Pädagogische Psychologie. Beltz, Weinheim, pp 297–322

Leutner D, Schmeck A (2014) The generative drawing principle in multimedia learning. In: Mayer RE (ed) The Cambridge handbook of multimedia learning. Cambridge University Press, Cambridge, pp 433–448

Lindner M (2020) Representational and decorative pictures in science and mathematics tests: do they make a difference? Learn Instr 68:1–11

Mautone PD, Mayer RE (2001) Signaling as a cognitive guide in multimedia learning. J Educ Psychol 93:377–389

Mayer RE (ed) (2005) Cambridge handbook of multimedia learning. Cambridge University Press, Cambridge

Mayer RE (2014) Cambridge handbook of multimedia learning, 2nd edn. Cambridge University Press, Cambridge

Mayer RE (2020) Multimedia learning, 3rd edn. Cambridge University Press, Cambridge, MA

Mayer RE, Fiorella L (2014) Principles for reducing extraneous processing in multimedia learning: coherence, signaling, redundancy, spatial contiguity, and temporal contiguity principles. In: Mayer RE (ed) Cambridge Handbook of multimedia learning, 2nd edn. Cambridge University Press, Cambridge, pp 279–315

Mayer RE, Moreno R (1998) A split-attention effect in multimedia learning: evidence for dual processing systems in working memory. J Educ Psychol 90:312–320

Mayer RE, Pilegard C (2014) Principles for managing essential processing in multimedia learning. In: Mayer RE (ed) The Cambridge handbook of multimedia learning, 2nd edn. Cambridge University Press, New York, pp 316–344

Mayer RE, Sims VK (1994) For whom is a picture worth a thousand words? Extensions of a dual-coding theory of multimedia. J Educ Psychol 86:389–401

Müller J, Stender A, Fleischer J, Borowski A, Dammann E, Lang M, Fischer HE (2018) Mathematisches Wissen von Studienanfängern und Studienerfolg. [Mathematical knowledge and study success of study beginners.] Zeitschrift für Didaktik der Naturwissenschaften 24:83–199

Niegemann HM, Domagk S, Hessel S, Hein A, Hupfer M, Zobel A (2008) Kompendium multimediales Lernen. [Compendium for multimedia learning]. Springer, Heidelberg

Nielsen T, Angell C, Grønmo LS (2013) Mathematical competencies and the role of mathematics in physics education: a trend analysis of TIMSS advanced 1995 and 2008. Acta Didactica Norge 7(1) (Art. 6). ISSN 1504-9922

Ohle A (2010) Primary school teachers' content knowledge in physics and its impact on teaching and students' achievement. Logos, Berlin

Opfermann M (2008) There's more to it than instructional design: the role of individual learner characteristics for hypermedia learning. Logos, Berlin

Opfermann M, Schmeck A, Wienand A, Leutner D (2014) The use of decorative pictures in elementary school: are they really as bad as their reputation? In: Paper presented at the 7th international cognitive load theory conference (ICLTC). Taipei, Taiwan

Oser FK, Baeriswyl FJ (2001) Choreographies of teaching: bridging instruction to learning. In: Richardson V (ed) Handbook of research on teaching, 4th edn. American Educational Research Association, Washington, pp 1031–1065

Paatz R, Ryder J, Schwedes H, Scott P (2004) A case study analysing the process of analogy-based learning in a teaching unit about simple electric circuits. Int J Sci Educ 26(9):1065–1081

Paivio A (1986) Mental representations: a dual coding approach. Oxford University Press, Oxford

Park B, Flowerday T, Brünken R (2015) Cognitive and affective effects of seductive details in multimedia learning. Comput Hum Behav 44:267–278

Park J, Chang J, Tang KS, Treagust DF, Won M (2020) Sequential patterns of students' drawing in constructing scientific explanations: focusing on the interplay among three levels of pictorial representation. Int J Sci Educ 42:677–702

Qasim SR, Kieseler J, Iiyama Y, Pierini M (2019) Learning representations of irregular particle-detector geometry with distance-weighted graph networks. Eur Phys J C 79(7):608. https://doi.org/10.1140/epjc/s10052-019-7113-9

Rau M (2017) Conditions for the effectiveness of multiple visual representations in enhancing STEM learning. Educ Psychol Rev 29:717–761

Reyer T (2004) Oberflächenmerkmale und Tiefenstrukturen im Unterricht - exemplarische Analysen im Physikunterricht der gymnasialen Sekundarstufe I. [Surface structures and deep structures in instruction: exemplary analysis of lower secondary physics instruction]. Logos, Berlin

Rieber LP (2000) Computers, graphics and learning. Online: http://homepage.ufp.pt/lmbg/reserva/livro_graficos%20e%20interface.pdf

Scheiter K, Wiebe E, Holsanova J (2008) Theoretical and instructional aspects of learning with visualizations. In: Zheng R (ed) Cognitive effects of multimedia learning. IGI Global, Hershey, PA, pp 67–88

Schmeck A, Mayer R, Opfermann M, Pfeiffer V, Leutner D (2014) Drawing pictures during learning from scientific text: testing the generative drawing effect and the prognostic drawing effect. Contemp Educ Psychol 39:275–286

Schmidt WH, Cogan L, Houang R (2011) The role of opportunity to learn in teacher preparation: an international context. J Teach Educ 62(2):138–153. https://doi.org/10.1177/0022487110391987

Schnotz W (2002) Towards an integrated view of learning from text and visual displays. Educ Psychol Rev 14:101–120

Schnotz W (2005) An integrated model of text and picture comprehension. In: Mayer RE (ed) Cambridge handbook of multimedia learning. Cambridge University Press, Cambridge, pp 49–69

Schnotz W (2014) Integrated model of text and picture comprehension. In: Mayer RE (ed) Cambridge handbook of multimedia learning, 2nd edn. Cambridge University Press, Cambridge, pp 72–103

Schraw G, Dunkle ME, Bendixen LD (1995) Cognitive processes in well-defined and ill-defined problem solving. Appl Cogn Psychol 9:523–538

Schwamborn A, Mayer RE, Thillmann H, Leopold C, Leutner D (2010) Drawing as a generative activity and drawing as a prognostic activity. J Educ Psychol 102:872–879

Schwamborn A, Thillmann H, Opfermann M, Leutner D (2011) Cognitive load and instructionally supported learning with provided and learner-generated visualizations. Comput Hum Behav 27:89–93

Schwan S, Dutz S, Dreger F (2018) Multimedia in the wild: testing the validity of multimedia learning principles in an art exhibition. Learn Instr 55:148–157

Simonyi K (2003) Kulturgeschichte der Physik. Harri Deutsch, Thun, Frankfurt a. M. 1995. ISBN 3-8171-1379-X

Sweller J (2005) Implications of cognitive load theory for multimedia learning. In: Mayer RE (ed), The Cambridge handbook of multimedia learning (pp 19–30). Cambridge University Press

Sweller J (2010) Element interactivity and intrinsic, extraneous, and germane cognitive load. Educ Psychol Rev 22(2):123–138

Sweller J, van Merriënboer JJG, Paas FWC (1998) Cognitive architecture and instructional design. Educ Psychol Rev 10:251–296

Sweller J, Ayres P, Kalyuga S (2011) Cognitive load theory. Springer, New York

Tabbers HK, Martens RL, van Merriënboer JJG (2004) Multimedia instructions and cognitive load theory: effects of modality and cueing. Br J Educ Psychol 74:71–81

Tsui C, Treagust DF (2013) Introduction to multiple representations: their importance in biology and biological education. In: Treagust D, Tsui C (eds) Multiple representations in biological education. Springer, The Netherlands, pp 3–18

Ullrich M, Schnotz W, Horz H, McElvany N, Schroeder S, Baumert J (2012) Kognition-spsychologische Aspekte der Bild-Text-Integration. [Cognitive and psychological aspects of text-picture-integration]. Psychologische Rundschau, 63:11–17

Van Hout-Wolters B, Schnotz W (2020) Text comprehension and learning from text. Taylor & Francis, New York

Van Meter P, Garner J (2005) The promise and practice of learner-generated drawing: literature review and synthesis. Educ Psychol Rev 17:285–325

Weidenmann B (1993) Informierende Bilder. [Informational pictures.] In: Weidenmann B (ed) Wissenserwerb mit Bildern [Knowledge acquisition with pictures.]. Hans Huber, Bern, pp 9–58

Won M, Yoon H, Treagust D (2014) Students' learning strategies with multiple representations: explanations of the human breathing mechanism. Sci Educ 98:840–866

Chapter 8
Physical–Mathematical Modelling and Its Role in Learning Physics

Gesche Pospiech and Hans E. Fischer

Abstract Emphasising the role of mathematics in describing the universe, Galileo Galilei (1564–1641) pointed out that mathematics is the language of physics (Galileo in Il Saggiatore, 1623). At the latest since the time of Isaac Newton (1643–1727), the interplay of mathematics and physics as subjects with different nature has grown increasingly important and crucial for doing physics as a science and, vice versa, theoretical physics sometimes provides suggestions for developing mathematics. In this chapter, mathematics is seen as a prerequisite for understanding physics because mathematics contributes greatly to the theoretical development of physics concepts in theoretical physics and to their empirical validation in experimental physics (see chap. 1). The development, description and processing of physics concepts by means of mathematical modelling are called physical–mathematical modelling. Physical–mathematical models are used for logical and structural integration of mathematics to describe the necessary laws and functional aspects of space–time relationships. It must nevertheless be taken into account that several empirical studies in the field of education have shown that knowledge of mathematics alone is not sufficient for good understanding and modelling of physics concepts, because mathematics has different functions in physics compared with those in mathematics it-self. This is especially so because the solutions to physics tasks do not only consist of applying mathematical methods to solving the physics tasks as mathematical tasks, but also of modelling the solution process on the basis of physics concepts. Recent research results have shown that students' learning processes—for example, their learning of physics concepts, their understanding of the epistemological development of physics, their ability to use multiple representations and their understanding of physical–mathematical modelling—appear to be correlated. Therefore, physics educators and physics teachers are recommended to take into consideration modelling-centred instruction to support and improve their students' understanding of physics concepts. In order to ensure that physics lessons and learning processes are designed and performed

G. Pospiech (✉)
TU Dresden, Dresden, Germany
e-mail: gesche.pospiech@tu-dresden.de

H. E. Fischer
Universität Duisburg-Essen, Duisburg, Germany

© Springer Nature Switzerland AG 2021
H. E. Fischer and R. Girwidz (eds.), *Physics Education*, Challenges in Physics Education,
https://doi.org/10.1007/978-3-030-87391-2_8

appropriately, it is important to address the specific features of theoretical modelling in physics and to teach physical–mathematical modelling explicitly. To describe or analyse physical–mathematical modelling, we present in this chapter several models, each of which focuses on one aspect of this complex process. Keeping the education of physics teachers in mind, we focus on the relationship between both modelling for physics teachers at university and for teaching the subject at school. In the course of this chapter, approaches to teach physical–mathematical modelling are presented and discussed on both the school and university levels (Geyer and Kuske-Janßen in Mathematics in Physics Education. Springer, Cham, pp. 75–102, 2019; Kanderakis in Sci Educ 25(7):837–868, 2016; Lehavi et al. in Mathematics in Physics Education. Springer, Cham, pp. 335–353, 2019).

Keywords Modelling · Mathematics in physics education

8.1 Introduction

Theoretical models have been developed in the natural sciences, particularly in physics, to describe natural systems in space and time as mediators between theory and natural systems (Morgan and Morrison 1999). Examples for theories are Newton's laws for describing phenomena in the meso-world, the theory of general relativity for describing the interaction of matter and space and quantum mechanics for dealing with objects beyond classical physics. In these theories, it becomes clear that mathematical models only acquire their meaning in a theoretical framework of physics and that, vice versa, the theory developed in mathematics has its effects on the perception and interpretation of the world from a physics point of view. As an example, we could take the prediction of electromagnetic waves in Maxwell's electromagnetism or the incompatibility of the quantum behaviour with a classical deterministic description of the world. Generally, processes in physics cannot be described and their development in space and time cannot be predicted without mathematics. According to Feynman (1965), "Mathematics is a language plus reasoning; it is like a language plus logic. Mathematics is a tool for reasoning" (p. 40). Therefore, physical–mathematical modelling is an important part of physics education at university as well as in schools, since it is at the centre of physics concepts and serves to describe processes in physics. The generation of a suitable description or a prediction in this context is a task whose solution essentially requires physical–mathematical modelling. The related activities should support students to understand physics concepts or to solve (specifically) designed physics tasks or problems. A systematic approach to physical–mathematical modelling might also be helpful for scientists who often apply similar routines intuitively or are constrained by research conditions. In the following sections, we focus on this complex process.

In order to describe a complex situation in nature, the space–time relations and states of matter/energy must first be described by suitable physics concepts. Physicists should be able to use these concepts to predict specific space–time relationships or states of observed matter or energy in order to describe or predict the behaviour of observed physical objects as a whole. Appropriate physical–mathematical modelling then plays a crucial role in describing the different physics concepts and showing their similarities. This can be explained, for example, with the example of the *Gaussian curvature K* in any point of a sphere $K = 1/r^2$. In physics, K always comes into play when spherical symmetries are used to explain the space distribution of fields. Point sources (e.g. light and sound sources) and their propagating intensities or masses or charges and their equipotential surfaces can be described with the function $f(r) = 1/r^2$. Also, *differential equations* play a predominant role in physics mostly to describe processes and states in space and time. Examples are all the equations of motion (including oscillations), the Laplace and Poisson equations in electrostatics, magnetostatics and in stationary heat conduction, the Helmholtz equation for natural oscillations and natural modes and the Schrödinger equation for temporal change of the quantum mechanical state of a non-relativistic system. The exponential function is universally used in decay and growth processes in all natural sciences; and in the complex number space, the trigonometric functions are connected to the exponential function by means of Euler's formula. In this sense, generic physical–mathematical models can be used for organising physics space–time relations in general and, if taught explicitly, should help students to understand the related concepts in particular. Further examples, which might be also relevant to teaching on the school level, consist of the modelling of complex bodies by the concentration of their mass in the centre of gravity of the body. The non-measurable instantaneous velocity of a body can be described theoretically by the differential quotient of distance and time and the behaviour of optical systems by the simplified representation of light as a geometrical ray.

Results arise from the interplay between physical–mathematical modelling and empirical results through the control of variables and parameters that are assumed theoretically and made measurable experimentally. In such a case, it is necessary to consider the functional relations between physics quantities and the variables and parameters that describe these quantities. Niss (2010) called this procedure *anticipation*: When modelling the non-mathematical situation, the possible mathematical options (models) must be considered. This process of creating physical–mathematical models is at the core of the method of physics. It is used to describe abstract processes studied in theoretical and experimental research in physics such as the analysis of elementary particles, predictions of quantum field theory (baryon spectrum), laser dynamics, topological insulators and many more. The analysis of these processes requires very complex mathematical elements (e.g. systems of partial differential equations), which often can only be processed using advanced numerical or statistical methods. Nevertheless, this method of physical–mathematical modelling is also needed to describe everyday situations or less advanced physics problems where the required physics and mathematical concepts and tools are within reach of students. In the modelling process, the qualitative verbal and written formulation of physics concepts can be an essential component, as it occurs in the learning process of students at every level of education. However, describing relationships in physics requires not only qualitative knowledge of physics concepts and relationships, but also mathematical knowledge at the learner's level. In many cases, only when the functional mathematical relationships are denoted and understood correctly can the physics phenomenon or situation be described adequately. For instance, the instantaneous velocity of an object is the mathematical derivative of its position with respect to time; the corresponding mathematical function and the concept of velocity are closely related but not identical. In physics, it is impossible to reduce a time interval to zero. This difference of interest between physics and mathematics proves to be extremely difficult for students inhibiting their learning process (Galili 2018). Moreover, to interpret or apply the mathematical model, it is necessary to explain the corresponding physics concept and to investigate the effects of different physics quantities expressed by mathematical variables in order to make predictions about the space–time behaviour of the observed natural process. In this process, students have to be able to formulate the physics concepts, their relation to mathematics and the physics meaning of mathematics statements in their own words.

Mastering the necessary mathematical methods and techniques and being able to blend them with physics concepts become a particular challenge not only for learning physics of students from primary school to university and but also for their teachers. They want to teach their students the use of mathematics in physics appropriately in each case, which is a specific challenge of teaching physics. The physical–mathematical logic that physicists need to understand and develop their concepts and their findings must be extended to an instructional logic of teaching physics. Teaching physics involves understanding and integrating concepts of students' learning processes in physics learning to design learning opportunities at school or university. To understand the instructional logic of teaching physics from a physical–mathematical modelling perspective, physics teachers should distinguish a technical aspect that comprises the algorithmic procedures and calculations and a structural aspect that emphasises the manifold connections of physics and mathematical concepts and merges these in suitable instructional approaches (Pietrocola 2008). In the following sections, we discuss these aspects in detail based on an analysis of mathematical modelling and physical–mathematical modelling.

8.2   Mathematical and Physical–Mathematical Modelling

According to Greefrath, Kaiser, Blum and Borromeo Ferri (2006), mathematical modelling serves for solving tasks from the extra-mathematical world. Therefore, mathematical elements are mapped to processes or situations in extra-mathematical systems to construct mathematical models as a description of these processes and situations using mathematical concepts and mathematical representations. Niss (2010) described this mapping as a *translation* between the two domains (extra-mathematical and mathematical), which always must include the awareness of the amalgamation of mathematics and physics as an important characteristic of the modelling process. In this process, Niss (2010, p. 54) placed a special focus on mathematisation, which he characterised, by two subprocesses:

- The first subprocess consists of the translation of extra-mathematical objects and their relations into appropriate mathematical objects and their relations.

- The second subprocess consists of the extra-mathematical questions that have to be translated into mathematical questions.

In order to perform these two subprocesses, mathematical representations must be constructed and used to describe a specific situation. Niss (2010) gave some examples that extra-mathematical objects and relations can be taken from different domains, including physics. We start with an example before taking on a general perspective.

Figure 8.1 shows the everyday situation of cooling a beverage with ice cubes. Students are asked: How many ice cubes are needed to cool the beverage by 10 K?

In order to answer this question, the physical–mathematical concept of constructing an energy balance must be applied and simplifying assumptions of the complex physics situation must first be made in a way that they do not prevent a meaningful solution to the task. Therefore, idealised experimental conditions (e.g. no energy losses) are assumed in order to get the values for a mathematical treatment of the situation. If the experimental side conditions can be neglected with good reason, physicists speak of idealisation. Idealisations are well suited for mathematically estimating the behaviour of physical objects in complex everyday problems.

The picture is complex and lacks an initial description of the phenomenon from a physics point of view. Therefore, in the first subprocess of modelling, the appropriate physical–mathematical components of the situation must be identified. The energy balance can be modelled with the two involved masses m_{ice} and m_{water}, the specific heat capacities of their materials c_{ice} and c_{water}, the specific heat of fusion H_{fus} and the melting temperature T_m of the ice, the initial temperature of the ice T_1, the

Fig. 8.1 Cooling a beverage with ice cubes

initial temperature of the beverage T_2 and the final temperature of the mixture T_{mix}, respectively. These must be classified as variables or parameters. The energy balance leads to the algebraic equation:

$$m_{ice} \cdot c_{ice} \cdot (T_m - T_1) + m_{ice} \cdot H_{fus} + m_{ice} \cdot c_{water} \cdot (T_{mix} - T_m)$$
$$= m_{water} \cdot c_{water} \cdot (T_2 - T_{mix})$$

The second subprocess is to identify the suitable variable that corresponds to the answer of the extra-mathematical question and to solve for it. This step leads to the evaluation of an algebraic term. Since the question is how many ice cubes are needed, the equation has to be solved for the variable m_{ice}, which gives:

$$m_{ice} = \frac{m_{water} \cdot c_{water} \cdot (T_2 - T_{mix})}{c_{ice} \cdot (T_m - T_1) + H_{fus} + c_{water} \cdot (T_{mix} - T_m)}$$

Since T_{mix} is not known, but the difference between T_2 and T_{mix} is given, the equation must be changed into:

$$m_{ice} = \frac{m_{water} \cdot c_{water} \cdot (T_2 - T_{mix})}{c_{ice} \cdot (T_m - T_1) + H_{fus} + c_{water} \cdot (T_{mix} - T_2) + c_{water} \cdot (T_2 - T_m)}$$

This equation has to be checked for plausibility. From general considerations, it would be expected that $m_{ice} \sim m_{water}$. This proportionality is fulfilled by the equation and hence seems appropriate.

In this example, the extra-mathematical situation is represented as an algebraic expression, which is at the same time a result of modelling in physics and a mathematical model. In general, however, physical situations can be described with multiple representations such as pictures, diagrams, graphs, experimental settings or digital simulations (Linder 2013). In the example above, these different representations could be the following: the temperature curve could be shown in a diagram or be numerically simulated. Alternatively, an energy flux diagram could be used or an experiment could be done. All these representations contribute to the derivation of a physical–mathematical model in a specific way. Multiple representations in general and in physics in particular are described in Chapter 7, and experiments and digital media in Chapters 10 and 11, respectively. According to Kind, Angell and Guttersrud (2017, pp. 25–26), "Physics, in particular, makes use of mathematical representations and uses mathematics to describe phenomena, construct models and solve problems. Physics students, however, do not always see physics formula and equations as modelling tools". Indeed, understanding that a mathematical equation or a graph is representing a phenomenon in physics is very challenging for many students.

More generally, the solution of a task or problem in physics with the help of mathematics is difficult for students. A guideline, for example, a modelling cycle, can help them to approach the solution process of a problem systematically (for mathematics, see Borromeo Ferri 2006). Modelling cycles are the starting point for the systematic design of modelling lessons. Brand (2014, p. 6ff) summarised the state of research and contrasted commonly used modelling cycles. In principle, mathematical modelling cycles connect the extra-mathematical and the mathematical domains. They can comprise cognitive activities (understanding, reducing, mathematising, mathematically working, interpreting and verbalising) and metacognitive activities (planning, regulating, validating and solving) (Schukajlow 2011, p. 84). This chapter mainly deals with physical–mathematical modelling in connection with solving tasks because tasks play a major role in physics teaching at school and university. The difference between task and problem is addressed in Chapter 9. In summary, for tasks, the goal is unknown, whereas the solution process should be known, while for problems, the goal is known, whereas the solution process has to be developed. In the previous years, several cycles for solving tasks using physical–mathematical modelling have been developed, each focusing on a different aspect of this multi-faceted process. In the following sections, we present some of these models and highlight their similarities and differences.

8.2.1 Model 1: Routine to Solve Physics Tasks

We first describe the solution process of a physics task shown in Fig. 8.2 with the example from the previous section before analysing it in general.

Example The *physics task* "Cooling a beverage with ice cubes" (see Fig. 8.1) described above is given as an everyday situation and a question. In the first step, the *description of the question,* the non-mathematical situation, must be analysed including a stepwise identification of simplifying assumptions and identification of suitable physics descriptions (e.g. masses and temperatures) and *physics concepts/theory* (concept of energy conservation). It must then be recognised that the appropriate mathematical model to describe the process in physics is an equilibrium

Fig. 8.2 Model 1: The sequence of a solution process for physics tasks according to Schukajlow (2011, p. 84)

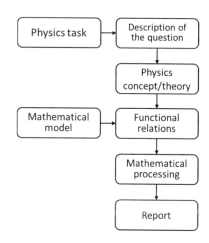

equation. With the general concept of energy conservation, the *functional relation*, that is, a relationship between the initial and the final state, can be derived. To identify the individual energy contributions, a visual representation of the energy flows can be constructed. For *mathematical processing*, the derived equation is solved for the required quantities. In the *report*, the result is interpreted and discussed. The report contains the reflection of the boundary conditions and a plausibility check. In this example, the proportionality of m_{ice} and m_{water} is plausible, and hence, the solution of the derived equation can be accepted as a correct result.

General description of Model 1. Students should be enabled to analyse a physics task by themselves. For this purpose, they should describe the problem to be analysed in their own words and identify the appropriate physics concepts. According to Vygotsky (2012), individuals are only able to use new concepts and ideas alone after learning to use them on a social level. Therefore, students have to know the *physics concepts* addressed by the task and the teacher should initiate a discussion in the classroom in order to identify the mathematical models needed to describe it. The teacher, classroom/group discussions and literature research (Internet, etc.) can support this process. It can be expected that students' understanding of the respective physics concepts after these activities will also be improved.

To elaborate a *mathematical model* in physics, the student's need a goal (to solve the task), have to know physics concepts and the involved variables, and they have to develop ideas about the task-related relationships and constraints (functional relations, complexity, missing variables, parameters, etc.), and quality measures for research at university or expectations of the result at school, respectively.

The necessary *functional relations* and variables in particular must be formulated on the student level. They should be in line with the expected mathematical resources of the type of school and grade level and, if necessary, be adapted during the lessons to the concrete needs of the students.

Solving a physics task implies the application of *mathematical methods* and *operations (processing)*. Therefore, the level of mathematical procedures must match the students' abilities, and digital tools can be used when appropriate. To complete the solution process, the task should be described and critically reflected upon by the class, moderated by the teacher who could identify the limitations of the solution. To this end, students should present the results of analysis, including the plausibility/validity check of the individual solutions.

8.2.2 Model 2: The Integrated Physical–Mathematical Model

In the aforementioned Model 1 on solving tasks, we have described that in the mathematical modelling part, students need two different types of mathematical abilities. The first is the ability to apply mathematical operations, for example, to rearrange or solve equations. The second ability is called the structural ability to find a suitable mathematical expression in relation to a process in physics and/or use this functional relationship for solving a task. According to Karam et al. (2012), understanding the relationship between physical modelling and mathematical modelling is one of the most difficult requirements in physics lessons at school and university. According to Ivanjek et al. (2016), many students have difficulties to transfer knowledge from mathematics lessons to physics lessons and vice versa. Cognitive development of this relationship implies the identification of physics concepts, their mathematical description and their mathematical processing. This cognitive process is modelled sequentially by Model 1 (see Fig. 8.2). To analyse this process in detail, Uhden et al. (2012) developed Model 2 described Fig. 8.3. It focuses on the structural aspects of the process of mathematisation, that is, the combination of physical and mathematical concepts, such as the mathematical derivation of an equation based on physics principles, the mathematical modelling of processes in physics or the identification of non-obvious analogies and common mathematical-physical elements in different processes in physics (Uhden et al. 2012).

Figure 8.3 presents Model 2 in a generic way, implying that it must be detailed for a given problem. The arrows point to the right (see middle box in Fig. 8.3) in the direction of a more intensive use of mathematics, whereas the arrows pointing to the left reflect the physicists' interpretation of the mathematical steps. This indicates that the mathematisation, as well as the interpretation, belong to the physical–mathematical modelling process. Both directions are interlaced. Elementary steps are represented by shorter arrows, complex methods by longer arrows (Uhden et al. 2012). The aspect of mathematical operations is included as "technical mathematical operations" as described above. This model was derived theoretically and then validated in a laboratory study, in which students of grades 9 and 10 (aged 15–16 years) worked on specific physical–mathematical tasks that were intended to encourage them to use structural skills and to explicitly reflect on the meaning of mathematical expressions (Uhden 2016). The link to everyday situations was not analysed in detail in the study, but it appeared to be a possible source of modelling difficulties (see Sect. 8.3).

In explaining how the model is to be applied, we use the aforementioned example (see Fig. 8.1) and focus on the central box in Fig. 8.3. The following steps in mathematisation can be identified with the arrows in the central box:

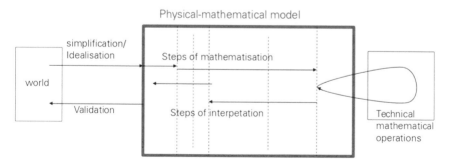

Fig. 8.3 Model 2: An integrated physical–mathematical model (Pospiech 2019, p. 11; modified from Uhden et al. 2012). The left border of the central box marks the interface between the world and the physical–mathematical model

1. Identifying the relevant quantities and processes together with the describing equations and functional relations.

   (a) specific heat of ice and water, melting heat of ice, masses of ice and water
   (b) invoking the heat equation $Q = m \cdot c \cdot \Delta T$

2. Setting up the balanced equation by identifying which energy transfers take place.
3. Solving for the searched quantity m_{ice}.
4. Interpreting the final equation (e.g. that if the temperature difference is given, the quotient of the masses of ice and water is constant).
5. Checking if only differences of temperature appear.

In this way, Model 2 can be used to analyse the complexity of physics tasks for students in detail. This is carried out by identifying every necessary step, each represented by an arrow pointing to the right or to the left in the central box of Fig. 8.3. In addition, the teacher is reminded to identify the necessary pre-knowledge of the students and to analyse which different mathematical elements, or physicists' interpretations thereof, have to be taken into account for finding the solution.

In contrast to Model 1, Model 2 focuses on representing the process of physical–mathematical modelling as a whole.

8.2.3 Model 3: Consideration of Activities in the Modelling Process

Whereas Model 1 represents a general procedure for working on tasks with physical–mathematical modelling and Model 2 represents the detailed blending of physical and mathematical modelling steps, other models focus on the individual sequence of task-solving activities of the students. One of such models (Model 3) is depicted in Fig. 8.4. It refers in particular to the interpretation and validation of the result and the critical reflection of modelling during a task-solving process as a whole and upgrades the models 1 and 2 to a cycle. Model 3 involves the consideration of the possibility of varying the process or repeating some parts of it as often as necessary. In addition, the goal orientation of the task-solving process is explicitly related to the modelling process.

Since students tend to not follow the theoretical modelling cycle consistently when solving physics tasks, Model 3 can serve as an orientation for them during their solving process. To solve the above task of cooling a beverage with ice cubes, the students could run the cycle several times. Perhaps they first only think about how to determine the mixture temperature, and then consider the latent heat of the melting ice, and finally they remember that the temperature of the ice cubes might initially be lower than the melting temperature. They must also control the sign of the energy transfer between the beverage and the ice cubes.

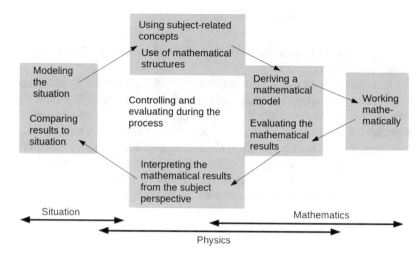

Fig. 8.4 Mosdel 3 focuses on activities in problem solving. The intertwining of mathematics and physics is made visible by the arrows at the bottom

8.3 Mathematical Tools on Different School and Grade Levels

It is well known that the successful solution of a task requires knowledge of both: mathematical operations, such as arithmetic operations or finding the derivative of a function, and mathematical elements, such as terms, functions, derivatives or integrals. The following sections discuss what mathematical knowledge students of different school types and grades have at their disposal for solving physics tasks by physical–mathematical modelling. It should be noted that mathematical knowledge cannot simply be applied in physics without deeper reflection. First, the conventions differ with regard to the rules for the writing and the order of evaluation of mathematical expressions (notation) (see also Heck and Buuren 2019). Even more important is that physical–mathematical terms and operations are laden with meaning in the context of physics. As an example, the equation $F = ma$ has to be interpreted within the framework of Newtonian mechanics or in the case of electrical current, the equation $R = \frac{U}{I}$ has to be analysed with regard to limiting cases, that is, if, for example, the value of I becomes very small. Therefore, students must combine mathematical syntax with the meaning of physics concepts in order to be able to model the physics situation successfully.

8.3.1 Basic Mathematical Elements

This section describes how the meaning of physics concepts is related to the mathematical syntax.

At first, *numbers* as basic mathematical elements are needed to compare and estimate measuring values including their *units*. In most examples from elementary physics, values and units of velocity, temperature, pressure and distances are present in everyday life but the everyday understanding of these concepts often differs significantly from that of the corresponding physics concepts. Therefore, already on this basic level and even before any calculation, it has to be taken into account that in physics not only the already developed concepts but also their internal signs, icons, symbols and values carry a certain meaning. In addition, there is a nested problem with units. In physics, equations for the units of the measuring values that are actually independent are often integrated into the respective algebraic operation used in the function or equation. Sometimes, this is indicated by square brackets, but often not. As a first (technical) validation when solving a task, it is always necessary to control whether the units have been changed correctly in the solution process, whether the values are of the right order of magnitude and so on.

The second basic point is that *algebraic operations* always carry specific physics meaning, as explained by the *basic ideas* in the context of mathematics by Hofe and Blum (2016), and analysed in the context of physics by Sherin (2001). Most mathematical operations in physics result in a change of meaning. For example, addition can mean adding not only a quantity but also, for example, *adding a change* such as a change of energy $(E + dE)$ or *an influence* such as the addition of forces $(F_1 + F_2)$. Multiplication can be interpreted as *adding the same number several times*. In physics, multiplication can lead to new quantities such as work (if force is taken as constant) or momentum, and therefore, to a change or extension of the physics meaning. In mathematics, division is defined as the inverse operation of multiplication. In physics, division may carry an additional meaning of splitting up or of distributing. Accordingly, also division can lead to new physics quantities and meanings. Two examples are: (1) Pressure p is defined as $p = F/a$, force F distributed over a surface a; (2) under the consideration of boundary conditions, electrical resistance R is defined as $R = U/I$ and the voltage U is required to maintain a certain electrical current I in a conductor. In physics functions or equations, the results of algebraic operations must therefore be transferred into new physics meanings.

Using *fractions* creates an additional learning difficulty that cannot completely be explained by the complexity of physics concepts in which they are used: From mathematics education, it is known that it is difficult for students up to the secondary school level to calculate with fractions. This mathematical problem occurs in physics lessons in a more intensive form, because changing equations to solve a task usually requires the *use of mathematical symbols* instead of numbers (Torigoe and Gladding 2011). This difficulty becomes even more problematic by the fact that the standardised use of symbols is different in mathematics and physics. Furthermore, in physics, the physical and mathematical meanings of the symbols are intertwined. To solve a physics problem, the mathematical symbols in a function or equation must always be thought together with their physics meaning. In order to understand and solve physics tasks, therefore, students must not only know the arithmetical operations, but it is also essential that they have understood the physics concepts represented in the functions and equations in order to be able to solve them (see "structural aspects" in Sect. 8.1). Furthermore, the allocation of the same symbols in different contexts with different meanings, which is trivial for physicists, may be confusing for students. This also applies to the conventions for naming the elements of a functional relationship in mathematics and physics, for example, $f(x)$ *versus* $s(t)$, and even for drawing diagrams (see Chap. 7 and Heck and Buuren 2019).

In the following sections, we look at particular problems in dealing with functions and equations in the classroom. For example, $8 = 5 + x$ is an equation and $f(x) = 5 + x$ is a function.

Functions are needed to describe functional relations. The difference between a function and an equation consists of the fact that, in a function, the value of one variable, for example, t, is assigned to another variable, for example, $f(t)$. It thus indicates a relationship between dependent and independent variable quantities, which must be identified at the beginning of a task-solving process. Furthermore, the solution sets depend on the set of numbers from which the independent variable originates and on the function itself. For integers or real numbers, the solution set can even be infinite if no restriction is defined. Often, the sets of numbers are even changed by the functional relation, for example, for $f(x) = 1/x$, if $x \in Z$, then values $f(x) \in R$. This is relevant to solution sets from a mathematics point of view but mostly not relevant for school physics.

In an equation, on the other hand, the expressions on the right and left sides of the *equals* sign are the same in terms of equivalence and equalisation, giving the same amounts and units or defining a new quantity (Heck and Buuren 2019). The equation thus results in a fixed solution set which is the same for both sides.

As an example from physics, we consider the quadratic function in polynomial form, $f(x) = ax^2 + bx + c$, as the mathematical basis for the description of projectile motion. The individual symbols herewith each get a meaning in physics: the symbol x could represent position or time, depending on the context, the symbol a is the inverse of a length or acceleration, and so on. In related tasks, the student can be asked for the throw distance or the maximum throw height or the time a projectile needs for the flight. These tasks can only be solved if the physics concept of the oblique throw is understood, if the needed equations can be applied and interpreted and if the symbols of the function are assigned their physics meanings. During the calculation, knowledge about mathematical operations, such as handling fractions, is needed. Therefore, structural as well as technical skills are needed.

The most important mathematical function that students encounter in physics lessons in their first years on the lower secondary level is *proportionality* which is introduced in mathematics as a linear relation between two numerical quantities. Which of the variables in such a relationship becomes the independent or the dependent variable depends on the situation in physics. Other functions used in physics, sometimes before their introduction in mathematics lessons, are quadratic functions, trigonometric functions and exponential functions. For example, the beginning of the teaching of differential calculus can be illustrated already on early class levels as *difference quotient* or *exponential progressions* with the halving/doubling of the number of events in certain periods. Both functional relationships describe processes, which may be socially relevant and at the same time relevant to physics, such as the measurement of instantaneous velocity or atomic decay or growth processes in pandemic developments. In addition, with digital media, students can sometimes handle complex functions without knowing their mathematical background in detail. Dealing with the structure of mathematics and physics clarifies the nature of each of these sciences respectively. Beyond the mathematical methods described above, also approximate methods or estimations can be important for physics, such as scaling, dimensional analysis and orders of magnitude calculations.

8.3.2 *Advanced Mathematical Elements*

Students in high schools should be able to apply advanced mathematical operations— such as *differentiation* and *integration* with basic functions—to learning physics. This is the prerequisite for a deeper understanding of fundamental physics concepts such as velocity or work. Herewith, it has to be kept in mind that these methods have their roots in geometrical considerations before they were translated into calculus. Using both views might support modelling and can help students to understand their application and meaning in physics.

The *derivative of a function* characterises the behaviour of a function near a given point. The basic idea is to approximate a function sufficiently well by a linear function, the derivative. In physics, the derivative often is approximated by the quotient of differences (Martinez-Torregrosa et al. 2006). An important example for school physics is the slope of a linear function. It is used to characterise processes in physics because the steeper (or flatter) the gradient, the higher (or lower) the speed, for example. In physical–mathematical modelling, the interpretation of the derivative as rate of change is an important way of thinking because this idea allows students to consider processes step by step.

Preliminary stages of *integral calculus* go back to antiquity. For example, Archimedes calculated volumes of cones or spheres by breaking the bodies down into thin cylindrical discs and thus calculating their volumes as the sum of the volumes of the discs. Its formalisation led to the modern integral concept in mathematics. The idea often used in physics is to assume a variable quantity to be constant on small intervals and then to add it up. As an example, we take the concept *work* which is defined as the path integral of the force. The idea of summation allows students to proceed intuitively in simple cases, for example, if work is to be calculated (piece-wise) with a constant force, this is already possible in lower grades of school. In this case, the students do not need to fully master the integral concept but just the basic idea: to compose an irregular area from smaller regular ones such as squares whose area can be calculated and the number of squares counted for finding the irregular area. Furthermore, this idea is fruitful in many contexts, such as distance as integral over velocity or electric charge as integral over electric current in the case of charging or discharging a capacitor.

8.4 Students' Strategies and Difficulties in Physical–Mathematical Modelling

In the previous sections, we have pointed out that the relationship between mathematics and physics, especially in the context of physical–mathematical modelling, which is characterised by a complex interplay of syntax and semantics. Therefore, modelling requires on the side of students their technical and structural skills. As structural abilities are connected to conceptual understanding in both mathematics and physics, students usually show great difficulties in modelling, which go far beyond the usual "students do not master the mathematical rules of calculation" (see, for example, Carli et al. 2020; Ivanjek et al. 2016). There are recent intensive research efforts to explain the lack of success of many students in physical–mathematical modelling and ultimately in task solving. In the following sections, the solution strategies on the one hand and specific modelling difficulties of students in secondary schools, on the other hand, are addressed.

8.4.1 Strategies in Solving Tasks with Physical–Mathematical Modelling

According to the models from Sect. 8.1 on students' modelling when solving tasks, mathematics and physics have to be related to each other during the modelling process. It is to be expected that the solution strategies used by students will depend on their knowledge of mathematics and physics but they often either work in a mathematical mode or in a physical mode (Erickson 2006). Some studies also show that this knowledge is correlated with epistemic beliefs about the role of mathematics in physics (Domert et al. 2007; Greca and Ataide 2017). There are beliefs of students that emphasise the operational role of mathematics in physics and beliefs that support a structural role. Accordingly, students' strategies to solve tasks that were observed in studies are either—more formal and calculus-oriented but neglecting the physics meaning of mathematical elements—or those that combine mathematical and physics concepts with mathematics being considered as a reasoning tool. These latter strategies seem to be more promising for success in problem solving and for supporting the understanding of physics concepts (Bing and Redish 2009; Greca and Ataide, 2017).

When evaluating the strategies of the students, it has to be taken into account that in simple tasks an extended modelling approach is not necessary. This is also why some students—with an adequate understanding of the role of mathematics in physics and a high level of modelling ability—use the simplest method leading to the solution. For example, the *plug and chug strategy* (i.e. they look for the right equation and add the numbers for calculation without referencing to physics concepts; see also Tuminaro and Redish 2007) is often applied by students to solving tasks. Efforts have therefore been made to promote physical–mathematical modelling among Norwegian grammar school students with appropriately demanding tasks (e.g. Angell et al. 2008). Nevertheless, the results show that many students have only basic knowledge and competencies in mathematical modelling (Guttersrud and Angell 2014). The ability to reason in physics seems to be closely connected to the ability to handle different representations and mathematical elements such as graphs and quadratic equations. On the whole, it seems to be helpful to students in their problem-solving process if they know that a modelling cycle such as that described by Model 1 contains several steps and on which step they are. It can be observed that students do not usually follow a modelling cycle rigidly, but approach the given problem flexibly and jump individually between the different steps (Borromeo Ferri 2006). On the other hand, to follow the modelling cycle systematically does not guarantee success in solving a task. This observation might be attributed to the problem-solving difficulties described in the next section.

8.4.2 Specific Difficulties in Problem Solving

There are only few studies that examine students' difficulties in physical–mathematical modelling with the aim of task solving (see, for example, Meli et al. 2016; and for examples that focus on mathematics, Ibrahim et al. 2017; Niss 2010). For physics in particular, Uhden (2016) conducted a laboratory study based on Model 2 with the focus on the central box as shown in Fig. 8.2, that is, the interplay of physics and mathematics. For examining the specific difficulties of physical–mathematical modelling, tasks were developed to ensure that students do not just use their routine procedures for solving them. The tasks induced students to find their own strategy for integrating physics and mathematical concepts. Four main areas of students' difficulties concerning operational and structural aspects of the relation between physics and mathematics have been identified (Uhden 2012, 2016).

1. *Difficulties to understand physics functions and equations*
   The following characteristics of functions and equations must be considered when solving physics tasks:

   (a) Any physics quantity is the product of a numerical value with a unit. Numerical values and units have to be worked out in mathematical functions in order to arrive at solutions that are meaningful in physics and correctly processed mathematically. In addition, there are constant quantities and values without units that play a general role (e.g. the gravitational constant, the Boltzmann constant or the elementary charge) or a substance-related role (e.g. density, friction or specific electrical resistance).

(b) The interaction of mathematical operations and physics meanings makes it difficult for students to understand functions for solving physics tasks (see Chap. 9). According to Uhden (2016), for example, time-dependent variables, such as $v(t) = a{\cdot}t$, the product of acceleration a and time t, is often interpreted as a being dependent on time or vice versa. Therefore, in the case of functions, students should be encouraged to first identify and assign the independent and dependent quantities (variables). Already the identification often causes problems simply because of different notations in mathematics and physics, for example, $f(t)$ instead of $f(x)$.

(c) To solve equations in a task, it is necessary to know about the mathematical rule of balance between both sides of the equation (each mathematical operation has to be applied on both sides). Physics adds the units and the dependence between values and units. Equations in physics often contain both, quantities and units in one equation, and sometimes, the units are written in square brackets. A change of units, for example, from km/h to m/s, results in a factor that affects the magnitude of the quantity velocity (v [km/h] $= v$ [km/3600 s] $= v$ [1000 m/3600 s]). This is also the case when the units are written in a separate equation, for example, when the result of an equation conversion is to be checked. A preferred solution strategy is, if necessary, to first process the equation mathematically with symbols as quantities. Then, if necessary, numerical values and units are inserted to generate numerical results. Sometimes, equations in physics contain quantities and values, which refer to different physical objects such as in the case of static friction force $F_S = \mu_S{\cdot}F_N$. The static friction coefficient μ_S and therefore the static friction force F_S depend on both materials in contact on each other, and F_N is the normal Force: for steel on steel, $\mu_S = 0.2$, and for steel on wood $\mu_S = 0.5$ (without unit).

(d) There is a vague belief that the position of a term in an equation determines the importance of the term for the situation or that proportional dependencies are more important than indirectly proportional ones (Uhden 2016).

(e) There are difficulties for an adequate understanding of proportionality, often even in relation to basic mathematical concepts (Uhden 2016).

2. *Schematic-technical approach.* It can be observed that students try to solve tasks by using a schematic approach and superficially fall back on memorised knowledge. For example, they try to apply equations they know from similar contexts without checking their validity for the new task. One example in Uhden's (2012, p. 133) study was the speed-time law for linear uniform motion, which was applied to solving an accelerated motion task.

3. *Interferences with the students' everyday experience.* Examples are inadequate generalisation of special cases without consideration of the limits of physical

models, of insight into the role of idealisations (see Fig. 8.1) or of the abstract meaning of symbols.

4. *Basic physical or mathematical difficulties.* There are additional factors such as unstable knowledge of mathematical or physics concepts. Therefore, a wrong result might be accepted or a correct result might not be accepted. Sometimes, the mathematical concept is seen as stronger than the physics concept, or vice versa, so that one concept overrides the other.

5. The *confusion of concept definition versus concept image* addresses the difference between students' cognitive pre-concept, which mostly contradicts the physics concept taught at school or university. The cognitive conflict should be confronted with an adequate offer of the aimed physical–mathematical model (Lee and Yi 2013; Tall and Vinner 1981).

These five areas of weaknesses indicate that there is great confusion in students' problem-solving process regarding the interaction between physics and mathematics. If the mathematical operations were not correctly related to the physics concepts required in the task, the calculated results were neither adequately interpreted nor critically questioned.

8.5 Teaching Physical–Mathematical Modelling

Making the connection between physics and mathematics systematically and explicitly clear to students should be a central aim of physics teaching. According to Tursucu et al. (2017), this is important both for students at school and those at university. Developing abilities for physical–mathematical modelling is a complex process that does not simply involve applying learned material to solving tasks. Rather, this process—that requires time and practice to combine the different concepts from mathematics and physics (Roorda et al. 2015)—should also be aimed at in physics teaching regardless of whether mathematics and physics are taught in an integrated way or separately. In the following sections, different ways of explicitly promoting the link between physics and mathematics from a physics teaching perspective are presented.

8.5.1 *Understanding Terms and Equations*

A *term* in mathematics is a structure of numbers and variables that are correctly linked. Terms can be calculated if numbers are used for the variables where a single number (taken as a constant) is already a term. Because the word *formula* is used rather colloquially in mathematics and physics, we use the word *equation* in the following discussion. An equation connects two terms with an equal sign. Whereas we are allowed to insert any numbers into terms, we sometimes get untrue statements for equations if we insert the wrong values into variables. For example, in the equation $3 \times = 9$, we are only allowed to insert 3. Therefore, an equation has a certain solution set that depends on the equation and the physics concept. On the absolute temperature scale, for example, there is no negative number. In physics, solutions of tasks that result in negative absolute temperatures or other solutions of equations that contradict physics concepts are therefore not acceptable. In these constellations, the physical prerequisite does not correspond to mathematically possible solutions. It also has to be considered that equations may have different connotations. They can represent a definition (such as units), a general principle (the continuity principle of classical mechanics or energy conservation) or a law for a special case (such as $F = q \cdot E$ or $p = r \cdot g \cdot h$). According to Karam and Krey (2015); Kim et al. (2018), it is helpful for students to be aware of these different roles when solving physics tasks, especially in the light of Karam and Krey's research findings that students and, even physics student teachers, routinely apply equations to solve physics tasks without such awareness. That students are not sufficiently aware of the underlying physics concepts is a finding often discussed by physics educators and physics teachers (Uhden et al. 2012).

An equation can be seen as a very dense representation of an empirically generated or theoretically derived functional relation in physics. It represents a condensed representation of physics concepts and allows the prediction of space–time processes (Burkholder et al. 2020). Generally, the symbols of an equation in physics have fixed meanings. For the solution of a physics task or problem, the elements of an equation can therefore be easily remembered or converted and might reduce the cognitive load during the solving process (see Chap. 7). In addition, equations may serve as a means of communication (Krey 2012, p. 55). The dense packing of information can be unpacked, for example, with the help of different representations on different levels according to a model developed by Kuske-Janßen (2020) (see Fig. 8.5).

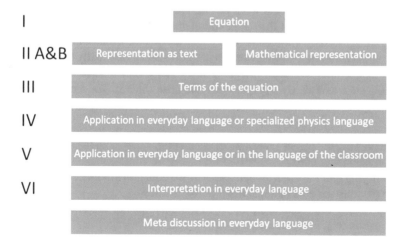

Fig. 8.5 A model for working with an equation (developed from Kuske-Janßen 2020, p. 102)

This model defines several levels of awareness and unpacking of a physics equation. It prototypically describes the functional treatment of equations and functions in classroom teaching and systematises the teaching of physical–mathematical relationships. It can be used, for example, to derive the meanings of equations defining or containing instantaneous velocity, acceleration, force and so on. Levels I–III in Fig. 8.5 serve to make the students aware of the meanings of the symbols and the arithmetic operations. On levels IV–VI in Fig. 8.5, the dense mathematical representation is translated into a specialised physics language or a language used in class respectively (see Chap. 13). Therefore, on these levels, mathematical representations are transferred into verbal representations of different levels of differentiation. This model thus describes in a systematic way how the mathematically represented physics concepts can be qualitatively explained by referring to the students' level of cognitive and concept development with the help of verbal representation and possibly an everyday reference. Everyday examples could be used to clarify the meaning and possible consequences of an equation and contribute to motivation in using the physics concept (see Chap. 13). It has to be considered that everyday references cannot lead to physics equations, and physics equations cannot be derived from specialised or everyday language (i.e. we can go from levels I–III to levels IV–VI but we do not arrive at the equation from levels IV–VI). The transformation between these representations is especially important and necessary to help students understand the equations they need for solving a given task and for organising their modelling process. The different levels do not have to be used in a fixed sequence, but should be addressed in teaching physics and oriented to the individual student's needs. In addition, it might also help students to understand how equations relate to everyday or laboratory situations in the classroom.

In a related way, Bagno et al. (2008) gave a suggestion for the use of equations in class with the examples of motion with constant acceleration and the basic law of Newtonian dynamics. The goal of the proposed learning activity *interpretation of an equation* was that students work out the meaning of the symbols in the related equation, the conditions of their application and the related physics concepts. This activity was guided by a worksheet and realised using the Think-Pair-Share method. In this method, the students first work on the worksheet individually (Think) and then discuss their results with their partner (Pair). Finally, they discuss the worksheet in the whole class and clarify any questions that arise (Share). The worksheet guiding the activity *interpretation of an equation* contains the following steps (Bagno et al. 2008):

1. Identifying the physics meaning of each of the symbols in the equation and the corresponding units.
2. Verifying that the units on both sides of the equation are identical.
3. Explaining the area of application of the equation.
4. Visualising the relationship between the quantities or the components of the equation (e.g. in a graph).
5. Exploring special cases or limiting cases.

6. Explaining the meaning of components of the equation (e.g. if there are several summands).
7. Explaining the meaning of the equation in one's own words.

As described above, these tasks are then reviewed with a partner and discussed in class. This activity showed several effects (Bagno et al. 2008): Most students reported to have the subjective feeling of having understood better the topic of the activity, the meaning of the law of motion with constant acceleration. Some also stated to have acquired a new learning strategy or that they would apply the method to understand other equations. Only a few students declared not to have learned anything from this activity. An analysis of the worksheets handed in by the students showed that the crucial step for learning success seemed to be the final discussion in class (Share).

8.5.2 Developing Physical–Mathematical Tasks

Learning tasks (see Chap. 9) that promote the ability of students to interrelate physics and mathematics may be an important contribution to introduce physical–mathematical modelling in physics classrooms and/or to increase students' related abilities. For example, when determining the instantaneous velocity in solving secondary school tasks, physical–mathematical modelling can be well combined with experimental tasks, so that students can find connections between relevant physics quantities and velocity as a mathematical function of distance and time to plan and understand a self-performed experiment for measuring, for example, the velocity of a bicycle. From a speed measurement in the school yard to understand the function of a speed metre on a bicycle, students should develop the physics concept of instantaneous speed by reducing the time interval when measuring the average speed in their experiment. The mathematical analogy is the transition from $v = d/t$, where d is the overall distance travelled, to $v = \Delta x/\Delta t$, where $x(t)$ is the position as a function of time and finally to $v = dx/dt$, where the instantaneous velocity v is defined by taking the limit $\Delta t \rightarrow 0$. In the lower secondary school physics, we have to stop at the *difference quotient* because the *differential* is usually taught in the upper secondary school. This process can be supported by scaffolding of the teacher from explaining the problem over designing an experiment to the graphical support of the limit value consideration.

A procedure according to Model 2 from Sect. 8.1 allows teachers to support their students by a graded mathematical approach for them to better understand the relationship between the significance of instantaneous velocity in physics and the required mathematical structures. In the case of instantaneous velocity, physics concepts whose mathematical foundations have not yet been covered in mathematics can thus be taught on the secondary school level. The tasks can be set in a way that students' mathematical and physics understanding support each other when they are solving these tasks (Roorda et al. 2015).

Already a slight modification of textbook tasks could have positive effects on students' understanding (Uhden 2012). Providing a realistic situation, but not to give directly numerical values, can avoid routines and lead to an improvement of the solution strategies, for example, by making sketches for problem analysis and estimating realistic values of quantities. If numbers are used, they should be meaningful to students and, if possible, come from their own field of experience.

Also in the case of teaching physics concepts, the focus should first be on conceptual understanding. The teacher should encourage students to start from a process in physics needing explanation, either with their own or with one of the teacher's questions, identify the associated physics concepts and, with the teacher's assistance, discuss the mathematical implications. A scaffolding approach (Zavala 2019) avoids going to routine procedures too rapidly, and it helps the student to approach the offered physics concept step by step and adapted to the individual learning process (Bagno et al. 2019).

There are only very few studies on how students process mathematical elements in problem solving. Generally, for success in problem solving it seems important what students think they are expected to do and what their picture of the situation is. In the process of problem solving, the unfolding of the physics situation and its mathematical description connected to mathematical sense making depends on the underlying assumptions and concepts. It seems that the required favourable combination of symbolic and conceptual reasoning could be reached by suitable instruction (e.g. Kuo et al. 2020; Redish 2017).

8.5.3 Interdisciplinary Teaching

An additional approach to enhance physical–mathematical modelling in a way that it supports physics learning consists of interdisciplinary teaching mathematics and physics. In this context, the teacher's beliefs are also important, but we focus in this section on the possible effects on the students' learning. In an example of teaching from the mathematics perspective, a connecting link is represented by the concept of function, which is also important for other sciences (Michelsen 2006). Already, early in the history of physics education research, the integration of physics (or science) and mathematics was considered as fruitful but only relatively few studies actually measured the learning success of students or the change in their attitude towards science (Berlin 1991).

There are numerous approaches on different levels to teach physics and mathematics in an integrated way. These are expected to lead to a more positive attitude towards science and to a better understanding of concepts and to increased availability and better application of problem-solving skills. Most of these approaches relate to mechanics in physics and to calculus in mathematics. According to Yeatts and Hundhausen (1992), higher performing students on the university level seemed to benefit, but not lower performing students. Similar observations were made for students in grade 7 (Friend 1985). The study was conducted in a 2×2 design (integrated mathematics-science instruction versus non-integrated standard instruction and a class with significantly above-average performance in mathematics and reading versus a class with average performance) to examine grade 7 students' attitudes towards science and their knowledge in science. The unit of instruction covered electricity and heat with teaching spanning over ten weeks. The high-performing integrated-taught students showed a significant advantage over the non-integrated-taught students in their performance on the knowledge test. Among the average students, the integrated-taught students showed an improvement in attitudes towards science. However, overall results showed that only about 100 students were involved so that no generalization could be made.

After the Trends in Mathematics and Science Study (TIMSS) in 2000, a programme was launched in Austria to promote integrated teaching in mathematics and sciences. Rath (2006) described some projects of this programme concerning the integration of mathematics and physics. The study showed in particular, how an integrated physics and mathematics education programme in grades 9 and 10 (age 15–16) could look like, for example, by introducing functions in mathematics along with teaching kinematics in physics, by relating vectors with force and momentum and by connecting trigonometric functions with oscillations and waves. This programme was implemented with one class in grade 10. The results were compared with those of a traditionally taught class. The data on the interest of students for concrete topics was collected with help of a short questionnaire and their problem-solving strategy in an open-ended problem was observed; and interviews were also conducted with selected students. The findings of this small case study give the impression that teachers can gain not only additional insight into students' difficulties and adjust their explanation but also get to know the methods, similarities and differences between the different subjects. Students who were taught with this integrated approach could better understand the physics concepts and better transfer the learned functional relationships to other physics settings.

Other examples were given, among others, by Michelsen (2015), Mäntylä and Poranen (2019) who proposed to model everyday problems in a physical–mathematical manner, considering both aspects in an integrated perspective. Michelsen (2015) focused on geodesic problems, distance in kinematics, or exponential growth in various contexts as examples and pointed out that mathematisation does not have to be accompanied by a fully developed academic mathematical syntax and the related mathematical concepts. The modelling process could be carried out, for example, in an 8th-grade class with an experiment on radioactivity or the lever ratios on a bicycle. Data collection and modelling are supported by appropriate experimental designs, by digital devices and by processing the functional relationships (see Chap. 11). Depending on the goal of the lesson, digital tools can support students' learning process. For example, the effort for data collection and processing can be minimised in order to focus on the reconstruction of physics concepts.

Nguyen and Krause (2020) provided suggestions for the training of physics teachers by joint efforts of mathematics and physics education researchers. In addition to experimentation, modelling was highlighted by Nguyen and Krause as a particularly fruitful way in physics and mathematics education for teachers to mutually transfer from one to the other—and benefit from each—subject.

8.5.4 Explicit Teaching of Modelling

According to Galili (2018), physical–mathematical modelling requires an understanding of the concepts of physics and mathematics and abilities (referred to as structural abilities in Sect. 8.1.2) to associate them. His conclusion is that teaching physical–mathematical modelling explicitly in physics at school contributes to reach science literacy and meaningful learning. A central component of structural abilities is that students not only learn equations by heart but also are able to explicitly relate mathematical elements and physics concepts and discuss their relationships. Such learning activities must therefore be explicitly encouraged in physics teaching. According to Lehavi et al. (2017), four patterns of corresponding activities reported and used by teachers could be identified. These teachers' reports were supported by classroom observations in which several lessons in grade 9 were videotaped and afterwards discussed with the teacher (Lehavi et al. 2015):

1. The **exploration pattern:** Possible consequences of a mathematical description are analysed, for example, in terms of limits of approximation, extreme cases, limits of validity and so on. An example was about force that must be applied to move an object on a table against friction. Students should analyse the function in terms of the limiting cases (e.g. when the denominator approaches zero).
2. The **construction pattern:** Suitable mathematical tools are chosen for describing and analysing physics situations in an inductive or deductive way. An example is the derivation of Snell's law from experimental data.
3. The **application pattern:** Students apply known laws to solve physics problems using mathematical tools to develop physical–mathematical models and use mathematical representations. For example, students must relate the mathematically obtained solutions of a given task to the concrete situation and interpret them in physics.
4. The **broadening pattern:** The physical–mathematical model is placed in a larger context and discussed from a top–down perspective using general principles, analogies or symmetries (guiding concepts). Examples include applying conservation laws such as conservation of momentum or finding analogies (such as harmonic motion in different contexts), where mathematics helps students by organising their expertise.

The four activity patterns map the course of a teaching sequence (see Chap. 4). First, the concept to be learned is made accessible, then constructed with the help of a prototype, then applied to an example in a task with the prototype and finally, transferred to other similar situations.

Often, instruction is characterised by the third pattern, although learning through one's own experience (1st pattern), constructing concepts (2nd pattern) and transferring learning to similar contexts represent the entire learning process.

In each pattern, graphical representations may be applied (e.g. Doorman and Gravemeijer 2009; Carrejo and Marshall 2007; Opfermann et al. 2017; see Chap. 7). For example, data obtained from experiments are represented graphically by using digital devices. From a graphical representation, an algebraic description is developed and the graphical and algebraic representations are related to each other and must be interpreted.

A far more extended approach for explicitly teaching physical–mathematical modelling—with use of experiments, multiple representations and meaningful learning activities—was implemented in upper secondary schools in Norway with nearly 300 students (Angell et al. 2008). The modelling activities were systematically integrated in the lessons during a school year. This approach gave hints that for students' learning success also attention to their learning strategies has to be paid.

An important research method in physics are numerical simulations. They are used to evaluate physical–mathematical models and to match models and empirical data. In the school context, graphical simulation systems are available for this purpose, which relieve students of their programming effort but still require their understanding of the physics concepts involved and their mathematical description. There are numerous suggestions for using such graphical modelling tools to develop understanding of physics concepts (see Chap 11), for example, tools provided by Buuren and Heck (2019).

8.6 Conclusions

Physics is inconceivable without a mathematical description of its processes and concepts. Accordingly, physical–mathematical modelling is the core of physics and its methodology. Therefore, mathematics and mathematisation in connection with physics play a significant role in physics education on all school levels, at university and also in the in-service education of teachers. In this chapter, we have focused on the treatment of physical–mathematical modelling and the related difficulties of students at school and university. A detailed analysis shows, on the one hand, the diversity of the roles of mathematics in physics and, on the other hand, that mathematical knowledge cannot be applied in physics without a specific transformation process. The complexity of this process is reflected in the various models that have been developed to describe specific aspects of mathematisation and physical–mathematical modelling. Knowing such models can support physics teachers in planning and structuring their instruction, as well as in analysing their students' learning processes and consequently adjusting support for them. In presenting the models in this chapter, we have detailed different aspects of physics–mathematical modelling, both from the perspective of task solving and from that of analysing tasks and delivering instruction. While students often use mathematical tools in a superficially technical way, their structural abilities—which consist of closely linked physical–mathematical concepts and mathematical elements and their connections in a variety of ways—often play only a minor role in classroom learning.

We have also outlined which mathematical elements and structures are relevant to learning physics on the different educational levels from a physics perspective and pointed out the possible pitfalls. These show up in difficulties that students have in developing physical–mathematical models. If physics teachers look at the mathematical elements and forms of representation in detail, they should find that mathematical operations themselves have already incorporated different physics meanings. When working on physics tasks with mathematical methods, learners have to take all these different aspects into account and link them appropriately. It is not surprising that problems arise in this process, and the reasons are manifold ranging from deficits in students' understanding physics concepts to their technical errors in mathematical procedures. Various sources of error interact and lead to unfavourable solution procedures. It is not yet fully understood how to prepare students efficiently for understanding physical–mathematical modelling. However, there are some suggestions for teaching strategies. Developing students' structure-building abilities can be achieved in a physics lesson by addressing the following details: (1) students' own analysis of the related equations in detail, (2) students' systematic verbalisation of symbols and their meanings in the equations, (3) students' own modelling processes guided by the teacher and (4) students' use of digital, graphical media (see Chap. 11). In this context, we have described several possible teaching strategies without being able to identify specific methods, which, according to empirical investigations, would promise more success and would therefore be preferable. This chapter aims to help teachers become aware of the teaching goal of physical–mathematical modelling. Nevertheless, much more research is needed to better understand possible teaching and learning strategies in the field of physical–mathematical modelling.

Acknowledgements We would like to thank Carl Angell (University of Oslo) and Igal Galili (Hebrew University of Jerusalem) who carefully and critically reviewed this chapter.

References

Angell C, Kind PM, Henriksen EK, Guttersrud Ø (2008) An empirical–mathematical modelling approach to upper secondary physics. Phys Educ 43(3):256–264. https://www.uio.no/studier/emner/matnat/fys/nedlagte-emner/FYS2150L/v09/undervisningsmateriale/physics_education.pdf

Bagno E, Berger H, Eylon B-S (2008) Meeting the challenge of students' understanding of formulae in high-school physics: a learning tool. Phys Educ 43(1):75–82

Bagno E, Berger H, Magen E, Pollingher C, Lehavid Y, Eylon B (2019) Starting with physics: a problem-solving activity for high-school students connecting physics and mathematics. In: Pospiech G, Michelini M, Eylon B-S (eds) Mathematics in physics education, Springer, Cham, Switzerland, pp 317–331. https://doi.org/10.1007/978-3-030-04627-9_14

Berlin DF (1991) Integrating science and mathematics in teaching and learning: A bibliography. ERIC Clearinghouse for Science, Mathematics, and Environmental Education, Columbus, Ohio. http://eric.ed.gov/?id=ED348233

Bing T, Redish E (2009) Analyzing problem solving using math in physics: epistemological framing via warrants. Phys Rev Phys Educ Res 5:020108. https://doi.org/10.1103/PhysRevSTPER.5.020108

Borromeo Ferri R (2006) Theoretical and empirical differentiations of phases in the modelling process. ZDM Math Educ 38(2):6–95. https://doi.org/10.1007/BF02655883

Brand S (2014) Erwerb von Modellierungskompetenzen: empirischer Vergleich eines holistischen und eines atomistischen Ansatzes zur Förderung von Modellierungskompetenzen [Acquisition of modelling competencies: empirical comparison of a holistic and an atomistic approach to promoting modelling skills]. Springer Spektrum, Wiesbaden, Germany. https://doi.org/10.1007/978-3-658-06679-6

Burkholder E, Blackmon L, Wieman C (2020) Characterizing the mathematical problem-solving strategies of transitioning novice physics students. Phys Rev Phys Educ Res 16(2):020134. https://doi.org/10.1103/PhysRevPhysEducRes.16.020134

Buuren O, Heck A (2019) Learning to use formulas and variables for constructing computer models in lower secondary physics education. In: Pospiech G, Michelini M, Eylon B-S (eds) Mathematics in physics education, Springer, Cham, Switzerland. https://doi.org/10.1007/978-3-030-04627-9_8

Carli M, Lippiello S, Pantano O, Perona M, Tormen G (2020) Testing students ability to use derivatives, integrals, and vectors in a purely mathematical context and in a physical context. Phys Rev Phys Educ Res 16:010111. https://doi.org/10.1103/PhysRevPhysEducRes.16.010111

Carrejo DJ, Marshall J (2007) What is mathematical modelling? Exploring prospective teachers' use of experiments to connect mathematics to the study of motion. Math Educ Res J 19:45–76

Domert D, Airey J, Linder C, Kung RL (2007) An exploration of university physics students' epistemological mindsets towards the understanding of physics equations. Nord Stud Sci Educ 3(1):15–28. https://doi.org/10.5617/nordina.389

Doorman LM, Gravemeijer KPE (2009) Emergent modelling: discrete graphs to support the understanding of change and velocity. ZDM Math Educ 41(1–2):199–211. https://doi.org/10.1007/s11858-008-0130-z

Erickson T (2006) Stealing from physics: modeling with mathematical functions in data-rich contexts. Teach Math Appl 25(1):23–32. https://doi.org/10.1093/teamat/hri025

Feynman R (1965) The character of physical law. The MIT Press, Cambridge

Friend H (1985) The effect of science and mathematics integration on selected seventh grade students' attitudes toward and achievement in science. School Sci Math 85:453–461. https://doi.org/10.1111/j.1949-8594.1985.tb09648.x

Galileo G (1623) Il Saggiatore

Galili I (2018) Physics and mathematics as interwoven disciplines in science education. Sci Educ 27(1):7–37. https://doi.org/10.1007/s11191-018-9958-y

Geyer M-A, Kuske-Janßen W (2019) Mathematical representations in physics lessons. In: Pospiech G, Michelini M, Eylon B-S (eds) Mathematics in physics education, Springer, Cham, Switzerland, pp 75–102. https://doi.org/10.1007/978-3-030-04627-9_4

Greca IM, Ataíde ARP (2017) The influence of epistemic views: about the relationship between physics and mathematics in understanding physics concepts and problem solving. In: Greczyło T, Dębowska E (eds) Key competences in physics teaching and learning, Springer International Publishing, Cham, Switzerland, pp 55–64. https://doi.org/10.1007/978-3-319-44887-9_5

Guttersrud Ø, Angell C (2014) Mathematics in physics: upper secondary physics students' competency to describe phenomena applying mathematical and graphical representations. In: Kaminski W, Michelini M (eds) Teaching and learning physics today: challenges? Benefits? Proceedings of selected papers of GIREP-MPTL-ICPE-conference 2010 in Reims, France, Università degli Studi di Udine, Udine, Italy, pp 84–89

Heck A, Buuren O (2019) Students' understanding of algebraic concepts. In: Pospiech G, Michelini M, Eylon B-S (eds) Mathematics in physics education, Springer, Cham, Switzerland. https://doi.org/10.1007/978-3-030-04627-9_3

Ibrahim B, Ding L, Heckler AF, White DR, Badeau R (2017) Students' conceptual performance on synthesis physics problems with varying mathematical complexity. Phys Rev Phys Educ Res 13:010133. https://doi.org/10.1103/PhysRevPhysEducRes.13.010133

Ivanjek L, Susac A, Planinic M, Andrasevic A, Milin-Sipus Z (2016) Student reasoning about graphs in different contexts. Phys Rev Phys Educ Res 12:010106. https://doi.org/10.1103/PhysRevPhysEducRes.12.010106

Kanderakis N (2016) The mathematics of high school physics. Sci Educ 25(7):837–868. https://doi.org/10.1007/s11191-016-9851-5

Karam R, Krey O (2015) Quod erat demonstrandum: understanding and explaining equations in physics teacher education. Sci Educ 24(5):661–698. https://doi.org/10.1007/s11191-015-9743-0

Karam R, Pietrocola M, Pospiech G (2012) The complex road to mathematization in physics instruction. In: Bruguière C, Tieberghien A, Clément P (eds) E-Book proceedings of the ESERA 2011 conference, Science learning and Citizenship. Strand, Lyon, France, pp 112–117

Kim M, Cheong Y, Song J (2018) The meanings of physics equations and physics education. J Korean Phys Soc 73:145–151

Krey O (2012) Zur Rolle der Mathematik in der Physik: wissenschaftstheoretische Aspekte und Vorstellungen Physiklernender [On the role of mathematics in physics: epistemologic aspects and conceptions of physics learners]. Logos, Berlin

Kuo E, Hull MM, Elby A, Gupta A (2020) Assessing mathematical sense making in physics through calculation-concept crossover. Phys Rev Phys Educ Res 16:020109. https://doi.org/10.1103/PhysRevPhysEducRes.16.020109

Kuske-Janßen W (2020) Sprachlicher Umgang mit Formeln [Verbal handling of formulas]. Logos, Berlin

Lee G, Yi J (2013) Where cognitive conflict arises from? The structure of creating cognitive conflict. Int J Sci Math Educ 11(3):601–623. https://doi.org/10.1007/s10763-012-9356-x

Lehavi Y, Bagno E, Eylon B-S, Mualem R, Pospiech G, Böhm U, Karam R (2015) Towards a PCK of physics and mathematics interplay. In: Fazio C, Sperandeo Mineo RMS (eds) The GIREP MPTL 2014 conference proceedings, Università degli Studi di Palermo, Palermo, pp 843–853

Lehavi Y, Mualem R, Bagno E, Eylon B-S, Pospiech G (2019) Taking the phys–math interplay from research into practice. In: Pospiech G, Michelini M, Eylon B-S (eds) Mathematics in physics education, Springer, Cham, Switzerland, pp 335–353. https://doi.org/10.1007/978-3-030-04627-9_15

Linder C (2013) Disciplinary discourse, representation, and appresentation in the teaching and learning of science. Eur J Sci Math Educ 1:43–49. https://doi.org/10.30935/scimath/9386

Mäntylä T, Poranen J (2019) Combining physics and mathematics learning: discovering the latitude in pre-service subject teacher education. In: Pospiech G, Michelini M, Eylon B-S (eds) Mathematics in physics education, Springer, Cham, Switzerland, pp 247–266. https://doi.org/10.1007/978-3-030-04627-9_11

Meli K, Zacharos K, Koliopoulos D (2016) The integration of mathematics in physics problem solving: a case study of Greek upper secondary school students. Can J Sci Math Technol Educ 16(1):48–63. https://doi.org/10.1080/14926156.2015.1119335I

Michelsen C (2006) Functions: a modelling tool in mathematics and science. ZDM 38(3):269–280

Michelsen C (2015) Mathematical modeling is also physics—interdisciplinary teaching between mathematics and physics in Danish upper secondary education. Phys Educ 50:489

Morgan MS, Morrison M (1999) Introduction. In: Morgan MS, Morrison M (eds) Models as mediators. Perspectives on natural and social science, Cambridge University Press, Cambridge, pp 1–9

Nguyen PC, Krause E (2020) Interdisciplinarity in school and teacher training programs. In: Kraus S, Krause E (eds) Comparison of mathematics and physics education, Springer Spektrum, Wiesbaden, Germany, pp 15–35. https://doi.org/10.1007/978-3-658-29880-7_2

Niss M (2010) Modelling a crucial aspect of students' mathematical modelling. In: Lesh R, Galbraith PL, Haines CR, Hurford A (eds) Modelling students' mathematical modelling competencies: ICTMA, 13. Springer, US, Boston, MA, pp 43–59

Opfermann M, Schmeck A, Fischer HE (2017) Multiple representations in physics and science education: why should we use them? In: Treagust D, Duit R, Fischer HE (eds) Multiple representations in physics education. Springer, Cham, Switzerland, pp 1–22

Pietrocola M (2008) Mathematics as structural language of physical thought. In: Vicentini M, Sassi E (eds) Connecting research in physics education with teacher education, International Commission on Physics Education, New Delhi

Pospiech G (2019) Framework of mathematization in physics from a teaching perspective. In: Pospiech G, Michelini M, Eylon B-S (eds) Mathematics in physics education, Springer, Cham, Switzerland, pp 1–33. https://doi.org/10.1007/978-3-030-04627-9_1

Rath G (2006) Auseinandergelebt?—Physik und Mathematik [Living apart? Physics and mathematics]. Plus Lucis 1:9–13

Redish EF (2017) Analysing the competency of mathematical modelling in physics. Springer, Cham, Switzerland

Roorda G, Vos P, Goedhart MJ (2015) An Actor-oriented transfer perspective on high school students' development of the use of procedures to solve problems on rate of change. Int J Sci Math Educ 13(4):863–889. https://doi.org/10.1007/s10763-013-9501-1

Schukajlow S (2011) Mathematisches modellieren, vol 6. Waxmann Verlag, Münster

Sherin BL (2001) How students understand physics equations. Cogn Instruct 19(4):479–541. https://doi.org/10.1207/S1532690XCI1904_3

Tall D, Vinner S (1981) Concept image and concept definition in mathematics with particular reference to limits and continuity. Educ Stud Math 12(2):151–169. https://doi.org/10.1007/BF0 0305619

Torigoe ET, Gladding GE (2011) Connecting symbolic difficulties with failure in physics. Am J Phys 79:133. https://doi.org/10.1119/1.3487941

Tuminaro J, Redish E (2007) Elements of a cognitive model of physics problem solving: epistemic games. Phys Rev Spec Top Phys Educ Res 3:020101. https://doi.org/10.1103/PhysRevSTPER.3.020101

Turşucu S, Spandaw J, Flipse S, Vries MJ (2017) Teachers' beliefs about improving transfer of algebraic skills from mathematics into physics in senior pre-university education. Int J Sci Educ 39:587–604. https://doi.org/10.1080/09500693.2017.1296981

Uhden O, Karam R, Pietrocola M, Pospiech G (2012) Modelling mathematical reasoning in physics education. Sci Educ 21(4):485–506. https://doi.org/10.1007/s11191-011-9396-6

Uhden O (2012) Mathematisches Denken im Physikunterricht—Theorieentwicklung und Problemanalyse [Mathematical reasoning in physics education: theory development and problem analysis]. Logos, Berlin

Uhden O (2016) Verständnisprobleme von Schülerinnen und Schülern beim Verbinden von Physik und Mathematik [Students' comprehension problems in connecting physics and mathematics]. Zeitschrift für Didaktik der Naturwissenschaften 22(1):13–24. https://doi.org/10.1007/s40573-015-0038-4

Vom Hofe R, Blum W (2016) "Grundvorstellungen" as a category of subject-matter didactics. J Math-Didakt 37(1):225–254

Yeatts F, Hundhausen J (1992) Calculus and physics: challenges at the interface. Am J Phys 60:716–721

Zavala G (2019) The design of activities based on cognitive scaffolding to teach physics. In: Pietrocola M (ed) Upgrading physics education to meet the needs of society. Springer International Publishing, Cham, Switzerland, pp 169–179

Chapter 9
Physics Tasks

Hans E. Fischer and Alexander Kauertz

Abstract In both university and school classrooms, tasks are an important element of teaching physics. They are used to organise students' learning processes and to diagnose students' learning progress for teachers' diagnosis and guidance of their students' learning progressions and performance measurement. In the classroom, tasks are opportunities for interaction, especially communication, between students and between teachers and students. In *learning situations*, students' abilities and competences can be specifically developed and diagnosed with learning tasks. A teacher can use task sequences to scaffold the students' learning processes and keep the focus on learning goals; task sequences help teachers to shape his or her teaching in the classroom, respecting the abilities and needs of the students. In *performance measurement* situations, a single test task demands a certain student ability to apply a physics concept to solving the task. The abilities needed to solve the task should be identifiable without overlapping with other abilities, if possible, in order to decide why a student could not solve a task. A complete test in turn depicts the entire area of interest (e.g. abilities in mechanics at the end of grade 10) for which also other abilities are also to be measured, such as reading ability or cognitive ability in general. The quality of tasks can be assured by task characteristics with already researched effects on learning processes or learning outcomes. The characteristics described in the chapter can help teachers, among other things, to develop tasks from task collections or textbooks in order to improve them and adapt them to the teachers' planned learning goals. The criteria come from many different fields of research and are selected and described from a physics education perspective. When experts and teachers discuss learning or test items, for example, in the context of test item development for class tests or final exams, they can systematically vary the difficulty and variety of the items by using the described characteristics as a toolbox in order to create valid testing instruments (see Chap. 16) or implement successful teaching

H. E. Fischer
Universität Duisburg-Essen, Essen, Germany
e-mail: hans.fischer@uni-due.de

A. Kauertz (✉)
Universität Konblenz-Landau, Landau, Germany
e-mail: kauertz@uni-landau.de

© Springer Nature Switzerland AG 2021
H. E. Fischer and R. Girwidz (eds.), *Physics Education*, Challenges in Physics Education,
https://doi.org/10.1007/978-3-030-87391-2_9

(see Chap. 4). This chapter describes both the functions of tasks to structure teaching and create learning tasks (learning opportunities) and diagnostic or test tasks as well.

9.1 Characteristics of Tasks in Physics Lessons

Tasks are typical elements of physics lessons. The most general description of a task is given by the definition, that is, the oral or written instructions to do something. In physics lessons, we usually limit the use of tasks to activities for calculating, naming or explaining something in order to reproduce, apply or transfer physics knowledge or concepts to solving the tasks. In education, in general, a more differentiated concept of tasks is used; according to (Klauer 1987, p. 15) it is a "… combination of a stimulus component with a response component. The stimulus component consists of a certain content that is presented in a certain way. The response component consists of the action to be performed on the stimulus component" (translated by authors of this chapter). For students, the stimulus or offer of the teacher must be directed either to organise their learning activities or to elicit their response by activating the related knowledge and abilities to answer a test item.

According to Renkl (1991) and Villarroel et al. (2020), a text-based task in a physics lesson or class test consists of several requirements, each of which should be solvable by the students. Reading ability is necessary for reading the text; students should also have the ability to develop a plan for the solution, which requires their ability to apply physics concepts and mathematical operations, and in the end, the solution must be critically reflected and the results communicated. Physics tasks are generally independent of the form of presentation used and thus can be presented orally, in writing, or graphically or as a mixture of several forms of representation (see Chap. 7). However, oral tasks play a special and varied role, for example, those with teacher prompts and student responses, and written tasks have use for testing and for specific phases in learning in and out of the lesson.

In the following sections, we aim to structure different characteristics of stimuli and responses of a task and their processing by students for the task in a goal-oriented selection. Research results already exist on how some of these characteristics influence student processing of learning tasks and their chances of solving these tests or performance tasks. This leads to an appropriate selection of tasks, which is important for teachers to control the quality of their teaching and to vary the difficulty of a task. Fitting the level of difficulty to students' abilities is a precondition for cognitive activation and support for the students (internal differentiation) in a lesson and for an adequate diagnosis. The difficulty-based sequencing of tasks allows planning, structuring and organising teaching as required by the teaching goals and the cognitive prerequisites of the students. The development of a good task and adapting it for varying conditions is a creative act and usually requires an iterative process of trial, informative error and redesign before the task works well as desired. However, there are many task collections in textbooks, which provide a good basis for the application of selected tasks in teaching contexts. A systematic and critical analysis of these tasks helps to select, adapt or vary tasks if necessary.

This chapter explains how tasks can be processed by learners and proposes categories with which they can be assessed and constructed for the learners' specific needs and gives advice on their selection and design. For this purpose, we first distinguish learning tasks and performance tasks as well as tasks and problems. Then, we look at the structure of a task and establish a connection between its formulation or design and the process of task processing by learners. We then describe the embedding of a learning task and a complex problem solution in the physics classroom and a situation of performance measurement. We next clarify effects that can result from the combination of tasks. We conclude with recommendations for the design and use of tasks in lessons and tests.

9.1.1 Learning Tasks and Performance Tasks

In teaching physics, a clear distinction between learning and achievement situations is essential and an important feature for the quality of teaching. Students must be able to rely on when they have the *right to make mistakes* (in learning situations) and when they should better avoid them (in achievement situations). (Neumann et al. 2013; Osborne et al. 2016). Accordingly, learning and performance must be distinguished. *Learning tasks* have the function of initiating and controlling learning processes and learning progressions. Test tasks applied at school are a subcategory of performance tasks and so are oral and practical exams or interviews.

Learning tasks are designed to fit into certain phases of a learning process, for example, they offer various approaches to activate previous knowledge and student conceptions, demand exchange with others to encourage creative solution ideas or to compare results. The focus lies on the operationalisation of the learning goal. Students should be able to follow the operational instructions to achieve the goal. A learning task offers many opportunities for further activities that should lead to new cognitive constructions, which are necessary to develop students' learning processes in the physics classroom.

Performance tasks have the function of checking the availability and application of knowledge, abilities and competencies. Very superficially, competence can be defined as the ability to apply specific knowledge to solve tasks or problems. A detailed description of competence models in general and in physics education can be found in Shavelson and Stern (1981), Leutner et al. (2017), and Weßnigk et al. (2017).

Organising learning and assessing performance, the two different functions of tasks require different designs of tasks according to the standards of their quality. As we will see later in this chapter when we look at the task-solving process, the design of a task has an intended influence on the learning process.

In the case of *learning tasks*, a strict classification of abilities needed for solving the tasks is not possible. Learning tasks are made to offer learning opportunities for students and, therefore, should be appropriately and attractively formulated in order to allow students to find the goal of the offer. Therefore, learning tasks usually consist of complex requests for actions that imply several steps that can be identified according to individual needs and abilities of students (see Task 1). The range between high and low complexity and difficulty of the task and the offer of different possibilities of its solutions must contain potential for differentiation to address the different abilities of different students in a class. For example, theoretical and experimental solutions and individually adapted solution aids can be offered.

In the case of *performance tasks*, the focus is on the clarity of the demand for limiting and avoiding all processes and approaches that are not helpful for the specific solution. The solution, in turn, is characterised by clear criteria that are directly related to the knowledge, abilities or competences to be tested. Although difficult to implement, a performance task or part of it should relate to only one ability; otherwise, it is not possible to clarify the cases when it could not be solved. This is important not only for the teacher to be able to use the tasks to assess students' performance fairly, but also to provide the clearest possible feedback as to whether a particular knowledge, ability or competence has been sufficiently acquired by the students or needs to be improved in classroom teaching or learning by specific students. For example, if the students cannot read the text of a task correctly or if they are not able to solve an equation mathematically, the teacher will not get any information about his or her physics knowledge or ability. Therefore, a teacher should ensure as far as possible that the wording of his or her tasks is appropriately formulated and that the required mathematical procedures should be manageable by all students in the class. It has to be noted here, however, that in every class there is a distribution of abilities, willingness and motivation with regard to learning physics and that therefore a distribution of the difficulties of the tasks of a test and of student performance cannot be avoided. In principle, this applies to any sample, be it a class in a class test or the group of all sixteen year olds in a large-scale assessment (see Chap. 16).

Learning and performance phases often alternate in the classroom, so it is important for students to get clear guidance from the teacher as to which phase they are in. In learning phases, students should know that mistakes are not only allowed but mostly productive on their way to the newly learned physics concepts. In tests, however, mistakes should be avoided.

9.1.2   Tasks and Problems

Before we take a closer look at tasks and their characteristics, we have to distinguish between tasks and problems. In everyday life, the term *problem* is usually used when a task cannot be solved (or requires high cognitive effort to solve it). In learning psychology, tasks and problems have fundamentally different characteristics, which, according to Doyle (1983), can be judged with three categories. (1) Problems aim for the procedure to reach a given result, and tasks aim for the result of a known solving procedure. (2) The solution process requires different resources, such as knowing the necessary physics concepts for solving tasks and knowing solving procedures for solving problems. (3) Problems and tasks require specific steps of action, such as generating hypotheses and new solving processes for problems and applying known mathematical procedures and solving routines for tasks.

However, a *task* is clearly structured as step-by-step instructions with explicit working invitations and often with a question on the production of a calculation result, and a *problem* gives only few hints that focus on the initial situation and the final state. The specification of a dynamic problem invites students to explore and model the situation first and then find a solution in a reasonable number of trials.

For complex tasks and problems, for example, conducting an experiment or applying a physics concept, there is at least one assumption regarding the appropriate experimental procedure or the physics concept, which can serve as an initial input for the development of the experiment or concept application. The solution of the task must be found, and the process for solving the task should be known.

Because the ability to solve problems and tasks depends on a learning process, for individual students, physics problems can become tasks, and tasks can have the characteristics of a problem. Therefore, a teacher should be able to provide the appropriately adapted learning support for each student.

The first example can be applied as a task or a problem. After teaching about the superposition and independence of motion, the teacher could pose the following task (see Fig. 9.1) to offer students the opportunity to apply both concepts.

Task 1 can be solved experimentally or theoretically. *Experimental problem solving* succeeds qualitatively quickly if the physics concepts are known, but also without knowledge of the physics concepts by trial and error as a dynamic problem, with a moderate number of trials. For the *theoretically based solution* of the task,

Task 1:
The toy car first rolls down a ramp and then falls freely to the ground. Its acceleration is due to the force of gravity.
Questions:
1. As task: Find the position at which the toy car falls through the centre of the ring that drops from the ceiling of the room as shown in the drawing below. The ring is cut off when the toy car leaves the ramp. Note the superposition principle and the independence principle of motion.
2. As problem: Where to place a ring (see picture) and when to cut the thread that holds it to make the toy car fall through the ring?

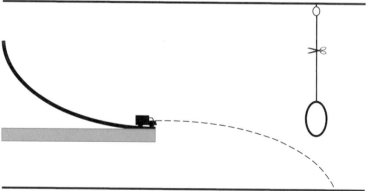

Fig. 9.1 Toy car rolls down a ramp and falls to the ground (Task 1)

the horizontal throw must be divided into two partial movements according to the superposition principle (principle of independence), the movements in x-direction and in y-direction do not influence each other. However, this only works if other forces, such as air friction, are not taken into account. The direction in which the solution should go depends on the teacher's goals.

It is plausible that the difficulty of a problem (and a task) depends on students' knowledge. Accordingly, in a test, each student receives a score for each task that corresponds to his or her ability in solving the task (Atkinson and Renkl 2007; Renkl 1991). If the physics concept is not known, students often use the heuristic method of trial and error, a method that is also used in physics research when nearly nothing is known about the object under investigation. If students remember the preparation in the previous lesson, they should be able to develop a hypothesis and design the experiment accordingly; after a few attempts, the solution should succeed. Question 1 is an easy task for students who can apply the superposition principle to this case. They would also immediately come to the right solution in an experimental setting. Students who do not know the superposition principle must develop an experimental solution strategy to answer Question 2. These students should therefore know how to proceed with a physics experiment. The ability of the students influences not only the goal of the lesson in which the task/problem is to be solved, but also the character of the task/problem. The teacher's requirement and the students' abilities decide the characteristics as a task or as a problem. According to Stigler and Hiebert (2004), helpful embedding of a task in the learning situation, for example, through detailed experimental instructions, can also turn a problem into a task (see Task 1). However, a task cannot turn into a problem because, different from problems, the target states of tasks are not known. In addition, if a student does not know the possible way to solve a task and the format of the task is open ended, it is mostly unlikely that the task can be solved only by chance or trial and error. However, the multiple-choice task format allows the solution of a task by chance. In the following sections, we use the term task for all questions that the teacher can ask in physics lessons, because it is the more general term from the psychology of learning perspective.

9.1.3 Structure of a Task

Starting from the basic assumption that learning is influenced by an interaction between a teacher and his/her students, a task can be considered as part of their communication. In classroom communication, the teacher tries to make an offer (a task as a request for response) which the students can use to clarify the physics content presented by the teacher and, in the best case, learn something new (Fig. 9.2).

According to the four-sides model of communication, the offer of the sender (teacher) contains information about the matter in question, but also includes a request for action of the student. The following example illustrates the situation in which the communication is embedded to provide additional factual information that is not an explicit part of the offer as indicated in Task 2 (see Fig. 9.3).

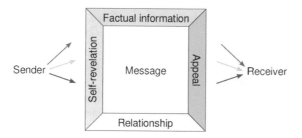

Fig. 9.2 Four-sides model or four-ear model of Schulz von Thun (1981). Graphic by Jazzy Julius—translated from http://en.wikipedia.org/wiki/File:Vier-Seiten-Modell_de.svg, which was based on https://www.schulz-von-thun.de/die-modelle/das-kommunikationsquadrat and the public domain (https://commons.wikimedia.org/w/index.php?curid=7685255)

Task 2:
Determine the falling distance of a freely falling body after t = 2 s.

Fig. 9.3 Task for additional information needed by the student

The text of Task 2 is an offer for the student; the solution is expected to be $s = 20$ m. The question contains the information that a body falls freely along a certain distance and the request to calculate this distance. From physics lessons, it is clear that the equation $s = \frac{1}{2} g\, t^2$ is useful and for g the local acceleration due to gravity, or its rounding to 10 m/s^2, that is common in physics lessons, can be assumed. Although the task looks very simple, several skills are required to solve it. For example, the text must be understood, and therefore, students' reading ability cannot be taken for granted, since technical terms (falling distance, free, etc.) and an equation as mathematical representation (see Chap. 10) are used. The teacher's responsibilities include not only ensuring that the given equation must be available and correctly interpreted by the students, but also that they have both the required mathematical and physics knowledge because it must be recognised what the equation as a whole and the individual symbols mean in physics. Arithmetical operations are needed when calculating the results and so is physic knowledge, when interpreting and indicating the results, including the correct units.

For all the tasks in this example, the following five-point structure can be applied:

1. Task text (contains information on the subject content, the context, which may not be subject specific and its embedding in the situation).
2. Question to the students.
3. Solution process.
4. Result.
5. Answer.

Just as not every communication is successful, a task that has been set in the classroom may not lead to the desired processing in student learning. According to Fig. 9.2, not only the clarity of the offered task is important for successful communication, but also its appropriate interpretation by the students (as receivers). The teacher (as sender) should therefore anticipate the students' interpretation, which should be considered, when designing the offer. This means that a task is not good in itself, but it should fit in well with particular learners in a particular situation. When using tasks from task collections or textbooks in their lessons, teachers must therefore take into account the planned learning goal, information about the abilities required to solve the tasks and the abilities of their own students when assessing and modifying the tasks.

When students fail in solving a task, it is because either the text of the task has not been interpreted correctly by them or their knowledge and ability are not sufficient. The knowledge that can be implicitly assumed by the teacher can, for example, refer to the meaning of the technical terms and the correct equation for the solution. In Task 2 (see Fig. 9.3), it is quite possible that the students do not conclude that an accelerated motion is from the formulation *free falling body*, that is, the information is not interpreted correctly. The probability that learners make the connection between free fall and accelerated motion increases if accelerated motion is the subject of the lesson in which the task is for the students to solve (for priming effect, see Becker et al. 2020). However, when students make this connection, they may still not know or be able to look up the equation and thus not be able to solve the task because the implicit information is not present.

Since the text of learning tasks and test tasks has to be interpreted by the learners and even implicit information cannot be avoided for every student, the fundamental individual differences must be taken into account when constructing and applying the tasks. In learning tasks, this difficulty can be taken into account by communicative adaptation of the tasks to the needs of the students.

In interactive communication, as is common in physics classes, the teacher can respond to the misunderstanding and adapt his responses in case of failure. In the case of classroom tasks, this can be done by means of aids offered by the teacher. For example, to explicate the implicit information (use the equation $s = \frac{1}{2} g\, t^2$), the teacher can mention it explicitly and give the further information to use the value 10 m/s^2 for calculating with g to clarify the context. In addition, the original task from a textbook can be reformulated linguistically such as *calculate the height of fall when g and t are given and use the equation for free fall*.

In contrast, test tasks should be formulated in such a way that no additional information or even correction is required. However, the students' different interpretations of the text caused by the formulation of the tasks cannot be completely avoided. This is especially true for class tests because, in most cases, individual requirements are represented by only one task and no reduction in reliability can be achieved by several tasks on the same physics concept. In test situations where aids are rather unusual, the ambiguity of a text is problematic. It should always be considered whether information cannot be perceived or interpreted correctly or whether all the information needed to solve the task is available, for example, about the equations used. Therefore, test items in particular must be formulated clearly and unambiguously and must leave as little room as possible for different interpretations. If during a class test, it becomes apparent that individual students are having problems in understanding the text, and advice for individual students must always be given to all students in order not to disadvantage anyone.

In scientific tests, multiple-choice items requiring the same implicit information improve the reliability of the test instrument because the probability that the information has not been correctly interpreted decreases with the number of different formulations (see Chap. 16).

In the following section, the solution process is discussed in more detail in order to open up the possibility of adapting tasks at possible critical points in classroom teaching and learning.

9.1.4 Solution Process of a Physics Task.

The solution process of a task can be divided into different steps and student activities (see Fig. 9.4).

Model of the Task Situation

The meaningful reading of a task leads to a context or situation-based mental model (Rebello et al. 2007). The developed model is referred to in the following as the *situation model*. In many subject areas, especially in physics, the verbal representation of the situation model can be transferred into a pictorial representation. To explain a physics concept, physicists always need a piece of paper and a pen. As a starting situation, one could imagine a block falling from the sky in Task 2. However, the situation model already contains initial ideas about connections and processes. It contains physics assumptions, which ideally represent physics concepts, but in case of doubt, it also contains everyday ideas (for student ideas, see Chap. 14) and predictions derived from them for the spatial and temporal progression of the situation. For most people, it should be clear, for example, that the colour of a falling toy car is not relevant for the fall distance. However, size and weight might not be disregarded in an everyday observation, so they would probably be part of the situation model of students less competent in physics. To describe physics situations, learners use such situation models, but they cannot necessarily describe them verbally. Sometimes, it is very difficult to describe situation models verbally when they refer to physics concepts constructed as mathematical representations (e.g. the Maxwell equations or the general relativity theory). However, the representations of the situation model should help to solve the physics problem. In order to develop situation models that are useful for task solving, teachers therefore should know about those situation models based on everyday knowledge and the learners' way of thinking and working in general. In addition, the teacher should know about task

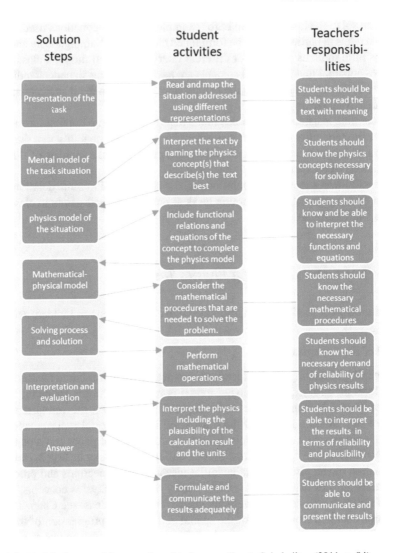

Fig. 9.4 Model of sequential processing of tasks according to Schukajlow (2011, p. 84)

processing as well as about conceptual and declarative knowledge of the students and their communicative abilities.

Teaching physics in a particular class always needs the students' specific physics knowledge and conceptions about the required, possible and actual physics content taught. *The required* is what the curriculum demands as to which concepts have to be taught in an age group. The level at which the acquisition of this physics knowledge is possible depends on the average individual cognitive development of the students. Younger students, for example, do not yet have very pronounced abilities to construct and apply mathematical relationships. The actual knowledge has to be assessed by the teacher himself or herself because it refers to the individual level of knowledge of the students in his or her class, and it should be assessed by a diagnostic test in the current lesson (see Chap. 15). Therefore, curricular content, age-dependent cognitive abilities and potentials of students and methods of diagnostics should be taught in teacher education.

The actual knowledge can be taken into account by adequately designing the tasks and preparing the students for finding their solutions, for example, by using only already known terms in the task text (see Chap. 13). Clear references to the students' knowledge must be established by explicitly naming the physics concepts just taught. For example, the expressions *free fall* should be named as a synonym for *accelerated motion,* and also *acceleration due to gravity* should be included at central points of the task text. The very fact that a task is set in physics lessons suggests to the students that certain physics rules and meanings must be taken into account when forming the situation model. For example, reading the term *force* in the task should indicate that thinking about *interaction* instead of *impulse* is a better reference for solving the task. When looking for energy conversions, the students may need to think about an energy flow chart that identifies the work needed to transform the energy form at the transformation points. For energy flow diagrams, the thickness of the arrows can also indicate the corresponding proportions of the transformed energies and the energy losses of the process. Additional difficulties arise when the same task is set outside the physics classroom in an unknown context (cf. situated learning and inert knowledge; see Chaps. 6, 8, 12 and 14). Usually, only students who have advanced knowledge in physics can make connections between everyday situations and physics concepts in terms of analogies or structural equality and apply this knowledge to construct their situation models.

Physics and Physical–Mathematical Models of the Situation

The formation of the situation model as physics model or physical–mathematical model is influenced by the selection of the representation used in the task (see Chap. 7). In addition to text that is presented orally or written, mathematical symbols, photos, (schematic) drawings, sketches and other representations of physics concepts and their combinations are common. Digital media in schools bring new forms of representation to physics lessons, for example, the use of films, animations and inter-active applications is already widespread (see Chaps. 7, 10 and 11). In the near future, virtual realities and augmented reality may be added as new representations (e.g. information superimposed on a live camera image using overlay technology). Forms of representation serve different sensory channels, which in turn can favour or hinder certain forms of knowledge retrieval. The more realistic the forms of representation in a task are used (e.g. photos or films), the more information is presented on a visible and audible level. The more abstract the information of the task is presented (as math-ematical symbols or text), the more the interpretation and conceptual knowledge is necessary to construct the situation model, because this is associated with a higher cognitive effort (Baumert et al. 2010; Cohn 2019). The combination of different forms of representation creates multimedia effects. Furthermore, multiple represen-tations can result in the *redundancy effect*—redundant information in text and images unnecessarily ties up memory capacity; the *split-attention effect*—switching between image and text requires memory capacity; the *expertise-reversal effect*—experts are prevented from discovering relevant information by an oversupply of information (Mayer 2008, 2015).

Text and Meaning

Based on the situation model, students have to interpret the oral prompts to think about and act on the written questions of the task, as well as the given information. As described in detail in Chap. 13, linguistic principles become relevant in such a case. In the context of tasks, demands connected to a certain action are called *command verbs*. Command verbs do not have unique significance, such as determining, solving, calculating, constructing and so forth. In everyday contexts, the term *determine* has the meaning of an independent decision, in the context of a physics task in particular, and it requires the calculation or measurement of a quantity. Therefore, learners in physics lessons have to learn about command verbs and their meaning in physics. Similarly, physics uses certain linguistic turns of phrase (syntax) to describe processes, actions and situations so that learners must first become familiar with, especially substantiations of, these processes and passive constructions. This is where language-sensitive teaching begins (see Chap. 13; Janík et al. 2020; van Dijk et al. 2020). Such teaching not only requires teachers' understanding of physics language, as well as their familiarity with technical terms and syntactic turns of phrases, but also their ability to make meaning of specific terms in physics. Teachers should also know the rationale for using physics language transparently for learners and show its advantages. In principle, it must be clarified in this context whether a text is appropriate for the target group (see Chaps. 4 and 13).

The physics language (terms and syntax) in a task is used to describe a situation and to communicate a request for thought and action. Therefore, it influences the selection and interpretation of the information during the processing of the task. Löffler (2016) described this process as transparency of information in the task and showed that, in so-called contextual tasks, the frequency of task solution can be influenced by this kind of transparency (see Chap. 13).

9.1.5 Contextual Tasks

Context-related tasks are tasks that provide information about the situation in an adequate language such as everyday language in everyday context or scientific language in a research laboratory. The text of the task in an everyday context contains (from the perspective of physics) irrelevant and implicit information about the situation, which can also have its emotional and affective characteristics. Accordingly, the construction of the situation model can be influenced and possibly even dominated by many other types of information and emotions that have nothing to do with the physics concepts addressed in the task. In addition, activating emotions and motivating students both have an effect on the their willingness to invest efforts and influences in their expectations in completing the task positively (self-efficacy expectation; Pozas et al. 2020). In addition, contextual tasks are an important element for the competence-oriented design of a lesson. They allow students to discuss the structuring of their reasoning and working processes when solving physics tasks, as well as the relevance of physics-related descriptions for everyday situations (Fischer 1998). This can create a precondition for the transfer of physics–typical ways of thinking and working to mostly very complex everyday problems.

The example of the free fall from above is a less contextualised task with a high degree of transparency from a physics point of view. The transparency could be further increased by explicitly naming the equation to be used or by specifying the acceleration due to gravity. It is neither motivating for most learners nor does it encourage them to engage with their ways of thinking and working in physics. Nevertheless, it can, for example, be used to illustrate and practise their use of physics-specific language or to test their physics knowledge related to the law of gravity.

In contrast, the following Task 3 (see Fig. 9.5) contains much different non-physics information, and it does not explicitly include the application of specific physics concepts. It is a much more contextualised task with only little transparency regarding physics.

Task 3:

The picture on the left shows the glacier mummy called Ötzi, who was found about 25 years ago in the Ötztal Alps in the border region of Italy and Austria. The man was deadly hit by an arrow about 5,300 years ago and immediately frozen in the glacial ice. Many objects were found in Ötzi's surroundings that he presumably carried with him. Among them was a mysterious grass mat (middle picture), whose purpose is still disputed among experts. One of the assumptions is that it could be a Stone Age sleeping mat; the picture on the right shows a modern mat.

Question:

Can such a thin mat protect body heat from the freezing cold ground? Should the Stone Age mat theory be discarded, or is there a scientific explanation that supports this assumption?

Fig. 9.5 Context task example: was it Ötzi's sleeping mat? ©Hohenstein Institute (Löffler 2016)

The steps described in Fig. 9.4 are necessary in order to arrive at a solution and an answer to the Task 3 question. The better the modelling of the situation (i.e. the identification of the physics elements and their appropriate interpretation) is, the higher the probability that the solution will succeed.

Role of Knowledge and Cognitive Processes in Contextualised Tasks

In successful task solving, students have first transformed the original situation model into a mathematically adequate physics model of the situation (see Fig. 9.4 and Chap. 8). The solution may be: air is trapped in small chambers in the tightly woven grass mat. However, the trapped air, like all gases, is a poor conductor of heat, so convection in the grass mat and the flow of thermal energy can be reduced. The theory of the Stone Age mat is therefore not to be discarded. In a physics unit on thermodynamics, this might be the starting point for designing related experiments to compare the thermal insulation of grass and foam mats.

The types of knowledge required to solve the problem also moderate the difficulty of tasks. According to Krathwohl (2002), cognitive processes and types of knowledge can be applied to describe learning goals, tasks and task solutions (see Table 9.1). The example of Task 3 can be classified as B.3 (applying conceptual knowledge) and Task 2 as A.3 (applying declarative knowledge). Declarative knowledge is also called factual knowledge. The experimental comparison of the mats needs C.3 in addition.

According to Miyake et al. (2000), solutions of tasks are not only influenced by the knowledge of the person performing the task but also by so-called *executive functions*, which are necessary for processing tasks successfully:

1. Shifting/switching (task, attention, strategy changes, etc.), for example, in Task 3 shifting the attention from the picture of the mummy to the physics content.
2. Updating (adjusting and monitoring working memory representations and processes), for example, in Task 3 linking the grass mat properties to those of the modern sleeping mat.
3. Inhibition (suppressing hasty, dominant, and/or automated responses), for example, in Task 3 to stop oneself from just using everyday knowledge and common sense, which seems adequate for such a context and going deeper into scientific thinking and reasoning which is more demanding and needs more effort.

In the case of more extensive and complex solutions of tasks in particular, so-called metacognitive abilities of solvers play an important role. They include setting goals, motivating oneself, making plans and implementing and evaluating the plans.

Role of Structure and Language in Contextualised Tasks

The application of learning strategies (metacognitive abilities) can be supported by structuring aids such as tables or flow charts and by structuring the task text according to the content. The different demands (e.g. naming the concept, determining the variables, solving an equation, checking the units, etc.) should be taken into account when structuring the work process (Kistner et al. 2015). When the teacher designs

Table 9.1 Taxonomy according to Krathwohl (2002, p. 216) for the classification of tasks and learning goals

Knowledge types	Cognitive processes					
	1. remembering	2. understanding	3. applying	4. analysing	5. evaluating	6. creating
A. Declarative knowledge			Task 2			
B. Conceptual knowledge			Task 3			
C. Procedural knowledge						
D. Metacognitive knowledge						

Teachers can use the table for their preparation

and formulates the task, he or she starts from his or her learning goal. But in addition, to give the students an orientation and make them concentrate on the task, he or she should point out in which form the answer should be given (e.g. "The number of answer options there are in a multiple-choice task." and "Do not forget whole sentences, units.", etc.) and which aids and representations are available for processing (only their own workbook, calculator, Internet, collection of formulas, etc.). The representations must also be carefully planned and realised. For example, care must be taken to ensure that students already have learned and therefore should know the mathematical representations (see Chap. 7) and the interpretations of the graphs and, if needed, other illustrations (Treagust et al. 2017).

However, working on one task cannot directly influence the cognitive processes necessary for solving similar tasks. The cognitive processes must be developed and practised over a longer period of teaching, and therefore, starting with a prototype and by practising the solution process through tasks with the same or similar structure can support learners in this process. Immediate and clear feedback on the solution attempts is also important. Students must be able to see whether their solution attempts need to be changed and adapted according to the teacher's suggestions (Broughton 1981; Vu et al. 2020). Direct help may also be possible in digital forms. Digital learning environments that automatically provide feedback and select subsequent tasks according to the learners' abilities seem to be suitable for such learning support (see Chap. 11; Euler et al. 2020; Ruiz-Mallén et al. 2020).

In addition, it is especially necessary with contextualised tasks for students to need support in reading their text. Students not only have to read the words and grasp the meaning of the text (with its own difficulty if it contains many special words and a complex syntax) to know what to do, they also need to differentiate between physics language and everyday language (e.g. to translate between both languages and understand which language is used for which purpose). This also applies to the understanding of the non-text-based representations used in the task text. In addition, organising and writing down the solution process must be taught. It is especially important to outline the work process on paper and separate it from the transcript to be handed in to the teacher or the presentation of the solution in a test or assessment.

Therefore, the teacher's linguistic support should start when students are reading the task, continue when they use different support and options of the working process and end when they are formulating an appropriate answer. As already described, much of the support can be provided in advance if language (especially the meaning of terms) is consciously used in the task. The most important features of support are summarised in the following questions, which the teacher can work through when constructing a task:

1. What do I want to be achieved with the task (e.g. learning task or test task)?
2. What physics knowledge do students need and what abilities do they need (e.g. laws, functions mathematical procedures)?
3. Which terms are important for the solving process?
4. Which syntax should be used?

5. Do formulations and complexity of the text meet the respective design standards?
6. Is the text appropriate for the respective age group/class?
7. Is it possible to anticipate and avoid or address conflicts with the students' prior understanding?
8. Which representations are necessary and which can help the students to understand the physics concept of the task?
9. Do the forms of representation I have used meet the design standards?
10. Are the students of my class able to understand the different forms of representations I have used?

Answering these questions means that the teacher needs to take into account the linguistic standards, the standards for constructing forms of representation and their applications. That also means that the everyday language standards and physics-related linguistic standards and norms must be part of teacher education at university, pre-service training in schools and in-service teachers' further education (see Chap. 13).

Summary of Skills and Abilities for Solving Tasks

As Table 9.1 suggests, solving a task follows several steps. Each of these steps requires a specific activity on the part of the student. In carrying out these activities, students need a wide range of skills and abilities, as we have described in the previous sections of this chapter. We now try to summarise these skills and abilities to give an overview of the intended requirements and unavoidable prerequisites of the student completing the task.

Below, we present a hierarchical system of five levels of skills and abilities that build on each other and are relevant to each of the seven activities. The system goes back to basic ideas of Krathwohl (2002), the discussion of thinking skills (cf. Lewis and Smith 1993; Tsaparlis 2020) and competence orientation (Kauertz et al. 2012). The five levels combine an increasing understanding of physics as a science with processes of increasing complexity necessary to acquire or apply that understanding. Each level provides clues as to how and why students cannot adequately perform the activity required in the solution process (see Table 9.2).

Both requirements, higher-order thinking skills and self-regulation skills, have to be developed in a longer process of working on physics tasks/problems where students receive feedback on their work with an explicit focus on the first three levels. Fourth and fifth levels are the target levels of competency-based teaching and addressed in competency tests. Tasks at these two levels encourage students' reflection on the broader understanding of physics during learning and require explicit responses to specific aspects of the first three levels in tests. Self-regulation as such is not physics specific; however, learning physics as a cognitive and formalistic approach to know the world relies heavily on students' abilities related to self-regulation.

Several studies on competence levels, higher- and lower-order thinking skills and Krathwohl's taxonomy (see Table 9.1) suggest that if students are able to perform

Table 9.2 Five levels of increasing understanding of physics as a science

On a basic or first level, students need to get access to the task, which could mean in inclusive settings that students with disabilities need support because they cannot read text or grab experimental materials. This also includes basic logical operations such as cause–effect relations, time sequences or identifying objects in their spatial and timely order. Such basic logical operations may be limited, for example, by limited attention spans, extreme emotional states or impairment of brain functions
On the second level, students need to read and write well enough in the language of the task, which could be challenging for students with non-native-speaking background and those with limited skills to perform basic mathematical operations (e.g. performing term transformations for school level; solving differential equations and integrals for university level) because of dyscalculia or inadequate practice
The third level is the level of declarative knowledge. This includes the knowledge of technical terms in physics, concepts of physics, functional relations and equations and the different representations of this knowledge. These different aspects of knowledge depend on each other; however, they might be differently stored in students' memories due to the situation where they were acquired, time and intensity of use, student's pre-knowledge or individual preferences. Students must be able to actively recall and use this knowledge based on what is given in the task and the learning or test situations. In learning situations, the teacher wants to introduce new knowledge and shows its relevance by embedding it into a solving process. In test situations, the teacher wants to evaluate, if his or her students' have the ability for and can apply it to, solving the task
The fourth level of skills or abilities is formed by physics-related higher-order thinking skills such as formulating, interpreting, analysing, applying or evaluating, as well as problem solving and critical thinking. Higher-order thinking occurs when a person needs newly learned concepts and interrelates and/or rearranges and extends these concepts to achieve a purpose or find possible answers in perplexing situations (for a general definition of higher-order thinking see Lewis and Smith 1993; Piaget 2013). This requires that students know and understand physics more broadly, that is, they can answer questions about what physics problems are aimed at, how physics work processes are structured, and what information another physicist needs to understand their answer and reasoning
The fifth level addresses students' self-regulation skills, that is, they need to be careful not to rush to an answer, but to structure and plan their work, set and evaluate intermediate goals, think carefully step by step, reflect critically on their own thoughts and check each partial solution for plausibility with the given information or other sources

at fourth and fifth levels, it is very likely that they will also perform well at the lower levels (first three levels). This means also that a task that causes unnecessary challenges at a lower level will hinder high-ability students who could solve the task under optimal task conditions at fourth and fifth levels. At the first level, these unnecessary challenges might be a poorly copied worksheet that can only hardly be read or provide additional verbal explanations that exceed the capacity of students' working memory. At the second level, such hindrance could be an unusual sentence structure or the use of an unnecessary amount of technical terms or foreign words. At the third level, this may concern detailed information students who have learned long time ago, and therefore, they cannot recall such information in the given time.

After detailed examination of the task-solving process, it becomes clear that task solving is associated with a variety of abilities that need to be developed, supported and promoted in the course of students' learning processes in the classroom. The most important abilities required to solve physics tasks are listed in Table 9.3. The absence of one of these abilities or knowledge claims may prevent the student from completing the task. The table gives a more detailed description of the abilities needed on the third to the fifth level and categorises them in four areas (comprehension, cognitive model, metacognition and knowledge).

It is important, especially in test situations, for the teacher to have clarity about the students' abilities needed to solve a certain test task. Ideally, each test task should focus on exactly one kind of ability; the other abilities should not play a decisive role to solve the task. Since the abilities and knowledge claims are not independent of each other, this is only possible to a limited extent in the real case but should be aimed at when developing tasks for tests and for learning. All tasks (items) of a test must represent the physics concept to be tested with all facets of ability and knowledge that have previously been theoretically and empirically established in the classroom.

Table 9.3 Abilities to solve physics tasks

1. Reading comprehension appropriate to the age and grade (Höttecke et al. 2018)
a. Knowledge of a language appropriate to physic (see Chap. 13)
b. Adequate interpretation of text related to the physics concepts used to solve the task
c. Reproduce, select, combine, add or transfer parts of the task text
d. Ability to handle devices for communication
2. Construction of a cognitive model of the task situation (Álvarez et al. 2020)
a. Activation of task-related declarative, conceptual and procedural knowledge (see Chap. 6)
b. Recognition of command verbs and their significance in physics such as describe, analyse or solve
c. Task-based selection and interpretation of the text
d. Using analogies or structural equality of different physics concepts (see Chap. 12)
3. Effective use of metacognitive processes (Akben 2020)
a. Task-related knowledge about the nature of the task and its requirements
b. Strategic knowledge that allows to evaluate the suitability of solutions for the task and to assess the effectiveness of alternative solutions
c. Knowledge about activating own knowledge, motivation and volition (see Chap. 2)
d. Application of physics standards and norms such as ISO standards, rules of scientific work, citation methods and forms of presentation
4. Knowledge about the epistemology of physics (Lederman et al. 2020)
a. Knowledge about distinguishing common sense predictions, expectations, explanations and judgments of plausibility and the interplay between theoretical and experimental physics (see Chap. 1)
b. Knowing a justificatory structure (objectivity, reliability and validity) and resolving conflicts on the basis of knowledge within the system (see Chaps. 16 and 17)
c. Knowing causal argumentations in terms of agents, parameters, experimental interventions, variables and the organisation of empirical work in physics (see Chap. 10)
d. Knowing a tendency to focus on characterisations of balancing and equilibrium such as the development of validity ranges of theories of micro-, meso- and macro-physics and key ideas such as conservation concepts or the general principle of relativity (see Chap. 5)

Of course, it is never possible to formulate each task appropriately for each student. In each class, there is a statistical distribution of the students' abilities, which affects the distribution of the accuracy of the answers. The goal must therefore be to influence the distribution for each learning group in such a way that the difference between the best and the weakest students is kept as small as possible. Even test anxiety of students can be taken into account by creating a relaxed atmosphere in the classroom during the test.

Teachers should therefore be guided by whether, in his or her opinion, the weaker students are able to understand the text and the representations in tasks and whether they should have the necessary knowledge to solve the tasks of a class test or an assessment. Nevertheless, a distribution of the task difficulties for the students in a class and the corresponding differences in their solutions can never be avoided. This leads to another aim of a test. It should be possible to assess the different abilities of the students with a high degree of discrimination. The difference between student abilities in one class should be justifiable in terms of abilities that have led to the failure to solve a task in an individual case (see Chap. 16). Only in this way can the results of a test be used to diagnose student abilities which are needed to plan further lessons (see Chap. 15).

9.1.6 Solving Tasks in Groups

So far, we have assumed what students have to achieve in order to solve tasks individually. In physics lessons, however, it is a common practice to perform learning tasks as partner and group work and to obtain a common result. Here, further abilities have to be added, they concern teamwork, including communication and cooperation. In the context of cooperation, metacognitive skills are important, especially with regard to goal setting and planning the joint approach. Finally, it is necessary that the abilities of the students involved in the process of developing situation models for task solving have a sufficient intersection as a common basis to maintain task-related communications and to plan and carry out coordinated and goal-oriented activities (Jin and Kim 2018; Wilson et al. 2018). These group work-dependent additional demands in turn require additional cognitive abilities and create additional risks of error, which must be considered in relation to the intended learning process and learning goal.

Therefore, cooperative tasks are particularly useful when these additional abilities are to be developed or practised or when they are in any other way functional for the task. They are functional when sharing of work is necessary to solve the task and in order to get to know typical working methods in physics, for example, experimental work or the development of theoretical approaches in a team. According to Nieswandt et al. (2020), group work demands the three areas of abilities which refer to the three areas: cognition (related to content or the problem to be solved), social interaction (based on social interactions and relations) and affection (referring to the emotional life of the working group). Based on these dimensions, students construct a *problem-solving space* that evolves and changes at every moment. Whether the task is successfully solved on the content dimension depends strongly on the social and affective dimensions. In addition, there is evidence of the need for social regulation processes regarding the available resources within the group.

The content dimension for (small) group work often relates to scientific enquiry, including experimental work. This kind of work combines theoretical abilities in physics explained above with knowledge about empirical research such as useful heuristics, measuring procedures and methods and measuring instruments necessary for solving the problem or task (see Chap. 10). These abilities should be related to declarative knowledge, as group members need to be able to share and discuss their knowledge, and their abilities need to be related to procedural knowledge, that is, routines need to be planned for carrying out at least part of the activities efficiently. Procedural knowledge needs much training to become automated to a certain degree. If the task requires complex problem solving—which is quite common in scientific enquiry tasks—the group also needs to be creative. Creativity involves the use of analogies typical for physics (e.g. Yerrick et al. 2003), such as identifying common structures of concepts in different areas of physics (e.g. heat and electric current, gravity and electric field). Analogies also refer to, for example, symmetries or extreme states of concepts. In negotiating knowledge and the creative parts of solving, students need to know and accept the physics way of communicating and reasoning in order to be efficient and successful in solving the tasks. These knowledge and acceptance are part of knowing the nature of science and to some extent the epistemology of physics.

Enghag and Niedderer (2008) distinguished two aspects of student responsibility for their learning process during task solving in a group. They argued that "(o)wnership refers to the importance and need for students to actually participate by discussions, choice, responsibility and decision taking;" (p. 631). However, the aspect of individual ownership comes from students' questions and ideas based on their experiences and motivation, group ownership is formed by their actions of choosing and controlling the activities. In addition, the Enghag and Niedderer identified three basic characteristics in social interaction for task-solving processes: *power and opportunity*, *management* and *learning processes*. Power and opportunity are a matter of how the task is approached by the group, and whether the group action is actually taken at all, and if so, what action is taken. It is a kind of gatekeeper for the rest of the process and depends on the social system within the class and the structure of the peer group and relations of the members working on the task. Management refers to the actual work process and the organisation of the group members in relation to the different activities required for solving the task. It is influenced by the distribution of self-regulation skills among the group members and whether they can use them under the given social relations. Learning processes describe how constraints, understandings, cognitive efforts, ideas and questions are considered in the solution process. They depend on how much each participant feels about his or her part of the group and feels responsible for its success and how much his or her questions and ideas are taken into account or at least considered by the group (social integration).

During the solution process in the group, students not only have been related to each other individually because of this social system, but they also have to deal with emotions regarding the task and the expressions of their interactions in the group in order to learn successfully (e.g. Boekaerts and Pekrun 2015). As Mänty et al. (2020) showed, negative emotions at the beginning or even before the solving process starts increase group-level regulation, whereas negative interactions during the process negatively affect students' emotions after the task, and finally, group-level regulation in negative interactions has consequences for further interaction in the classroom. Emotions are therefore an important and long-established factor affecting motivation to learn. Interpretation of the learning situation and the development of students' self-efficacy in turn lead to beliefs about what goals can be achieved in relation to physics and what contribution one can make to solving a particular task. Finally, negative emotions can lead to disengagement and then hinder successful group performance. Positive emotions, on the other hand, promote social interactions in the group. Emotional regulation, such as motivating oneself or each other, is therefore essential for successful group work. This regulation must take place both individually and on the group level (co-regulation). The task itself can also support emotional regulation to a certain extent, for example, by avoiding triggers for negative emotions such as fear, shame or language that is not adequate for the students. If students master all three dimensions proposed by Nieswandt et al. (2020), the solution process can proceed as described depending on the design and formulation of these dimensions, as well as the students' abilities and skills required for the solution. In the best case, the solution is shared and contributed by all members of the group.

The theories, models, ways of thinking and working, as well as the standards of physics, form the elements of knowledge that are necessary for the construction of the situation model, the adequate physics model, the physical–mathematical processing and the formulation of an answer to the question posed (see Table 9.2). Task 2 (see Fig. 9.3) is obviously a physics task, since it requires solvers to access corresponding physics knowledge. Nevertheless, Task 2 is very simple in its structure and not an example of an appropriate, competence-oriented learning task for physics lessons. In the following, we therefore look into the question of what design and application possibilities there are for tasks in physics lessons and go into more detail about test tasks (for more information, see Chap. 10).

9.2 Use of Tasks in Competence-Oriented Teaching

Competence, as defined by Weinert (2001), refers to students' cognitions, emotions and volitions. In order to optimise learning, learning tasks, as an essential element of physics teaching, must address these areas of competence. For example, for tasks that require problem-solving abilities, it might be necessary to plan and adapt graded help, openness and challenge levels to the respective needs of individual students or at least to the estimated needs of several competence levels (Klieme et al. 2008).

According to most national education standards in the world, physics lessons aim to develop certain abilities and skills of learners that they need in order to participate actively in shaping our society and to shape their own lives. Physics education at school should develop these specific abilities and skills of students using theories, concepts or models, their ways of thinking and working in physics. This includes, in particular, the theoretical–mathematical modelling of physics concepts (see Chap. 8).

In order to construct tasks appropriately, it is important to know the epistemological background of physics in relation to the students' cognitive development. The following section gives an example of this relationship.

Every observation already contains a theory, the minimum of which is a subjective theory that has proven itself in everyday life. Therefore, every observation is subjective because it actually depends on the individual observer. Such theories develop with age, at least at the beginning of human life. Young children in the first grade, for example, take the spherical shape of the earth as a disc, covered by a spherical vault of sky to which the sun is attached as a person. This is their observation of a certain phenomenon, and since they argue very egocentrically, they have no choice. Similarly, our Germanic, Greek, Egyptian or Indian ancestors perceived the world such that sometimes the sun was pulled across the sky in a horse-drawn cart. This was already a theoretical advance with which the movement of the sun could be described and explained. The system was described in spatial and temporal dimensions. Similarly, every scientific observation contributes to a new theoretical and empirical description of this phenomenon. Kant studied this development and concluded that an object that exists but is not described cannot be a phenomenon. Kant identified it as *das Ding an sich [thing-in-itself]*, an entity not directly accessible to human beings (Kant 1781). The phenomenon, however, enables individual access to the *thing-in-itself*. The observed is therefore not the thing-in-itself, but the phenomenon, which is always shaped by individual or general (scientific) theories. In this way, human beings gain the freedom of interpretation and, in the final analysis, the freedom of decision. For what applies to the observation of nature also applies to the observation of social relations. In social relations, according to Kant, human beings must make reasonable decisions.

When applied to solving tasks, the Kantian view means that each phenomenon to be worked on in a task is determined by the cognitive development and the competences of the student. If the instruction around the task has given all students access to the concept common in physics and prepared them for solving the task in the classroom, and if all students in the class have the competence to handle the physics concept equally well, the task is equally difficult for every student in the class. If the students have different competences (variance with respect to the task to be solved), which are common in school classes, the teacher should assume that they have different prerequisites for solving the task. One approach to guide this process is the *basic concept* construction proposed by Oser and Baeriswyl (2001) (see Chap. 4).

Provided that the students show adequate general reading competence (see third level in Table 9.2), it is nevertheless a complex situation for all of them to successfully solve physics tasks because this demands their competencies in other different areas. The application of physics concepts in different contexts not necessarily aims solely at solving the task or problem posed by applying mathematical operations but demands also a theoretical understanding of the situation that causes the problem or leads to the solution of the task. Physics concepts, which are usually referred to as expert knowledge, can serve as prototypes, which should be useful to learners also as applicable concepts in everyday life. This leads to an understanding of physics in its role for decision making and development in our society.

Learning physics should enable students to make reasonable decisions based on a broad understanding of physics (first and fourth levels in Table 9.2). Since every observation already contains at least a subject theory about phenomena of the world, it seems important to equip learners with the most reasonable ideas available for making certain observations. In the context of physics, phenomena are the combination of an observation with the underlying theory, which leads learners to identify the observation as such.

9.2.1 Competence Orientation and Tasks

Tasks—that are intended to address students' competence and its development—make transparent the content-related metastructures of physics (basic concepts/guiding ideas) and the physics-related competence to be learned by applying them to examples and reflecting on them. Tasks for competence development thus should build on each other in terms of learning theory and content logic (for the development of the enery concept see Neumann et al. 2013).

Experiments and experimental parts of learning tasks fulfil, among other things, the goal of developing physics phenomena from a physics perspective, allowing prototypical scientific working methods and clarifying concepts of a physics description of the world or an approach to describe the world. The learning environment that the teacher needs to provide for these learning situations to optimise the students' learning process should therefore be clearly structured and presented in a way that is accessible to them.

Learning tasks in physics lessons should contribute to this goal. Learning tasks that only involve the practice of using definitions and other declarative knowledge—such as formulas and laws or the insertion of numbers in equations—are therefore of limited relevance for modern physics education. However, they are very much in line with the historically developed tradition and can still be found in numerous task collections (especially at university level). Moreover, they only take into account the logic of physics and do not include the instructional logic necessary for teaching and learning physics and developing physics-related competencies.

Dealing with typical and learning-resistant everyday ideas of students is an example of a learning difficulty that cannot be overcome exclusively with the logic of physics (see Chap. 14). Learning tasks must be focused on students' learning processes, abilities and knowledge that are characteristics of a particular school type, age and grade level in order to achieve the learning objective.

However, traditional tasks are often designed without considering learning processes, and learners have hardly any chance to construct resilient concepts for solving them (Sinaga and Feranie 2017). In contrast to the traditional use of tasks, we therefore speak of a competence-oriented use of tasks, which should enable learners to acquire the abilities for solving physics tasks and for applying their new concepts in everyday situations as described above (Fauth et al. 2019; Kohl and Finkelstein 2005; Zlatkin-Troitschanskaia et al. 2015).

9.2.2 Change of Task Difficulties

As we have shown above, each task requires a wide range of abilities, many of which are related to physics concepts and physical–mathematical modelling and whose development is important for solving physics tasks and problems. The extent to which an ability of students must be developed in each case—for enabling them to solve a task successfully—is influenced by the design of the task. In order to describe corresponding variations in the design of tasks, various researchers developed analysis schemas to distinguish characteristics of tasks and their aspects systematically. These characteristics usually refer to formal criteria and aspects. They include openness and answer formats of the tasks, the content structures and central, required models or working methods and the structural design with forms of representation, complexity or required cognitive processes that are related to cognitively activating terms such as *explain, draw and calculate.*

As some of these criteria successfully describe the variance in task difficulties, they could be a helpful tool to adapt tasks to students' abilities, skills and needs. Across the findings of many researchers, the following are coherent (Kauertz 2008; Le Hebel et al. 2017; Liou and Bulut 2020; Prenzel et al. 2002; Stiller et al. 2016). Text length, the number of calculations and inferences increase task difficulties, whereas meaningful graphical representations, multiple-choice formats and the use of concepts already covered in class or their additional mention in the text make tasks easier. From these researchers' analyses, it is known that these difficulties are due to additional or higher skills or abilities, such as literacy or numeracy, that are needed beyond the intended skills required by the physics-specific demands of the tasks. However, these task features affect task difficulties in general. For each student, the individual difficulty additionally depends on his or her abilities. With the characteristics we have mentioned, the difficulties of the tasks, therefore, can only be changed on average for a class.

9.2.3 Open-Ended Experimental Tasks

The most challenging situation in terms of a *double diagnostic approach* is an open-ended experimental complex problem-solving task (e.g. Akben 2020; Lock 1990).

Such open-ended tasks are based on a fundamental idea in many curricula and standards—such as the next generation science standards in the USA (c.f. Calmer 2019), in Germany (KMK 2005; Kremer et al. 2012) or in Australia (Fraser et al. 2020)— and might be posed to students to solve in a typical competence-oriented physics lesson.

Open ended means that there is more than one reasonable way to solve the task and sometimes more than one acceptable solution (as demonstrated in Task 1 of this chapter). In open-ended tasks, the actual problem or question often needs to be identified or clarified before a task-solving process can be attempted. These types of tasks must have a minimum level of complexity to represent multiple solutions and options for accomplishing the task. Appropriate complexity increases the number of different ways of solving the task and the number of experiences that are allowed by the task. This provides solvers with the opportunity to learn from failures or dead ends during the solving process if they reflect upon their failures afterwards and if the solving process can take place in an angst-free atmosphere and a reliable social environment. These tasks also invite solvers to reflect upon and discuss the basic principles of working and thinking in physics, for example, aspects of the nature of physics as a science that become apparent in the solving process. It is due to the nature of open tasks that a wide variety of skills and knowledge types must be considered for solving the task. Therefore, when tasks of this kind are adequately constructed, students with different competences, knowledge types or interests can each acquire different competences from different aspects of the task. Task 1 at the beginning of the chapter can be solved, for example, experimentally or theoretically. In the case of an experimental solution, there are at least three different solutions with different attempts:

1. If the theory is known, the movement is decomposed into two partial movements according to the superposition principle (independence principle). The experimental solution is thus obvious.
2. If the theory is not known, the experiment is set up according to the trial-and-error principle.
3. In the theoretical solution, the vertical and horizontal solutions must be transferred separately to the ring and the toy car.

A special type of these open-ended tasks is complex analytical problem (Fleischer et al. 2010). In an analytical problem, the initial state and the final state are given, and the means must be found to transform the initial state into the final state. This is achieved by analysing the initial state and assuming plausible functional relationships between the given variables. In contrast to analytical problem solving, complex problem solving always requires an exploratory phase in which fundamental functional relationships within an unknown system must be systematically determined through interaction with the system. Furthermore, hidden dynamics within the system often lead to a fluidity of the situation, for example, certain variables change with time. The initial state is therefore not fixed, and the final state is often defined by a field of possible outcomes that are evaluated according to their compliance with several (normative) goals that sometimes logically contradict each other, so that some kind of balance has to be found.

The typical laboratory work is the equivalent of such a complex problem in physics. The underlying physics model of the experiment is usually well defined (at least at school), and the complex problem is solved when its transfer to the laboratory is successful, nevertheless, several aspects make the physics experiment a complex problem. First, there is the question of operationalising the variables in the model and thus deciding on the measurement process. Second, there is the elimination of systematic errors through calibration, adjustment and control variables (such as friction, torque of the toy car wheels or surrounding temperature). Third, increasing accuracy to minimise measurement error is also a complex problem and closely linked to operationalisation of the variables. Finally, in some cases—and for learning purposes, these cases can be specifically generated—new effects can occur, for example, the extreme states of the experiment, which are not covered by the original physics model on which the actual experiment is based. As another example, when examining pendulum oscillations, the accuracy depends on the amplitude set in the experiment. A further example can be found in Task 1, in which the choice of the accelerated object is important because the torque of a ball could influence the measurement, whereas the torques of the toy car's wheels are small enough to meet the demand for accuracy in this experiment. Exploring extreme states and boundary conditions of an experiment is often a great challenge for physicists; it has led to some new insights into understanding nature in the past. It is therefore to be expected that students need help to recognise and interpret these states.

This situation is even more challenging when considering the entire process of research and not just the isolated experiment in the laboratory. Finding plausible explanations for phenomena in physics (in Kant's sense) actually means going through a long iterative process of complex problem solving, constantly switching between theoretical considerations and experimental work to *explore* the phenomena, *develop* a model that can help explain the phenomenon, *calculate* the possible effects and *test* predictions of or hypotheses about the changes in the laboratory. This idealisation of the physics research process is undoubtedly demanding and extremely time consuming for students at school and university. However, it comes closest to the pedagogical goal of not only executing this research process with the students, but also of encouraging them to reflect on the process. It becomes even more important if this process is not only related to physics, but is also transferred to solving other natural and social problems in other natural and social sciences. This can demonstrate to the students the power and necessity of this approach, as it is taught in schools according to most of the national educational standards for physics.

It is obvious that this approach also demands a lot from the teacher. It requires the teacher to have a broad knowledge of the field of physics, a deep understanding of the physics research process, an awareness of a wide range of abilities and knowledge areas—related to the different ways that students might reasonably choose to solve the task—and finally, a sound pedagogical and instructional understanding of the situation, the students in the classroom and the educational goal. Striking the right balance in solving experimental problems—between allowing failure and avoiding frustration, encouraging success and taking control of a student's learning—cannot be standardised and certainly required much expertise through reflective experience and empathy. Therefore, open-ended complex experimental problem solving is certainly the most interesting but also the most demanding task a teacher can set for the class and for him/herself (see Chaps. 2, 3 and 4).

To reduce this complexity, it is common to focus on parts of this overall process within a task. However, by sequencing tasks, even such complex processes can be implemented in the classroom. By helping the students to see the overarching idea behind these sequences, the teacher can ensure that they can also learn about this process without having done it alone and in one piece.

9.2.4 Effects of Combining Tasks

In physics lessons, learning tasks are used to achieve certain learning goals with the students, but they also structure the classroom instruction. Therefore, they are usually not individual events, but planned and implemented ones by the teacher in a sequence to link already taught and newly developed concepts. They follow an instructional logic and key concepts, which should be oriented towards a specific learning goal and represent the learning processes of the respective target group of learners (Trendel et al. 2008). As described above, complex problem solving could have such a learning goal for the students and is used to organise the necessary experimental lesson structure with a task sequence. The sequence contains subtasks connected by the expected phases of development of the learning process. In response to student questions and answers, feedback is usually given by the teacher when the students are working on a task and during class discussions. These phases of the lesson can be used as a means of diagnosis for the teacher to narrow down the selection of the next task (for adaptive differentiation, see Chap. 15).

According to Pozas et al. (2020), the sequence of tasks can be expected to have effects that can be attributed to priming and framing. Priming means that the formulation of the previous task makes the associations related to its content more easily available. If the content changes, as is common when working on task sequences, accessing the necessary skills and knowledge to solve the next tasks is more difficult. Framing means that the expectation of the following task and its understanding is shaped by the previous task (Scheufele 1999; Scheufele and Tewksbury 2007). For example, failure in the first task lowers the expectation of success in the next task (Pozas et al. 2020). Both effects can be used specifically to guide learners' attention through the task sequence and to maintain motivation through a sense of achievement. In addition, the systematic design of tasks that demand new skills and knowledge can be facilitated through ingenious construction of the sequence, for example, with the help of a dialogue or whole-class discussion. Examples of this idea are task sequences that follow the idea of fading out, that is, where a model solution is first offered and then the level of detail and extent of help to solve the tasks is gradually reduced (Schmidt-Weigand et al. 2009). In international science tests, such as TIMSS or PISA, the combination of tasks has a different goal (Zlatkin-Troitschanskaia et al. 2020). For example, it is determined how physics competence can be modelled in a certain age group. Groups of tasks are then constructed to represent this model. The tasks in each group have to be solved with the same competencies in order to measure this competency more accurately (with greater reliability). The basic idea is that in a good test, all other competencies—that are to be measured to represent the model—are required to the same extent. How well this will be achieved is shown by the correlation between the tasks in terms of internal consistency (Clayson 2020) or the fit into a probabilistic test model, for example, the Rasch model (Mešić et al. 2019) (see Chap. 16).

We have seen in this chapter that the goals of the different phases of physics education and the goal of physics education as a whole can be achieved by tasks arranged in instructional logic sequences. Accordingly, tasks are strong tools for structuring physics teaching. Tasks offer the possibility of developing students' competencies and knowledge, but at the same time they are also tools for the teacher to organise lesson according to his or her learning goals in a lesson or even a sequence of lessons. The understanding of the processes involved makes it possible for the teacher to provide targeted learning support and to focus on specific the students' competences in performance measurements.

It is not easy to select and construct tasks that meet the expectations of respective learning goals, fit the heterogeneous prerequisites of the target group of students and form a meaningful offer by the teacher for individualised understanding of physics concepts. In the following concluding section, we suggest some measures to assess and ensure the quality of the tasks. We conclude the chapter with some guidance and suggestions on task construction and selection. We consider, on the one hand, the linguistic construction of tasks, and, on the other hand, we suggest guiding questions that can be used for the teacher to reflect on the selection and use of tasks. After all, as with any pedagogical decision, there is also no automatism in the use of tasks to arrive at an appropriate solution to the problem of this pedagogical challenge. The task quality criteria described in the following section help teachers to construct appropriate tasks.

9.2.5 Quality Assurance of Tasks

With the considerations presented in this chapter for the design and optimisation of learning tasks and tasks for class tests or tests, teachers can develop their own tasks for their lessons or optimise tasks from task collections or textbooks and adapt them to their own teaching objectives. From physics education research, we can suggest to physics teachers three measures that are standards in the research to ensure the quality of tasks and could be transferable to school settings:

1. Take a structured and deliberate approach: create a manual that includes descriptions of task features and guidance on how to structure tasks and task sequences,

using theoretical considerations such as those in this chapter and the teachers own reflective experiences with tasks in the school.

2. Expert evaluation and agreement: form a group of experts (e.g. the physics group in your school) to discuss, agree on and evaluate tasks created according to or matching the manual, using agreement in the group as an indicator of the quality of the tasks.

3. Trial and empirical evaluation: whenever possible, try out the tasks with a few students who match the target group as much as possible while covering the expected range of diversity and reflect with them on the task in terms of the cognitive, motivational and emotional implications of solving it.

These measures are based on the conviction that quality can only be assured in a structured way in a social expert group through reflected empirical evidence. This quality assurance includes the following:

- the correctness of all information, especially that in the solution (regarding physics and all other aspects of the task context);
- the assumptions and expectations about the students' abilities, motivations and knowledge required and their availability or accessibility for successfully completing and solving the tasks;
- the fit between the task and the target group, which is closely related to, but also focused on the non-intentional competencies and knowledge components of the task (e.g. literacy, numeracy, self-concept, self-regulation, etc.) and
- the relationship to other tasks within the learning sequence, and finally, in view of this, also to the contribution of the task to the educational goal of physics in your school.

In more detail, we now present some ideas on how the above-mentioned demands for accuracy and verifiability could be implemented.

A manual on task design can improve the assignment to the specified task characteristics, which can be varied in a targeted manner. The manual can be created by physics teachers themselves using the criteria suggested in this chapter and can be used and updated in the future task constructions. A detailed construction description of examples made by the teacher, taking into account all student difficulties, is particularly helpful. The kind of handbook can be checked by an evaluation of several experts (e.g. by the members of a physics department at school). A high quality is then shown by a high degree of agreement between these experts. The more clearly the task characteristics can be identified in the tasks, the better the agreement and the better the task is in terms of this quality aspect. The joint processing of the tasks by the physics teachers in a school increases the quality of all physics tasks in that school, and, if the characteristics are discussed with the students, this also increases the transparency of the teaching goals and performance assessment. In a similar way, the technical correctness and educational appropriateness of the teaching structure can be checked. In the literature of physics education research, there are numerous references to corresponding descriptions in different subject areas (see Chaps. 4, 10, 11 and 15). Experience from various studies shows, however, that although it is possible to formulate a sample solution in terms of the physics content, it is difficult to compare it with the results formulated by the students themselves. As the interpretation processes involved in assessing the technical correctness of student responses are complex, the level of expert agreement is usually not very high. The clearer the expected solution is, the simpler these coordination processes and the greater the agreement of the experts in assessing the correctness of the student's answer are.

A manual could include direct hints on how the task should look like or how the posed questions should be checked. Here are some examples of what a manual could include:

- Cognitive structuring is achieved, for example, by summarising, highlighting similarities and differences of individual pieces of information, meaningful structuring and outlining paragraphs.
- Linguistic simplicity is achieved by short and simple sentence structures with subject, predicate and object. Common words, active verbs and few substantiations also contribute to simplicity.
- Semantic redundancy is achieved by reducing information density and by repetition.

Brevity and conciseness in the text are achieved by a balance between comprehensibility of the content with as little text and unnecessary redundancy as possible, that is, a reasonable relation between information and reading effort. Conciseness includes structuring the task in a logically and technically correct way and structuring the task according to the logic of the assumed learning process and the connectivity of the physics concept to further physics lessons and alternative views.

Therefore, the students' respective age, their level of mathematical competence, the power of the task for cognitive activation, the teaching phase in the respective unit, the organisation of the intended learning process and the variability of the task related to individual learning processes should all be considered as in the following questions.

- Are the technical or foreign words used known to the learners and necessary?
- Is the information in the task text known to the learners or at least easy for them to understand? The offer (a task as a request for response) must be made in line with Vygotsky's zone of proximal development (Tudge and Winterhoff 1993; Vygotsky 1978). The competences needed for solving the task must be above the abilities of the learners so that the learning goals themselves can be developed but such differences do not appear trivial.
- Are all the representations necessary and do they need to be combined to come to an answer (e.g. text, graphics and tables)?
- Is the information in the text relevant for solving the task or does it unnecessarily increase the number of competencies (and thus the difficulty) needed to solve the task?
- Are realistic details given, especially in tasks that require mathematical skills for the solution, and have the mathematical skills been developed in advance (in previous physics lessons)?

9.2.6 Teacher's Role and Responsibility

To achieve the overall educational goal of physics education enabling students to participate in society and successfully manage their own lives, the teacher should ensure that they have the opportunities to experience physics and the other natural sciences and mathematics in school looking at the world from an enlightenment and anti-mystical perspective. This is especially important in times of the spread of mysticism and conspiracy beliefs in the so-called social media.

To achieve an independent and profound attempt to understand the functional relations of physics concepts and a probabilistic view of modern physics, students should take a view, which is, also relevant for social relations, at the end of their school and ideally be able and willing to work out complex open-ended problems independently. Their attempt in this regard should involve or require physics concepts and procedures, and related mathematical concepts, such as exponential functions, and the difference between probability and causality. This will be achieved by the teacher setting tasks to enable students' learning in and about physics, to promote the development of their motivation and self-concept regarding physics. In addition, the teacher should enable students' physics-related social activities and the transfer of their competences to understand experienced everyday phenomena such as the development of a pandemic, the health-related relevance of radiation or energy conversion and its problem for climate change. According to Markic and Abels (2014), this attempt can only be successful if these tasks and their sequence are as student-adaptive as possible; this means that all dimensions of student heterogeneity must be taken into account.

This is the challenge for the teacher. He or she must ideally select, create and evaluate tasks for each individual student in the class, as well as offer and evaluate support, moderate interactions, communication and argumentation and support processes of enquiry, learning and understanding (scaffolds). As suggested by Blömeke et al. (2015), the teacher must first analyse and decide which (type of) task fits the individual student's current learning situation, how the student approaches the task, what kind of support is needed and whether the next step or goal will be achieved through this task as intended (cf. Santagata and Yeh 2016). Even the way the teacher asks questions during an experiment of the students can influence the learning effect. For example, if the teacher asks a specific question about a procedure of an experiment, the students remember this specific content better; but if the question is more general about the procedure, the students remember the context better (Endres et al. 2020). Accordingly, the teacher needs to take the necessary actions based on the analysis and the learning goals and finally evaluate the students' activities.

As shown in Fig. 9.2, the teacher is responsible for ensuring that students are able and have the necessary knowledge to carry out all the steps of the solution process. This responsibility can be met in two ways. (1) The teacher can provide tasks that he or she has already solved and know (has assessed) that the students are very likely to have the competence to learn from the tasks or to solve them; or (2) he or she can use the task as a starting point for developing students' intended abilities. In a testing situation, the first way is preferred, for both fairness and diagnostic reasons. In learning situations, a combination of both ways is usually appropriate. If the students are invited to work on a task independently, the first way would be more pronounced; in a more teacher-led learning situation, the second way makes sense.

In a heterogeneous class, in particular, the teacher may have difficulties in finding tasks that enable all students to work successfully on one task or in a comparable way due to different individual competences. Thus, a "double diagnostic approach" is needed, that is, diagnosis of students' competencies and diagnosis of the competencies needed to solve the task. Since it is not possible to ensure that this diagnosis covers the full range of competences of, and needed by, the students, it is always necessary for the teacher to respond to a possible mismatch of students and tasks by offering help and support and by adapting the tasks to the students' abilities.

Summary

This chapter provides a theoretical and structured view of tasks that can help the teacher to create, select and evaluate tasks and to think about their use in teaching and testing. Tasks are essential for teaching physics and achieving the educational goals in learning physics defined in most national educational standards. We have discussed tasks as a means of communication between teachers and students and analysed the solution process from the point of view of necessary abilities, skills, knowledge, heuristics and so on (see Fig. 9.1). In doing so, we have identified intended and unintended demands on students and highlighted teachers' responsibilities. By looking more closely at open-ended experimental problem solving, we are able to clarify the need for a double diagnostic approach to tasks in order to identify students' potentials and requirements for their physics learning. Since all tasks are communication tools, they are usually never finished and perfect for each target group. For this reason, in particular, their quality needs to be assured, as they play such an important role in physics learning. Quality assurance of tasks can be achieved by drawing on the experience of physics education research and applying manual-based structured task features, reflective experiences and task quality indicators based on expert assessment and agreement, including empirical evidence, to the work of physics teaching groups in schools.

Acknowledgements We would like to thank Andrée Tiberghien (Université Lyon, ICAR) and Dirk Krüger (Freie Universität Berlin) for carefully and critically reviewing this chapter.

References

Akben N (2020) Effects of the problem-posing approach on students' problem solving skills and metacognitive awareness in science education. Res Sci Educ 50(3):1143–1165. https://doi.org/10.1007/s11165-018-9726-7

Álvarez V, Torres T, Gangoso Z, Sanjosé V (2020) A cognitive model to analyse physics and chemistry problem-solving skills: mental representations implied in solving actions. J Baltic Sci Educ Šiauliai 19(5):730–746. https://doi.org/10.33225/jbse/20.19.730

Atkinson RK, Renkl A (2007) Interactive example-based learning environments: using interactive elements to encourage effective processing of worked examples. Educ Psychol Rev 19(3):375–386. https://doi.org/10.1007/s10648-007-9055-2

Baumert J, Kunter M, Blum W, Brunner M, Voss T, Jordan A et al (2010) Teachers' mathematical knowledge, cognitive activation in the classroom, and student progress. Am Educ Res J 47:133–180

Becker M, Wiedemann G, Kühn S (2020) Quantifying insightful problem solving: a modified compound remote associates paradigm using lexical priming to parametrically modulate different sources of task difficulty. Psychol Res 84(2):528–545. https://doi.org/10.1007/s00426-018-1042-3

Blömeke S, Gustafsson J-E, Shavelson RJ (2015) Beyond dichotomies competence viewed as a continuum. Zeitschrift für Psychologie 223:3–13. Retrieved from https://doi.org/10.1027/2151-2604/a000194

Boekaerts M, Pekrun R (2015) Emotions and emotion regulation in academic settings. In: Corno L, Anderman EM (eds) Handbook of educational psychology. Routledge, London, pp 90–104

Broughton JM (1981) Piaget's structural developmental Psychology: III. function and the problem of knowledge. Hum Dev 24(4):257–285. Retrieved from http://www.jstor.org/stable/26764224

Calmer JM (2019) Teaching physics within a next generation science standards perspective. Pedag Res 4(4). Retrieved from https://doi.org/10.29333/pr/5868

Clayson PE (2020) Moderators of the internal consistency of error-related negativity scores: a meta-analysis of internal consistency estimates. Psychophysiology 57(8):e13583. https://doi.org/10.1111/psyp.13583

Cohn N (2019) Chapter four—visual narratives and the mind: comprehension, cognition, and learning. In: Federmeier KD, Beck DM (eds) Psychology of learning and motivation, vol 70. Academic Press, pp. 97–127

Doyle W (1983) Academic work. Rev Educ Res 53:159–199

Endres T, Kranzdorf L, Schneider V, Renkl A (2020) It matters how to recall—task differences in retrieval practice. Instr Sci 48(6):699–728. https://doi.org/10.1007/s11251-020-09526-1

Enghag M, Niedderer H (2008) Two dimensions of student ownership of learning during small-group work in physics. Int J Sci Math Educ 6(4):629–653. https://doi.org/10.1007/s10763-007-9075-x

Euler E, Gregorcic B, Linder C (2020) Variation theory as a lens for interpreting and guiding physics students' use of digital learning environments. Eur J Phys 41(4):045705. https://doi.org/10.1088/1361-6404/ab895c

Fauth B, Decristan J, Decker A-T, Büttner G, Hardy I, Klieme E, Kunter M (2019) The effects of teacher competence on student outcomes in elementary science education: the mediating role of teaching quality. Teach Teach Educ 86:102882. https://doi.org/10.1016/j.tate.2019.102882

Fischer HE (1998) Scientific Literacy und Physiklernen [Scientific literacy and physics learning]. Zeitschrift Für Didaktik Der Naturwissenschaften 4(2):41–52

Fleischer J, Wirth J, Rumann S, Leutner D (2010) Strukturen fächerübergreifender und fachlicher Problemlösekompetenz. Analyse von Aufgabenprofilen. Projekt Problemlösen [Structures of transdisciplinary and specialised problem solving. Analysis of task profiles. Project problem-solving] In: Klieme E, Leutner D, Kenk M (eds) Kompetenzmodellierung. Zwischenbilanz des DFG-Schwerpunktprogramms und Perspektiven des Forschungsansatzes [Modeling of competencies. Interim report of the DFG priority programme and perspective of the research approach], 56 edn. Beltz, Weinheim, Basel, pp 239–248

Fraser PR, Sidhu LA, Jovanoski Z, Hutchison WD, Tran TP, Arnold J (2020) Teaching university physics to students from different school systems: Australia's state-based education. J Phys Conf Ser 1643:012165. https://doi.org/10.1088/1742-6596/1643/1/012165

Höttecke D, Feser MS, Heine L, Ehmke T (2018) Do linguistic features influence item difficulty in physics assessments? https://doi.org/10.18452/19188

Janík T, Slavík J, Najvar P, Jirotková D (2020) The same and the different: on semantization and instrumentalization practices in the (maths) classroom. SAGE Open 10(3):2158244020950380. https://doi.org/10.1177/2158244020950380

Jin Q, Kim M (2018) Metacognitive regulation during elementary students' collaborative group work. Interchange 49(2):263–281. https://doi.org/10.1007/s10780-018-9327-4

Kant I (1781) Critik der reinen Vernunft [Critique of pure reason]. Retrieved from http://www.deutschestextarchiv.de/book/view/kant_rvernunft_1781?p=7

Kauertz A (2008) Schwierigkeitserzeugende Merkmale physikalischer Leistungstestaufgaben [Difficulty-generating characteristics of physical performance test tasks]. Logos, Münster

Kauertz A, Neumann K, Haertig H (2012) Competence in science education. In: Fraser BJ, Tobin K, McRobbie CJ (eds) Second international handbook of science education. Springer, Dordrecht, pp 711–721

Kistner S, Rakoczy K, Otto B, Klieme E, Büttner G (2015) Teaching learning strategies. The role of instructional context and teacher beliefs. [Strategievermittlung im Unterricht. Welche Rolle spielen Unterrichtskontext und Lehrerüberzeugungen?]. J Educ Res Online 7(1):176–197. https://doi.org/10.25656/01:11052

Klauer KJ (1987) Kriteriumsorientierte Tests [Criterium oriented tests]. Hogrefe, Göttingen

Klieme E, Hartig J, Rauch D (2008) The concepts of competence in educational contexts. In: Leutner D, Klieme E, Hartig J (eds) Assessment of competencies in educational contexts. State of the art and future prospects. Hogrefe & Huber Publishers, Göttingen, pp 3–22

KMK (2005) Bildungsstandards im Fach Physik für den Mittleren Schulabschluss: Beschluss vom 16.12.2004 [Educational standards for middle school physics: resolution approved by the Standing conference on 16 December 2004]. Luchterhand, München

Kohl PB, Finkelstein ND (2005) Student representational competence and self-assessment when solving physics problems. Phys Rev Spec Top Phys Educ Res1(1):010104. https://doi.org/10.1103/PhysRevSTPER.1.010104

Krathwohl DR (2002) A revision of Bloom's taxonomy: an overview. Theory Pract 41(4):212–218. https://doi.org/10.1207/s15430421tip4104_2

Kremer K, Fischer HE, Kauertz A, Mayer J, Sumfleth E, Walpuski M (2012) Assessment of standard-based learning outcomes in science education: perspectives from the german project ESNaS. In: Bernholt S, Neumann K, Nentwig P (eds) Making it tangible—learning outcomes in science education. Waxmann, Münster, pp 159–177

Le Hebel F, Montpied P, Tiberghien A, Fontanieu V (2017) Sources of difficulty in assessment: example of PISA science items. Int J Sci Educ 39(4):468–487. https://doi.org/10.1080/09500693.2017.1294784

Lederman NG, Abd-El-Khalick F, Lederman JS (2020) Avoiding de-natured science: integrating nature of science into science instruction. In: McComas W (ed) Nature of science in science instruction: rationales and strategies. Springer International Publishing, Cham, pp 295–326

Leutner D, Fleischer J, Grünkorn J, Klieme E (2017) Competence assessment in education: an introduction. In: Leutner D, Fleischer J, Grünkorn J, Klieme E (eds) Competence assessment in education: research, models and instruments. Springer International Publishing, Cham, pp 1–6

Lewis A, Smith D (1993) Defining higher order thinking. Theory Pract 32(3):131–137

Liou P-Y, Bulut O (2020) The effects of item format and cognitive domain on students' science performance in TIMSS 2011. Res Sci Educ 50(1):99–121

Lock R (1990) Open-ended, problem-solving investigations. Sch Sci Rev 71(256):63–72

Löffler P (2016) Modellanwendung in Problemlöseaufgaben - Wie wirkt Kontext? [Model application in problem solving tasks—how does context work?], vol 205. Logos, Berlin

Mänty K, Järvenoja H, Törmänen T (2020) Socio-emotional interaction in collaborative learning: combining individual emotional experiences and group-level emotion regulation. Int J Educ Res 102:101589. https://doi.org/10.1016/j.ijer.2020.101589

Markic S, Abels S (2014) Heterogeneity and diversity: a growing challenge or enrichment for science education in German schools? Eurasia J Math Sci Technol Educ 10(4):271–283. https://doi.org/10.12973/eurasia.2014.1082a

Mayer RE (2008) Applying the science of learning: evidence-based principles for the de-sign of multimedia instruction. Am Psychol 63(8):760–769

Mayer RE (2015) The Cambridge handbook of multimedia learning, 2nd edn. Cambridge University Press, Cambridge

Mešić V, Neumann K, Aviani I, Hasović E, Boone WJ, Erceg N et al (2019) Measuring students' conceptual understanding of wave optics: a Rasch modeling approach. Phys Rev Phys Educ Res 15(1):010115. https://doi.org/10.1103/PhysRevPhysEducRes.15.010115

Miyake A, Friedman NP, Emerson MJ, Witzki AH, Howerter A, Wager TD (2000) The unity and diversity of executive functions and their contributions to complex "frontal lobe" tasks: a latent variable analysis. Cogn Psychol 41(1):49–100. https://doi.org/10.1006/cogp.1999.0734

Neumann K, Viering T, Boone WJ, Fischer HE (2013) Towards a learning progression of energy. J Res Sci Teach 50(2):162–188. https://doi.org/10.1002/tea.21061

Nieswandt M, McEneaney EH, Affolter R (2020) A framework for exploring small group learning in high school science classrooms: the triple problem solving space. Instr Sci 48(3):243–290. https://doi.org/10.1007/s11251-020-09510-9

Osborne J, Henderson B, Macpherson A, Yao S-Y (2016) The development and validation of a learning progression for argumentation in science. J Res Sci Teach 53(6). https://doi.org/10.1002/tea.21316

Oser F, Baeriswyl FJ (2001) Choreographies of teaching: bridging instruction to learning. In: Richardson V (ed) Handbook on research on teaching, 4th edn. American Educational Research Association (AERA), Washington, pp 1031–1065

Piaget J (2013) Origin of intelligence in the child, vol 3. Routledge, London

Pozas M, Löffler P, Schnotz W, Kauertz A (2020) The effects of context-based problem-solving tasks on students' interest and metacognitive experiences. Open Educ Stud 2(1):112–125. https://doi.org/10.1515/edu-2020-0118

Prenzel M, Häußler P, Rost J, Senkbeil M (2002) Der PISA-Naturwissenschaftstest: Lassen sich die Aufgabenschwierigkeiten vorhersagen? [The PISA science test: can the difficulty of the tasks be predicted?]. [The PISA science test: can we predict the item difficulties]. Unterrichtswissenschaft 30(2):120–135. https://doi.org/10.25656/01:7682

Rebello NS, Cui L, Bennett AG, Zollman DA, Ozimek DJ (2007) Transfer of learning in problem solving in the context of mathematics and physics. In: Jonassen D (ed) Learning to solve complex scientific problems. Lawrence Earlbaum Associates, New York

Renkl A (1991) Die Bedeutung der Aufgaben- und Rückmeldungsgestaltung für die Leistungsentwicklung im Fach Mathematik. [The importance of task and feedback design for the development of performance in mathematics]. Universität, Heidelberg

Ruiz-Mallén I, Heras M, Berrens K (2020) Responsible research and innovation in science education: insights from evaluating the impact of using digital media and arts-based methods on RRI values. Res Sci Technol Educ 1–22. https://doi.org/10.1080/02635143.2020.1763289

Santagata R, Yeh C (2016) The role of perception, interpretation, and decision making in the development of beginning teachers' competence. ZDM Math Educ 48(1–2):153–165

Scheufele DA (1999) Framing as a theory of media effects. J Commun 49(1):103–122. https://doi.org/10.1111/j.1460-2466.1999.tb02784.x

Scheufele DA, Tewksbury D (2007) French abstract. J Commun 57(1):9–20. https://doi.org/10.1111/j.1460-2466.2006.00326_5.x

Schmidt-Weigand F, Hänze M, Wodzinski R (2009) Complex problem solving and worked examples. Zeitschrift Für Pädagogische Psychologie 23(2):129–138. https://doi.org/10.1024/1010-0652.23.2.129

Schukajlow S (2011) Mathematisches Modellieren: Schwierigkeiten und Strategien von Lernenden als Bausteine einer lernprozessorientierten Didaktik der neuen Aufgabenkultur. [Mathematical modelling: difficulties and strategies of learners as building blocks of a learning process-oriented methodology of the new task culture]. Waxmann, Münster

Schulz von Thun F (1981) Miteinander reden: Störungen und Klärungen. Psychologie der zwischenmenschlichen Kommunikation. [Talking to each other: Disruptions and clarifications. Psychology of interpersonal communication.] Rowohlt, Reinbek 1981. ISBN 3499174898

Shavelson RJ, Stern P (1981) Research on teachers' pedagogical thoughts, judgments, decisions, and behavior. Rev Educ Res 51(4):455–498. https://doi.org/10.3102/00346543051004455

Sinaga P, Feranie S (2017) Enhancing critical thinking skills and writing skills through the variation in non-traditional writing task. Int J Instr 10:69–84

Stigler J, Hiebert J (2004) Improving mathematics teaching. Educ Leadersh 61(5):12–17

Stiller J, Hartmann S, Mathesius S, Straube P, Tiemann R, Nordmeier V et al (2016) Assessing scientific reasoning: a comprehensive evaluation of item features that affect item difficulty. Asses Eval High Educ 41(5):721–732. https://doi.org/10.1080/02602938.2016.1164830

Treagust DF, Duit R, Fischer HE (eds) (2017) Multiple representations in physics education. Springer, Cham

Trendel G, Wackermann R, Fischer HE (2008) Lernprozessorientierte Fortbildung von Physiklehrern [Learning process oriented further education of physics teachers]. Zeitschrift Für Pädagogik 54(3):322–340. https://doi.org/10.25656/01:4354

Tsaparlis G (2020) Higher and lower-order thinking skills: the case of chemistry revisited. J Balt Sci Educ 19:467–483

Tudge JRH, Winterhoff PA (1993) Vygotsky, Piaget, and Bandura: perspectives on the relations between the social world and cognitive development. Hum Dev 36(2):61–81. https://doi.org/10.1159/000277297

van Dijk G, Hajer M, Kuiper W, Eijkelhof H (2020) Design principles for language sensitive technology lessons in teacher education. Int J Technol Des Educ. https://doi.org/10.1007/s10798-020-09622-w

Villarroel V, Boud D, Bloxham S, Bruna D, Bruna C (2020) Using principles of authentic assessment to redesign written examinations and tests. Innov Educ Teach Int 57(1):38–49. https://doi.org/10.1080/14703297.2018.1564882

Vu DP, Nguyen VB, Kraus SF, Holten K (2020) Individual concepts in physics and mathematics education. In: Kraus SF, Krause E (eds) Comparison of mathematics and physics education I: theoretical foundations for interdisciplinary collaboration. Springer Fachmedien Wiesbaden, Wiesbaden, pp 215–256

Vygotsky LS (1978) Mind in society: the development of higher psychological processes. Harvard U Press, Oxford, England

Weinert FE (2001) Concept of competence: a conceptual clarification. Hogrefe & Huber Publishers, Ashland, OH, US

Weßnigk S, Neumann K, Viering T, Hadinek D, Fischer HE (2017) The development of students' physics competence in middle school. In: Leutner D, Fleischer J, Grünkorn J, Klieme E (eds) Competence assessment in education: research, models and instruments. Springer International Publishing, Cham, pp 247–262

Wilson KJ, Brickman P, Brame CJ (2018) Group work. CBE—Life Sci Educ 17(1):fe1. https://doi.org/10.1187/cbe.17-12-0258

Yerrick RK, Doster E, Nugent JS, Parke HM, Crawley FE (2003) Social interaction and the use of analogy: an analysis of preservice teachers' talk during physics inquiry lessons. J Res Sci Teach 40(5):443–463

Zlatkin-Troitschanskaia O, Pant HA, Toepper M, Lautenbach C (2020) Modeling and measuring competencies in higher education. In: Zlatkin-Troitschanskaia O, Pant HA, Toepper M, Lautenbach C (eds) Student learning in german higher education: innovative measurement approaches and research results. Springer Fachmedien Wiesbaden, Wiesbaden, pp 1–6

Zlatkin-Troitschanskaia O, Shavelson RJ, Kuhn C (2015) The international state of research on measurement of competency in higher education. Stud High Educ 40(3):393–411. https://doi.org/10.1080/03075079.2015.1004241

Chapter 10
Experiments in Physics Teaching

Raimund Girwidz, Heike Theyßen, and Ralf Widenhorn

Abstract Experiments are an integral part of physics research and physics instruction. In physics research, an experiment is a reproducible empirical procedure for acquiring knowledge. Experimental physicists use experiments to gather empirical evidence by developing research questions, and designing and conducting appropriate experiments to answer these questions (see Chap. 1). They control and systematically alter parameters during data collection. An experiment requires comprehensive planning, precise data acquisition, analysis of the experimental data and their interpretation in the context of a theoretical framework. In research, the experiment primarily serves as a method to test assumptions about the outcomes of the experiment (hypotheses) in order to generate new insight and evidence (more differentiated considerations on the procedure and significance of experimentation in physics research are discussed in Chap. 5). In physics education, the experiment has many additional and different functions and serves a variety of goals. Teachers can use *experiments as tools* to convey new content knowledge, for example, by demonstrating a physics phenomenon or by investigating laws of physics quantitatively. In addition, experiments can be used to support students in learning experimentation *as a fundamental method* to establish and verify knowledge and to offer insights into processes of scientific inquiry. Furthermore, experimentation as a method has to be discussed in the context of *nature of science knowledge* and *nature of scientific inquiry*, which is discussed in detail in Chap. 5. Physics teachers should know and be able to use the various functions of experimental activities in physics learning. In this chapter, we highlight the diversity of experimental approaches in physics instruction, their different goals (see Sect. 10.1) and their designs (see Sect. 10.2). Finally,

---

R. Girwidz
Ludwig-Maximilians-Universität München, München, Germany
e-mail: girwidz@lmu.de

H. Theyßen (✉)
Universität Duisburg-Essen, Essen, Germany
e-mail: heike.theyssen@uni-due.de

R. Widenhorn
Portland State University, Portland, OR, USA
e-mail: ralfw@pdx.edu

© Springer Nature Switzerland AG 2021
H. E. Fischer and R. Girwidz (eds.), *Physics Education*, Challenges in Physics Education,
https://doi.org/10.1007/978-3-030-87391-2_10

269

we present recommendations for teaching from multiple instructional perspectives, including related psychological and pedagogical aspects (see Sect. 10.3).

10.1 Experimentation and Learning Goals

In physics education, experiments—unlike in research—are primarily a tool to support various goals, for example, to illustrate physics phenomena, to promote conceptual understanding, experimental design, or data analysis skills. Figure 10.1 shows an example of the use of an experiment as a teaching and learning tool. It indicates that Mrs. Lee is pursuing primarily a content-related learning goal: The students should learn about the elongation of springs by investigating quantitatively the relationship between force and elongation. The acquired content knowledge can be used in various contexts, for example, to explain the function of a spring scale. In order to achieve this content-related learning goal, Mrs. Lee uses an experiment as a tool by precisely specifying the setup, the performance of the measurement and the analysis of the experimental data. If Mrs. Lee's experimental setup is well designed, students will most likely obtain data that show a linear relationship between the applied force and the elongation of the spring, and then the mathematical representation of Hook's law (see Chap. 7).

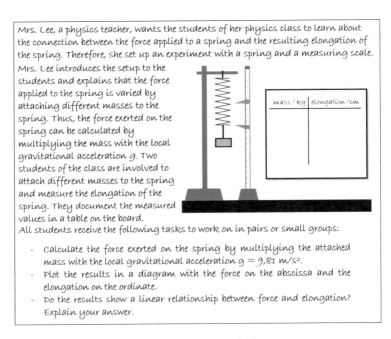

Mrs. Lee, a physics teacher, wants the students of her physics class to learn about the connection between the force applied to a spring and the resulting elongation of the spring. Therefore, she set up an experiment with a spring and a measuring scale.

Mrs. Lee introduces the setup to the students and explains that the force applied to the spring is varied by attaching different masses to the spring. Thus, the force exerted on the spring can be calculated by multiplying the mass with the local gravitational acceleration g. Two students of the class are involved to attach different masses to the spring and measure the elongation of the spring. They document the measured values in a table on the board.

mass / kg	elongation / cm

All students receive the following tasks to work on in pairs or small groups:

- Calculate the force exerted on the spring by multiplying the attached mass with the local gravitational acceleration $g = 9{,}81$ m/s².
- Plot the results in a diagram with the force on the abscissa and the elongation on the ordinate.
- Do the results show a linear relationship between force and elongation? Explain your answer.

Fig. 10.1 Example for the use of an experiment in a physics lesson

Experiments do not only serve as a tool for conveying subject-related content, for example, improving students' understanding of phenomena, their surrounding natural world or of physics concepts, but can also be used to support students' learning on three different levels of knowledge and understanding as follows:

- the phenomenological framework (e.g. showing the deflection of light through a prism);
- the physics laws (e.g. quantitatively investigating the correlation between the incident and refracted angle at an air-glass interface, leading to Snell's law);
- the explanatory theory (e.g. in this case the wave model of light and the index of refraction).

As an important scientific method, experimentation can also promote students' understanding of scientific inquiry and their ability to employ scientific methods. In physics education, the basic steps of knowledge acquisition through experimentation should be addressed, for example, how to plan and set up an experiment. Additionally, troubleshooting, experimental uncertainty, and boundary conditions should be discussed. Therefore, experimentation taught as a subject-specific method becomes at the same time a subject of teaching on a meta-level.

So far, two central learning goals that a teacher might want to achieve with experiments are discussed: the learning of subject-related content or the learning of scientific methods. According to several studies on experimentation in physics education or science education (e.g. Welzel et al. 1998; Ma and Nickerson 2006), these and other functions are frequently described:

- content learning, for example, about the phenomena or related physics laws;
- the acquisition of experimental skills, for example, how to collect and analyse experimental data (see above);
- promoting motivation and social skills, for example, effective teamwork; and
- learning about nature of scientific inquiry (see details in Chap. 5).

One challenge in using experiments in physics lessons is not to use experiments as an end in themselves. The learning goals associated with the lesson should guide the choice of the experiments and how the experiments are implemented and embedded in the lesson. It is important to note that one experimental activity cannot cover all of the objectives mentioned above at the same time. It is therefore important that a teacher is clear about the main learning goal of a lesson or for a specific phase in the lesson. In the following sections, we discuss for the first three learning goals (subject-related content, scientific method, motivation and social skills) how experiments might be used as a tool to reach these goals.

10.1.1 Supporting Physics Content Learning

Teaching and learning physics aims at building up an appropriate cognitive structure. At the pre-university level in particular, experiments should help students to experience phenomena and to apply and substantiate physics theories and knowledge (see Chap. 1).

In physics instruction, experiments show phenomena, can direct students' focus towards technical questions and help them to get insights about nature. Often physics phenomena cannot be presented verbally in a way that is nearly as illustrative and vivid as in an experiment. Thus, the experiment in physics education is an important conveyer of information from an educational point of view and can fulfil special communicative functions.

The benefits and effects of using experiments in teaching physics must always be considered in relation to the already described learning goals. The effectiveness of experiments in physics instruction will depend on the specific teaching–learning environment (see Sects. 10.2 and 10.3). Thus, teachers must know the range of applications to implement the experiments effectively in their physics courses. Therefore, the following selected examples show how experiments can be used as a tool to achieve various content-related instructional goals (see Fig. 10.2).

Making Physics Phenomena Visible

To visualise the phenomenon that electric currents are associated with magnetism, small compasses are placed around a vertically running wire. Without electric current, the compass needles align with the earth's magnetic field. With a strong current in the wire, they align in a circle around the wire, and the existence and spatial characteristics of the magnetic field are displayed.

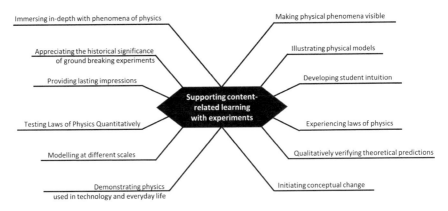

Fig. 10.2 Content-related instructional goals

Illustrating Physics Models

To illustrate the ray model of light, a narrow bundle of light, for example, from a laser is observed. The linear path of the light is visible in a darkened room when the air is enriched with chalk dust or water vapour. The observation motivates the description of light propagation by infinitely narrow linear bundles ("light rays") in the ray model.

Developing Student Intuition

Labudde (1993) used a team experiment to provide practical experience of acceleration for uniform circular motion. A circle of about 2 m radius was marked on the floor and the students lined up around this circle. The aim was to get a ball to roll on the marked circular path with short pushes of just the right amount of force towards the centre of the circle. The kinematic and dynamic treatment of the circular movement in class is based on the knowledge gained about the direction, strength and time sequence of the short pushes. Further experiences that allow for individual experiences are, for example, the buoyancy in water, air resistance or inertial forces.

Experiencing Laws of Physics

Students can use their senses to perceive physics laws of energy and power in electrical circuits (Muckenfuß and Walz 1992). For this purpose, students use a hand crank to operate an electrical generator (dynamo)—once idle and then under load with a light bulb. The devices are sized so that the higher power to operate the light bulb can easily be felt. This makes it immediately apparent that work is required to light-up the bulb. As another example, for kinematics, students can experience physics quantities such as position, velocity or acceleration by acting out different scenarios and experiencing mathematical and physical relationships with their own bodies (Dale et al. 2020).

Qualitatively Verifying Theoretical Predictions

Sound requires a medium for propagation. To verify the prediction derived from this statement that sound does not propagate in a vacuum, a bell is operated in a vacuum chamber. If the air is pumped out, the bell is not audible. The sound becomes louder as the air flows back into the chamber.

Initiating Conceptual Change

A widespread misconception about electric current is the idea of the consumption of current, for example, in a light bulb. Accordingly, in a corresponding electrical circuit, the measured current should be different on both sides of a light bulb. The conceptual change towards the viable concept that electric current is not consumed can be supported by a measurement with a clamp ammeter (Girwidz and Ireson 2007). The device, which measures the electric current through the magnetic field using Hall sensors, is simply passed over the wires and the light bulb allowing to quickly find the current at different places in the circuit.

Demonstrating Physics used in Technology or Everyday Life

In order to demonstrate the application of physics in everyday life, the operation of many devices can be shown in a simplified setup. For example, the temperature control in a water cooker can be demonstrated with a bimetallic switch. A more technical application can be shown with an experiment on the melting process in an induction furnace using a transformer with a circular melting channel on the secondary side.

Modelling at Different Scales

Transferred to smaller dimensions, astronomical relationships are easier to understand. To develop insights into the cause of lunar and solar eclipses, but also of the phases of the moon, these can be reproduced in modelling experiments using a light bulb, a globe and a ball. Students can carry spheres representing the earth or the moon around a light bulb representing the sun. This allows them to compare the phenomena observed from an inside and an outside perspective.

Testing Laws of Physics Quantitatively

Laws of physics, mostly represented mathematically as equations, describe relations between observable physics quantities (Lederman et al. 2002). Students can use experiments to test predictions derived from a law, for example, Ohm's law, Hooke's law (as shown in Fig. 10.1) or Snell's law, and thus verify their applicability or reveal discrepancies.

Providing Lasting Impressions

A strong impression of the atmospheric pressure can be gained by the implosion of a tin can. For this purpose, the can is filled with some water and heated. When the water boils, steam displaces the air from the can. The tin can is then sealed tightly and cooled down. As soon as the steam condenses, the can is compressed by the external atmospheric pressure. (Alternatively, a vacuum pump can be used.)

Appreciating the Historical Significance of Ground-Breaking Experiments

Some physics experiments are of special importance in the development of our views of the world. These experiments include, but are not limited to, the investigation of the following laws and phenomena: Basic laws of motion, law of gravity, Brownian motion, cathode rays, magnetic field of moving electric charges, law of induction, photoelectric effect, interference of light, speed of light, line spectra, resonance fluorescence, electromagnetic waves, black body radiation, X-rays, electron diffraction, natural radioactive decay, Rutherford's scattering experiments, pair annihilation and so on. Building on these experiments, exciting insights can often be gained into the complex and interconnected paths of *nature of scientific inquiry*.

Immersing in Depth with Phenomena of Physics

To provide the impulse to think in depth about the reversal (or non-reversal) of spatial directions in reflections, two words are cut out of coloured paper (here two templates

Fig. 10.3 Experiments on the reversal (or non-reversal) of spatial directions in reflections

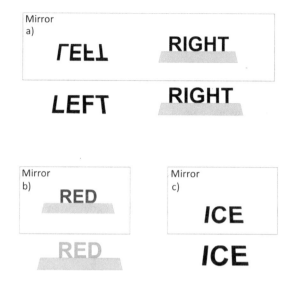

with the words LEFT and RIGHT; see Mirror a in Fig. 10.3). One cut-out (LEFT) is laid on the table, flat in front of an upright mirror, whereas the other (RIGHT) is placed upright. However, only the word RIGHT appears correctly in the mirror image. This observation can initiate further in-depth analyses of the rules behind the reversal of spatial directions. For example, the analysis of the word RED with different colours on the side facing the mirror (red) and the other side (green) (see Mirror b in Fig. 10.3) or with words such as ICE with symmetric letters that can be read in the mirror (see Mirror c in Fig. 10.3).

The examples shown above are mainly meant to give an overview of the multitude of physics content-related objectives pursued with experiments in the classroom. However, the learning objective achieved with an experiment is not determined by the experiment itself, but also by the way it is embedded in the lesson, for example, in which teaching phase it is used (see Sect. 10.2.4). For example, an experiment on Faraday's law of induction can be used to appreciate the historical significance of this experiment for the development of a comprehensive theory of electromagnetism. In another context, it can be used to make the phenomenon of electromagnetic induction visible or it can demonstrate the use of physics in technology and everyday life.

10.1.2 Supporting Experimental Skills

Various national educational standards require the acquisition of experimental skills, for example, the educational standards in Germany (Kultusministerkonferenz [KMK] 2005) or the Next Generation Science Standards in the USA (National Research

Council 2013). Among other skills, students should be able to plan, carry out and document experiments, evaluate the data obtained and assess the validity of empirical results (KMK 2005, p. 11). This should not result in the misleading impression that there is a single experimental method in science (see Chap. 5), which would imply one single sequence of steps which must be learned and worked through by students. To label classroom experiments as a model for current research methods in the natural sciences would paint a distorted, oversimplified picture. Nevertheless, experiments can and should demonstrate elementary steps of physics research. One goal of experimental education is to enable students to independently plan and perform the steps for solving experimental tasks in simple arrangements. Various models describe those skills. Even if these models differ in their granularity, the skills described can generally be associated with the phases of planning and conducting experiments as well as the evaluation of the collected data (for an overview, see Emden and Sumfleth 2016).

Many models of experimental skills are based on the Scientific Discovery as Dual Search (SDDS) model by Klahr and Dunbar (1988). This model assumes so-called *hypothesis* and *experimental spaces* containing possible hypotheses and experiments respectively. It also comprises three basic processes: the search in the hypothesis space for an experimentally testable hypothesis, the testing of the hypothesis by a suitable experiment from the experimental space and the evaluation of empirical findings with reference to the hypothesis.

The model by Nawrath et al. (2011) describes experimental skills, particularly those for teaching practice. The experimental process is divided into seven steps and seven skills associated with these steps, which are described (see Fig. 10.4) as follows:

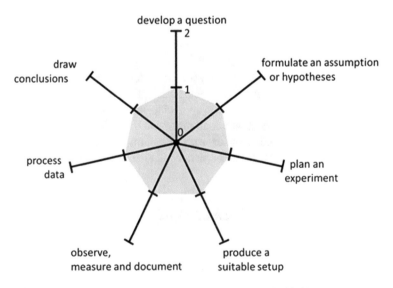

Fig. 10.4 Model of experimental skills according to Nawrath et al. (2011)

- Developing and defining physics questions that can be answered by experimentation.
- Formulating assumptions or hypotheses in relation to a physics question that can be tested experimentally.
- Planning a suitable experiment with which an assumption or hypothesis can be tested. This includes the basic experimental procedure, that is, how the effect to be investigated is produced, and which parameters are changed, measured or kept constant. This also includes the choice of material and a sketch or description of the experimental setup.
- Putting together a suitable setup for carrying out the planned experiment. This includes, on the one hand, building a setup that fits the planning, and on the other hand, an initial functionality test.
- Observing, measuring and documenting relevant parameters while carrying out the experiment. This includes, for example, the correct reading of experimental data, a suitable number and step size for data collection, as well as the clear, objective documentation of observations and experimental data as a basis for the subsequent analysis.
- Analysing experimental data. This includes, for example, calculations of parameters not directly measured or the presentation of the data in diagrams.
- Drawing conclusions that are based on experimental data or observations and address the hypothesis or the initially formulated assumption.

If teaching aims to promote experimental skills, this model can serve as a basis for planning tasks for physics instruction. The teacher can visualise on three levels (cf. Fig. 10.4), whether the respective skill should be secondary (0), relevant (1) or important (2) for the students to accomplish the task. In this context, a single task should not promote and challenge all the skills described in the model to the same extent, but should focus specifically on a single or a few skills, for example, on planning or building a functioning setup. The requirements in the other components of the model can be reduced by providing extra support or guidance. In other words, the task should not demand the highest level (level 2) in each component of the model component.

For example, if Mrs. Lee (cf. Fig. 10.1) wants to improve her students' experimental planning skills, she could specify the assumption that the elongation of the spring and applied force are linearly related. Then, the main task for the students could be to decide what experimental setup they can use to test this assumption, what quantities should be varied and measured, and to select suitable instruments for the experiment. Students can compare different ideas and evaluate them with regard to their fit to the given question and assumption, as well as to the practicability of the experimental setting. To reduce the demands in the subsequent phases of the experiment, Mrs. Lee could also conduct the experiment as a demonstration (cf. Sect. 10.2.1), and results can be analysed cooperatively among the students. A representation of this lesson plan in the model is shown in Fig. 10.5.

The experiment would be quite different if Mrs. Lee mainly aims at enabling students to practise data analysis (process data and draw conclusions; cf. Fig. 10.4).

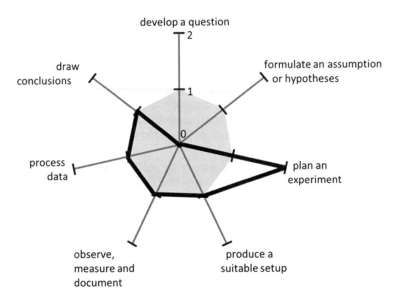

Fig. 10.5 Exemplary representation of the skills that an experimental task demands in the model by Nawrath et al. (2011)

In this case, she could reduce the other demands by demonstrating the experiment (as a hands-on experiment or video) and having all students analyse the results. The key challenges for the students are the calculation of the gravitational force for a given mass, the scaling and labelling of the axes, the entering of the experimental data into the diagram and the decision on whether or not the data relationship is linear. Alternatively, Mrs. Lee could put data acquisition into focus as a learning goal (observe, measure and document; cf. Fig. 10.4). In this case, she should reduce demands in the other phases of the experiment, for example, by analysing the data together with the students.

The skills in the model can be developed and practised separately. Further tasks can require and train the combined application of several skills that should enable students to use experiments to answer their own (or from the teacher provided) research questions.

Besides the skills described in the model, which are aligned with the temporal sequence of an idealised experimental cycle, there are further skills that are relevant throughout the entire experimental process. This also includes the identification and reduction of systematic errors and dealing with measurement uncertainties (cf. Heinicke and Riess 2012), both of which are relevant during the discussion and analysis of the data. For example, an appropriate design of the experimental setup can minimise systematic errors. Additionally, statistical uncertainties should be considered when planning the number of measurements.

When promoting experimental skills, the teacher should always remember to clearly formulate the addressed learning goals. In the examples given above, it should be clear to students that it is not primarily about learning about the behaviour of springs, but that they should expand or practise their experimental skills, for example, in planning experiments to test an (given) assumption or in dealing with measurement uncertainties.

In order to clarify the learning goals and to explicitly promote experimental skills, rules for the subject-specific methods can be formulated, worked out and practised in learning tasks, for example, Vorholzer et al. (2020) formulated the rule that a prediction about the outcome of an experiment needs to comprise a meaningful reason in order to be called a hypothesis (and not just an assumption).

10.2 Designing and Using Experiments in Physics Education

If the learning goal is clearly defined and shall be achieved with an experiment, the teacher has to make a number of additional decisions about the learning goal and the learners' prior knowledge and skills in order to implement the experiment appropriately in the lesson plan. For example, the teacher has to decide in which phase of the lesson the experiment should be used (see Sect. 10.2.4) and which digital tools are used to support experimentation (see Sect. 10.2.3). These and other considerations are presented in this section. First, we discuss in more detail whether the students should carry out the experiment themselves or if it should be demonstrated by the teacher.

10.2.1 Demonstrations and Lab Experiments

When student participation is considered, there is a continuum between a demonstration by the teacher and a lab experiment carried out by the students. The example of Mrs. Lee (cf. Fig. 10.1) shows a mixed form, in which the planning and setup is performed by the teacher, two students are involved in the measurement and the data analysis is carried out by all students in small groups.

Cognitive engagement of the students is intended in all implementations from teacher demonstrations to lab experiments. If the experiment is considered as a unit from defining the research question to drawing conclusions, it can be decided anew for each phase in which way the students are involved.

Besides the availability of equipment, classroom space and technical aspects, the following pedagogical questions should guide the teacher's choice between demonstrations and lab experiments:

- Is the active engagement of the students in the planning and performance of the experiment helpful to achieve the desired learning goal or is it likely to distract them from it?

- Do the students have the cognitive and technical skills to perform the respective parts of the experiment themselves?

There are important distinctions between demonstrations and lab experiments in the ease with which the teacher can guide and control the processes and results. In demonstrations, the process of experimentation can be more closely guided. The setup looks exactly as the teacher has planned it and important aspects can be highlighted, for example, to present phenomena in a particularly instructive manner. Through exemplary experimentation by the teacher, experimental skills can be promoted explicitly by explaining the procedure or implicitly through imitation effects (see Sect. 10.3.2).

Lab experiments, on the other hand, allow students to work on their own or in groups and provide the various avenues of active learning and teaching methods. Individualisation and differentiation are also easier to realise in lab experiments than in demonstrations. However, the more the students themselves contribute to the experiment, the less the teacher can control, for example, if a desired correlation between two parameters is apparent in the results. The knowledge acquisition in open lab experiments is often less structured and not always systematic. However, the students acquire subjective experiences with the experiment, which in turn may have other advantages in the learning process. Furthermore, the development of social skills such as cooperation and communication can be trained in partner and group work.

For the actual implementation of demonstrations or lab experiments, safety requirements must be taken into account. Demonstrations are necessary if experiments have to be conducted with great caution in order to avoid damage to experimental equipment or harm to students' health. Further organisational aspects that have to be considered are, for example, that limited availability of equipment or space for group work may make demonstration experiments a better option. Furthermore, lab experiments often require more time and effort for preparation and supervision from the teacher. The simultaneous support of several groups of students working on lab experiments has its limitations. Due to the organisational form, disciplinary issues are more likely to occur. The amount of instructional time needed for preparation, performance, analysis and discussion should be weighed against the instructional goals as well as the learning gains that can be achieved with a lab experiment.

The learning efficiency of demonstrations and lab experiments is still an area of research in physics education and teaching–learning psychology. Demonstrations usually provide strong guidance from the teacher. A meta-analysis by Alfieri et al. (2011) showed that discovery learning without support led on average to lower learning gains compared to direct instructional methods (e.g. Klahr and Nigam 2004). In this respect, lab experiments are particularly demanding with regard to the organisation and support of learning processes (see Sect. 10.3.2).

10.2.2   *Forms of Engagement with Experimental Activities*

When implementing experiments in physics instruction, teachers may initially think of hands-on experiments in which experimental equipment is used to investigate physics phenomena and, if necessary, collect experimental data. Such experiments can employ everyday material or may require specialised experimental setups for the detailed investigation of phenomena. Instruments (such as thermometers or voltmeters) provide data that cannot be acquired by human sensory organs only.

However, there are numerous other ways to obtain empirical data, for example, by evaluating videos of experiments. In addition, simulations are often used in physics instruction, which do not provide empirical but calculated data. Some alternatives to provide access to empirical or calculated data or observations in physics instruction are discussed in the following sections.

Remotely Controlled Experiments

In labs using remotely controlled experiments (Remote Labs), students can control the experimental procedure remotely via the Internet. The course of the experiment and the experimental data are usually captured by at least one camera and displayed on the user's monitor. Parallel to the camera image, digitally measured data can be displayed directly and accessed through the user interface. However, the data are not necessarily recorded by digital instruments. It is also possible to document data that can be observed, for example, on an analogue scale, via a camera. Just as in direct manipulation of the experimental setup, data will be recorded close to real time and anew in each run of the Remote Lab experiment (de la Torre et al. 2011, 2016). This is useful, for example, in experiments on radioactivity, as Remote Labs provide a method to capture the statistical processes of nuclear decays in a safe environment (Jona and Vondracek 2013). Remote Labs are also suitable for practising repeated measurements and error analysis, for example, implemented as homework without time-consuming repetitions in class. In addition, there are options to provide further assistance by means of augmented reality. Remote Labs can provide more flexibility for experimentation (anywhere and anytime) and an expanded range of attractive experiments (cost-effective and safe). However, measurement slots must be booked and respected, and maintaining the experimental setup may be difficult and time consuming.

Heradio et al. (2016) provided an overview of key literature, and articles by Lowe et al. (2013) and de la Torre et al. (2016) are also a good resource for teachers. Further background and possibilities are discussed in Chap. 11 on digital media.

Interactive Screen Experiments

In an interactive screen experiment (ISE), one takes multiple sequences of images of the experiment within certain range of experimental parameters. Afterwards, the images are combined so that selected parameters, for example, the voltage at a power supply, can be virtually manipulated on the images. The ISE shows the corresponding response of the experiment, for example, the change in brightness of a bulb or the recordings of experimental data for current and voltage (Kirstein and Nordmeier 2007). Mrs. Lee (cf. Fig. 10.1) could let the students vary the elongation of a spring within an ISE and measure the force accordingly (http://udue.de/hooke). Unlike the Remote Labs, the ISE is accessible to all students at the same time. However, one has to accept that the data displayed in the ISE have been previously generated. From an epistemological point of view, the ISE is closer to the actual experiment than to the simulation (see the following section), because it depicts actual physics effects and experimental data and no calculated data derived from underlying physics laws. Therefore, deviations from theoretically expected values can be observed in the ISE as soon as the approximations of a simplified model (that may be simplified for teaching purposes) are no longer valid. For example, the oscillation of a pendulum in the ISE is no longer well described by a sine wave if the amplitude is too large.

Videos

Videos of experiments are also used in physics instruction. The main difference between those videos and ISEs is that videos always display a linear sequence of an experiment. A variation specially developed for teaching purposes are *silent videos*. These show the performance of experiments without any comments or explanations. There are, however, time slots (freeze frames or delays) provided for adding verbal explanations. Users can easily add audio tracks (examples can be found at http://www.physikonline.net/springer/video/index.html). Silent videos can be used as a template for adding explanations. In this way, silent videos create an opportunity for teachers and students to practice using physics to explain phenomena. In pre-service and in-service teacher training, they can also be used for practising verbal support of demonstrations in simplified experimental settings (time-efficient and without equipment requirements).

Simulations

In simulations, devices are represented either in a photo-realistic or abstract form (e.g. as electrical symbols), trying to illustrate the experimental setup. As in hands-on experiments, selected parameters can be manipulated and the effects can be observed. However, in contrast to hands-on experiments and the variants mentioned above, the effects of parameter variations that can be observed or measured in simulations are not based on empirical data, but on calculations based on laws of physics. Therefore, simulations cannot replace experiments, rather they can help to compare calculated data with experimental data in order to test the validity of the laws that are used in the simulations.

For teaching, simulations can be used to illustrate physics concepts. The complexity of the simulation can be deliberately reduced compared to a corresponding experiment because the boundary conditions and the investigated parameters can be defined. For example, a simulation meant to investigate the angles of reflection and refraction might neglect the angle dependence of the intensity of light.

Additionally, systematic errors (see Chap. 16) can be excluded. Furthermore, simulations can be designed in such a way that they allow extended and easier parameter manipulations compared to an experiment. For example, Mrs. Lee (cf.

Fig. 10.1) could easily let her students explore the elongation of springs in a simulation for different springs and also on other celestial bodies by changing the spring constant or the gravitational acceleration (https://phet.colorado.edu/sims/html/masses-and-springs/latest/masses-and-springs_en.html).

Another possible use of simulations is to design and test experimental setups in advance. In physics research, this is common practice, especially when the setups are time consuming to build, expensive or hazardous. For educational purposes, such computer programs can be used in a similar way. For example, in electronics, there are simulation programs available with which an electric circuit can be simulated and tested in advance. They can also be used for troubleshooting or to practise circuit design in general. Students can find suitable computer programs online not only for learning electric circuits but also for learning mechanics or optics, for example, Yenka (https://www.yenka.com/) or Algodoo (http://www.algodoo.com/).

Virtual Labs and Virtual Reality Labs

Despite the many common features and in contrast to simulations, Virtual Labs and Virtual Reality Labs offer user interfaces that are visually designed to come close to authentic experimental situations. The aim is that the activities of the learners come as close as possible to working with real devices or a realistic working environment (Potkonjak et al. 2016), correspondingly, as follows:

- All objects and equipment should look and behave as much as possible like the real devices in the laboratory.
- The visualisations should appear very close to reality (e.g. also three-dimensional).
- Communication and collaboration between experimenting students should be possible.

The proximity of the representation to the real situation, however, always bears the danger that the students are no longer aware that the data obtained in a Virtual Lab or Virtual Reality Lab are stored data or data calculated from physics laws and that the observed "phenomena" are also presented on the basis of theories and/or a set of stored data.

Thought Experiments

Thought experiments allow for extrapolation into areas that are not so easily or not at all accessible in experiments. In addition, they often provide good training for scientific reasoning. This shall be illustrated by a thought experiment of Galileo (e.g. Segre 1980, p. 236). Galileo showed in an elegant proof of contradiction that the motion in free fall must be the same for all bodies: The argumentation contains three important considerations (see also Fig. 10.6).

At first, it is assumed that the heavier body B reaches the ground faster than the lighter body A (Sketch a). Then both bodies are connected (Sketch b). Since body A now slows down the faster body B, together they fall more slowly than body B alone. On the other hand, the combination of body B and A is heavier than body B alone and should therefore fall faster (Sketch c). Therefore, the assumption that the

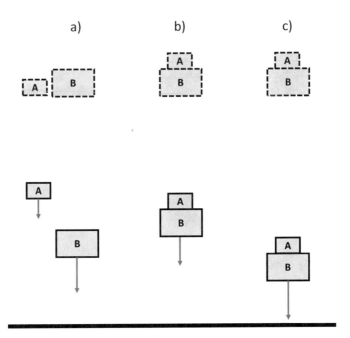

Fig. 10.6 Sketches to illustrate the thought experiment by Galileo

heavier body falls faster than the lighter one leads to a logical contradiction and must be wrong.

While Galileo's conclusions could also be tested with hands-on experiments, this does not apply to all thought experiments. For example, in the twin paradox, the speed of light cannot be realised experimentally. The results from the thought experiment are discussed under the assumption that the applied theory is correct (e.g. the time dilation). However, in a thought experiment, the empirical background is missing, and it is exclusively related to the theory.

The choice between the above options of how students can be engaged with experiments should primarily be based on the learning goals. If the goal is to visualise theoretical relationships, simulations are suitable and allow for exploration under variable conditions. However, when it comes to demonstrating the limitations of the applicability of physics laws, hands-on experiments or at least alternatives based on empirical data (e.g. ISEs) must be used. In addition, organisational aspects and safety issues can guide the decision. In hands-on experiments, in particular those with physics apparatus, the material costs can be high and safety is more difficult to guarantee than with the computer-based variants such as Remote Labs and ISEs. For example, experiments on radioactivity can only be carried out as lab experiments under strict safety guidelines.

10.2.3 Using Digital Tools

In various phases and forms of experimentation to support learning processes, several digital tools can be used, for example, to reduce student workload, visualise data or provide alternative approaches (see also Chap. 11 for digital media in physics instruction). Properly applied, they can facilitate routine work and make it easier for students to recognise the actual physics concepts. They also provide options for linking experimentation with other elements of scientific work. However, digital tools should always be used according to the learners' level of proficiency and in line with the instructional goals. For example, there are digital tools that automatically represent experimental data in diagrams. This can be very helpful to reduce the cognitive load, if the focus is on the interpretation of the data, but not, if the students are meant to learn about the scaling of diagrams. In the following section, the use of various digital tools is presented. Due to rapid technological development, these can only serve as examples to indicate the breadth of the spectrum of digital tools that can be used in physics teaching and learning.

In addition to the auxiliary function of digital tools, learning how to use modern digital technologies should be mentioned as an additional goal in connection with experimentation. Physics instruction can illustrate the use of digital technology in modern data acquisition and control technology in a way few other subject areas can. Not many other fields offer comparable perspectives and approaches on subjects currently referred to by keywords such as the *Internet of Things* or *Industry 4.0*. In any case, the fundamentals of data acquisition have to be covered during experimental instruction and a wide range of measurements, evaluations and digital output techniques are used. Students can thus experience and build up physics-specific and IT-specific skills in an integrative way.

Digital Measurement Devices

In addition to observation, the acquisition of experimental data is a central part of experimentation. Digital measuring devices are largely standard instruments today. For example, temperature is rarely measured with mercury or alcohol thermometers (unless one wants to teach about the expansion of liquids) or electric current is rarely measured with moving-coil instruments. Data acquisition systems from educational equipment manufactures and pocket calculators with graphics capability also offer interfaces to which several sensors for different physics quantities can be connected simultaneously. Smartphones and tablets already contain a camera and many sensors that can be used to measure physics quantities, and corresponding apps, for examples, *phyphox* (Staacks et al. 2018; https://phyphox.org), enable the targeted integration into experimental physics instruction. This can support the planning of experiments with a focus on data acquisition, because these devices can be used repeatedly. The aforementioned tools for data acquisition usually offer both tabular and graphical representations of the data. By means of visualisation, this supports a first cursory assessment of the data quality already during the measurement and facilitates the evaluation.

Microcontroller and Mini-Computer

Microcontrollers such as Arduino (Kinchin 2018) and mini-computers such as Raspberry Pi offer further options, which include not only the data acquisition but also the control and regulation of experiments. Today, microcontrollers and mini-computers are increasingly used in industry for the control and regulation of smaller production processes and machines. They can also be used in physics instruction to design experiments in modern and engaging ways. In addition, the great functional openness and flexibility of the devices not only allow them to be used for a few demonstrations but also to be adapted to various problem-solving tasks for lab experiments. This opens up solutions that are often closer to the reality of experimental work today than are those from common tools designed specifically for physics teaching.

With such devices, physics education can also highlight fundamental ideas about the "Internet of Things" or "Industry 4.0". Arduino "shields" offer a multitude of instruments, for example, also for measuring environmental data (see https://create.arduino.cc/projecthub/projects/tags/sensor?page=2; or https://howtomechatronics.com/arduino-projects/).

Spreadsheets for Data Analysis

With spreadsheet programs, experimental data can be displayed in tabular and graphical forms. The data can be entered manually or imported from data acquisition systems. Commonly, the spreadsheet calculation programs offer additional or more user-friendly tools for graphical evaluation than those provided by the software of the acquisition systems. Spreadsheet programs can also be used to model theoretical curves and compare them with experimental data (see examples in Liengme 2014). Additionally, online spreadsheet programs allow students to work collaboratively on data documentation and analysis.

Augmented Reality

This rather new technology provides additional support in the mental representation of the experimental setups. For example, a smartphone or special glasses can be used to view the experimental setup and procedure with their explanations overlaid on the image viewed by the user. Additional information, for example, force vectors for moving objects, can also be superimposed on the experimental setup in real time. This supports the linking of abstract concepts, such as force vectors, with real processes.

Tools such as those mentioned above serve to support and relieve cognitive strain when experimenting and to expand experimental and instructional possibilities. Nevertheless, when using them, it is important to note the following:

- The variety of tools with their specific operational requirements should not lead to an excessive cognitive load, which in the end might have a negative effect on learning. Thus, to be beneficial, the use of the digital tools has to be practised, and repeated use of the same or similar tools can be helpful.
- Even though students can use digital tools without fully understanding the mechanism with which these tools work, it might be important that at some point in their learning, students gather some insights into the functioning of those tools.

For instance, students should have the chance to learn how to create and analyse diagrams "by hand" before or after using software that calculates or creates diagrams more or less automatically from a data table.

10.2.4 Further Design Aspects

In addition to the educational decisions already discussed above, the teacher has to make numerous other choices about the use of experiments in physics instruction. These should be guided by the learning goals and the students' prior knowledge. In addition, as is always the case in physics instruction, one must consider the availability of equipment and safety aspects. Here are some examples of further classifications relevant for planning experimental physics instruction.

Single, Serial and Parallel Experiments

A classification of experiments, which is particularly relevant for demonstrations, is the distinction between single, serial and parallel experiments. In addition to a single experiment, in which a procedure is demonstrated without variation of parameters, series and parallel experiments can be distinguished.

A **series of experiments** puts together single experiments in a logical sequence. The aim can be to identify rules and regularities through systematic variations. For example, in an experiment using Newton's cradle, one ball is first deflected to hit the remaining four resting balls, then the experiment is repeated in a series with two, three and finally four deflected balls. The series of experiments is conceptually close to the sequence of measurements, in which one parameter is systematically varied and the effects of this variation on another parameter are investigated. However, the series of measurements is always quantitative and limited to the investigation of physics relationships. In a series of experiments, qualitative observations can also be made. Several parameters may be varied one after the other. The aim of a qualitative series of experiments is therefore to show which parameters have an influence rather than to quantify a specific relationship.

A **parallel experiment** shows processes at the same time next to each other and offers great opportunities for comparison. The effects of changing one parameter become immediately apparent. For example, in the experiment on stretching a spring (see Fig. 10.1), Mrs. Lee could mount several identical springs close to each other at the same height and attach linearly increasing masses. Then the linear elongation of the spring is immediately visible and could even be evaluated graphically in a photo. If the teacher additionally uses deflection rolls and pulls the springs upwards from the table (see Fig. 10.7), the upper edges of the springs can immediately be visualised by the students as the straight line in the elongation-versus-force diagram. The recording and documentation of experimental data could be omitted in favour of a clear presentation of the correlation. To make the correlation even clearer, Mrs.

Fig. 10.7 Setup of a parallel experiment with springs

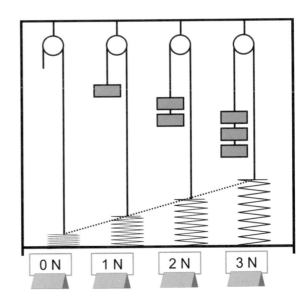

Lee could attach the masses one after the other to neighbouring springs. The students are likely to quickly recognise the pattern and predict the next elongation.

Qualitative Versus Quantitative

Data collection can be qualitative or quantitative. Quantitative experiments require the selection of suitable instrumentation, objective data acquisition, documentation, data processing and evaluation. Though qualitative experiments tend to focus on direct observations, they also require clear criteria for what is to be observed. A mere qualitative observation would be, for example, *"If I hang something on the spring, it elongates"*. Semi-qualitative experiments lead to correlations, for example, *"The heavier the mass I hang on the spring, the more the spring elongates"*. In a quantitative experiment, mass and elongation are measured. They lead to statements such as *"If I attach 20 g more to the spring, the spring elongates 2 cm further"* and can, for example, result in the determination of the spring constant.

Experiments as Part of Different Phases of Instruction

The learning goal, the learners' prior content knowledge and skills, the prior preparatory work in class and the overall methodological concept determine the phase of instruction in which an experiment is used as follows:

- Introductory experiments are meant to motivate, introduce a question, create awareness of a problem and give food for thought. Only basic knowledge is assumed, but careful observation is required.
- Experiments in the investigation phase are used to test hypotheses, to determine qualitative and quantitative relationships including the deliberate collection and

analysis of data. In particular, skills for precise work and for linking theory and practice are required here.

- Experiments to deepen or to validate understanding can reveal apparent contradictions, point out similarities or analogies and prepare knowledge transfer. They build on the detailed knowledge of a subject area.

In general, qualitative as well as quantitative experiments, demonstrations or lab experiments can be used in all phases of instruction. The fit of the experiment to the teaching phase depends on the embedding of the experiment, for examples, the instructions and the concrete formulation of the tasks.

10.3 Recommendations for Teaching

Experiments in physics teaching should provide more than a gathering of observable facts. Thus, it is important that learning material is well organised, structured, and that the information is appropriately sequenced and structured as scaffolder. Furthermore, it should not be overlooked that the instruction needs to first develop basic experimental skills.

In addition to subject-specific and content-specific requirements, general recommendations can be summarised for the use of experimentation in physics instruction. The recommendations have their foundation and draw from pedagogical and psychological perspectives (see Fig. 10.8). Though these recommendations do not originally focus on teaching physics, some of them can be transferred to suggestions relevant for teaching with experiments.

Fig. 10.8 Perspectives that contribute to recommendations for teaching

10.3.1 Recommendations from the Psychology of Learning

There are three aspects that require special attention:

1. Experimental activities must be combined with corresponding cognitive activities. A "hands-on but minds-off" effect (discussed by, for example, Osborne 2019; Furtak and Penuel 2019; Stiller et al. 2018) is likely to occur when working through recipes for experimental tasks. This should be counteracted by embedding experiments in well suited teaching strategies.

   According to Ausubel (1968), appropriate structuring of the learning content and the linking of new knowledge to the students' previous knowledge are essential factors for effective learning. Thus, the following questions should be examined:

   - Can the content be integrated into a meaningful context for learning? Application-related questions can provide guidelines for thinking, for example, "How can a cup of tea be brought quickly to a comfortable drinking temperature?" and "How can cooling below a certain temperature be prevented?"
   - Can the content be linked to the learner's existing concept knowledge and which additional support is suitable?
   - How familiar, clear and consistent are the representations, the applied symbols and terminology in the students' conceptual world?
   - Are important substeps recognisable as such for the learners?
   - Are the students able to recognise important relations in the experimental results and is it possible for them to deduce causalities?

2. Learning from experimental activities needs support from the teacher. Based on their analysis of constructivist, discovery, problem-based, experiential and inquiry-based teaching, Kirschner et al. (2006) highlighted the need for guidance. Alfieri et al. (2011) concluded from two meta-analyses with 164 studies that discovery learning needs some guidance, for example, through feedback, worked examples, scaffolding or additional explanations.

3. An experiment is not an isolated element of teaching but requires both adequate mental preparation, discussion of the results and reflection of the approach. In addition, the process and the result should be documented in various forms of representation (e.g. verbally, written and graphically).

According to Mayer (2004), teaching with lab experiments should adhere to basic guidelines as follows:

- Cognitive activities should take priority over behavioural activities.
- Structured and guided working is preferable to aimlessly exploratory learning (by trial and error).
- The design should be oriented towards a learning goal and should not lead to unstructured experimentation.

10.3.2   Recommendations from a Pedagogical Perspective

In demonstrations, students observe how the teacher or individual students conduct an experiment. The experimenter, explicitly or at least implicitly, acts as a role model. Imitative learning and role model effects are related in a complex way to social relationships. In fields new to the students, however, there is a strong tendency to adopt methods that have already been witnessed. A theoretical overview of observational learning can be found, for example, in Bandura (2004) and Long et al. (2010) (see also "scaffolding", e.g. Reiser and Tabak 2014). Some recommendations can be derived for lecture demonstrations. Thus, a teacher should pay attention to the following aspects when demonstrating an experiment:

- conduct experiments in an exemplary manner (appropriate, professional and targeted);
- show precise and careful work;
- point out safety guidelines and follow them explicitly (e.g. connect electrical circuits step by step towards the power supply—but switch power on only after a final accuracy check; shield open flames, use protective devices such as safety screens, safety goggles, etc.);
- show appropriate handling of instruments, for example, setting the measuring range of a multimeter to the highest range at the beginning, check operating conditions (e.g. avoid stray magnetic fields that could influence the results, follow handling guidelines of instruments, etc.);
- dispose consumables appropriately; and
- use correct technical language and terminology when describing the experimental setup, procedure and results.

10.3.3   Recommendations from Motivational Psychology

Recommendations for experiments can also be derived from the perspective of motivational psychology. For a broader background, articles written by Rheinberg et al. (2000) or by Heckhausen and Heckhausen (2008), are helpful. Motivational aspects should be considered, especially when there are no hands-on activities and thus an increased likelihood of students passively "consuming" an experiment. Therefore, the following recommendations from the literature are primarily aimed at increasing motivation through students' personal engagement and active participation in all essential thought and action processes:

- make the process interesting, build tension and do not verbally anticipate any observable effects;
- strengthen the individual relation to the outcome of an experiment, for example, let students make predictions about the course of the experiment;

- select contexts that have an explanatory value for applications in the everyday world of the students (e.g. motion detectors, brightness control in building technology, etc.);
- provide incentives through experience of success, for example, let students build functionally interesting circuits and setups that can be directly tested (e.g. "building your own electric motor using a magnet and wires"); and
- if possible, even in demonstrations, assign suitable tasks to the learners, for example, to read measured values as part of the work and have them documented on the board.

In addition to observing and, if possible, manually participating in the experiment, logging and documenting the experimental procedures are among the tasks that can be carried out by all students in parallel. This can also support a deeper cognitive processing of the presented phenomena and procedures.

10.3.4 Recommendations from Perceptual Psychology

A broad overview over fundamental perception principles is given by Winn (1993). The basic ideas provide an interesting background for deriving guidelines for observing experimental procedures.

Accurate observations are required in experimental work. The complex interplay between receiving and processing information sets the following requirements, for which students should be specifically trained:

- Ability to differentiate: This refers to how accurately different physics aspects are taken into account in an experiment to precisely capture a given situation, for example, when studying motion, velocity, acceleration, or the impact of different forces. It is important that students are clear about which parameters are being studied.
- Ability to discriminate: This includes treating certain parameters as less important or even neglecting them, for example, friction of an air track or unimportant externalities in an experimental setup. Abstracting secondary side effects is an important aspect for understanding physics processes and should be explicitly explained and justified.
- Ability to integrate: This refers to the ability to establish connections between different categories and characteristics and also to the ability to link prior knowledge with new information.

Furthermore, recording modalities such as data rate or limited memory size can be performance-limiting factors. However, various measures can facilitate the observation of experiments and put a focus on essential components. This is particularly important for demonstrations, where the students are further away from the setup and cannot interact directly with the experiment. Thus, for demonstrations in particular, the following guidelines apply:

1. Providing a clear view on the experiment

   - Easily readable and large display scales of the instruments should be used.
   - Small experimental setups can be shown magnified using video projection.
   - Instruments should be set up in such a way that important operating elements (e.g. important control buttons) are visible to all students.

2. Focusing on the essentials/highlighting important components

   - Only one experiment should come into focus at a given time.
   - If applicable, the important object that students need to observe should be in the centre, possibly highlighted in colour.
   - Clearly label devices that are important for the procedure.
   - Irrelevant aspects should be minimised (e.g. by using low friction carts).

3. Structuring of the experimental setup and procedure

   - Functional units/subunits can be spatially separated and structured by vertical and horizontal arrangements.
   - Hose and cable connections should remain short and clear, and electrical cables should be colour-coded according to their function.
   - Supply units and accessories can be positioned in the background, possibly covered and only indicated by a symbol (e.g. a symbol for the power supply unit).
   - Relevant time periods should be clearly distinguished (e.g. distinguish the settling process from stationary vibration states).
   - If possible, show time-structured processes also in a spatial order (e.g. from bottom to top, from back to front or from left to right).
   - Fast, complex processes may be shown several times and different focal points for observation may be specified. As an alternative or additional presentation, a slow-motion film can be offered.

4. Using a concise procedure

   - According to Schmidkunz (1992), the application of Gestalt psychology is characteristic of concise experimental setups. Thus, proximity, external similarity (e.g. colouring) or symmetry belong to superficial perceptual factors, which often suggest cognitive associations to a decisive degree (for a short English version of Schmidkunz's principles, see also Nehring and Busch 2018).

5. Offering guidance and multiple presentations.

   - For example, a schematic panel sketch of the experimental setup can highlight key components.
   - Various forms of representation, for example, presenting a realistic image of an electrical circuit together with a circuit diagram, stimulate rethinking processes and thus a more intensive mental examination of the facts.

10.3.5 Mastering Challenges and Providing Help

Unsurprisingly, an experiment can fail. When demonstrations fail, teachers quickly lose their role model function if students doubt their experimental skills. Therefore, experiments that may not be reliably working should be introduced as such. If comprehensible for the students, reasons for experimental challenges can provide valuable insight into the nature of experimental work. Discussing problematic experimental conditions can be very instructive. Troubleshooting can also be a meaningful joint task of teachers and students. However, the time for this should be limited. If necessary, an experiment can be repeated in the next lesson. To avoid troubleshooting during demonstrations, the classroom demonstrations should always be tested before the lesson with exactly the same setup that is to be used for the experiment in class. This applies even to seemingly trivial and familiar experiments.

The teacher usually notices a failure of a well prepared demonstration right away. This allows the teacher to react immediately. When students conduct experiments in groups, different difficulties can occur at different times and students might not be aware of these issues. Teachers cannot keep an eye on all groups at the same time and cannot always help instantly. Therefore, it is appropriate to anticipate potential difficulties and prepare suitable strategies. In order to anticipate issues, it is also helpful to carry out the experiment in advance and to pay attention to potential difficulties from the learners' perspective. Help can refer to conceptual underpinning of an experiment (e.g. explanations of technical details and terms), the methodical approach of an experiment (e.g. structuring the procedure and the systematic variation of experimental parameters), the equipment used, or to operating and safety instructions. Especially circuits with sensitive electrical equipment must be checked by the teacher before providing it to students. In general, experimental troubleshooting is one of the skills that should be trained with explicit exercises (see Sect. 1.2).

Acknowledgements We would like to thank Lori Shabaan (Liberty High School, Hillsboro, USA) and Claudia von Aufschnaiter (University of Gießen, Germany), for carefully and critically reviewing this chapter.

References

Alfieri L, Brooks PJ, Aldrich NJ, Tenenbaum HR (2011) Does discovery-based instruction enhance learning? J Educ Psychol 103(1):1–18

Ausubel DP (1968) Educational psychology: a cognitive view. Holt, Rinehart & Winston, New York

Bandura A (2004) Observational learning. In: Byrne JH (ed) Learning and memory: Macmillan psychology reference series (Psychology reference series, 2), pp 482–484. https://b-ok.org/book/737686/6ac635

Dale Z, DeStefano PR, Shaaban L, Siebert C, Widenhorn R (2020) A step forward in kinesthetic activities for teaching kinematics in introductory physics. Am J Phys 88:825 (2020). http://doi.org/10.1119/10.0001617

de la Torre L, Sánchez J, Dormido S, Sánchez JP, Yuste M, Carreras C (2011) Two web-based laboratories of the FisL@bs network. Hooke's and Snell's laws. Eur J Phys 32(2):571–584

de la Torre L, Sánchez JP, Dormido S (2016) What remote labs can do for you. Phys Today 69(4):48–53

Emden M, Sumfleth E (2016) Assessing students' experimentation processes in guided inquiry. Int J Sci Math Educ 14(1):29–54 (Online-first: 08/2014). http://doi.org/10.1007/s10763-014-9564-7

Furtak EM, Penuel WR (2019) Coming to terms: addressing the persistence of "hands-on" and other reform terminology in the era of science as practice. Sci Educ 103(1):167–186. https://doi.org/10.1002/sce.21488

Girwidz R, Ireson G (2007) The clamp-on ammeter: a tool for understanding electricity. Phys Educ 42:93–97

Heckhausen J, Heckhausen H (2008) Motivation and development. In: Heckhausen J, Heckhausen H (eds) Motivation and action. Cambridge University Press, Cambridge, pp 384–444. http://doi.org/10.1017/CBO9780511499821

Heinicke S, Riess F (2012) Missing links in the experimental work: student's actions and reasoning on measurement and uncertainty. In: Maurines L, Redfors A (eds) ESERA 2011 proceedings. Nature of science, history, philosophy, sociology of science

Heradio R, de la Torre L, Galan D, Cabrerizo FJ, Herrera-Viedma E, Dormido S (2016) Virtual and remotelabs in education: a biblio-metric analysis. Comput Educ 98:14–38

Jona K, Vondracek M (2013) A remote radioactivity experiment. Phys Teach 51(1):25–27

Kinchin J (2018) Using an Arduino in physics teaching for beginners. Phys Educ 53:063007

Kirschner PA, Sweller J, Clark RE (2006) Why minimal guidance during instruction does not work: an analysis of the failure of constructivist, discovery, problem-based, experiential, and inquiry-based teaching. Educ Psychol 42(2):75–86

Kirstein J, Nordmeier V (2007) Multimedia representation of experiments in physics. Eur J Phys 28:115–126

Klahr D, Dunbar K (1988) Dual space search during scientific reasoning. Cogn Sci 12(1):1–48

Klahr D, Nigam M (2004) The equivalence of learning paths in early science instruction: effects of direct instruction and discovery learning. Psychol Sci 15(10):661–667

Kultusministerkonferenz [KMK, The Standing Conference of the Ministers of Education and Cultural Affairs] (2005) Bildungsstandards im Fach Physik für den Mittleren Schulabschluss. Beschluss vom 16.12.2004 [Educational standards in physics for the middle school]. Luchterhand, München

Labudde P (1993) Erlebniswelt Physik [World of physics experience]. Dümmler, Bonn

Lederman NG, Abd-el-Khalick F, Bell RL, Schwartz RS (2002) Views of nature of science questionnaire: toward valid and meaningful assessment of learners' conceptions of nature of science. J Res Sci Teach 39(6):497–521

Liengme BV (2014) Modelling physics with Microsoft excel (R). https://iopscience.iop.org/book/978-1-627-05419-5

Long M, Wood C, Littleton K, Passenger T, Sheehy K (2010) The psychology of education: the evidence base for teaching and learning. Taylor & Francis Group. ProQuest Ebook Central

Lowe D, Newcombe P, Stumpers B (2013) Evaluation of the use of remote laboratories for secondary school science education. Res Sci Educ 43(3):1197–1219

Ma J, Nickerson JV (2006) Hands-on, simulated, and remote laboratories: a comparative literature review. ACM Comput Surv 38(3):Article 7

Mayer RE (2004) Should there be a three-strikes rule against pure discovery learning? Am Psychol 59(1):14

Muckenfuß H, Walz A (1992) Neue Wege im Elektrikunterricht: vom Tun über die Vorstellung zum Begriff [New ways of teaching electricity: from action through imagination to the concept]. Aulis-Verlag Deubner, Köln

National Research Council (2013) Next generation science standards: for states, by states. The National Academies Press, Washington, DC. https://doi.org/10.17226/18290

Nawrath D, Maiseyenka V, Schecker H (2011) Experimentelle Kompetenz - Ein Modell für die Unterrichtspraxis [Experimental competence: A model for teaching practice]. Praxis der Naturwissenschaften - Physik in der Schule 60:42–48

Nehring A, Busch S (2018) Chemistry demonstrations and visual attention: does the setup matter? Evidence from a double-blinded eye-tracking study. J Chem Educ 95(10):1724–1735

Osborne JF (2019) Not "hands on" but "minds on": a response to Furtak and Penuel. Sci Educ 103(5):1280–1283

Potkonjak V, Gardner M, Callaghan V, Mattila P, Guetl C, Petrović VM, Jovanović K (2016) Virtual laboratories for education in science, technology, and engineering. A review. Comput Educ 95:309–327

Reiser BJ, Tabak I (2014) Scaffolding. In: Sayer RK (ed) The Cambridge handbook of the learning sciences. Cambridge University Press, Cambridge, pp 44–62

Rheinberg R, Vollmeyer R, Rollett W (2000) Motivation and action in self-regulated learning. In: Boekaerts M, Zeidner M, Pintrich PR (eds) Handbook of self-regulation. Elsevier, San Diego, pp 503–529

Schmidkunz H (1992) Zur Wirkung gestaltpsychologischer Faktoren beim Aufbau und bei der Durchführung chemischer Demonstrationsexperimente [On the effect of Gestalt psychological factors in the construction and performance of chemical demonstrations]. In: Wiebel KH (ed) Zur Didaktik der Physik und Chemie: Probleme und Perspektiven. Vorträge auf der Tagung für Didaktik der Physik/Chemie in Hamburg, 1991 [On physics and chemistry education: problems and perspectives. Lectures at the conference for physics/chemistry education in Hamburg, 1991]. Leuchtturm, Alsbach, pp 287–295

Segre M (1980) The role of experiment in Galileo's physics. Arch Hist Exact Sci 23:227–252

Staacks S, Hütz S, Heinke H, Stampfer C (2018). Advanced tools for smartphone-based experiments: phyphox. Phys Educ 53(4):045009. http://doi.org/10.1088/1361-6552/aac05e

Stiller C, Stockey A, Wilde M (2018) Hands-off, minds-on? The pros and cons of practical experimentation? In: Finlayson OE, McLoughlin E, Erduran S, Childs P (eds), Electronic proceedings of the ESERA 2017 conference. Research, practice and collaboration in science education, Part 2. Dublin City University, Dublin, pp 332–342. ISBN 978-1-873769-84-3

Vorholzer A, von Aufschnaiter C, Boone WJ (2020) Fostering upper secondary students' ability to engage in practices of scientific investigation: a comparative analysis of an explicit and an implicit instructional approach. Res Sci Educ 50:333–359. http://doi.org/10.1007/s11165-018-9691-1

Welzel M, Haller K, Bandiera M, Hammelev D, Koumaras P, Niedderer H et al (1998) Teachers' objectives for labwork: research tool and cross country results. Working paper 6 from the European project labwork in science education (Targeted Socio-Economic Research Programme, Project PL 95-2005). http://www.idn.uni-bremen.de/pubs/Niedderer/1998-LSE-WP6.pdf

Winn W (1993) Perception principles. In: Fleming M, Levie WH (eds) Instructional message design: principles from the behavioral and cognitive sciences. Educational Technology Publications Englewood Cliffs, New Jersey, pp 55–126

Chapter 11
Multimedia and Digital Media in Physics Instruction

Raimund Girwidz and Antje Kohnle

Abstract Research on media-based learning has led to many new insights and applications in the last decades. However, the technological developments in digital media are diverse and dynamic. For effective education training and instruction, it is important to know theoretical foundations and how they can be applied in instruction. For these reasons, this chapter focuses on theoretical models and guidelines that are valid in the long-term for the learning of physics with digital media. These include insights into multiple modalities, multiple representations and interactivity. Aspects of learning theories are used as a basis for considering concepts and concrete examples of the use of digital media and multimedia in physics instruction. We discuss theories of anchoring of knowledge and situated learning, mental models, supplantation theory and coherence formation, multiple representations, cognitive flexibility, knowledge structuring, the use of simulations, guided discovery learning and learning physics with online resources and tools.

11.1   Introduction: Multimedia and Digital Media

Mayer (2002, 2014a, b) defined multimedia learning through the multiple forms of presentation (both verbal and visual) used. *Multimedia learning* occurs when a student constructs mental representations from the presented verbal and pictorial information. Word-based information may include printed or spoken text, and pictorial information may include both static and dynamic images such as pictures, drawings, diagrams, video and animations. Multimedia can enhance learning, but also poses particular challenges. On the one hand, the Internet offers an enormous range of resources and information; on the other hand, it is not always easy to identify useful learning materials from the wealth of information. A second aspect is the

R. Girwidz
Ludwig-Maximilians-Universität München, München, Germany
e-mail: girwidz@lmu.de

A. Kohnle (✉)
University of St Andrews, St Andrews, United Kingdom
e-mail: ak81@st-andrews.ac.uk

© Springer Nature Switzerland AG 2021
H. E. Fischer and R. Girwidz (eds.), *Physics Education*, Challenges in Physics Education,
https://doi.org/10.1007/978-3-030-87391-2_11

multimedia nature of the materials: Information is presented via different channels and in different representations, often interactively. Thus, teaching must also ensure that learners can process and relate information in these various forms.

Specific strengths of multimedia include *multiple modalities* (integration of different sensory inputs such as auditory and visual), *multiple representations* (representations in different formats, e.g., as images, text and mathematical formulas) and also the interactive use of these resources (especially with adapted challenges and feedback). These strengths are the starting point of our considerations. From a didactical point of view, however, multiple modalities, multiple representations and interactivity only characterize the "surface structure" of the teaching and learning environments, describing the human–computer interface. The "deep structure," that is, the factors that impact learning effectiveness, is thus not necessarily specified. For example, multiple representations can be helpful because they illustrate different aspects of a concept; but they can also lead to the learner being overwhelmed by the flood of information. For these reasons, we discuss theories of multimedia learning as well as specific examples in the following sections.

This chapter is structured as follows: Sect. 11.2 introduces theories of learning with multimedia. Section 11.3 discusses different types of computer programs for physics instruction. Section 11.4 has a central role in showing how learning theories translate into practice in applications of multimedia learning for different instructional goals. Section 11.5 discusses simulations and guided discovery learning. Finally, Sect. 11.6 describes physics learning with online resources. Figure 11.1 gives a graphical overview of the sections of this chapter.

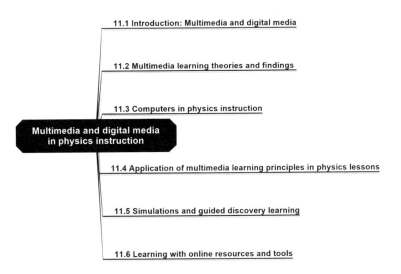

Fig. 11.1 Overview of the sections of this chapter

11.2   Multimedia Learning Theories and Findings

This section discusses theoretical considerations and guidelines for learning with multiple representations and modalities. We first discuss the models of Mayer (2001, 2014a, b), and Schnotz and Bannert (2003). This is followed by a conceptual framework for the functions of multiple representations and considerations of cognitive load theory.

11.2.1   Mayer's Theory of Multimedia Learning

Based on his cognitive theory of multimedia learning, Mayer (2001, 2014a, b) developed multimedia design principles based on *three key assumptions*:

- *Two information channels*: Incoming visual and auditory information is processed in working memory according to its modality in two separate cognitive channels: images in the visual channel and words in the auditory channel. This assumption is based on Paivio's (1986) dual coding theory (Clark and Paivio 1991).
- *Limited capacity*: Processing capacity is limited in each channel. This assumption is based on Baddeley's (1992) theory of working memory and Chandler and Sweller's (1991) cognitive load theory.
- *Active processing*: Learning requires active processing of information. Learners select meaningful information from the material presented, convert it into a coherent mental representation and establish links with their prior knowledge. This assumption is based on Wittrock's (1974, 1989) generative theory of learning.

A central idea is that multimedia environments with suitably designed visual and verbal information can promote the deep processing needed for meaningful learning. Figure 11.2 shows Mayer's (2001, 2014b) model of multimedia learning which describes a multistage process.

For visual and verbal information, there are initially two parallel processing paths based on the dual coding principle. The first step is to pay attention to the visual and

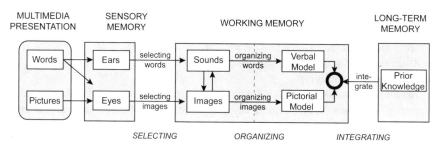

Fig. 11.2 Cognitive theory of multimedia learning, from Mayer (2014b, p. 52)

verbal information, and to select (personally) relevant terms and aspects for further processing in working memory ("selecting").

The sensory information is processed separately in pictorial and verbal representations in working memory. The selected text and image representations, which are still strongly based on the sensory input, are organized into coherent pictorial and text-based mental representations ("organizing"). This results in verbal and pictorial internal models.

In the final step, the verbal and pictorial models are integrated with each other and with prior knowledge ("integrating"). Learning thus requires engaging in all three of these cognitive processes of selecting, organizing and integrating (see Chap. 7).

Design Principles for Multimedia Applications

Based on this model and a series of empirical studies, Mayer (2009, 2014a, b) formulated the following *twelve design principles* for multimedia applications:

- *Coherence principle*: Extraneous text, images and auditory information should be omitted. Include only information that is necessary for understanding and learning the relevant ideas.
- *Signaling principle*: Cues that highlight essential facts or important organizational information can improve learning. An example is an introduction to clarify the goals and structure of a text.
- *Redundancy principle*: Learning is improved if visuals are explained using audio narration only and not also via on-screen text. If a single source of information can fully explain a situation, no supplementary information is needed. Redundant information can even interfere with and reduce learning.
- *Spatial contiguity principle*: Pictorial information and accompanying text should be placed in close proximity to one another. Separating related representations spatially is less effective than integrating them.
- *Temporal contiguity principle*: Learning is improved when visual information and audio narration are presented simultaneously and not sequentially.
- *Segmenting principle*: Breaking down a lesson into individual sections and allowing students to access them at their own pace is preferable to a continuous learning unit.
- *Pre-training principle*: It is better to discuss the necessary basic definitions, terms or concepts up front prior to using these ideas in new content.
- *Modality principle*: Learning is improved when graphical representations are combined with audio narration rather than on-screen text.
- *Multimedia principle*: Learning is improved when using words and graphics rather than just words alone.
- *Personalization principle*: It is preferable to use a conversational style of writing (using first- and second-person language) rather than formal texts in the third person.
- *Voice principle*: Audio narration should use a friendly sounding human voice and not a computer-generated voice.

- *Image principle*: A picture of the speaker on the screen has no demonstrable positive effect.

These effects have been demonstrated in short multimedia sequences dealing with simple cause-and-effect situations, such as how bicycle pumps or car brakes work. Complex and highly connected content may need further research to verify the effectiveness of these principles or adapt and modify their details. In addition, students bring their prior knowledge, experiences and preferences to the learning environment. Thus, these *design principles may need to be tailored* to students' needs, and their relative importance may vary for different groups of learners.

11.2.2 Schnotz and Bannert's Integrated Model of Text and Picture Comprehension

Text and images are fundamentally different forms of representation. Text organizes information sequentially; images provide information in an integrated/simultaneous form. This leads to different cognitive processing and organizing of information from text and images.

Schnotz and Bannert (2003) fundamentally distinguished between descriptive (text-based) and depictive (pictorial) representations. Descriptive representations are based on symbol structures, whereas pictorial representations are based on analog structure mapping. A descriptive representation uses symbols (such as the letters of the alphabet) that are understood on the basis of conventions. However, they bear no resemblance to the content they represent. On the other hand, pictorial representations have common structural characteristics with the content they represent. Structural characteristics (such as size relations) are inherent in the representation.

Given the structural differences between descriptive and depictive representations, even the selection of information will involve different processes. *The integrated model of text and picture comprehension* (Schnotz 2014; Schnotz and Bannert 2003) combines these ideas into a cognitive model of multimedia learning.

Figure 11.3 illustrates the model. There are *different branches for processing text and pictures/diagrams*. When reading a text (left side of Fig. 11.3), organizational processes first result in the construction of a mental representation of the text's syntactic form, which is still strongly oriented toward the text's surface structure. This form is transformed further through semantic processing into a structured propositional representation of the semantic content. Accordingly, when looking at an image (right side of Fig. 11.3), a visual mental representation is first created based on visual perception. This representation is developed further into a mental model through analog structure mapping. This mental model can then also be linked to a corresponding propositional representation.

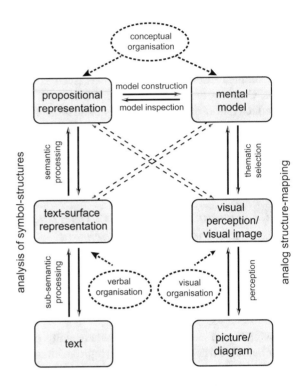

Fig. 11.3 Integrated model of text and picture comprehension (Schnotz 2014, p. 79)

11.2.3 Multiple Representations in Physics

Multiple representations (text, images, graphs and formulas) can illuminate different aspects of physics concepts, especially in the case of complex content. Using multiple representations can ensure that learners do not just relate concepts to a single particular representation but can also flexibly support learners in abstracting invariant structures from models and processes.

Savelsbergh et al. (1998) noted that experts can use internal representations more flexibly than novices do when solving problems. Kozma (2003) also identified differences between experts and novices in their use of multiple representations. Whereas experts use multiple representations in a targeted manner, novices have difficulties in using and integrating them, often focusing only on surface features of the representations. Novices also seem to be more restricted to a particular representation, whereas experts make use of multiple representations and flexibly switch between them. Experts use multiple representations for different purposes, for example, to support content-related reasoning and for communication with other scientists. Kozma's work relates to chemistry education, but is likely to be transferable to physics education. In summary, experts use multiple representations flexibly and appropriately, whereas novices tend to focus on surface features.

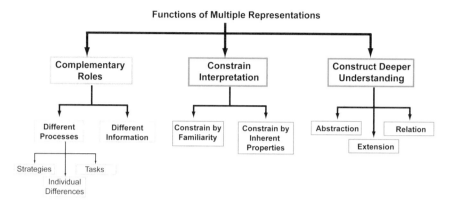

Fig. 11.4 A taxonomy of functions of multiple representations (from Ainsworth 2006, p. 187)

Multiple representations are particularly useful in the following ways:

- Specific information can best be conveyed in a particular representation.
- Different representations can be used to illuminate the same concept in different ways in the learning process.
- Having multiple representations available for reasoning implies more flexible use of knowledge.

For the planning of lessons, it is important for physics teachers to know the various functions that multiple representations can play in instruction. Ainsworth (2006) provided a systematic taxonomy (see Fig. 11.4), describing three key functions:

- Different representations can *complement* each other. This can apply to both their information content and the processes they support.
- Combinations of different representations can help a learner to interpret and understand an unfamiliar representation based on their familiarity with the other one (*constraining* function). Constraining can make use of inherent properties of the representation and the underlying physics content.
- Abstraction, extension or the integration of information from multiple representations can support the *construction* of deeper understanding.

Further information on the use of multiple representations can be found in the book *Multiple Representations in Physics Education* (Treagust et al. 2017).

11.2.4 Cognitive Load Considerations

The density of information in multimedia and the need to deal with multiple representations can easily overload learners, especially in the introductory phase. Learning environments should allow students to understand the different representations and be able to work with them without cognitive overload. Thus, teachers should be aware of strategies to reduce cognitive load.

Cognitive Load Theory

Baddeley (1992) developed a model of working memory that includes three components:

- The central executive, which controls attention.
- The visuospatial sketch pad, short-term storage for visual material.
- The phonological loop, short-term storage for verbal acoustic material.

Working memory is limited, with only about three to seven "chunks" (subjectively perceived as a unit) being simultaneously available in working memory (Miller 1956; Baddeley 1990).

Cognitive load theory (Chandler and Sweller 1991; Sweller 1994; Paas et al. 2016) emphasizes the limitations of working memory capacity as an important factor to consider when designing lessons. Due to the limited working memory capacity, presented information must be structured in such a way that cognitive load does not become too high. Each additional allocation of cognitive resources reduces the fraction available for learning. Cognitive load theory distinguishes between extraneous, intrinsic and germane cognitive load.

Extraneous cognitive load refers to working memory load due to external factors. The way in which information is presented can cause a higher cognitive load than that for the understanding of the content (Leung et al. 1997). Unusual notation in equations can create a high cognitive load because symbols and content need to be processed at the same time. In addition, interactions between the learners and the multimedia resource need to be considered. Thus, the load of discovery learning with interactive animations can also become too high due to the need to coordinate tasks in collaborative learning (Schnotz et al. 1999).

Intrinsic cognitive load of content comes from its complexity in relation to the prior knowledge of the learners. Physics content can itself cause a high load if many details must simultaneously be present in working memory in order to understand a situation.

Germane cognitive load refers to the learning-related, resulting cognitive load that arises from the interplay of intrinsic and extraneous cognitive load. Successful learning is only possible if extraneous and intrinsic cognitive loads still leave capacity for the learning process (see Fig. 11.5).

Fig. 11.5 Cognitive overload hinders learning

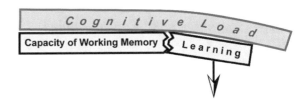

Factors Relevant to Cognitive Load

Sweller (2002) described various *cognitive load effects* that are particularly relevant to visualizations. Similar descriptions can also be found in Mayer (2002, 2014a, b), Paas and Sweller (2014), and Paas et al. (2016). Some of these effects are as follows:

- *Split-attention effect*: Cognitive load is reduced when relevant text is integrated into graphical representations and learners do not have to focus on different places to connect text with graphics (e.g., angles between vectors or the values for masses can be specified directly on the objects).
- *Modality effect*: Combining visual material and spoken text (e.g., when a diagram is combined with narrated text) facilitates information processing compared with visual material and written text.
- *Redundancy effect*: Additional or redundant information may interfere with learning as it increases extraneous cognitive load.
- *Element interactivity effect*: Cognitive overload can occur when content elements are strongly connected and cannot be considered independently of one another. The effect does not occur with isolated information elements.
- *Imagination effect*: Asking learners to imagine a procedure or concept can facilitate learning compared with studying information about this procedure or concept, as it reduces extraneous cognitive load.

Self-directed multimedia learning can also easily lead to cognitive overload. In particular, the following process-oriented questions also add to cognitive load: "What are the next steps?" and "What information do I need and where can I find it?" In addition, technical or operational aspects also increase cognitive load, for example, "Which program options do I have to choose?".

Approaches to Limiting Cognitive Load

We discuss here three general approaches to help limit the density of information and thus reduce cognitive load in multimedia learning environments.

The first approach is to design simulations according to the *single concept principle*. Here, the focus is on a single content element, term or concept at a time. This is particularly useful in introductory sequences and for introducing new ideas. Examples from classical wave theory are shown in Figs. 11.6 and 11.7 (program available at http://www.physikonline.net/springer/multimedia). Wave phenomena can be considered individually in a sequence from simple to increasingly complex situations.

A second approach is to create a user interface that can be navigated by learners at their own pace. The interface controls allow users to individually adjust the information density and cognitive load.

A third approach is to use both visual and auditory channels, for example, the use of visuals and narrated text rather than written text. Using *multiple modalities* can expand working memory capacity and thus enhance learning.

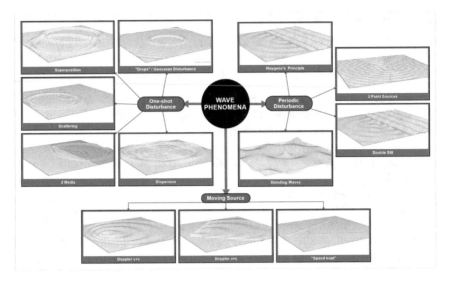

Fig. 11.6 A selection of basic wave phenomena arranged in a clickable map, following the single concept principle (see http://www.physikonline.net/springer/multimedia/mindmap)

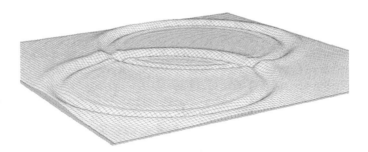

Fig. 11.7 The superposition of two overlapping circular waves

Supplantation and Coherence Formation

Media can be used as a tool to illustrate a process that learners cannot yet realize independently. A missing cognitive function of the learner is demonstrated externally by the medium. Salomon (1979) called this *supplantation*. A missing cognitive function can thus be supplanted by an external medium, which bridges a gap between the task requirement and learner capabilities.

An example is the dynamic linking of different descriptions as shown in Fig. 11.8. This figure shows the oscillation of a spring in a realistic depiction (right) and how the corresponding $y(t)$ displacement graph builds up with time (left). The horizontal arrow highlights the connection at the current time point between the displacement of the physical oscillator and the abstract graphical representation.

Fig. 11.8 Animation illustrating the relationship between a physical oscillator and a y(t) displacement graph

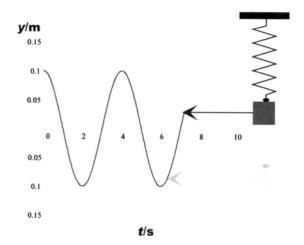

For deep understanding, it is necessary for learners to *create connections within and between multiple representations*. Only such integration of different representations makes it possible to develop a coherent knowledge structure (coherence formation). Seufert (2003) investigated these integration processes that are needed for coherence formation and distinguished two types: identifying relevant elements and the relations between these elements within a representation (intra-representational coherence formation), and across different representations (inter-representational coherence formation).

The demanding cognitive process of *coherence formation* can be supported by several approaches (Seufert and Brünken 2006). This includes support for recognizing related elements using color highlighting, dynamic linking (e.g., when a formula or the values in a table are changed, the corresponding graphical representation also changes), "text image hyperlinks" (e.g., if a relevant concept in a text is clicked, the corresponding part of an image is highlighted) or additional descriptions of the relations between representations. The most appropriate support for coherence formation depends on the prior knowledge of the learners and the learning objectives (Seufert 2003; Seufert and Brünken 2006).

11.3 Computers in Physics Instruction

11.3.1 Categories of Computer Programs

Computer programs for physics instruction can be classified in a variety of ways focusing on different aspects. The following classification is based on their roles in the learning process, as different roles require different methodological approaches.

Thus, knowledge of the different functions allows the targeted selection of programs for physics lessons.

The standard design of *training programs* consists of: (1) presenting the task, (2) registering student input, (3) evaluating student input and providing feedback and (4) transitioning to the next task (in a potentially adaptive way based on student performance). Such training programs allow factual knowledge and procedures to be practiced individually in a differentiated manner.

Traditional *computer coaches* initially present information, which is followed by questions testing understanding. The program provides feedback on students' answers and based on their responses leads to sections which give them further information or possibly repeat the previous content. "Intelligent computer coaches" aim to provide content for different learning phases that is adapted to students' current knowledge and performance levels.

Cognitive computer tools include word processing systems (with spell-checking) and computer algebra systems (e.g., calculating integrals or plotting functions). They facilitate routine work and thus increase capacities for more in-depth investigations.

Simulations are intended to reconstruct selected aspects of reality. They work on the basis of mathematical models of phenomena. Users can control displayed elements and vary parameters within the model. In this way, relationships between parameters and their influence can be identified under simplified assumptions. Knowledge and skills to control complex systems can also be trained.

Computer-based modeling tools offer more possibilities to adapt and vary parameters compared to simulations. Here, the underlying model assumptions (e.g., the equations of motion) can also be varied, so that causal relationships of the model behavior can be investigated, and students can pursue their own ideas.

Using computers as *data collection and presentation tools*: Computer-based data collection is helpful if many readings have to be recorded in a short time or if the time periods are large, and if an automatic system control for the acquisition of data is necessary. Computer programs can also be used to analyze and present data in ways that reduce cognitive load and support learning, for example, by presenting acceleration vectors in place of tables of durations and distances. Programs that allow collected data to be processed and graphically displayed in real time are particularly useful. Another use of computers is for the automatic control of systems using measurements to provide corrective feedback, for example, simple robots with optical sensors. Mobile phones also offer many data collection possibilities via their inbuilt sensors (e.g., for accelerometers, see Sect. 11.6.5).

Remote labs are suitable for experiments that require extensive calibration or complex procedures and are expensive or potentially dangerous. Experiments are remotely controlled over the Internet, and the setup can be observed via webcams. Data can be read off measuring instruments or recorded digitally. In addition, there is the possibility in the design to use multimedia technology to overlay information or graphics. This *augmented reality* approach offers interesting didactical perspectives in terms of multiple representations and modalities (see Sect. 11.6.6). An example of the use of augmented reality for the deflection of electron beams in electric and magnetic fields can be found at http://www.physikonline.net/springer/electronbeam.

11.3.2 The Role of the Teacher

The teacher is an important part of the multimedia learning environment. Comi et al. (2017) emphasized that the effectiveness of information and communications technology (ICT) in the classroom depends on teachers' specific practices and details of the implementation and pointed to more studies being needed to elucidate the conditions under which students benefit from ICT. Koh and Chai (2016) studied teachers' ICT lesson design discussions. Their study underlined the need to consider knowledge of lesson design (which comprises knowledge about design processes and the management of design) as part of teacher ICT professional development. Jen et al. (2016) described the gap between technological pedagogical content knowledge and its application in actual teaching practices and emphasized the importance of practical experiences of using technology in science teacher education.

Issing (2002) devised general didactical planning principles for the development of multimedia learning environments. Figure 11.9 shows a graphical adaptation of these ideas that can provide guidance to teachers on how to embed multimedia applications in a lesson, and in which phases additional resources or aids should be included.

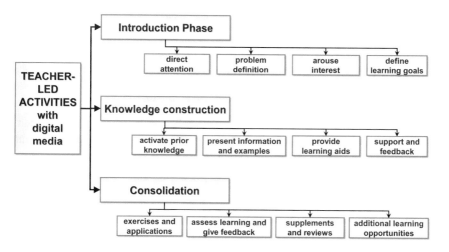

Fig. 11.9 A schematic diagram to assess roles of digital media in different learning phases, and what supplements are useful (adapted from Issing 2002).

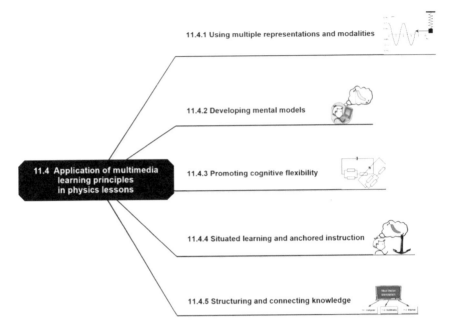

Fig. 11.10 Overview of the principles discussed in Sect. 11.4

11.4 Application of Multimedia Learning Principles in Physics Lessons

This section discusses further theoretical approaches to the question of how multimedia can be useful in learning processes. The focus is on multifaceted learning objectives for connected content. As shown in Fig. 11.10, this section covers multiple representations and modalities, supporting the development of mental models, promoting cognitive flexibility, situated learning and anchored instruction, as well as structuring and connecting knowledge. Examples show how the general principles can be translated into concrete applications.

11.4.1 Using Multiple Representations and Modalities

Systems using multiple modalities make use of multiple senses. The simultaneous use of acoustic and visual information can illustrate different aspects of a physics phenomenon and relate perceptions to graphical representations. In what follows, we describe applications from introductory acoustics. A sound card, microphone and loudspeaker are ubiquitous in today's PCs and mobile devices. With the corresponding software, functionality is then available that goes beyond a standard frequency generator and storage oscilloscope. Thus, a suggestion for teaching is to provide students the option to create their own investigations using multiple modalities and representations to reinforce learning.

The following selection of experiments *referring to acoustics* can be implemented with various computer programs. Further information can be found at http://www.physikonline.net/springer/acoustics or http://www.compadre.org/. In the following examples, the freely available program *Audacity* by the Audacity Team (https://www.audacityteam.org/) was used.

This selection starts with introductory experiments that are standard examples in physics lessons. Using suitable software, their implementation is straightforward, and they are even suitable as home experiments. This is followed by experiments that could not be realized without the use of a computer.

Relationships Between Amplitude and Volume, Frequency and Pitch

One can start by finding simple *qualitative relationships* between amplitude and volume, and between frequency and pitch. These experiments facilitate the connection between the sensory experience and the graphical representation of sinusoidal notes (see examples on the Web site of Audacity Team at https://www.audacityteam.org/).

Note, Sound, Noise and Bang

Different *types of sound* (see Fig. 11.11), such as a sinusoidal note, a note of a musical instrument, a noise and a bang, can be recorded and analyzed. The auditory

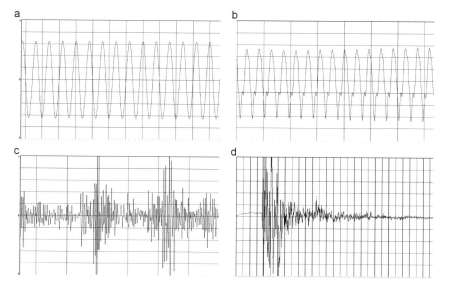

Fig. 11.11 Waveforms of a sinusoidal note (**a**), note played by a musical instrument (**b**), noise (**c**) and bang (**d**)

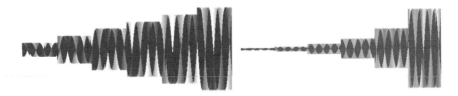

Fig. 11.12 Amplitudes of pressure variations in the sound samples at http://www.physikonline. net/springer/acoustics

impressions can then be related to particular types of waveforms and later also to their characteristic frequency spectra.

Quantitative Analysis of Sounds

For quantitative considerations, it is advantageous if the program does not only allow the analysis of sounds, but also the generation of defined note sequences. The synthesis of acoustic events allows students to apply and test acquired knowledge in a goal-oriented manner.

The sound samples shown in Fig. 11.12 can give interesting insights into perceived *sound level*. In the example on the left, the graph shows a stepwise linear increase in amplitude. However, the example on the right gives the perception that the sound level increases linearly with time. Here, however, the graph shows an amplitude that increases exponentially with time.

The apparent contradiction is resolved by the *law of Weber and Fechner*, according to which the perceived sound level is proportional to the logarithm of sound intensity.

The law of Weber and Fechner not only approximates perceived sound level, but also the perception of brightness and the *perception of pitch*. Examples can be found on the accompanying Web site http://www.physikonline.net/springer/acoustics. In all of these cases, a logarithmic scale is useful. Semitone steps, that is, the increase of the frequencies by a factor of $\sqrt[12]{2}$, appear on a logarithmic scale as equidistant steps. Thus, also for pitch, a logarithmic scale best suits our acoustic perception.

These examples illustrate the strength of multiple modalities in facilitating links between *acoustic perceptions,* and *the mathematical and graphical descriptions*.

Superposing Notes and Fast Fourier Transform

A first application task can be to create sound samples by *superposing different notes*. The results can then be played back and heard immediately and connected with graphical representations of the pressure variations. If one starts from a fundamental tone and then overlays it with higher harmonics of different amplitudes, their significance for the sound experience becomes clear.

Conversely, the sounds of different musical instruments can be analyzed via a Fourier decomposition (*fast Fourier transform/FFT*) and viewed as frequency spectra (see Fig. 11.13). From a didactical point of view, the possibility of relating various graphical representations with the acoustic perception is again of particular use.

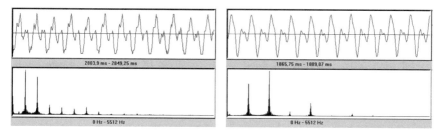

Fig. 11.13 Waveforms (top) and frequency spectra (bottom) of a piano (left) and flute (right) tone

An interesting extension task is *modifying sound*, for example, to filter particular frequencies of the spectrum (e.g., via high-pass or low-pass filters) in a given or self-recorded sound and to investigate their effects.

Experimenting with Sound Effects

Many acoustic programs offer special sound effects, often via a simple user interface. Thus, students can quickly *create impressive sound effects*, for example, when recording their own voice. The following effects can be realized with a wide range of acoustic programs.

- An echo can be implemented by varying sound intensity with time delay.
- For stereo recordings, a sound source moving transversely to the listener can be simulated by reducing the volume continuously on one channel and simultaneously increasing it on the other channel.
- A continuous frequency shift in time, combined with an increase or decrease in volume, can be used to simulate sound sources that appear to move toward or away from the listener.
- With filters, for example, a high-pass or band-pass filter, voice can be modified ("Mickey Mouse voice") or the sound of an antiquated gramophone with shellac records can be created (also for current music pieces).
- If piano sounds are played backward in time, this sounds like a harmonium.

Sound samples are available at http://www.physikonline.net/springer/acoustics. In these examples, the possibility to switch easily between different modalities and representations and to use multiple representations can help build a bridge between theory and perception.

11.4.2 Developing Mental Models

Mental models provide a theoretical explanatory framework for the use of pictorial analog representations. This theory can be helpful for the development and use of multimedia applications.

Mental models are *analogical, cognitive representational* forms of complex relationships (Gentner and Stevens 2014).

However, media can only indirectly support the development and extension of mental models by offering information in the form of *external representations*. However, visualizations can offer important cues that *promote the formation of mental models*. In the following, the term is used in accordance with early work (see, in particular, Johnson-Laird 1980; Forbus and Gentner 1986; Seel et al. 2000): Mental models denote a representation of knowledge that allows for a mental simulation of physics processes and procedures. Corresponding examples would be ideas about the particle model of an ideal gas and the behavior of charge carriers at a *pn* junction or in the transistor effect.

Media to Support the Development of Mental Models

Simulations allow students to explore relationships between quantities and can thus support the development of mental models. For example, the PhET interactive simulation "States of Matter" supports the development of a particle model of matter (https://phet.colorado.edu/sims/html/states-of-matter-basics/latest/states-of-matter-basics_en.html). As another example, Fig. 11.14 shows the interactive simulation "Interferometer experiments with photons, particles and waves" (https://www.st-andrews.ac.uk/physics/quvis/simulations_html5/sims/photons-particles-waves/photons-particles-waves.html) that allows students to compare and contrast the measurement outcomes for these three types of objects when sent through the same interferometer setup (Kohnle et al. 2014). Students can easily change the type of particle and experimental setup and collect data under idealized conditions. This allows students to use their prior knowledge of classical

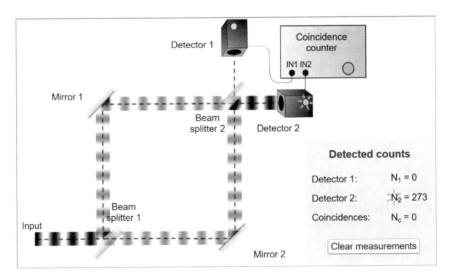

Fig. 11.14 A screenshot from the simulation "Interferometer experiments with photons, particles and waves"

objects to determine how photons (quantum objects) are similar to and how they are different from classical objects and thus promotes the development of a photon model in introductory quantum physics.

Combining multiple representations and *connecting models to real-world depictions* are further strategies to support the development of students' mental models and to ensure that they do not limit their investigation to surface features of a given visual depiction.

11.4.3 Promoting Cognitive Flexibility

Cognitive Flexibility and Multiple Knowledge Representation

Cognitive flexibility denotes the ability to apply knowledge flexibly in different situations. This includes the ability to spontaneously *reorganize and assemble knowledge* in response to changing situations and contexts (Spiro and Jehng 1990; Spiro et al. 2004). If necessary, preexisting knowledge can be flexibly reassembled in a format that is best suited to a specific situation (Spiro et al. 2003). Prerequisites are both appropriate representational forms of knowledge and adequate knowledge retrieval processes. Knowledge which is to be applied in many ways must be organized and represented in many ways. Achieving these goals has implications for instruction.

Cognitive flexibility theory emphasizes the importance for knowledge acquisition that learners consider the same content at different times, in newly constructed contexts, under different objectives and from different conceptual perspectives (Spiro et al. 2003). Accordingly, for complex areas of knowledge, especially so-called "ill-structured domains", *considering multiple examples with different perspectives* and relating them to one another is important. Hypermedia systems are well suited to replicate the complex paths needed for such knowledge acquisition.

In order to promote cognitive flexibility, *content should be presented in various representational forms* and integrated into different contexts. This also facilitates the development of efficient information retrieval processes for subsequent problem solving. Instruction should thus aim for the following:

- Activities that allow learners to process content using multiple representations.
- Learning materials should avoid oversimplifications and support learners in making connections between contexts.
- Teaching should consider multiple cases and emphasize the construction (rather than transmission) of knowledge.
- Sources of information should be highly interconnected and not isolated from each other.

A strong initial simplification of the learning content which is maintained for too long a time can be problematic in terms of interfering with learners' later attainment of complex understanding.

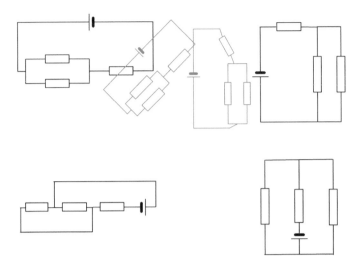

Fig. 11.15 Computer animations showing transformations of circuit diagrams. The circuit diagrams shown on the left- and right-hand sides are equivalent for the top and bottom circuits

Restructuring

Flexibility in applying knowledge implies not only mastering different representations, but also flexibly working within a single representation. For example, electrical circuits can be displayed in a variety of ways. Learners may initially not realize that the circuit diagrams shown on the left- and right-hand sides of Fig. 11.15 are equivalent.

A computer program which is based on ideas from Härtel (1992) can restructure the circuits in steps and thus make the connections between them apparent (see Fig. 11.15). Further examples can be found at http://www.physikonline.net/springer/multimedia/circuits.

Connecting Different Representations

At the more advanced level, learners can create connections between different representations themselves if multiple representations are presented simultaneously. As an example, Fig. 11.16 shows several representations of the electric field of a dipole. With the corresponding html5 program (available at http://www.physikonline.net/springer/multimedia/electricfield), the electric field of a configuration of point charges created by the user is displayed in different representations. Such use of multiple representations is not intended for the introductory phase of instruction.

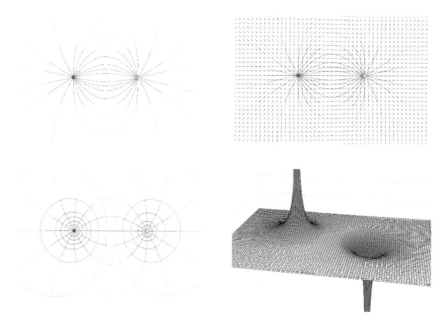

Fig. 11.16 Screenshots from the program "E-Field" showing multiple representations of the electric field of a dipole

11.4.4 Situated Learning and Anchored Instruction

Constructivist learning theories point to particular potential for the use of new media (1) in the inclusion of *authentic* physics and social contexts in the learning process and (2) the support of *communication* and *collaboration* between learners.

Researchers in *situated learning* (Brown et al. 1989; Lave 1988; Lave and Wenger 1990) emphasized the need to bridge the gap between theoretical learning in the classroom and real-life applications of knowledge. There is thus a need to avoid "inert knowledge" that is learned but not available in authentic contexts. According to situated learning theory, learning is a function of activities, context and cultural environment.

Herrington and Oliver (1995) described the following *key characteristics of instructional design for situated learning*:

- Provide authentic contexts that reflect how knowledge is used in real life.
- Carry out authentic activities.
- Provide access to expert performances and modeling of processes.
- Provide diverse roles and multiple perspectives.
- Support collaborative knowledge construction.
- Provide coaching and scaffolding in critical phases.
- Stimulate reflection to enable abstractions to be formed.
- Promote articulation that allows tacit knowledge to be made explicit.

- Provide integrated assessment of learning within the tasks.

Multimedia programs, video sequences and the Internet can contribute to appropriate contexts for situated learning.

However, the inherent *complexity of many real-life situations* must be taken into account. Sandberg and Wielinga (1992), among others, stressed that this complexity can also make the learning process more difficult. On the other hand, oversimplification is also problematic for learning (Spiro et al. 2003). Here again, it becomes clear that the old didactical principles are still important for new media, particularly, the rule to progress from simple to complex.

Anchored Instruction

Anchored instruction is an implementation of the theoretical postulates of the situated cognition approach using multimedia elements (Cognition & Technology Group at Vanderbilt 1993; Bransford et al. 1990). The intent is to provide application-oriented "anchors" (macrocontexts) as focal points for subsequent teaching and learning.

In anchored instruction, it is essential that the content (knowledge of concepts, theories and principles) has meaning and personal value for the learner. Knowledge is not seen as an end result, but as a tool for tackling (subjectively) relevant questions. Anchors highlight in which situations certain knowledge is useful to apply and provide a starting point for integrating information from different areas of knowledge. The anchoring of knowledge in complete, realistic problems is intended to make the development of specific, but also transferable problem-solving skills more effective (Goldmann et al. 1996).

An example of acoustic anchoring using multimedia programs can be found in Sect. 11.4.1. Here, physics knowledge can also have a *personal relevance for students* who are interested in music. The digital recording and further processing of music pieces or students' own voice and their addition of various sound effects are now directly possible with standard equipment.

11.4.5 Structuring and Connecting Knowledge

Structuring knowledge means the organization and connection of cognitive elements.

De Jong and Njoo (1992) analyzed 32 learning processes and identified two important categories: integrating (structuring new knowledge) and connecting (linking new knowledge to existing knowledge). *Structured and organized knowledge is also necessary for problem solving* (Reif 1981). Above all, a hierarchical knowledge structure influences memory retrieval. Key general concepts can control access to relevant specific details. Thus, van Heuvelen (1991) emphasized the need to help students construct a knowledge structure around physics principles and to see how a small number of unified concepts form the basis for diverse applications.

Knowledge must be connected in order to be useful for problem solving. Clark (1992) distinguished between *vertical connections* (e.g., the assignment to a general principle) and *horizontal connections* (connections to similar knowledge structures, e.g., via analogies). The latter are of particular interest in relation to transfer. Graphics that visualize relationships between concepts can support students in acquiring structural knowledge and can also promote recall and retention of learning content (Beisser et al. 1994). New media offer tools that can support the visualization and construction of relevant structures.

Maps and Charts

Mind maps and *concept maps* are structured and organized representations of key ideas and their relations (e.g., also as text image combinations). Concept maps represent a domain of knowledge via core concepts and ideas, depicted by nodes and their connections as a propositional structure. *"Reference maps"* (e.g., maps constructed by experts) aim to map knowledge structures. They aim to offer a cognitive framework and facilitate access to knowledge. Cañas et al. (1999) considered such maps as an opportunity for an elegant, intuitive representation of a field of knowledge. Unlike maps, *charts* do not start from a central concept. They tend to be more vertically organized and can thus clearly show hierarchical structures (see Fig. 11.17).

In addition to their use for teaching and learning purposes, concept maps can also be implemented with the aim of investigating individual students' cognitive structures. For example, Schaal et al. (2010), and Fischler et al. (2001) described concept maps as a tool for education research.

However, a differentiated approach is always needed. One cannot assume a direct one-to-one mapping of external structural representations onto internal representation (and vice versa). According to Jonassen and Wang (1993), it is not sufficient to depict knowledge structures in order to improve learning of structural knowledge. The active application of knowledge, stimulated by cognitive processing tasks and learning goals, seems to be essential.

Computer-Based Mapping

Figure 11.17 shows a concept map for heat transport and a schematic diagram of a *site map* which is automatically created by standard software to give a hierarchical list of Web pages with links between them. The structural similarity is obvious.

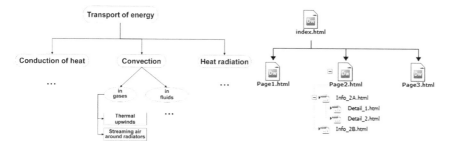

Fig. 11.17 A concept map for heat transport and a schematic diagram of a site map

Fig. 11.18 A mind map showing electricity from different types of power plants

Images can be used to further *illustrate mind maps and concept maps* (see Fig. 11.18).

Maps organize and structure content in a different manner to texts. Concepts are not sequentially ordered; they are placed side by side or below one another. Relationships and propositional structures are visualized graphically. Mind maps are thus also suitable for combining text-based and pictorial thinking, combining analytical and associative creative work and providing structural aids. In addition, the nodes can be attractively designed with images and furthermore linked to Web sites (see Fig. 11.18).

As suggestions for teaching, the depth of the hierarchical structure can be extended and linked to Internet addresses (see the right-hand side of Fig. 11.18), so that the information content can be expanded and additional explorations can be motivated. Pre-prepared hypertext or hypermedia structures can be explored (see, for example, http://hyperphysics.phy-astr.gsu.edu/hbase/hframe.html), offering new possibilities to assist teaching. In addition, it is also easily possible for learners to create their own concept and mind maps with modern computer programs. Even inserting different levels of hierarchy and linking to Internet addresses is straightforward (see also further suggestions in Sect. 11.6).

The following properties are useful for *characterizing concept maps* when introducing them in the classroom, but also for analyzing concept maps for assessment purposes (see also Cañas et al. 2016):

- *Structure type*: What are the criteria for structuring knowledge?
- *Structuring breadth*: How comprehensively is the topic covered?
- *Structuring depth*: How detailed is the topic covered?
- *Density of connections*: How complex and dense are the links in the concept map?
- *Clustering*: Which substructures can be identified?
- *Navigation*: How can the knowledge elements be accessed?

11.5 Simulations and Guided Discovery Learning

Computer simulations provide a model of real-world systems or phenomena based on a limited number of variables and their relations. Simulations respond to students' input according to predetermined rules that are derived from the model. Parameters can also be individually modified to investigate their impact within the model.

11.5.1 Simulations for Physics Instruction

Wieman et al. (2010) described particular advantages of simulations which can motivate their use:

- Simulations can provide alternatives where laboratory equipment is not available or not practical to implement.
- Simulations allow for "experiments" that would not be possible in the classroom (e.g., simulation of the greenhouse effect and simulation of a parachute jump).
- Variables can be easily changed in simulations in response to students' "what-if" questions.
- Simulations can visualize the invisible and establish relationships between different length scales.
- Students can run simulations on their own devices outside of class to repeat and expand experiments seen in the classroom.

As examples, Fig. 11.19 shows two simulations which also run on mobile devices and are thus directly available to students.

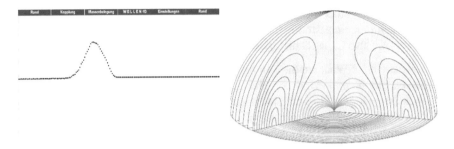

Fig. 11.19 Two simulations (both available at http://www.physikonline.net/springer/multimedia): (Left) Interactive wave machine. Using the mouse or finger, the chain can be deflected at an arbitrary point and the spread of the disturbance can be observed. (Right) The propagation of electromagnetic waves due to an oscillating electric dipole is displayed via dynamic field line images

11.5.2   Guided Discovery Learning with Digital Media

It is an appealing approach to use simulations and multimedia programs to create Internet-based guided discovering learning. Alfieri et al. (2011) defined *discovery learning* as learning in which the target information is discovered by learners themselves, framed by the task and the materials provided. Interaction with the material, the manipulation of variables and testing applications of logical principles open up possibilities for learners to identify important structures themselves, to deduce basic causalities and thus to learn in ways that give better results. According to de Jong (2011), relevant cognitive processes in guided discovery learning include:

- Orientation in terms of analyzing the domain and identifying main concepts and variables.
- Hypothesis formation.
- Testing hypotheses experimentally (e.g., via a simulation).
- Drawing conclusions.

In addition, there are actions to improve the management of the work procedure (especially in the areas of planning and evaluation) and to reflect on the procedure and the knowledge acquired.

The need and extent of meaningful assistance and scaffolding are intensely debated. For example, Kirschner et al. (2006) highlighted the need for guidance in discovery learning, problem-based learning, experiential learning and inquiry learning by referring to numerous empirical findings. They point to the limited capacity of working memory as the main reason why guidance is needed (see Sect. 11.2.4). In particular, they mentioned the use of worked examples or worksheets with descriptions and guidelines for the different phases of the problem-solving process as possibilities for more guided learning.

Alfieri et al. (2011) conducted two meta-analyses with 164 studies. They found that unassisted discovery learning was less effective than guided discovery, and that supporting learning through feedback and worked examples, offering guidance on how to proceed and giving elicited explanations were helpful and effective in discovery learning.

Further ways to support inquiry with simulations are the use of contrasting cases to support the induction of underlying structure (Chase et al. 2010) and driving questions to promote exploration (Chamberlain et al. 2014).

For work in lower secondary school education, Zhang and Quintana (2012) found positive results by implementing a tool to organize and structure a targeted search in digital media. The tool led to a more efficient, continuous, metacognitive and focused online inquiry.

11.5.2.1 Teaching with Simulations

The interplay of simulations with the content, the students and the teacher is decisive for instruction.

Wieman et al. (2010) discussed guidelines for inquiry-based activity design for their outstanding PhET simulations. They stressed that general strategies for effective teaching must be taken into account and, therefore, specifically repeat the following eight general *guidelines*:

1. Define specific learning goals.
2. Encourage students to use sense-making and reasoning.
3. Activate and build on students' prior knowledge.
4. Link to real-world situations and experiences.
5. Support productive collaboration.
6. Do not overly constrain student exploration.
7. Require reasoning in multiple representations.
8. Help learners to monitor their own understanding (see also PhET Interactive Simulations, https://phet.colorado.edu/en; Podolefsky et al. 2013).

Studies have found that interpreting visual information in simulations can be challenging. Lopez and Pinto (2017) found student difficulties relating to the spatial distribution and interconnection of the visual elements, the relevance and semantics given to visual elements, connection of multiple representations and interpretation of dynamic information. One approach to supporting student understanding of visual information is to get students to draw themselves (without simulation support) representations they will later see in the simulation for a simple case. In a second activity phase, students then work with the simulation and answer further questions (Kohnle et al. 2020). This approach builds on the literature of active construction of knowledge (Chi 2009) and drawing to learn (Ainsworth et al. 2011).

Simulations can enhance various educational settings. Wieman et al. (2010) described the use of simulations in the following contexts:

- Use by the teacher in class (as simple animations to show physics processes, especially for visualizing invisible phenomena).
- Use for testing physics ideas and reinforcing concepts.

- Use as interactive lecture demonstrations with the option to vary parameters.
- Use as collaborative in-class activities.
- Use as homework activities.
- Use to simulate laboratory activities.

It is important to note that simulations are tools that by themselves do not ensure effective instruction. They must be integrated into a suitable instructional design. Eysink and de Jong (2012) emphasized the key role of the depth of cognitive processing in student learning.

11.6 Learning with Online Resources and Tools

The following Web sites are examples of useful starting points for further exploration:

- ComPADRE resources: www.compadre.org
- PhET interactive simulations: www.phet.colorado.edu
- MERLOT online learning materials: www.merlot.org
- Phyphox mobile phone experiments: www.phyphox.org
- PhysPort research-based resources: www.physport.org
- Go-Lab online laboratories and inquiry learning spaces: www.golabz.eu.

11.6.1 Challenges of Internet Search

Despite the potential of the Internet for education, targeted search on the Internet also poses challenges. There are several reasons for this:

- There is no central coordination, no structured overview.
- There are errors, no checking of content or peer review.
- The structure of documents varies widely, ranging from short texts or graphics to entire books or databases.
- Internet presentations are often not designed to be complete or give an overview of a topic.
- The Internet is changing constantly with time, so that content and addresses can change. The continued existence of individual Web pages is not guaranteed.

If one does not quickly find a Web site that offers a well-prepared didactical presentation of a topic, independent learning via the Internet will rarely be effective and goal-oriented.

11.6.2 Organizing Information and Structuring Knowledge

Modern computer applications for mind mapping and concept mapping make it easy to compile, organize and explain information from Internet sources in a graphical format. In this way, small (content-related) sections of the Web can be structured, broken down and designed into learning paths (see Sect. 11.4.5). In particular, students can also create their own overviews.

From a didactical perspective, mind maps are interesting as *cognitive tools*, that is, as tools that help the learner to engage with the content *more closely, effectively* and *economically*. The following aspects are particularly relevant to Internet searches in a school environment:

- Mind maps document the current state of work; progress is directly visible in the graphic.
- No mind map is perfect right from the start. However, changes and corrections are easy to implement using the computer interface, and the graphic remains clearly laid out.
- The mapping programs allow students to easily create their own Web sites with an attractive graphic design. The possibility of creating outputs of their work in this form can be motivating from a psychological perspective (perception of self-efficacy).

11.6.3 Activity Design for Internet Search

It is rare that students already possess sophisticated techniques to use information on structural relationships effectively. Meaningful learning from hypertext structures usually requires externally scaffolded and mediated learning tasks.

Before starting online work and using search engines, students should compile search terms that are as accurate as possible. This makes the search more structured and targeted. In addition, the work with mind maps and concept maps can be integrated into different tasks and thus adapted to student performance and objectives. Some suggestions for student tasks are as follows:

- Work through a mind map with Internet links generated by the teacher.
- Design a mind map with images, links and accompanying texts.
- Create extensions and additions to a given mind map (e.g., elaborate and deepen an existing map).
- Extract relevant keywords from a predetermined list (e.g., from the table of contents of a textbook, compile relevant keywords, arrange them clearly in a graphic and link them to Web pages).
- Create an overview of the current teaching content for the school server, which is constantly updated.
- Compile additional links to teaching material, and sort and structure them by themes.

- Brainstorming in a first phase of a project and creating a target map. This map is then elaborated and linked to Web pages in collaborative group work.

ComPADRE (www.compadre.org) offers various search options for instructional materials for physics. It is possible for instructors to create their own compilations of materials and make them available to students.

11.6.4 E-Learning

E-learning denotes a form of learning supported by digital media. The electronic learning materials are multimedia-based, linked together and structured to enable learners to interact with the learning environment, the instructors and members of a learning group.

E-learning requires good preparation and an overall plan that takes the following dimensions into account:

- The learning *content*, that is, considerations of factual knowledge, conceptual knowledge, processes and procedures, must be coordinated with the multimedia presentation of the content.
- Structures to promote and support cognitive *processes* related to the learning of the content.
- Learning in a *social context* with interactive procedures supported by digital media must be prepared.

Alonso et al. (2005) described the following components for effective self-contained learning units:

- Presentation of the learning content that will be developed. This presentation is responsible for motivating and providing guidance to learners about the knowledge they are to acquire.
- Learning objectives that indicate the results of the learning, briefly describing the tasks that learners will be able to perform.
- Preparation and multiple presentations of the required knowledge.
- Learning tasks that train and support the desired skills.
- Practice to consolidate what was taught and put it into practice for real cases.
- Discussions and group activities to drive collaborative learning.
- Conclusions to reinforce and recall key points and focus learners' attention on the learning objectives.
- Feedback to learners on their performance.

Blended Learning

Blended learning aims to link virtual learning spaces and in-class instruction by using the strengths of both forms and by compensating disadvantages. Virtual learning spaces are organized as e-learning classes, e-mentoring groups, Web-based learning modules or online communities, and use email, knowledge and literature databases, e-workbooks, audio and video streams, as well as Web 2.0 technologies. In-class instruction makes use of conventional forms such as lessons, workshops, tutorials or laboratory experiments.

Strengths of in-class instruction allow for:

- Social and personal contacts (with teachers and between learners) as well as collaborative group work.
- Holistic communication (with nonverbal cues, which are important, e.g., for a teacher training course).
- A more direct treatment of problems and difficulties in understanding.
- Direct discussions and agreements.
- Secure and unambiguous individual proof of performance via assessments.

Strengths of e-learning allow for:

- Time- and location-independent learning.
- Individualized learning (in terms of learning pace, duration, environment, but also objectives).
- Linking of the instructional information units.
- Networking of different departments and locations (also internationally).
- Incorporating the learning material into different scenarios with an interdisciplinary character.
- Multimedia applications of the learning content (e.g., illustrating complex ideas through animations and simulations).
- Providing different media (image, video, sound, animation and text), in different task formats (with practical or theoretical emphasis, games, group work, individual work, etc....), which can accommodate the preferences of different types of learners.
- Techniques that facilitate access to information in databanks and electronic libraries and provide additional search functions.
- Dynamic and up-to-date content.
- New forms of collaboration and communication between learners and teachers, but also between learners or teachers themselves (e.g., in virtual discussion forums), including bringing in experts.
- Assessment of learning outcomes with direct feedback.

In particular, the use of Web 2.0 technologies is increasingly implemented in e-learning, which allows for further activities.

11.6.5 Physics with Mobile Devices

New Mobility for Digital Media

Mobile devices offer particularly *simple, flexible* and *location-independent* possibilities for

- Information searches on the Internet.
- Documenting ideas via image, sound and video.
- Data exchange and communication.
- Use of so-called cognitive tools to facilitate work, such as a dictionary, calculator and notepad.

These functionalities make mobile devices interesting for physics learning.

New Availability of Sensors and Modern Data Acquisition Technology

For physics lessons, mobile devices are also interesting as data acquisition devices with attractive applications. A wide range of modern sensors are integrated into these devices.

Sensors usually include:

- Acceleration and inclination sensor.
- Gyroscope.
- Magnetic field sensor.
- Brightness sensor/light meter for built-in cameras and for adjusting the display brightness.
- Color sensor.
- Microphone.
- Proximity sensor that turns off the display and touch functions based on a distance measurement (e.g., when holding the smartphone next to a head).

Further sensors are already installed in some devices:

- Temperature sensor.
- Humidity sensor.
- Barometer.
- Gesture sensor that detects hand movements, for example.

This makes it increasingly possible to transform standard demonstration experiments into student experiments, with students using their own devices. In addition, the new mobility also facilitates experiments outside the classroom and in everyday life.

Examples of applications using sensors in mobile phones:

- Mechanics (Kuhn and Vogt 2013; Hochberg et al. 2014).
- Acoustics (Klein et al. 2014; Hirth et al. 2015; Thoms et al. 2018).
- Waves (Müller et al. 2016).

- Radioactivity (Kuhn et al. 2014).
- Biological physics (Thoms et al. 2017; Thoms et al. 2019a, b).

Fortunately, the range of easily available and affordable application programs (apps) for mobile devices is increasing rapidly. In combination with the comprehensive information available via the Internet, application-related and interdisciplinary topics become accessible, for example, the physics of sports and health education.

Modern Communication Skills Training

Another interesting application using smartphones was proposed by Rath and Schittelkopf (2011) and tested in class: Students use their mobile phones to make video recordings of experiments, which they also narrate with descriptions and physics explanations. The ease of use and availability of working with your own device has particular appeal. However, Rath and Schittelkopf also pointed to the issue that the focus on the physics phenomenon is not automatically ensured, and that technology and implementation aspects can easily become the focus of discussions. In this respect, it is important from the outset to plan for targeted, content-oriented tasks.

Learning Physics with Mobile Devices

Recent research shows that learning with mobile devices as experimental tools can initiate active learning and thus foster learning and motivation. Hochberg et al. (2018) discussed the use of smartphones as experimental tools in secondary school physics. According to their studies, students developed higher interest and curiosity exploring mechanical oscillations with smartphones' internal acceleration sensors compared with traditional experiments. Hochberg et al. (2020) also showed that students were more cognitively engaged and had a lower cognitive load by learning the content using video motion analysis with mobile devices. This led to better conceptual understanding compared to a control group where students learned the same content with comparable traditional experiments.

Becker et al. (2020a, b) showed that exploring the content of uniform motion with video motion analysis fosters students' conceptual understanding and reduces their extraneous cognitive load compared with learning this content with traditional experiments. Video motion analysis is also an effective tool in introductory physics courses, and its positive effects on learning and motivation have been shown (e.g., Klein et al. 2017, 2018). Kuhn and Vogt (2015) also found successful learning effects with mobile devices in the context of acoustics.

11.6.6 Augmented Reality in Physics Education

Combining virtual and real experiments has the potential to promote learning processes (de Jong et al. 2013; Jones and Sharma 2019). Augmented reality (AR) can help to implement this combination or at least offer additional context-specific information. On the one hand, learners interact with real objects (i.e., experimental materials) and, on the other hand, engage with virtual objects (e.g., measurement data in different visualizations) integrated into the real experimental environment (Azuma 1997). Thus, AR can be used to connect virtual information and real components in space and time (in accordance with the contiguity principle), which makes it a promising technology for teaching and learning (Ibáñez and Delgado-Kloos 2018). This technology has been implemented, for example, by Fidan and Tuncel (2019), Garzón et al. (2019), and Hung et al. (2017). However, whether and to what extent such a synchronization of information sources enhances learning depends on the educational goals and the corresponding functions of virtual and real components (Altmeyer et al. 2020; Kuhn et al. 2016; Liu et al. 2020; Thees et al. 2020; Strzys et al. 2018).

Augmented reality, especially in the context of physics experiments, includes extensive elements of multimedia learning: Visual information (e.g., a real setup of electrical circuits consisting of cables, resistors and power supply or computer representations of data) and verbal information (on worksheets or narrated explanations of measurement data) are combined to assist knowledge construction. According to Santos et al. (2014), augmented reality can support a promising transformation of traditional learning environments into modern multimedia settings by integrating additional external representations into an authentic environment. Theoretical concepts for multimedia learning can be applied with the aim of creating effective novel learning environments.

11.7 Conclusions

In addition to the goal of enriching discipline content with multimedia-based learning opportunities, e-learning also offers the option to train students in digital literacy and the competent use of new media and communication possibilities.

Digital media allow for various ways to enhance and transform classical teaching techniques. For example, the SAMR model from Puentedura (2013) describes:

- Substitution: Direct replacement of tools without functional changes.
- Augmentation: Direct replacement of tools with functional improvement.
- Modification: Redesign of tasks with significantly new elements.
- Redefinition: Creation of new tasks that were previously inconceivable.

However, it is essential to implement technological tools based on didactical principles and findings from educational psychology. Classical learning theories remain valid and need to be taken into account. Nevertheless, research in the field of media-based learning has provided important new findings.

Mobile devices greatly improve students' abilities to access learning resources and offer opportunities to carry out experiments directly with their own devices. These opportunities open up additional perspectives for lifelong learning.

While the focus here was on multimedia applications, the medium alone does not ensure that learning objectives are achieved. The approaches presented in this chapter are intended to allow the development and implementation of materials based on theoretical considerations and empirical findings. The aim is to use media effectively and in a goal-oriented way, where other teaching means would make the objectives more difficult to achieve.

An overview of further articles in teacher education journals, especially in non-English languages, is available from Girwidz et al. (2019) and the Multimedia Physics Teaching and Learning (MPTL) 2019 Conference (Girwidz and Thoms 2021).

Acknowledgements We would like to thank Ian Lawrence (Institute of Physics, London, UK) and Jochen Kuhn (Technische Universität Kaiserslautern, Germany) for carefully and critically reviewing this chapter.

References

Ainsworth S (2006) DeFT: a conceptual framework for considering learning with multiple representations. Learn Instr 16:183–198

Ainsworth S, Prain V, Tytler R (2011) Drawing to learn in science. Science 333(6046):1096–1097

Alfieri L, Brooks PJ, Aldrich NJ, Tenenbaum HR (2011) Does discovery-based instruction enhance learning? J Educ Psychol 103(1):1–18

Alonso F, López G, Manrique D, Viñes JM (2005) An instructional model for web-based e-learning education with a blended learning process approach. Br J Edu Technol 36(2):217–235

Altmeyer K, Kapp S, Thees M, Malone S, Kuhn J, Brünken R (2020) The use of augmented reality to foster conceptual knowledge acquisition in STEM laboratory courses—theoretical background and empirical results. Br J Edu Technol 51(3):611–628

Azuma RT (1997) A survey of augmented reality. Presence Teleoperators Virtual Environ 6(4):355–385

Baddeley AD (1990) Human memory: theory and practice. Lawrence Erlbaum, Hillsdale, NJ

Baddeley A (1992) Working memory. Science 255:556–559

Becker S, Gößling A, Klein P, Kuhn J (2020a) Using mobile devices to enhance inquiry-based learning processes. Learn Instr 69:101350

Becker S, Gößling A, Klein P, Kuhn J (2020b) Investigating dynamic visualizations of multiple representations using mobile video analysis in physics lessons: effects on emotion, cognitive load and conceptual understanding. Zeitschrift Für Didaktik Der Naturwissenschaften 26:123–142

Beisser KL, Jonassen DH, Grabowski BL (1994) Using and selecting graphic techniques to acquire structural knowledge. Perform Improv Q 7(4):20–38

Bransford JD, Sherwood RD, Hasselbring TS, Kinzer CK, Williams SM (1990) Anchored instruction: why we need it and how technology can help. In: Nix D, Spiro R (eds) Cognition, education and multimedia: exploring ideas in high technology. Erlbaum Associates, Hillsdale, NJ, pp 115–141

Brown JS, Collins A, Duguid P (1989) Situated cognition and the culture of learning. Educ Res 18(1):32–42

Cañas AJ, Leake DB, Wilson DC (1999) Managing, mapping, and manipulating conceptual knowledge. In: Proceedings of the AAAI-99 workshop on exploring synergies of knowledge management and case-based reasoning. AAAI Press, Menlo Park

Cañas AJ, Reiska P, Novak JD (2016) Is my concept map large enough? In: International conference on concept mapping. Springer, Cham, Switzerland, pp 128–143

Chamberlain JM, Lancaster K, Parson R, Perkins KK (2014) How guidance affects student engagement with an interactive simulation. Chem Educ Res Pract 15(4):628–638

Chandler P, Sweller J (1991) Cognitive load theory and the format of instruction. Cogn Instr 8:293–332

Chase CC, Shemwell JT, Schwartz DL (2010) Explaining across contrasting cases for deep understanding in science: an example using interactive simulations. In: Gomez K, Lyons L, Radinsky J (eds) Proceedings of the 9th international conference of the learning sciences (ICLS 2010)—Volume 1 full papers. International Society of the Learning Sciences, Chicago, IL

Chi MT (2009) Active-constructive-interactive: a conceptual framework for differentiating learning activities. Top Cogn Sci 1(1):73–105

Clark RE (1992) Facilitating domain-general problem solving: computers, cognitive processes and instruction. In: de Corte E, Linn MC, Mandl H, Verschaffel L (eds) Computer-based learning environments and problem solving. Springer, Berlin, pp 265–285

Clark J, Paivio A (1991) Dual coding theory and education. Educ Psychol Rev 3:149–210

Cognition & Technology Group at Vanderbilt (1993) Anchored instruction and situated cognition revisited. Educ Technol 33(3):52–70

Comi SL, Argentin G, Gui M, Origo F, Pagani L (2017) Is it the way they use it? Teachers, ICT and student achievement. Econ Educ Rev 56:24–39

de Jong T (2011) Instruction based on computer simulation. In: Mayer RE, Alexander PA (eds) Handbook of research on learning and instruction. Taylor & Francis, New York, NY, pp 446–466

de Jong T, Njoo M (1992) Learning and instruction with computer simulations: learning processes involved. In: de Corte E, Linn MC, Mandl H, Verschaffel L (eds) Computer-based learning environments and problem solving. Springer, Berlin, pp 411–427

de Jong T, Linn MC, Zacharia ZC (2013) Physical and virtual laboratories in science and engineering education. Science 340(6130):305–308

Eysink THS, de Jong T (2012) Does instructional approach matter? How elaboration plays a crucial role in multimedia learning. J Learn Sci 21(4):583–625

Fidan M, Tuncel M (2019) Integrating augmented reality into problem based learning: the effects on learning achievement and attitude in physics education. Comput Educ 142:103635

Fischler H, Peuckert J, Dahncke H, Behrendt H, Reiska P, Pushkin DB et al (2001) Concept mapping as a tool for research in science education. In: Behrendt H, Dahncke H, Duit R, Gräber W, Komorek M, Kross A, Reiska P (eds) Research in science education—past, present, and future. Springer, Dordrecht

Forbus KD, Gentner D (1986) Learning physical domains: toward a theoretical framework. In: Michalski RS, Carbonell JG, Mitchell TM (eds) Machine learning: an artificial intelligence approach, vol 2. Morgan Kaufmann Publishers, Los Altos, pp 311–348

Garzón J, Pavón J, Baldiris S (2019) Systematic review and meta-analysis of augmented reality in educational settings. Virtual Reality 23(4):447–459

Gentner D, Stevens AL (eds) (2014) Mental models. Psychology Press, New York

Girwidz R, Thoms LJ (2021) Physics education with multimedia applications in non-English teacher-oriented journals: an analysis of 491 articles about multimedia in physics education. J Phys Conf Series 1929(1):012035. IOP Publishing

Girwidz R, Thoms LJ, Pol H, López V, Michelini M, Stefanel A et al (2019) Physics teaching and learning with multimedia applications: a review of teacher-oriented literature in 34 local language journals from 2006 to 2015. Int J Sci Educ 41(9):1181–1206

Goldmann SR, Petrosino AJ, Sherwood RD, Garrison S, Hickey D, Bransford JD, Pellegrino JW (1996) Anchoring science instruction in multimedia learning environments. In: Vosniadou S, de Corte E, Glaser R, Mandl H (eds) International perspectives on the design of technology-supported learning environments. Lawrence Erlbaum Associates, Mahwah, NJ, pp 257–284

Härtel H (1992) Neue Ansätze zur Darstellung und Behandlung von Grundbegriffen und Grund-größen der Elektrizitätslehre. In: Dette K, Pahl PJ (eds) Multimedia, Vernetzung und Software für die Lehre (English: New approaches to the representation and treatment of basic concepts and quantities in electricity. In: Dette K, Pahl PJ (eds) Multimedia, networks and software for teaching). Springer, Berlin, pp 423–628

Herrington J, Oliver R (1995) Critical characteristics of situated learning: implications for the instructional design of multimedia. Paper presented at the Australian Society for Computers in Learning in Tertiary Education Conference 1995 (ASCILITE95). https://researchrepository.mur doch.edu.au/id/eprint/7189/

Hirth M, Kuhn J, Müller A (2015) Measurement of sound velocity made easy using harmonic resonant frequencies with everyday mobile technology. Phys Teacher 53(2):120–121

Hochberg K, Gröber S, Kuhn J, Müller A (2014) The spinning disc: studying radial acceleration and its damping process with smartphone acceleration sensors. Phys Educ 49(2):137–140

Hochberg K, Kuhn J, Müller A (2018) Using smartphones as experimental tools—effects on interest, curiosity and learning in physics education. J Sci Educ Technol 27(5):385–403

Hochberg K, Becker S, Louis M, Klein P, Kuhn J (2020) Using smartphones as experimental tools–a follow-up: cognitive effects by video analysis and reduction of cognitive load by multiple representations. J Sci Educ Technol 29(2):303–317

Hung YH, Chen CH, Huang SW (2017) Applying augmented reality to enhance learning: a study of different teaching materials. J Comput Assist Learn 33(3):252–266

Ibáñez MB, Delgado-Kloos C (2018) Augmented reality for STEM learning: a systematic review. Comput Educ 123:109–123

Issing LJ (2002) Instruktionsdesign für Multimedia. In: Issing LJ, Klimsa P (eds) Information und Lernen mit Multimedia und Internet. Lehrbuch für Studium und Praxis (English: Instructional design for multimedia. In: Issing LJ, Klimsa, P (eds) Information and learning with multimedia and the Internet. Textbook for study and practice). Beltz, Weinheim, pp 151–175

Jen TH, Yeh YF, Hsu YS, Wu HK, Chen KM (2016) Science teachers' TPACK-practical: standard-setting using an evidence-based approach. Comput Educ 95:45–62

Johnson-Laird PN (1980) Mental models in cognitive science. Cogn Sci 4:71–115

Jonassen D, Wang S (1993) Acquiring structural knowledge from semantically structured hypertext. J Comput-Based Instr 20(1):1–8

Jones KA, Sharma RS (2019) An experiment in blended learning: higher education without lectures? Int J Digit Enterp Technol 1(3):241–275

Kirschner PA, Sweller J, Clark RE (2006) Why minimal guidance during instruction does not work: an analysis of the failure of constructivist, discovery, problem-based. Experiential Inquiry-Based Teach Educ Psychol 42(2):75–86

Klein P, Hirth M, Gröber S, Kuhn J, Müller A (2014) Classical experiments revisited: smartphones and tablet PCs as experimental tools in acoustics and optics. Phys Educ 49(4):412–418

Klein P, Müller A, Kuhn J (2017) KiRC inventory: assessment of representational competence in kinematics. Phys Rev Phys Educ Res 13:010132

Klein P, Kuhn J, Müller A (2018) Förderung von Repräsentationskompetenz und Experimentbezug in den vorlesungsbegleitenden Übungen zur Experimentalphysik - Empirische Untersuchung eines videobasierten Aufgabenformates (English: Promotion of representational competence and experiment relevance in exercises accompanying lectures on experimental physics—empirical investigation of a video-based task format). Zeitschrift für Didaktik der Naturwissenschaften 24(1):17–34

Koh JHL, Chai CS (2016) Seven design frames that teachers use when considering technological pedagogical content knowledge (TPACK). Comput Educ 102:244–257

Kohnle A, Bozhinova I, Browne D, Everitt M, Fomins A, Kok P et al (2014) A new introductory quantum mechanics curriculum. Eur J Phys 35:015001

Kohnle A, Ainsworth S, Passante G (2020) Sketching to support visual learning with interactive tutorials. Phys Rev Phys Educ Res 16:020139

Kozma R (2003) The material features of multiple representations and their cognitive and social affordances for science understanding. Learn Instr 13:205–226

Kuhn J, Vogt P (2013) Smartphones as experimental tools: different methods to determine the gravitational acceleration in classroom physics by using everyday devices. Eur J Phys Educ 4(1):16–27

Kuhn J, Vogt P (2015) Smartphone & co. in physics education: effects of learning with new media experimental tools in acoustics. In: Schnotz W, Kauertz A, Ludwig H, Müller A, Pretsch J (eds) Multidisciplinary research on teaching and learning. Palgrave Macmillan, Basingstoke, UK, pp 253–269

Kuhn J, Molz A, Gröber S, Frübis J (2014) iRadioactivity—possibilities and limitations for using smartphones and tablet PCs as radioactive counters. Phys Teacher 52(6):351–356

Kuhn J, Lukowicz P, Hirth M, Poxrucker A, Weppner J, Younas J (2016) gPhysics—using smart glasses for head-centered, context-aware learning in physics experiments. IEEE Trans Learn Technol 9(4):304–317

Lave J (1988) Cognition in practice: mind, mathematics, and culture in everyday life. Cambridge University Press, Cambridge, UK

Lave J, Wenger E (1990) Situated learning: legitimate peripheral participation. Cambridge University Press, Cambridge, UK

Leung M, Low R, Sweller J (1997) Learning from equations or words. Instr Sci 25(1):37–70

Liu R, Wang L, Lei J, Wang Q, Ren Y (2020) Effects of an immersive virtual reality-based classroom on students' learning performance in science lessons. Br J Edu Technol 51(6):2034–2049

Lopez V, Pinto R (2017) Identifying secondary-school students' difficulties when reading visual representations displayed in physics simulations. Int J Sci Educ 39(10):1353–1380

Mayer RE (2001) Multimedia learning. Cambridge University Press, New York

Mayer RE (2002) Multimedia learning. Psychol Learn Motiv 41:85–139

Mayer RE (2009) Multimedia learning. Cambridge University Press, New York, NY

Mayer RE (2014a) Introduction to multimedia learning. In: Mayer RE (ed) The Cambridge handbook of multimedia learning. Cambridge University Press, Cambridge, UK, pp 1–24

Mayer RE (2014b) Cognitive theory of multimedia learning. In: Mayer RE (ed) The Cambridge handbook of multimedia learning. Cambridge University Press, Cambridge, UK, pp 43–71

Miller GA (1956) The magic number seven, plus or minus two: some limits on our capacity for processing information. Psychol Rev 63:81–97

Müller A, Hirth M, Kuhn J (2016) Tunnel pressure waves—a smartphone inquiry on rail travel. Phys Teacher 54(2):118–119

Paas F, Sweller J (2014) Implications of cognitive load theory for multimedia learning. In: Mayer RE (ed) The Cambridge handbook of multimedia learning. Cambridge University Press, Cambridge, UK, pp 27–42

Paas F, Renkl A, Sweller J (2016) Cognitive load theory: a special issue of educational psychologist. Routledge, London

Paivio A (1986) Mental representations: a dual coding approach. Oxford University Press, New York

PhET Interactive Simulations: https://phet.colorado.edu/en (09.06.2020)

Phyphox (2018) https://phyphox.org (09.06.2020)

Podolefsky NS, Moore EB, Perkins KK (2013) Implicit scaffolding in interactive simulations: design strategies to support multiple educational goals. http://arxiv.org/pdf/1306.6544

Puentedura R (2013) SAMR and TPCK: introduction to advanced practice. http://hippasus.com/res ources/sweden2010/SAMR_TPCK_IntroToAdvancedPractice.pdf

Rath G, Schittelkopf E (2011) Mobile@classroom Handyclips im Physikunterricht (English: Mobile@classroom handyclips in physics lessons). Praxis der Naturwissenschaften - Physik in der Schule 60(7):12–14

Reif F (1981) Teaching problem solving—a scientific approach. Phys Teacher 19:310–316

Salomon G (1979) Interaction of media, cognition and learning. Jossey-Bass, San Francisco

Sandberg J, Wielinga B (1992) Situated cognition: a paradigm shift? J Artif Intell Educ 3:129–138

Santos MEC, Taketomi T, Sandor C, Polvi J, Yamamoto G, Kato H (2014) A usability scale for handheld augmented reality. In: Proceedings of the 20th ACM symposium on virtual reality software and technology. ACM, New York, pp 167–176

Savelsbergh ER, de Jong T, Ferson-Hessler MGM (1998) Competence-related differences in problem representations: a study of physics problem solving. In: van Someren MW, Reimann P, Bushuzien HPA, de Jong T (eds) Learning with multiple representations. Pergamon, Amsterdam, pp 263–282

Schaal S, Bogner FX, Girwidz R (2010) Concept mapping assessment of media assisted learning in interdisciplinary science education. Res Sci Educ 40:339–352

Schnotz W (2014) Integrated model of text and picture comprehension. In: Mayer RE (ed) The Cambridge handbook of multimedia learning. Cambridge University Press, Cambridge, UK, pp 72–103

Schnotz W, Bannert M (2003) Construction and interference in learning from multiple representation. Learn Instr 13:141–156

Schnotz W, Böckheler J, Grzondziel H (1999) Individual and co-operative learning with interactive animated pictures. Eur J Psychol Educ 14(2):245–265

Seel NM, Al-Diban S, Blumschein P (2000) Mental models and instructional planning. In: Spector JM, Anderson TM (eds) Integrated and holistic perspectives on learning, instruction and technology. Springer, Dordrecht, pp 129–158

Seufert T (2003) Supporting coherence formation in learning from multiple representations. Learn Instr 13:227–237

Seufert T, Brünken R (2006) Cognitive load and the format of instructional aids for coherence formation. Appl Cogn Psychol 20(3):321–331

Spiro RJ, Jehng J (1990) Cognitive flexibility and hypertext: theory and technology for the non-linear and multidimensional traversal of complex subject matter. In: Nix D, Spiro R (eds) Cognition, education, and multimedia: exploring ideas in high technology. Erlbaum, Hillsdale, NJ, pp 163–205

Spiro RJ, Collins BP, Thota JJ, Feltovich PJ (2003) Cognitive flexibility theory: hypermedia for complex learning, adaptive knowledge application, and experience acceleration. Educ Technol 43(5):5–10

Spiro RJ, Coulson RL, Feltovich PJ, Anderson D (2004) Cognitive flexibility theory: advanced knowledge acquisition in ill-structured domains. In: Ruddell RB (ed) Theoretical models and processes of reading, 5th edn. International Reading Association, Newark, DE, pp 602–616

Strzys MP, Kapp S, Thees M, Klein P, Lukowicz P, Knierim P et al (2018) Physics holo. lab learning experience: using smartglasses for augmented reality labwork to foster the concepts of heat conduction. Eur J Phys 39(3):035703

Sweller J (1994) Cognitive load theory, learning difficulty and instructional design. Learn Instr 4:295–312

Sweller J (2002) Visualisation and instructional design. In: Ploetzner R (ed) Proceedings of the international workshop on dynamic visualizations and learning. Knowledge Media Research Center, Tübingen, pp 1501–1510

Thees M, Kapp S, Strzys MP, Beil F, Lukowicz P, Kuhn J (2020) Effects of augmented reality on learning and cognitive load in university physics laboratory courses. Comput Hum Behav 108:106316

Thoms L-J, Colicchia G, Girwidz R (2017) Phonocardiography with a smartphone. Phys Educ 52(2):23004

Thoms L-J, Colicchia G, Girwidz R (2018) Audiometric test with a smartphone. Phys Teacher 56(7):478–481

Thoms L-J, Collichia G, Girwidz R (2019a) Real-life physics: phonocardiography, electrocardiography, and audiometry with a smartphone. J Phys Conf Ser 1223:012007

Thoms LJ, Colicchia G, Watzka B, Girwidz R (2019b) Electrocardiography with a smartphone. Phys Teacher 57:586

Treagust DF, Duit R, Fischer HE (eds) (2017) Multiple representations in physics education. Springer, New York

van Heuvelen A (1991) Learning to think like a physicist: a review of research-based instructional strategies. Am J Phys 59(19):891–897

Wieman CE, Adams WK, Loeblein P, Perkins KK (2010) Teaching physics using PhET simulations. Phys Teacher 48(4):225–227

Wittrock MC (1974) Learning as a generative process. Educ Psychol 11(71):87–95

Wittrock MC (1989) Generative processes of comprehension. Educ Psychol 24:345–376

Zhang M, Quintana C (2012) Scaffolding strategies for supporting middle school students' online inquiry processes. Comput Educ 58:181–196

Chapter 12
Instructional Explanations in Physics Teaching

Christoph Kulgemeyer and David Geelan

Abstract Explaining physics is at the core of physics teaching. Teachers explaining science matter is surely a standard situation in the science classroom. Such verbal attempts with the goal to engender students' understanding have been described as 'instructional explanations'. When planning an instructional explanation as a part of their science teaching teachers might wonder about different questions:

- What is 'explaining', what are 'instructional explanations'—and is that not a form of outdated, teacher-centred instruction?
- What are the characteristics of effective instructional explanations? Can instructional explanations be effective at all or is it always the better choice to motivate students for self-explanations?
- Is there a 'general rule of thumbs' to decide as a teacher what and when to explain?
- How can I plan an effective instructional explanation and how can I integrate it into my teaching?
- There are so many explaining videos on physics topics—can I rely on them? How can I integrate explaining videos into effective science teaching?

The chapter will address all of these questions by referring to a framework of effective instructional explanations based on empirical studies from science education and instructional psychology. The results of these studies will be described in the form of core ideas of effective instructional explanations and their integration into a learning process. We will discuss circumstances under which a physics teacher might rely on instructional explanations—and we will describe when to better avoid them. Finally, we will apply the core ideas on explaining videos (e.g. from YouTube) and discuss how to use them in science teaching (e.g. flipped classroom, students producing their own explaining videos).

C. Kulgemeyer (✉)
Universität Paderborn, Paderborn, Germany
e-mail: christoph.kulgemeyer@upb.de

D. Geelan
The University of Notre Dame Australia, Fremantle, Australia
e-mail: david.geelan@nd.edu.au

© Springer Nature Switzerland AG 2021
H. E. Fischer and R. Girwidz (eds.), *Physics Education*, Challenges in Physics Education,
https://doi.org/10.1007/978-3-030-87391-2_12

337

12.1 Introduction

It is a common situation in science teaching: A teacher wants to introduce a new concept in physics such as Newton's third law. Of course, several decisions must be made about how to teach the new concept. These decisions might include the question of whether students should mainly work on their own to *discover* the concept, or whether the teacher will present the content structure. The first option would include a lot of *self-explaining* on the part of the students, whereas the second option would incorporate *instructional explanations* given by the teacher. Instructional explanations, in this case, can be understood as a teacher's verbal talk (combined with an appropriate use of tools like visualizations, representations forms, demonstrations, …) made with the intention of enabling students to develop an understanding and make sense of a concept or principle.

To decide which way to go, a teacher might think about a number of different questions, for example:

1. What is *explaining*? What are *instructional explanations*? Isn't that a form of outdated, teacher-centred instruction?
2. What are some of the criticisms of explaining in science teaching?
3. What are the characteristics of effective instructional explanations? Can instructional explanations be effective at all? Or is it always a better approach to motivate students to develop self-explanations?
4. Is there a *general rule of thumb* to decide as a teacher what and when to explain?
5. How can I plan effective instructional explanations and integrate them into my teaching?
6. There are so many explanation videos on physics topics—can I rely on them? How can I integrate explanation videos into effective science teaching?

These questions and more are dealt with in the following sections in this chapter, based on empirical research in science education and instructional psychology (see Fig. 12.1). It is important to highlight the fact that the focus of this chapter is on

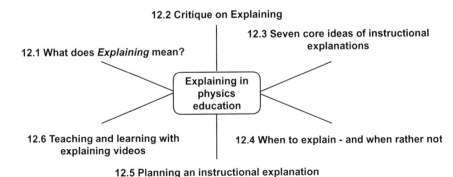

Fig. 12.1 Overview of the chapter

verbal explanations—how textbooks should include written explanations is beyond its scope. There are, however, a number of publications on this topic (e.g. Chambliss 2002; Mamiala 2002; Smolkin et al. 2013; in German: Kulgemeyer and Starauschek 2014).

12.2   What Is Explaining and What Are Instructional Explanations?

To clarify the term *instructional explanation*, we need to define the concept of *explaining*. In this section, we do so by describing differences between and commonalities of three (seemingly) similar concepts:

- Explanation and argumentation
- Scientific explanations and instructional explanations
- The process of explaining and its (temporary) product, the explanation

As mentioned above, an instructional explanation of a concept can be understood as a mainly verbal attempt on the part of a content expert to enable a novice to come to understand and make sense of the concept. Explaining is not limited to verbal language and should by all means be supported by media, representational forms such as diagrams, equations, videos and animations/simulations, experiments and various other features. The intention of the action of *explaining* is the most important part: An instructional explanation has the goal that someone seeks to support someone who attempts to understand a concept. That is the main difference between explanation and argumentation: The latter primarily aims at convincing someone to share or adopt a view or position. On a structural level, argumentation and explanation share many similarities (e.g. Osborne and Patterson 2011) but the goals of each action are fundamentally different.

Instructional explanations are a special kind of explanation and differ from scientific explanations in a number of ways (Geelan 2012). Scientific explanations are the kind given by scientists in formal communications such as in scientific journal articles. They are characterized by specific forms and conventions of communication in science.

For physics, probably the most important form of scientific explanation is the *deductive-nomological* explanation (Hempel and Oppenheim 1948), also known as the *covering-law* model of explanation. Other forms of explanations occur, for example, inductive-statistical explanations, teleological explanations or narrative explanations, but these are of less importance for school physics (with the very important exceptions of quantum and nuclear physics). The core idea of a covering-law scientific explanation is to find a logical connection between a phenomenon and an underlying law. The ideal case would be a law that allows prediction of the phenomenon as soon as all of the relevant initial conditions are known. It is important to highlight that scientific explanations should fulfil two quality criteria: (1) they should be logical and (2) they should explain the occurrence of various phenomena with the combination of a set of specified initial conditions and a single (or, in some cases, more than one) general physics law.

Instructional explanations, however, have a different character. Treagust and Harrison (1999) were among the first to point out the importance of this difference for science teaching. The main quality criterion for an instructional explanation is that it focuses on and effectively addresses the needs of an addressee (*explainee*) and makes it possible that someone makes sense of the information. Logic is certainly helpful for attempts to meet this goal, but it is not a sufficient condition. Just as important are characteristics of the explainee such as prior knowledge, possible misconceptions or areas of interest. There are even studies suggesting that explanations that primarily focus on an explainee's prior knowledge are more likely to be successful than those following a logical content structure (Nathan and Petrosino 2003).

The difference between scientific explanations and instructional explanations can, however, also be found in their structure. In the case of scientific explanations, one could say that the general law explains the occurrence of a phenomenon. In the case of instructional explanations, it is often the case that the general law is, at least in one sense, something mandated by a syllabus or curriculum that students are required to come to understand and to be able to demonstrate their understanding in an assessment task. Explaining, in this case, mostly means showing that this general law has the power to explain different phenomena that appear to be very different at first sight. Thus, in science teaching, explaining often works in the reverse order to scientific explanations: It is not that the law explains the occurrence of the phenomena, but that different phenomena are used to illustrate a law. These phenomena play the role of examples; and the choice of examples has a major influence on the comprehensibility of an explanation.

Finally, the process of explaining needs to be distinguished from its product, the explanation. Intending to *engender comprehension* (Gage 1968), we must make an important but somewhat disillusioning statement: There is no such thing as an ideal explanation for everyone. Explaining is an active, interactive and communicative process of developing candidate explanations, delivering them to explainees, testing for their comprehension and gaining their feedback, and then further modifying the explanation with the goal of enhancing the explainees' comprehension. Explanations play the role of recurring *products* in this process. Diagnosis and feedback lead to a gradually improved mental model in the mind of the explainer with respect to these explainees' needs and what can help them to construct meaning and understanding of the target information. How this process should be designed is described in Sect. 12.5. First, however, we need to understand why there are criticisms of the use of explaining in science teaching and why these criticisms highlight some very important issues in relation to explaining in science classrooms.

12.3   Criticisms of Explaining in Science Teaching

Instructional explanations by teachers have a little bit of a shady reputation. One important reason for this may well be that instructional explanations and scientific explanations are often mixed up by explainees, and instructional explanations are inappropriately judged using the standards appropriate for scientific explanations. It is a common misunderstanding that instructional explanations are simply scientific explanations that are reduced in complexity and somehow *covered with a shiny surface* that makes them understandable to explainees. We want to make it very clear that this is not the case. This misunderstanding is also present in the research literature.

There were studies that used the term *explaining* deprecatingly and equate it with a strong teacher-centred teaching or merely a presentation of scientific facts with no interaction (e.g. Chi et al. 2001). Indeed, explaining in this sense—of the attempted direct transmission of content knowledge from the mind of the explainer to that of the explainee—is not part of good teaching at all. One message should be clear: It is a misunderstanding to think that the secret of good explaining is finding a maximal understandable representation of content matter and that this will lead an explainee to understand an explanation automatically. This so-called *transmissive view* of explaining contradicts fundamental research findings about how learning works as the construction of meaning by learners based on their existing knowledge and their new experiences.

Explaining, however, can be understood based on constructivism (Geelan 1997) as well. The goal of explaining should be seen as the provision of experiences—including words from the teacher—increasing the likelihood that an explainee can construct meaning from an explanation. We can even show empirically that teachers who equate explaining with lecturing explained significantly less efficiently compared to teachers who acknowledged that explaining is a process targeting an explainee's knowledge construction (Kulgemeyer and Riese 2018).

It has been shown many times that learning works if students construct meaning based on their preconditions, most importantly their prior knowledge; this concept is at the core of constructivism itself. That insight may seem to contradict any justification for instructional explanations and support always having students construct their own understanding based on their experiences (in physics, either past experiences they can recall or the special category of experiences we call *experiments*)—but only at first sight. There is no need for students to just passively receive an instructional explanation. Good explaining is rather designed to avoid this. Good explaining aims at activating students and supports them to build mental models of the explained matter.

One might wonder whether there is another reason for the bad image of instructional explanations. Students might be activated cognitively during an explanation but cognitive activation should not be confused with physics activity. Physics activity can nearly always be observed in self-organized learning environments, whereas cognitive activation cannot be directly observed and must be inferred. Studies even showed that physics activity is not supportive for students' achievement but cognitive activity is (e.g. Skuballa et al. 2018). Indeed, the same quality criteria apply to self-organized learning environments and to instructional explanations. Cognitive activation is the most important of such criteria. In a nutshell, we can safely assume that instructional explanations in science teaching are often confused with teacher-centred lecturing even though they fundamentally differ from lecturing in a very important way: If they are designed properly, instructional explanations are interactive, aim at diagnosing misunderstanding and, at the same time, provide structures for the development of complex concepts that most students probably cannot simply discover on their own.

Interestingly, teachers' instructional explanations are not perceived as poor teaching by their students. For example, Wilson and Mant (2011) found that the ability to explain well is the one attribute in which exemplary teachers differ most from average teachers from their students' point of view. However, these exemplary teachers do not mention their explaining skills as a positive feature themselves. The reason, again, might be that they confuse interactive explaining and a teacher-centred presentation of scientific facts.

We also want to highlight the fact that students also explain regularly in science teaching, for example, in cooperative learning or when they present results of their work to the class. It might be important to teach them how to explain. Modelling on the part of teachers, as well as explicit attention to the skills of explaining, is an important part of that teaching; and attention to students' explaining skills may even increase the efficiency of cooperative learning (Kulgemeyer and Schecker 2013).

To sum up, criticism of teacher explanations is justified if explaining is understood as meaning nothing more than teacher-centred lecturing. This kind of poor and one-way explanation should be avoided. Explaining just makes sense in science teaching if the focus is shifted from the content structure of the discipline to the explainee's needs, and if explaining is performed as an interactive process in which learners and teachers both actively participate.

12.4   What Makes Instructional Explanations Successful? Seven Core Ideas of Explaining for Understanding

Much empirical work deals with efficient instructional explanations, mostly within psychology but also in science education research. In a critical meta-analysis of this research, Kulgemeyer (2019) identified seven core ideas for developing and delivering efficient and effective instructional explanations. Which of these core ideas is most relevant for which topic in physics teaching is determined by the type of content to be taught and the characteristics of the explainees (Geelan 2020).

12.4.1 Core Idea 1: Focus on the Explainee and Adapt to Prior Knowledge and Interests

The empirically best justified claim in relation to instructional explanations is probably also the most convincing: In order to be successful, instructional explanations should be oriented to the needs of the learners (Wittwer and Renkl 2008). Needs include interests, misconceptions and other characteristics of thinking. In particular, prior knowledge on the part of the explainee is crucial. That is, somehow, not surprising. Of course, someone who is unfamiliar with the terms and mathematics of quantum physics will have problems following a mathematical explanation of the Franck-Hertz experiment. The challenge, therefore, is to find a common ground for communication, to express things differently and to adapt to the explainee's prior knowledge. That is a tough challenge, indeed!

It may be more surprising that empirical research shows as that it is not a promising strategy to explain as simply as possible and to avoid all mathematics and physics diagrams. That is because an *expertise reversal effect* (Kalyuga 2007) has also been found in research on instructional explanations. That means that some explanations that are comprehensible for novices are unnecessarily complicated for experts and, therefore, less effective. An explanation for this effect comes from cognitive load theory (e.g. Sweller 1988). Including a summary, for example, can be helpful for novices, whereas this is an unnecessary repetition for experts. Unnecessary information is cognitive load that prevents the explainee's cognitive capacities from dealing with the relevant parts of the explanation. This effect is the same for the explanations offered in textbooks: If explanations follow criteria for text comprehensibility they become more comprehensible for novices, and at the same time less comprehensible for learners with higher prior knowledge (McNamara and Kintsch 1996).

That is another reason why there is no such thing as an ideal explanation for everyone. It very much depends on the individual's state of knowledge as to what becomes unhelpful cognitive load and what is of necessary help to increase comprehensibility. It also shows that two parts of a teachers' professional knowledge are of importance when it comes to good explaining: (1) profound knowledge about which knowledge can be expected from a learner (including knowledge of possible misconceptions and the curriculum) and (2) diagnostic skills to refine their assumptions about this knowledge. Good teachers, indeed, can diagnose learners' state of knowledge accurately and vary their explaining approach accordingly (Duffy et al. 1986). Regarding the first point—teacher knowledge about the students—it is important to highlight that studies showed that explainers often think the explainee has a much higher prior knowledge than is actually the case (e.g. Nickerson 1999). Furthermore, explainers with a high level of content knowledge often expect a much too high content knowledge level on the part of their explainees. This leads to an unexpected effect: Sometimes, experts with very high content knowledge explain less effectively than people with lower content knowledge because they do not adapt well to the explainees' needs (Kulgemeyer 2016). Studies also showed that pedagogical content knowledge (PCK)—particular knowledge about misconceptions and instructional strategies—mediates the influence of content knowledge (CK) on explaining quality. In other words, PCK makes CK useful in explaining (e.g. Kulgemeyer and Riese 2018) and can mitigate the issue where a high level of content knowledge on the part of the explainer makes the explanation less effective.

Besides meeting explainees' cognitive prerequisites, adaptations of the explanation to their interests are also important (Kulgemeyer and Schecker 2009). Among other things, this consideration—what explainees will find interesting—heavily influences the choice of examples to illustrate the topic being explained.

The idea of adapting the explaining activity to the explainees' prerequisites, however, is very abstract. How do we reach this goal? The next section describes four means towards adaptation: a handy toolbox that is helpful in successful adaptation.

12.4.2   Core Idea 2: Use Means for Adaptation

Adaptation to the learners' prerequisites and characteristics requires means to reach this goal. Profound knowledge of this toolbox is essential for good explainers. There are, basically, four areas that allow adaptation (Kulgemeyer and Schecker 2009, 2013):

1. Language level
2. Examples and analogies
3. Level of mathematization
4. Forms of representations and demonstrations.

Explainers can imagine these four means as tools that, when used skilfully, allow them to find the correct adjustment for all of the explainees' prior knowledge, misconceptions and interests. Indeed, an individual adjustment applies only to each individual, and it is even more of a challenge to explain to a whole class and to find a good balance. The core part of the process of explaining is to adjust in relation to each of these four means in order to reach a better adaptation of the explaining process (and the explanation product) to the explainees.

Adapting the language level to explainees' needs might mean replacing technical terms with everyday language but that depends on explainees. It might also mean that it is better to use technical terms for some explainees. The pragmatic rule of thumb for instructional explanations at school is to use a limited set of (new) terms and to critically evaluate which terms are really needed. Yager (1983) found that many secondary school science textbooks introduce more terms in the new vocabulary in a year than foreign language learning textbooks at a comparable level of schooling. For the German language, there is research showing that numerous technical terms in physics textbooks are used only once (Merzyn 1994). Whether or not a technical term is needed should be evaluated from a curricular perspective with consideration of connectivity to the following topics and a physics point of view about correctness, as well as in terms of the technical vocabulary mandated by the syllabus.

There is danger in a strategy that can be observed regularly on the part of inexperienced explainers. If they realize that explainees did not understand, they change the concepts they use for the explanation. This also leads to a whole new set of technical terms. Kulgemeyer and Tomczyszyn (2015), for example, videotaped pre-service science teachers explaining to students why blowing up an asteroid changes the trajectory of all of its pieces. When confronted with student misunderstandings, they reacted by changing the focus concept and talked about forces or energy instead of momentum. In general, this is not supportive and does not aid comprehension. Explainers should stick to the chosen concept except if they realize that the concept itself cannot be understood by the learners, in which case it may be necessary to revise some necessary concepts from those used earlier in the course or earlier in the explainees' education.

Adapting examples and analogies to the explainees' prerequisites is quite similar to adapting the language level to their needs and is probably one of the most important parts of explaining in general. Examples in studies (e.g. Duit and Glynn 1995; Clement 1993) showed how an explained concept applies and supports individuals' understanding of how a general concept solves a range of problems that appear very different at first sight. Analogies (and also metaphors and even models) bridge the gap between the new, unfamiliar topic and an area that has already been understood. Duit and Glyn, as well as Clement, showed that something new behaves (in parts) like something already known. Here, sometimes explainers use anthropomorphisms, which are analogies between physics and human behaviour. For example, an ideal gas is sometimes illustrated by a moving crowd. Anthropomorphisms are sometimes judged critically because free will is limited to humans whereas physics applies to inanimate matter. They can, however, also support understanding because human behaviour is familiar to all learners. The same goes for any kind of models: The limitations and the point where the analogy breaks down—with the features of the familiar concept no longer map to those of the concept being explained—have to be made clear.

The level of mathematization also needs to be adapted. No doubt, mathematical expressions of laws require verbal explanations as well. However, physics theories require efficient mathematical expressions as well: To avoid mathematics, in general, would be a mistake and shows a dubious picture of physics from a nature of science point of view. Geelan (2013), in a study observing experienced teachers, described the *move to mathematics*, in which teachers typically began with qualitative explanations or demonstrations and then moved to equations and calculations during a lesson. Drawing on an appropriate level of mathematics for student knowledge is also important. In some contexts in secondary school physics, some students within the class will have studied calculus, whereas others have not; and therefore, explaining needs to be adapted to the knowledge and skills of all students.

Forms of representations and demonstrations are helpful tools for adaptation if they support verbal information in the sense of dual coding. Forms of representations can include graphs of any kind, diagrams, photos, animations and simulations or even videos. Demonstrations have an important place in instructional explanations if they show an explained example as a model.

12.4.3 Core Idea 3: Highlight Relevancy and Use Prompts

Empirical research shows that the perceived relevance of an instructional explanation influences student achievement. This is especially the case if an instructional explanation is performed by the teacher in response to a misunderstanding on the part of a student or after a topic has been dealt with in class. Perhaps surprisingly, in this case, students tend to perceive further explanation as redundant and irrelevant (Acuña et al. 2011). The following simple approach is a promising candidate to solve this problem. In a study by Sánchez et al. (2009), it was sufficient to highlight at the beginning of the explanation of a topic that the explanation deals with a common misunderstanding and the teacher has chosen to present it because many people hold misconceptions about the topic. That reminds both teachers and students that teaching strategies can and should be chosen to deal with misconceptions and promote conceptual change.

The most important strategy of a successful explanation—(a) to show that an explanation is relevant to the students and (b) to signal which part of an explanation is particularly important—is called *prompting*. Prompts are explicit signals (Diakidoy et al. 2003) such as "this point is especially important because many people understand this wrongly" or "many people think about energy as something like a material—but is that appropriate?" Roelle et al. (2014) showed that these kinds of prompts affect students' achievement. The reason is probably that prompts affect the explainees' perceived relevance of the explanation, which in turn leads to their cognitive activation instead of passive listening.

12.4.4 Core Idea 4: Give It a Structure

The structure of an explanation has been discussed especially in the context of scientific explanations but structure is important for instructional explanations as well. Firstly, a clear structure helps to make relationships explicit for both explainers and explainees. It is very helpful to have a clear picture of an explained topic in order to build a coherent mental model. There is even research that examines the effects of different structures (e.g. Seidel et al. 2013). For example, two structures are compared directly: starting an explanation with a presentation of the general law and later illustrating it with examples (rule–example structure) or, vice versa, starting with examples and later deducing the general law from them (example–rule structure). The efficiency of these structures depends on the goal of the explanation. If the goal is the acquisition of content knowledge, an explanation with the rule–example structure seems to be superior (Seidel et al. 2013). However, if the acquisition of practical skills is the goal, an explanation with an example–rule structure outperforms one using a deductive approach (Seidel et al. 2013). What does this mean for teaching physics? If the learning goal is Newton's third law, it is appropriate to start with an explanation explicitly and later show different examples that illustrate how the law works. If the learning goal is to learn how to solve specific problems (e.g. by using a force or an energy approach), it is appropriate to start with a worked example and later present the underlying principle that justifies the approach. It is also important in this case, however, that the general principle is mentioned explicitly! Just explaining the steps required for the solution of the problem is not sufficient because it would only help students to understand the solution of this particular task but not help them to learn how every problem of this kind can be solved–they do not learn flexible conceptual knowledge and my even confuse the example (or analogy) with the concept.

That, by the way, does not mean that an explanation cannot start with an example if it aims for content knowledge acquisition. An example at the beginning might even help to increase the interest of students. However, the general law should not be deduced from this example; the law should still be given explicitly and later illustrated. The teacher may say, "What we just saw can be explained with Newton's third law—and this is what the law looks like".

12.4.5 Core Idea 5: Explain Precisely and Coherently

An explanation should be coherent and minimal, meaning that it should stick to what is important and leave out irrelevant details (Anderson et al. 1995). Coherence and minimalism are related to one another. Coherence as a characteristic of good explaining has been researched, especially in the context of textbooks (e.g. Wittwer and Ihme 2014). The underlying thought is that it helps students to build a mental model of an explained topic if the connection between elements of the topic is clear. That includes good connection of sentences, for example, by using the same terms in two sentences instead of using pronouns or synonyms. It already requires a certain knowledge of a topic to understand that two different words signify the same term and have the same meaning. For novices, two different words would appear as if they had two different meanings and it requires their cognitive capacities to clarify that: cognitive capacities that would have been useful in order to understand the subject matter! Also, connections between sentences can be highlighted by using connectives such as *because*. That helps explainees to understand what the phenomenon of interest is and to what general law it is connected.

Avoiding irrelevant details in an explanation is important because it helps explainees to focus their cognitive capacities on its relevant parts. Digressions should also be avoided (Renkl et al. 2006). Novices simply cannot know which part of an explanation is relevant and which part is extra information. In a nutshell, an explanation should be minimal and show the explained topic clearly. Later, new information may be added, especially if students ask for it.

Some researchers suggested that a good explanation should be built like a good story (e.g. Ogborn et al. 1996), in that it has a beginning that sets up expectations, a middle part that complicates them and an ending that resolves them. That really can have an advantage because a good story leads the audience to sympathize with the protagonist, increasing the perceived relevance of the explanation. It might, however, seduce the explainer into giving too much irrelevant information and that would result in reverse effects. Explainers should be careful to focus on the most salient details of their explanations.

12.4.6 Core Idea 6: Explain Concepts and Principles

Renkl et al. (2006) suggested that an explanation should address only concepts and principles. There is certainly something worthwhile in that suggestion. Understanding a principle means realizing that superficially very different problems can be reduced to the same idea and explained by the same approach. Explaining physics as a science is very efficient in doing so. There is also empirical research that supports this view. Dutke and Reimers (2000) showed that it is more effective to explain a principle than to show the solution of a problem step by step.

Knowing principles is an important goal of physics education. Therefore, if teachers want to use instructional explanations, they should do so, especially if their learning goal is for content knowledge acquisition rather than the development of skills. It is obvious, anyway, that, for example, experimental skills cannot be acquired by instructional explanations but must be developed in laboratory or other experiences.

12.4.7 Core Idea 7: An Explanation Should Be Embedded in Teaching

The last core idea refers to the notion that explaining is a process and not a product. An instructional explanation (the product of an explaining act) should be embedded appropriately in teaching. Explanations are not teaching in themselves, but they are just a part of teaching. Teaching includes diagnosis of understanding as an important part of the development of the explanation and as a prerequisite for further adaptation. Instructional explanations also should not replace cognitive activities. Learning is something only the learner can do; therefore, it cannot be replaced by an instructional explanation even if the explanation has an extraordinary quality. There is research on the question of the circumstances under which self-explanations (i.e. students engaging directly with physics phenomena or their representations and seeking to construct understanding for themselves) are more efficient than instructional explanations for constructing understanding. In principle, it is easier to reach high cognitive activation with self-explanations. That is why some researchers regard instructional explanations merely as a supportive means towards the process of self-explaining. However, there is an important problem with self-explanations. Students tend to interrupt self-explaining activities when they think they have understood a topic.

However, that is not always the case: They sometimes just think they have understood the topic, even though they have not (for the so-called *illusion of understanding*, see for example Chi et al. 1989). The well-known Dunning–Kruger effect is the fact that novices are not well positioned to make judgements of their own understanding (Kruger and Dunning 1999). After instructional explanations given by a teacher, it might also be the case that students' understanding of a topic is more superficial than they think, but the teacher has a certain level of control over this situation. He or she can diagnose their understanding and react to it accordingly. In particular, topics that include common misconceptions might be vulnerable to an illusion of understanding. A lot of common misconceptions about scientific concepts are very effective in predicting everyday phenomena and self-explanations might even strengthen these misconceptions rather than leading to conceptual change.

The clear advantage of self-explanations is their potentially high cognitive activation. However, instructional explanations can also reach high cognitive activation. To do so, they should be embedded in ongoing cognitive activities and not replace them (Wittwer and Renkl 2008). But how? A good instructional explanation initially leads to learners being able to memorize and comprehend the content of the explanation itself (declarative knowledge), for example, Ohm's law. Being able to use this knowledge to solve novel problems—that have not been part of the content or context of the explanation—requires of learners to use a different kind of knowledge or more flexible conceptual knowledge. Instructional explanations are not very useful tools in terms of providing this kind of knowledge (Kulgemeyer 2018). That is why instructional explanations should always be followed by learning tasks that require learners' autonomous application of the explained concept to solve new problems (Altmann and Nückles 2017). Studies showed that the quality of these learning tasks is at least as important for achievement as the instructional explanation itself (e.g. Webb et al. 2006). The criteria for well-designed learning tasks that foster cognitive activation can be found in Chap. 9 of this book.

Cabello and Topping (2018) proposed another promising approach to embed instructional explanations successfully into teaching. They propose looking for synergies between learners' and teachers' ideas by contrasting the students' ideas with the explanation. This indeed might lead to students' questioning a teacher's explanation and an improved co-construction of knowledge.

12.5 When Should I as a Teacher Explain and When Should I Avoid It?

As mentioned above, explaining has often been seen as an ineffective and outdated form of teaching. From the core ideas and empirical research outlined in this chapter, it follows that this is not always true. There are conditions under which instructional explanations are promising tools for learning and also conditions under which a teacher should rely on different forms of teaching.

Two *starting conditions* should be fulfilled if a teacher wants to use instructional explanations effectively:

1. Instructional explanations should focus on laws or principles. That is a starting condition from a physics point of view. Even if the occurrence of phenomena is being explained or if the goal is to show how a certain type of problem can be solved, the underlying principle should be stated and connected to the explanation.
2. There are prerequisites of the learners that determine whether instructional explanations offer a promising approach to teaching a particular concept. These are starting conditions from a pedagogical content knowledge point of view. It seems to be the case that, given a high level of content knowledge on the part of the students, self-explanations are the better choice. Instructional explanations have their place at the beginning of a teaching sequence when principles are first being introduced. This is especially the case if there are many common misconceptions in relation to the explained principle. Considering this starting condition, we suggest a side note here: It seems odd that there is a tendency in the educational system in which the higher the content knowledge level that the learners already have (e.g. university majors compared to high-school students), the more the teacher shifts towards teaching forms that contain instructional explanations.

An example that fulfils both starting conditions is already given in the introduction of this chapter: the introduction of Newton's third law. It is a principle that can be illustrated by many examples from everyday experience. There are many known misconceptions on this law (e.g. it is often confused with an equilibrium of forces) which makes it difficult to develop this principle through self-explanations. When Newton's third law is being introduced in physics, instructional explanations are a promising tool.

Skills in solving problems by calculating the amounts of different forms of energy present before and after a process is an example of something that should probably not be the subject of an instructional explanation. When it comes to solving a certain kind of problem or task, learning with worked examples is the better choice. If a verbal explanation was offered in this context, it would not aim at supporting the development and comprehension of a concept (except for the goal being an explanation of the conservation of energy), but rather at supporting skills for calculation, corrections of common calculation errors and perhaps ways to conduct calculations more efficiently in an examination. It is also likely that such an explanation would be performed after conservation of energy as a principle has been introduced, when a certain level of prior knowledge could then be expected. Both starting conditions for instructional explanations are not fulfilled in that case.

12.6 A Guide to Planning Instructional Explanations

The seven core ideas and the two starting conditions can now be summarized to form something like a guide to planning and conducting instructional explanations. The guide contains guidelines based on the current state of research; however, future research might lead to further refinement. Figure 12.2 shows this guide as a representation of the process of explaining in science teaching and all its steps.

Of course, the process starts with consideration of the starting conditions. Instructional explanations have their place in a teaching sequence if a new principle is to be introduced that is too complex for self-explanations and likely to contain misconceptions. The learners should also have low prior knowledge of this content.

After these considerations, the explaining starts. Planning an explanation is based on a mental model about the learners' prerequisites and the content structure. An explainer should have sound ideas about the learners he or she wants to address, especially those about their state of content knowledge and possible misconceptions. An explainer should also know the content structure well enough to identify the important parts that the explanation should contain. This mental model serves as an initial orientation for the explanation but the explanation is further refined iteratively in the process of explaining, seeking feedback from or testing for comprehension of the explainees and then, this is followed by further explaining. For example, the part of the explainer's mental model that is about the explainees' state of content knowledge is further refined after the explainer realizes that certain points of the explanation have

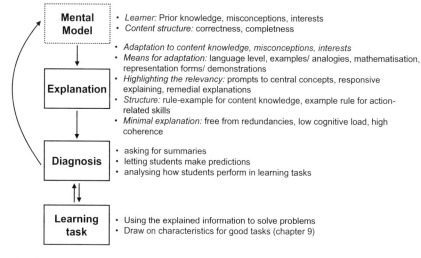

Starting conditions
- A new principle to be introduced is too complex for self-explanations;
- The students are likely to have low prior knowledge.

Mental Model
- *Learner:* Prior knowledge, misconceptions, interests
- *Content structure:* correctness, completeness

Explanation
- *Adaptation to content knowledge, misconceptions, interests*
- *Means for adaptation:* language level, examples/ analogies, mathematisation, representation forms/ demonstrations
- *Highlighting the relevancy:* prompts to central concepts, responsive explaining, remedial explanations
- *Structure:* rule-example for content knowledge, example rule for action-related skills
- *Minimal explanation:* free from redundancies, low cognitive load, high coherence

Diagnosis
- asking for summaries
- letting students make predictions
- analysing how students perform in learning tasks

Learning task
- Using the explained information to solve problems
- Draw on characteristics for good tasks (chapter 9)

Fig. 12.2 Explaining as a process in physics teaching

not been understood. It is worthwhile to acknowledge that it is highly challenging for a teacher to build such an accurate mental model. Many resources influence this mental model, and it is particularly important for it to hold both profound content knowledge and pedagogical content knowledge (Kulgemeyer and Riese 2018).

In a narrow sense, what follows after checking that the topic meets the initial conditions and developing a mental model—about both the explainees' prerequisites and the content—is a first attempt to explain the content in the explanation. The seven core ideas are very important for developing and refining an explanation. Adaptation to the learner's prerequisites can be achieved by using the four adaptation levels of (1) language, (2) examples/analogies, (3) level of mathematization, and (4) forms of representations/demonstrations. Also, the most important parts of the content should have been identified and prompts should be integrated such that they can be used to make clearer to the students what these important points are. Students, for example, should be aware of why the explained concept is relevant either to physics itself or to everyday experiences (or to passing a physics exam, but arguably this is a less motivating kind of relevancy); they should also know what kind of problems can be explained using this concept. It is important to involve students actively in the explanation, for example, by addressing them directly with *you* or by using questions they need to think about. These kinds of prompts and questions require sound preparation; it is very challenging for teachers to integrate them spontaneously.

Also, the structure of the explanation requires preparation. High coherence is important. The focus should be to avoid the use of synonyms and to connect the law and the examples with *because* or other causal connectors. For the acquisition of content knowledge, a modified rule–example structure is promising because it might help to increase the explainees' interest:

1. Using an example that shows why the principle is relevant and prompting that this example can be understood through understanding the following principle.
2. Naming the principle explicitly and explaining it.
3. Testing the principle by applying it to various examples and by explaining the first example.

The last important core idea includes using little or no digression and repetition, even in the form of a summary. It is recommended for teachers to bring along their notes of the planned explanation into the classroom. Even very experienced explainers are not as effective in ad hoc explanations as they could be in well-planned ones (Kulgemeyer and Schecker 2013).

After the initial explanation, there necessarily follows a diagnosis of understanding. Some tools for diagnosis could be to ask the students for summaries or to predict a phenomenon based on the explained principle. This diagnosis helps to refine teachers' mental model about the learners' needs and leads to a better adapted explanation. Teachers check formally and informally for both engagement (i.e. cognitive activation) and comprehension on the part of their students. Before moving onwards, it is important to ensure that all students have an understanding of the topic.

Ideally, the diagnosis is strongly connected to the following phase, the learning task. This should require students of using the explained principle in a self-guided way to solve new problems or to explain a new example. The criteria for learning tasks should be considered (see Chap. 9 of this book). How the students perform in this learning task can also serve a diagnostic purpose, leading to a better refined mental model and, finally, to a better adapted explanation.

According to Kulgemeyer and Schecker (2013), the following ten rules make the explained core ideas more approachable, when the explainer:

1. prepares the explanation (that means that during teaching, teachers should sometimes just say that they will answer this question in the following lesson),
2. illustrates the verbal language with visual forms of representations,
3. involves the students in the explanation by addressing them directly (which sometimes turns an instructional explanation into a dialogue),
4. asks frequently whether or not the students can follow the explanation,
5. answers questions briefly and precisely,
6. uses examples or analogies to connect the new topic with the content already familiar to the students,
7. considers the students' prior knowledge, misconceptions and interests,
8. highlights the relevant parts,
9. gives enough opportunities for students to ask questions and
10. follows a sound structure (e.g. rule–example structure).

Also helpful in teacher education is the use of the rubric of explanations for formative assessments developed by Cabello and Topping (2018). It allows direct feedback that might be very supportive for pre-service teachers.

12.7 Explanation Videos in Physics Teaching

An increasingly important special case of instructional explanations are explanation videos. They are accessible online on different platforms—most prominently YouTube—and there are numerous explaining videos for all common topics of school physics. Students watch these kinds of videos for several reasons, including entertainment and preparation for exams, but explaining videos are also increasingly part of formal learning (Wolf and Kratzer 2015). For example, in flipped classrooms, teachers give students access to videos (or even produce them for their students), assigning to students the task of watching them at home and answering related questions. Later, in the classroom, the students work on learning tasks concerning the explained topic, supported by the teacher. Producing explaining videos together with the students can also be a powerful tool for learning, especially at the end of a teaching sequence. Furthermore, producing an explanation video is an interesting tool not *just* because it requires students to understand the physics content and edit the video accordingly but also because it requires teachers' considerations about the explainees—an important part of science communication competence (Kulge-meyer and Schecker 2013). Having the students think about the audience to whom their explanation is directed—their peers, younger students, parents or members of society more broadly—involves them in a metacognitive process of modelling the thinking of others (see Table 12.1).

It is, however, an interesting question whether or not the quality criteria for good explaining can simply be applied to explaining videos. Of course, the medium is important and, for example, the cognitive theory of multimedia learning (Mayer 2001) surely adds to them valuable ideas. The most important difference between explanation videos and instructional explanations is certainly the problem of adaptation and audience. Whereas good instructional explanations are interactive and enable the explainers to revise their mental model about the learners and change their explanation attempts accordingly, explanation videos (not unlike explanations written in textbooks) are a static product. That makes it even more important to know the audience very well before producing a video, as well as when teachers produce videos together with their students. It is an important step to take enough time to write a profile of the audience and a script.

Kulgemeyer (2018) adapted the core ideas for instructional explanations to the specific context of explanation videos and developed a framework for potentially effective explanation videos. This framework has successfully been used to develop and analyse explanation videos (see Table 12.2).

Kulgemeyer and Peters (2016) used a similar framework to analyse explanation videos on YouTube and found that none of the quality criteria available on YouTube (e.g. Likes, Clicks, Dislikes) predicted explaining quality appropriately. It may not be very surprising, but these superficial measures are not helpful for teachers if they want to find a good video for their students. Furthermore, research suggests that the developers of YouTube videos do not aim for high explaining quality because that is not what makes a video successful (in terms of clicks)—the popularity of a

Table 12.1 Potential use of explanation videos in science classroom (Kulgemeyer 2018; Wolf and Kulgemeyer 2016)

		Producer of explaining videos	
		Teachers	Students
Explainee	Teachers	Teachers learn from other explaining experts how to explain a particular topic (e.g. good examples)	Teachers use students' explanation videos as an alternative form of assessment
	Students	Students learn from expert explainers, giving them access to an additional explaining approach compared to their own teacher's Teachers develop explanation videos for their own students and use them in flipped classroom settings	Students produce explanation videos; teachers assist them by giving feedback on (a) scientific correctness and (b) appropriateness of videos for a particular group of interest; and learning opportunity for (a) content knowledge and (b) communication skills

Table 12.2 Framework for effective explanation videos (Kulgemeyer 2018)

Factors	Feature	Description
Structure	Rule–example, Example–rule	If the learning goal is factual knowledge, the video follows the rule–example structure If the learning goal is a routine or procedural knowledge, the video follows the example–rule structure
	Summarizing	The video summarizes the explanation
Adaptation	Adaptation to prior knowledge, misconceptions, and interest	The video adapts the explanation to a well-described group of addresses and their potential knowledge, misconceptions, or interests. To do so, it uses the "tools for adaptation"
Tools for adaptation	Examples	The video uses examples to illustrate a principle
	Analogies and models	The video uses analogies and models that connect the new information with a familiar area
	Representation forms and demonstrations	The video uses representation forms or demonstrations
	Level of language	The video uses a familiar level of language
	Level of mathematization	The video uses a familiar level of mathematization
Minimal explanation	Avoiding digressions	The video focuses on the core idea, avoids digressions and keeps the cognitive load low. In particular, it avoids using too many "tools for adaptation" or summaries
	High coherence	The video connects sentences with connectors, especially *because*
Highlighting relevancy	Highlighting relevancy	The video highlights explicitly why the explained topic is relevant to the explainee
	Direct addressing	The explainee is getting addressed directly, e.g. by using the second-person singular instead of the passive voice
Follow-up learning tasks	Follow-up learning tasks	The video describes learning tasks the explainees can engage with to actively use the new information after the video

(continued)

Table 12.2 (continued)

Factors	Feature	Description
New, complex principles	New, complex principle	The video focuses on a new scientific principle that is too complex to understand by self-explaining, e.g. because there are frequent misconceptions

channel among students depends on very different factors. Even more, developers of explanation videos quite often probably have misconceptions about teaching and learning, or about the content, themselves and are not aware of the quality criteria for explaining.

One result, however, is encouraging: Teachers do not need to watch all of the videos on a topic. A first criterion to sort out videos could be a short look at the comment section. Kulgemeyer and Peters (2016) found that videos in which there is a content-related discussion in the comments have a higher probability of being good in terms of explaining quality. The reason might simply be that users need to have understood something about the content to discuss it. In any case, the number of such comments correlates with explaining quality. Comments that just briefly praise the video are not a good indicator. These kinds of comments tend to depend on the popularity of the channel and not on the quality of a single video.

Explanation videos and instructional explanations given by teachers are interesting to compare. The former often appear as a modern part of science teaching whereas the latter, as mentioned above, have the image of outdated instruction. However, regarding adaptation, instructional explanations even have advantages, whereas explanation videos surely are more powerful for learning with animations and multimedia. The intention of both is also the same: An explanation video is basically a filmed instructional explanation. Their place in teaching, therefore, differs just slightly. This chapter is based on empirical studies on their place in teaching. It certainly is worth it to revive teacher explanations in the sense of interactive, constructivist and communicational attempts. Placing them at an adequate place in a teaching sequence and performing them with high quality is very challenging for teachers and should be part of teacher education. Just as important are the skills to develop explanation videos for and together with students. That might even lead to an updated *culture of explaining* in science classroom where science teachers are aware of how instructional explanations and explanation videos work, what they can accomplish and where their limitations are. If researchers acknowledge that explaining is a part of teaching anyway (and sometimes even a powerful tool), they should acknowledge as well that teaching how to explain is crucial. Teachers are very welcome to train their own explaining skills and to use the seven core ideas given in this chapter to critically reflect on their attempts!

12.8   Additional Literature

Geelan, D. (2012). Teacher Explanations. In B. Fraser, K. Tobin & C. McRobbie (Eds.), Second International Handbook of Science Education (pp. 987–999). Dordrecht: Springer.

> This paper gives an excellent overview of research in science education on instructional explanations of science teachers. It completes the paper of Wittwer and Renkl (2008) by adding an educational perspective.

Kulgemeyer, C. (2019). Towards a framework for effective instructional explanations in science teaching, Studies in Science Education, https://doi.org/10.1080/030 57267.2018.1598054

> A critical meta-analysis of the research on instructional explanations, including work from instructional psychology and science education research. The seven core ideas are developed in this paper and it includes the most recent studies.

Wittwer, J. & Renkl, A. (2008). Why Instructional Explanations Often Do Not Work: A Framework for Understanding the Effectiveness of Instructional Explanations. Educational Psychologist 43(1), 49–64.

> A review paper on research on instructional explanations, drawing on studies mostly from psychology. To this date the probably most influential paper in this field of research. The depth of argumentation still is unmatched and it is a highly recommended paper for anyone who wants to learn more about instructional explanations. Science education and science teaching, however, play no role.

Acknowledgements We would like to thank Valeria Cabello (Pontificia Universidad Católica de Chile) and Lilia Halim (National University of Malaysia), who not only carefully and critically reviewed this chapter but also contributed valuable thoughts that helped to improve the quality of this chapter significantly.

References

Acuña SR, García Rodicio H, Sánchez E (2011) Fostering active processing of instructional explanations of learners with high and low prior knowledge. Eur J Psychol Educ 26(4):435–452. https://doi.org/10.1007/s10212-010-0049-y

Altmann A, Nückles M (2017) Empirische Studie zu Qualitätsindikatoren für den diagnostischen Prozess [Empirical studies on quality criteria for the diagnostic process]. In: Südkamp A, Praetorius A-K (eds) Diagnostische Kompetenz von Lehrkräften: Theoretische und methodische Weiterentwicklungen [Diagnostic competence of teachers: theoretical and methodological developments]. Waxmann, Münster, pp 134–141

Anderson JR, Corbett AT, Koedinger KR, Pelletier R (1995) Cognitive tutors: lessons learned. J Learn Sci 4:67–207

Cabello VM, Topping K (2018) Making scientific concepts explicit through explanations: simulations of a high-leverage practice in teacher education. Int J Cogn Res Sci Eng Educ 6(3):35–47

Chambliss MJ (2002) The characteristics of well-designed science textbooks. In: Otero JC, León J, Graesser AC (eds) The psychology of science text comprehension. Lawrence Erlbaum Associates, Mahwah, NJ, pp 51–72

Chi MTH, Bassok M, Lewis MW, Reimann P, Glaser R (1989) Self-explanations: how students study and use examples in learning to solve problems. Cogn Sci 13:145–182

Chi MTH, Siler SA, Jeong H, Yamauchi T, Hausmann RG (2001) Learning from human tutoring. Cogn Sci 25:471–533

Clement J (1993) Using bridging analogies and anchoring intuitions to deal with students' preconceptions in physics. J Res Sci Teach 30(10):1241–1257

Diakidoy IN, Kendeou P, Ioannides C (2003) Reading about energy: the effects of text structure in science learning and conceptual change. Contemp Educ Psychol 28:335–356

Duffy G, Roehler L, Meloth M, Vavrus L (1986) Conceptualizing instructional explanation. Teach Teach Educ 2:197–214

Duit R, Glynn S (1995) Analogien – Brücken zum Verständnis [Analogies—bridges to understanding]. Naturwissenschaften im Unterricht. Physik 6(27):4–10

Dutke S, Reimer T (2000) Evaluation of two types of online help for application software. J Comput Assist Learn 16:307–315

Gage NL (1968) The microcriterion of effectiveness in explaining. In: Gage NL (ed) Explorations of the teacher's effectiveness in explaining. Stanford Center for Research and Development in Teaching, pp 1–8

Geelan DR (1997) Epistemological anarchy and the many forms of constructivism. Sci Educ 6(1–2):15–28

Geelan D (2012) Teacher explanations. In: Fraser B, Tobin K, McRobbie C (eds) Second international handbook of science education. Springer, Dordrecht, pp 987–999

Geelan D (2013) Teacher explanation of physics concepts: a video study. Res Sci Educ 43:1751–1762. https://doi.org/10.1007/s11165-012-9336-8

Geelan D (2020) Physical science teacher skills in a conceptual explanation. Educ Sci 10(1). http://doi.org/10.3390/educsci10010023

Hempel C, Oppenheim P (1948) Studies in the logic of explanation. Philos Sci 15(2):135–175

Kalyuga S (2007) Expertise reversal effect and its implications for learner-tailored instruction. Educ Psychol Rev 19:509–539

Kruger JM, Dunning D (1999) Unskilled and unaware of it: how difficulties in recognizing one's own incompetence lead to inflated self-assessments. J Pers Soc Psychol 77:1121–1134

Kulgemeyer C (2016) Impact of secondary students' content knowledge on their communication skills in science. Int J Sci Math Educ 16(1):89–108

Kulgemeyer C (2018) A framework of effective science explanation videos informed by criteria for instructional explanations. Res Sci Educ (Advance online publication). http://doi.org/10.1007/s11165-018-9787-7

Kulgemeyer C (2019) Towards a framework for effective instructional explanations in science teaching. Stud Sci Educ (Advance online publication). http://doi.org/10.1080/03057267.2018.1598054

Kulgemeyer C, Peters C (2016) Exploring the explaining quality of physics online explanatory videos. Eur J Phys 37(6):1–14

Kulgemeyer C, Riese J (2018) From professional knowledge to professional performance: the impact of CK and PCK on teaching quality in explaining situations. J Res Sci Teach 55(10):1393–1418. https://doi.org/10.1002/tea.214571418

Kulgemeyer C, Schecker H (2009) Kommunikationskompetenz in der Physik: Zur Entwicklung eines domänenspezifischen Kompetenzbegriffs [Science communication competence in physics: On the development of a domain-specific concept of competence]. Zeitschrift für Didaktik der Naturwissenschaften 15:131–153

Kulgemeyer C, Schecker H (2013) Students explaining science: assessment of science communication competence. Res Sci Educ 43:2235–2256

Kulgemeyer C, Starauschek E (2014) Analyse der Verständlichkeit naturwis-senschaftlicher Fach-texte [Analysing the comprehensibility of science texts]. In: Krüger D, Parchmann I, Schecker H (eds) Methoden in der naturwissenschaftsdidaktischen forschung. Springer, Heidelberg, pp 241–253

Kulgemeyer C, Tomczyszyn E (2015) Physik erklären - Messung der Erklärensfähigkeit angehender Physiklehrkräfte in einer simulierten Unterrichts-situation [Explaining physics—measuring physics teachers' explaining skills in a simulated teaching situation]. Zeitschrift für Didaktik der Natur-wissenschaften 21(1):111–126

Mamiala LT (2002) Teachers' and textbooks' use of explanations in school chemistry and students' perceptions of these explanations. Ph.D. Thesis, Curtin University of Technology, Perth, Australia

Mayer RE (2001) Multimedia learning. Cambridge University Press, New York, NY

McNamara DS, Kintsch W (1996) Learning from texts: effects of prior knowledge and text coherence. Discourse Process 22(3):247–288

Merzyn G (1994) Physikschulbücher, Physiklehrer und Physikunterricht [Physics textbooks, physics teachers, and physics teaching]. IPN, Kiel

Nathan M, Petrosino A (2003) Expert blind spot among preservice teachers. Am Educ Res J 40(4):905–928

Nickerson R (1999) How we know—and sometimes misjudge—what others know: imputing one's own knowledge to others. Psychol Bull 125(6):737–759

Ogborn J, Kress G, Martins I, McGillicuddy K (1996) Explaining science in the classroom. Open University Press, Buckingham

Osborne JF, Patterson A (2011) Scientific argument and explanation: a necessary distinction? Sci Educ 95(4):627–638

Renkl A, Wittwer J, Große C, Hauser S, Hilbert T, Nückles M, Schworm S (2006) Instruk-tionale Erklärungen beim Erwerb kognitiver Fertigkeiten: sechs Thesen zu einer oft vergeblichen Bemühung [Instructional explanations and the achievement of cognitive abilities: six hypotheses on a failing attempt]. In: Hosenfeld I (ed) Schulische Leistung. Grundlagen, Bedingungen, Perspektiven (205–223). Waxmann, Münster, Germany

Roelle J, Berthold K, Renkl A (2014) Two instructional aids to optimise processing and learning from instructional explanations. Instr Sci 42:207–228

Sánchez E, García Rodicio H, Acuña SR (2009) Are instructional explanations more effective in the context of an impasse? Instr Sci 37:537–563

Seidel T, Blomberg G, Renkl A (2013) Instructional strategies for using video in teacher education. Teach Teach Educ 34:56–65

Skuballa I, Dammert A, Renkl A (2018) Two kinds of meaningful multi-media learning: is cognitive activity alone as good as combined behavioral and cognitive activity? Learn Instr 54:35–46

Smolkin LB, McTigue EM, Yeh YFY (2013) Searching for explanations in science trade books: what can we learn from Coh-metrix? Int J Sci Educ 35(8):1367–1384

Sweller J (1988) Cognitive load during problem solving: effects on learning. Cogn Sci 12(2):257–285

Treagust D, Harrison A (1999) The genesis of effective science explanations for the classroom. In: Loughran J (ed) Researching teaching: methodologies and practices for understanding pedagogy (pp 28–43). Routledge, Abingdon, VA

Webb NM, Nemer KM, Ing M (2006) Small-group reflections: parallels between teacher discourse and student behavior in peer-directed groups. J Learn Sci 46(4):426–445

Wilson H, Mant J (2011) What makes an exemplary teacher of science? The pupils' perspective. School Sci Rev 93(342):121–125

Wittwer JO, Ihme N (2014) Reading skill moderates the impact of semantic similarity and causal specificity on the coherence of explanations. Discourse Processes 51:143–166

Wittwer JO, Renkl A (2008) Why instructional explanations often do not work: a framework for understanding the effectiveness of instructional explanations. Educ Psychol 43(1):49–64

Wolf K, Kratzer V (2015) Erklärstrukturen in selbststellten Erklärvideos von Kindern [Explaining structures in students' self-produced explanation videos]. In: Hugger K, Tillmann A, Iske S,

Fromme J, Grell P, Hug T (eds) Jahrbuch Medienpädagogik 12 [Yearbook media pedagogics 12]. Springer, Berlin, pp 29–44

Wolf K, Kulgemeyer C (2016) Lernen mit Videos? Erklärvideos im Physikunterricht [Learning with videos? Explanation videos in physics teaching]. Naturwissenschaften Im Unterricht Physik 27(152):36–41.

Yager RE (1983) The importance of terminology in teaching K–12 science. J Res Sci Teach 20(6):577–588

Chapter 13
Language in Physics Instruction

Heiko Krabbe, Karsten Rincke, and Robert Aleksov

Abstract In this chapter, we focus on human language which can be used to express complex, abstract ideas in physics. It discusses the question of how language relates to scientific thinking. Does language determine thinking or thinking determine language? Are language and thinking the same thing or independent, albeit closely related? An argument in favour of a separate language module is that people acquire their native language intuitively and without explicit guidance, whereas abstract mathematical thinking, for example, does not develop spontaneously to such an extent. However, the native language is based on everyday experiences, whereas the understanding and communication of scientific concepts seem to require a specialised, abstract technical language for science, or scientific language, that must be learned and taught. In order to distinguish everyday language and scientific language as language registers (Halliday and Martin 1993), we discuss various factors in this chapter, for example, the contextual integration and the cognitive challenge of the object of communication, spatial, temporal and social proximity or distance of the communication partners, as well as the use of a particular vocabulary and conceptually written means of expression. This leads to a model that, in addition to everyday language and scientific language, also allows the interlinking of general academic language (language of schooling), scientific jargon and the language of teaching and learning in physics lessons. Within this model, we discuss whether the development of scientific language should be understood as a refinement of everyday language or as learning a new foreign language. In the first case, everyday language would appear as a deficient mode of scientific language, whereas in the second case, both registers are attributed an independent communicative function. Building on this, we consider the communicative function of scientific language for the acquisition

H. Krabbe (✉)
Ruhr-Universität Bochum, Bochum, Germany
e-mail: heiko.krabbe@rub.de

K. Rincke
Universität Regensburg, Regensburg, Germany
e-mail: karsten.rincke@physik.uni-regensburg.de

R. Aleksov
Universität Duisburg-Essen, Essen, Germany
e-mail: robert.aleksov@uni-due.de

© Springer Nature Switzerland AG 2021
H. E. Fischer and R. Girwidz (eds.), *Physics Education*, Challenges in Physics Education,
https://doi.org/10.1007/978-3-030-87391-2_13

of physics concepts, but conversely, also how physics content supports the formation of linguistic competences. For linking language learning and subject learning, a framework model of the Common European Framework of Reference for Language (Thürmann 2012) is presented and explained using the example of the lab report in physics education. This chapter shows that the promotion of language should be an inherent part of physics teaching.

13.1 Human Language and Thinking

It took centuries of human thinking to provide Isaac Newton with the basis for the development of classical mechanics and transformations in further centuries for this physics to take the form of today's physics. More and more people are trying systematically to describe the inanimate world with physics theory to predict the results of observable processes. In physics, this is not about representing the experience of nature in formulae; physics is rather about using abstract physical–mathematical models (for representations, see Chap. 7) to predict natural processes. With the help of these multiple representations, the models can be manipulated to apply them to new empirical experiences, to test them in experiments and to negotiate them in social processes. As shown above, physics teaching has many goals, one of the most important of which is to teach these theoretical physical–mathematical models. Since the theoretical models do not jump out of observations or experiments, and since a theoretical model is not sufficient to justify its validity on its own, a link must be established between models and experiments (see Chap. 5).

To understand this process of empirical research, physics educators and researchers need to consider human language as the most important medium for explaining and negotiating these models. What is special about human language and its use for communication compared to the communication of animals or plants? As scientists assume today, animals do not speak complex sentences and the language of plants has no phonology. However, they communicate anyway. Animals organise their social life by communicating in different ways. Trees in a forest also communicate with other plants, for example, through their roots, not only with their conspecifics but also with their heterospecifics. They form a complex and large network around their locations. Communication among animals or plants is not trivial. Biologists are just beginning to describe these processes in more detail as they found that the boundary between human and animal or plant communication is not sharp. What then is special about the language that humans speak?

According to current research, the special aspect of human language is the potential to express complex abstract ideas. In humans, this potential is developed to such an extent that we can distinguish right from wrong, we can see ourselves as part of a nation, or that we have religions and philosophies. This potential is also the precondition for being able to look at our surroundings with different individual mode of experiencing the world. We can experience it from the point of view of an artist, or through the eyes of a philosopher, a lawyer, or even a scientist. Since we can develop abstract ideas and communicate them with our language, an apple, for example, means many things: it is a symbol of temptation in paradise; it is a symbol of healthy eating; it can also be a symbol of harmful economics when apples are imported from foreign countries and brought to our table by plane; or a symbol of a company. If humans were unable to develop and communicate abstract ideas, an apple would be just something to eat and more or less tasty.

Moreover, humans can think about this complex, diverse and differentiated situations and can plan and manage communication about them. This asks the question of how language and thoughts are related: Are they identical or do they have to be distinguished? If they have to be distinguished, how do they influence each other?

About the Relationship Between Language and Thought

At first glance, it seems clear that spoken and written language plays an important role in the learning of physics as it is used to form, communicate or store thoughts. If one thinks about it more carefully, one notices that the relationship between language and thoughts is manifold. One of the great controversies of twentieth-century linguistics was how to describe this relationship. The following approaches have been used in linguistic research:

1. **Language and thinking are the same**. As behaviourism accepts only the observable as the basis for theory, language is seen as the sum of individual, conditioned speech habits and part of behaviour that can be learned. Whereas language is observable, thinking is not and therefore only speech is seen as real (see Chap. 1; Watson 1930). Radical supporters of behaviourist learning theories therefore reject the existence of thinking as an independent human ability and identify thinking with speaking. Thinking is reduced to *subvocal speaking* (Anderson 2014).

2. **Language determines thinking**. The founder of linguistic determinism in the 1950s was Benjamin Lee Whorf, who conceived of language as the basic material of human thought (Whorf and Carroll 1956). Linguistic determinists assume that thinking is completely determined by the way humans talk. The moderate form, known as linguistic relativity, is influenced mainly by the work of Whorf (1963) who assumed that if languages can be used to express an issue in different ways, especially with different grammatical structures, this is based on different thinking and a different worldview (Langenmayr 1997).

3. **Thinking determines language**. The attempt that thoughts are transmitted and expressed by language needs the assumption that thoughts are different from language. There is much evidence to support this view. According to Anderson (2014), the development of language in young children occurs later than their development of cognition and there is also much evidence that the phrase structure of language can be seen as a model of how information is encoded in the brain.

4. **Thought and language are independently existing modules**. This claim, known as the modularity hypothesis, is based primarily on the work of Chomsky (1980) and Fodor (1983; for an overview, see Anderson 2014). It states that the

ability to speak is based on a language module that operates in interaction with, but independently of, the rest of the cognitive system. Evidence is provided by the observation that there are people with linguistic deficits but no cognitive deficits (and vice versa).

Language acquisition is a scientific field in which the modularity hypothesis has been intensively discussed. The observation that people learn their language of birth without systematic instruction and without being aware of the difficult grammatical structures of a language was seen as evidence that the human brain needs to be prepared for language learning by means of a language module. This is supported by the fact that other complex cognitive skills, such as mathematical reasoning, are not acquired without systematic instruction. At birth, the human brain seems to be prepared differently for speaking than for calculating. Langenmayr (1997) gave an overview of various attempts to clarify the relationship between language and thought empirically, which is summarised as follows:

> The relationship between language and thought must be modelled as reciprocal. Linguistic determinism, that is, a full dependence of thought on language, has little plausibility. Linguistic relativity—for example, the idea that different linguistic modes of expression correspond to different rather than the same thought processes—can hardly be rejected. Effects on the lexical, even (to a lesser extent) on the tonal level, as well as on the grammatical level, can be demonstrated. (p. 224; translated by the authors)

This assumption does not directly say what effects can be expected from the quality of knowing and memory performance if subject-related communication is systematically practised in the classroom. However, in teaching practice, it is clear that the meanings of terms used in everyday life tend to be based on everyday concepts and those of subject-related expressions tend to be based on the concepts in a particular subject. Although it sounds consensual, this is a challenging thesis. It says that content knowledge tends not to be communicated in everyday language!

Factors to Distinguish Everyday Language and Technical Language

A first characterisation of everyday language and specialised technical language (e.g. scientific language) is given by the distinction of basic interpersonal communication skills (BICS) and cognitive academic language proficiency (CALP) defined by Cummins (1981, 2008). Considered in this distinction are the cognitive demand and contextual support involved in particular language tasks or activities (see Fig. 13.1).

For situated *everyday communication*, only basic language skills with low cognitive demands are needed and usually acquired in the out-of-school language learning process without special training. On the BICS level in Fig. 13.1, meaning is constructed through the context of a situation and one's own, cultural and social experiences (Cummins 2000). *Academic language*, on the other hand, is more likely to be used to communicate abstract cognitive models detached from everyday contexts. This is accompanied by a higher cognitive effort, because academic language develops only with the will or need to express and communicate more abstract ideas and ways of thinking. Accordingly, academic language is not acquired automatically but has to be fostered in conjunction with academic content. To reach

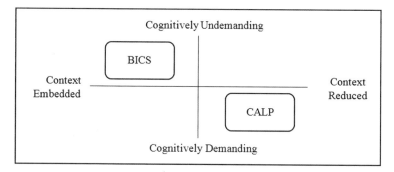

Fig. 13.1 Contextualization and cognitive effort for language learning according to Cummins (2000, p. 68)

this, Cummins argued that one should focus primarily on tasks that are related to the context and cognitive demands. Since what is "related to the context" or "cognitive demands" depends on learners' characteristics such as their prior knowledge or interest (Cummins 1981), what is offered has to be adapted, accordingly, for teaching and learning. However, Cummins did not make a distinction between general academic language and the *technical language* of a subject such as scientific language.

Habermas (1978) denoted and separated *colloquial (everyday) language*, specialised technical language and academic language in the following way. Academic language differs from colloquial language in the discipline of written expression and in a more differentiated vocabulary that includes specialised knowledge. On the other hand, it differs from the technical languages because it is basically open to everyone who can acquire orientation knowledge through general school education. According to Habermas, the separation of these conceptions of language takes place through the discipline of expression (written or oral), the domain-specific vocabulary, the communication partners and the degree of specialisation of the knowledge to be communicated.

Halliday's (1978) *register theory* offers a functional distinction between *everyday language, academic language and technical language*, which relates language to its social context. A *register* is characterised by certain morphological, syntactic and lexical properties that enable a person to communicate and construct meaning in a particular situation (cf. Halliday 1993). The given situation (e.g. a particular school context) determines a certain linguistic register, which is defined by the three categories *field, mode and tenor* (cf. Halliday and Martin 1993).

The category *field* refers to the situational setting in which the communication takes place (e.g. topic, content, subject matter or genre). *Mode* describes the medium of communication (e.g. written or oral, spontaneous or prepared) and its particular linguistic characteristics. *Tenor* describes the social relationship and the relationship between communication partners (e.g. personal closeness, distance or hierarchy).

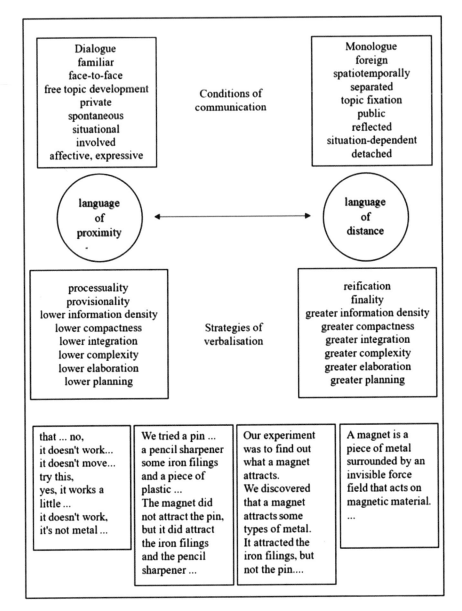

Fig. 13.2 Language of distance and proximity according to Koch and Oesterreicher (1985)

Koch and Oesterreicher (1985) also made a similar distinction between mode and tenor. They distinguished between medial and conceptual orality and written-ness. On the medial level, realisation is meant either in spoken form (phonic code) or written form (graphic code). This corresponds roughly to Halliday and Martin's (1993) mode, whereas the conceptual level corresponds to tenor. Koch and Oester-reicher conceived of conceptual orality as the language of proximity and conceptual writtenness as the language of distance. However, this demarcation is not dichoto-mous, but there exists a continuum spanning between these two poles by way of the interaction of a variety of parameters concerning the conditions of communication and strategies of verbalisation (see Fig. 13.2). In group work of students, for example, one finds dialogic speaking with presumably high familiarity of the partners. What is said will tend to be preliminary, unplanned and less elaborated. This is an example of language of proximity. A well-prepared teacher's lecture, on the other hand, is monological, fixed on a topic, reflective and concerned with objectivity. The teacher uses a comparatively compact language with deliberately designed sentences and a greater density of information, which would be equivalent to writing on a blackboard. This corresponds to the language of distance.

Modelling Different Language Registers for Science Learning and Understanding

In summarising these approaches, everyday language, academics language and scientific language can be interrelated as shown in Fig. 13.3.

Everyday language and academic language[1] both belong to the public field, but differ in their tenor. In contrast, academic language and technical language are similar in tenor, but refer to different fields (Pineker-Fischer 2017). This means that they use similar functional linguistic resources such as connectors, adverbial clauses or pronouns to establish coherence, but different domain-specific vocabulary. However,

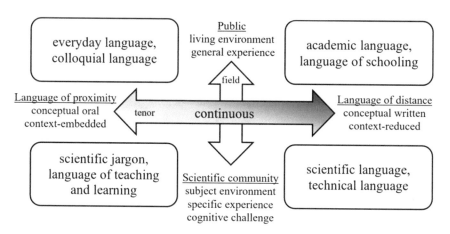

Fig. 13.3 Language registers and their interrelationships

---

[1] We use the term *academic language* following Cummins (2008). Schleppegrell (2004) calls this register the *language of schooling*.

scientific language is more than just technical vocabulary, because it also has certain grammatical forms and genres on sentence and text levels, which also need to be learned in their functionality. That is, because sciences such as physics consist of highly interconnected subject content that deals with abstract objects and topics (cf. Bernstein 1999). The complexity and abstractness of the subject matter require more extensive language skills (cf. Uesseler et al. 2013), which are not acquired and practised in everyday life, but only in science lessons (cf. Tajmel 2017).

Corresponding to everyday language, there is also a language of proximity in the scientific field, for example, in private discussions between experts. This specialist jargon differs from everyday language primarily in the use of specific terms and phrases (Montgomery 1989; von Polenz 1981). In dialogues with experts' direct counterparts, the language used becomes much less precise because there is a common basis on which knowledge is communicated and understanding negotiated. If necessary, the experts would be able to switch to a more precise, conceptually written or even mathematically symbolic language to clarify misunderstandings.

Discussions in physics lessons also use a less precise conceptual oral language of teaching and learning (Cazden 2001). It is neither possible nor desirable for teachers and students to talk about physics exclusively in scientific language because instruction and learning requires a personal relation with empathy and proximity. According to Lemke (1990), "learning science means learning to talk science" (p. 1) which does not mean that science lessons should be dominated by teacher talk, but the teacher should encourage genuine communication and greater application of discussion in science lessons. Typically, such talk involves evaluating observations for consistency with current ideas, recognising that earlier ideas may be wrong, modifying them using evidence, and coordinating all appropriate elements of knowledge into a coherent model (Keys 1997). Such reasoning, which is first used to search for understanding and meaning of terms and concepts, requires a language of teaching and learning that also needs proximity.

It is important to note that scientific jargon and language for teaching and learning are not the same register, although they are shown in the same box in Fig. 13.3 (cf. Bullock et al. 2019). Their difference lies in the communication partners. Experts in the field, who already have a firm grasp of the physics concepts, can therefore communicate less precisely when speaking in scientific jargon. For example, the statement "this mass has a weight force of 5 Newtons" is not a problem between experts because both are aware that they cannot attribute a force to the mass because forces are the interaction between two partners (in this case the earth and the mass). In contrast, novices in the field, who may also use everyday concepts, speak the language of learning. In this case, the same statement can be taken as an indication that the learners have not yet developed a concept of interaction. This shows how easy it is for misunderstandings to arise in physics lessons when the teacher speaks in scientific jargon but the students understand him or her on the basis of their everyday concepts (see Chaps. 6 and 14).

13.2 On the Relationship Between Everyday Language and Technical Language in Learning

In the following sections, we address the problem of promoting the learning of scientific language together with the learning of physics.

To structure the field, it is important that we distinguish between our concepts of how languages are learned and our conception of what kind of teaching can promote this learning. The first idea is part of a theory of learning; the second is part of a theory about teaching. Accordingly, the first has to do with the processes that take place in an individual's brain, and the second has to do with the environment, which can influence such cognitive processes of the individual. We first address learning theory and then suggest how a physics classroom should be equipped to support language learning.

Everyday Language and Scientific Language

How technical language and everyday language are related is discussed differently in linguistics. The two most prominent approaches are outlined in this section.

From the first point of view, technical language develops from everyday language. In the development process, everyday language is being refined until it meets the requirements of technical language. In describing the process, it is assumed that everyday language is imprecise, but technical language is precise. To develop technical language means, primarily, to express oneself in a particularly precise way. Accordingly, everyday language is assumed to be imperfect, whereas technical language, or scientific language in particular, is assumed to be a more developed linguistic register. However, the nuances of meaning that can be communicated in everyday language sentences cannot be identified as a lack of precision, but rather as an important communicative element that leads to adequate actions. Muckenfuß (1995) illustrated the difficult assessment of precision with the following example: "Which describes the situation most accurately: The soup is lukewarm. Or: The soup has a temperature of 32.5 °C?" (p. 247). A guest who wants to complain about his soup in a restaurant is more likely to be understood correctly with the first sentence than with the second one, which would seem absurd to the waiter or cook, although it seems more precise to a physicist. Precision in the sense of success in communication and activity is therefore not a privilege of using scientific language. Furthermore, Muckenfuß emphasised that scientific language is not precise everyday language. Rather, it is the language of abstract ideas in which we speak about theoretical models, that is, about something that does not have to be materially present. The special thing about scientific language, then, is not its particular precision compared to other linguistic registers, but its function, for example in physics, of communicating more or less precisely about theoretical models.

This leads to the assumption that a technical language, such as scientific language, does not emerge from the ongoing refinement of an everyday language. According to Vygotsky (2012), Rincke (2010, 2011) argued, on the basis of empirical data, that learning a technical language is similar to learning a foreign language. One advantage of this view is that everyday language no longer appears as a deficient mode of a technical language because both registers can be improved and both have their particular communicative functions.

Rincke adapted the so-called interlanguage approach of Selinker (1969, 1972). Learners of a foreign language show an interlanguage that is influenced by both their mother tongue and the target language. Rincke interpreted the language of students in physics lessons as a scientific interlanguage that is systematically influenced by everyday language and by the technical language of physics as the target language.

For teaching practice, it can be deduced that teachers in physics lessons should explicitly deal with everyday language and scientific language and address their differences. Not only should they clarify which words are connected with which objects in the sense of designation, they should also clarify in their teaching, especially in very abstract physics lessons, which connections of the new physics terms to be learned (subjects, predicates, prepositions etc.), communicate what the abstract physics models mean, and work out their differences for students' everyday understanding.

Supporting Language Learning and Physics Learning

Language plays a central role in learning scientific concepts and scientific working methods. Knowledge, skills and competences are conveyed through language and thus generated and integrated into pre-existing knowledge structures (von Glasersfeld 1987; Vygotsky 2012). Therefore, it is important to also support the development of language in subject learning. Christie (1981) identified three different learning goals that relate to different functions of language in teaching situations: learn through a language, learn about a language and learn how to use a language.

When *learning through language* the language serves the subject (physics) by negotiation and clarification of the meanings of technical terms and physics concepts. Basic language skills in the language of instruction are assumed for this purpose. *Learn about a language* refers to the subject matter as a resource for learning a language. The focus is on linguistic structures, such as grammar, word formation and vocabulary, with the aim of understanding the use and limitations of the language. This can be related to content-based language teaching and bilingual education, which targets, for example, learners who are studying a second language and need to learn to use it in speaking, reading and writing. *Learn how to use a language* aims for the nuanced application of language for different purposes and contexts. Here, for example, this is about native speakers who have to learn to communicate scientific content to different addressees in an appropriate way. Christie suggested that learners of technical languages should master competencies in each of these three areas. Norris and Phillips (2003) distinguished a *basic sense*, such as the ability to argue and inform, and a *derived sense*, such as understanding scientific concepts as two components that are, however, closely related and equally crucial for scientific teaching and learning processes and knowledge construction.

The Council of the European Union (2008) promotes concepts of multilingual teaching by offering non-linguistic subjects in foreign languages (CLIL—Content

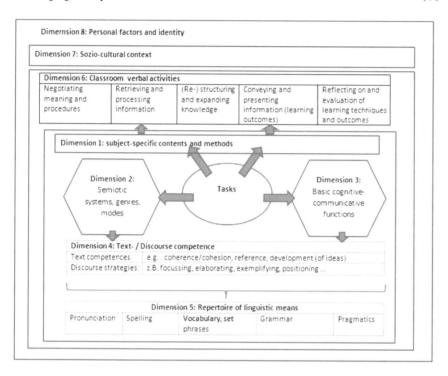

Fig. 13.4 Structured frame for academic discourse competence taken from Thürmann (2012)

and Language Integrated Learning). Working on the Common European Framework of Reference for Languages (CEFR), Thürmann (2012), and Vollmer and Thürmann (2013) suggested the following structured frame for academic discourse competencies as shown in Fig. 13.4.

In this model, the determination of language competencies is initiated by the first three dimensions:

The first dimension is to consider subject-specific content and methods and establishes their relationship to the goals of the subject, such as learning physics concepts and scientific research methods.

The second dimension refers to genres (text types) and semiotic systems, modalities (e.g. media written or conceptual written) and related conventions. Each subject has an inventory of genres and forms of representation that are common for teaching at school. For instance, Wellington and Osborne (2001) defined four major genres of science:

1. Reports that list properties, classify, describe functions and processes, or decompose a whole into its parts;
2. explanations drawing logical conclusions;
3. experimental reports describing procedures, observations, and results; and
4. expositions and argumentations to take a position.

Concerning the specific semiotic system, there are certain forms of representation such as illustrations, diagrams (e.g. circuit diagram) and symbols (e.g. force arrows, numbers and different mathematical representations).

The third dimension is about basic cognitive-linguistic rules, such as discourse rules, which are especially important for the development of language skills. Dalton-Puffer (2013, 2016) presented seven cognitive discourse functions (CDF) which are used to express elementary acts of thinking about subject matter in the classroom and were derived from educational curriculum theory and linguistic pragmatics (see Table 13.1). These CDF types can be marked in tasks by various performative verbs.

This is essentially a matter of the connection between language and thinking, since it is argued that learning science and learning the language of science cannot be separated (Halliday and Martin 1993; Martin and Rose 2003). From the perspective of content-based language teaching, this assignment of language patterns and discourse functions on the macro level has pedagogically the advantage that language activities in different subjects can be linked and language skills can be promoted more sustainably across subjects. From the subject perspective, however, these discourse functions may have to be specified more precisely on a micro level in order to represent how scientists are thinking and promote their science learning.

The fourth dimension is about strategies for text or discourse competence which can be derived from the above-mentioned determinants (Dimensions 1, 2 and 3). Examples are directed reading activities or text procedures.

The **fifth dimension** is to identify appropriate linguistic means in the production of oral or written statements and coherent texts.

Table 13.1 Cognitive discourse functions taken from Dalton-Puffer et al. (2018)

CDF type	Underlying basic communicative intention	Performative verbs
Classify	I tell you how we can cut up the world according to certain ideas	Classify, compare, contrast, match, structure, categorize, subsume
Define	I tell you about the extension of this object of specialist knowledge	Define, identify, characterize
Describe	I tell you details of what I can see (also metaphorically)	Describe, label, identify, name, specify
Evaluate	I tell you what my position is vis a vis X	Evaluate, judge, argue, justify, take a stance, critique, comment, reflect
Explain	I tell you about the causes or motives of X	Explain, reason, express cause/effect, draw conclusions, deduce
Explore	I tell you something that is potential (i.e. non-factual)	Explore, hypothesize, speculate, predict, guess, estimate, simulate
Report	I tell you something external to our immediate context on which I have a legitimate knowledge claim	Report, inform, recount, narrate, present, summarize, relate

In **the sixth dimension,** certain communicative (verbal) activities and classroom interactions can be distinguished. As shown in Fig. 13.4, and Vollmer and Thürmann (2013) generally differentiated between the following:

1. Negotiation of meanings and procedures by work instructions and recommendations, hints, assistance and corrections, requests for help and support, clarification and planning processes, or constituting meanings (often in the form of short initiate-response-feedback cycles).
2. Information retrieval by extracting and compiling relevant information from instructional text, textbooks, teaching materials, authentic factual text or from experiments.
3. Knowledge processing (declarative as well as procedural) by retrieving information from memory and relating it to new experiences and findings, making comparisons and evaluations, and examining, extending, refining, restructuring existing concepts and schemas, and developing appropriate questioning attitudes or solutions to problems.
4. Presentation of experiences, work results and learning outcomes results, that is, to express one's own understanding and position in a coherent and professionally appropriate manner and explain them if necessary. The proof of understanding in the classroom, written tests and exams is also part of this.
5. Meta-reflection of learning techniques and outcomes to optimise them for future tasks by checking whether selected strategies, methods, behaviours are goal-oriented and effective and how they can be improved if necessary. This includes reflecting on the subject as a whole and the use of research findings.

Whether and to what extent students accept and adopt the offered learning opportunities depends very much on the sociocultural context in which they grow up (Dimension 7)—such as being member of certain groups, their value systems and behavioural patterns—and on personal factors (Dimension 8) such as age, knowledge of the world, language biography, motivation and willingness to invest their effort in learning.

13.3 Writing of the Lab Report

In the following, we illustrate the benefit of this model for physics education using *the lab report* as the example for a typical text type in physics education (Dimension 2). As such, the lab report is a prototype for the recurring task of conducting an experiment, documenting and communicating the results obtained from it. Its purpose is to provide a concise account of the experimental procedures, reduced to the essentials, which allows later replication of the observations and examination of the conclusions drawn from them. In contrast to lab notes, which record the experimental procedure and results in their temporal sequence, the lab report reflects the final results in a logical order (Krabbe 2015). In its sequential structure, we assume a logical linear sequence of experimental steps, which distinguishes between stating a research question sometimes with a hypothesis, the description of the setup, of the procedure and of the observations, and the evaluation in the form of explanations or justified generalisations (Krabbe 2015). The first column in Table 13.2 gives an overview of the typical sections of the lab report. A study showed that students found it difficult to distinguish between the individual sections of the lab report and to understand the hierarchy of procedure, observation, explanation and justification, and thus the outline of the lab report even after instruction (Ricart Brede 2014).

Exploratory and Explanatory Experiments in the Lab Report

With regard to subject content (Dimension 1), we choose the phenomenon of charging an electroscope by induction (see Table 13.2). Furthermore, we make a distinction between exploratory and explanatory experiments as subject-specific methods. *Exploratory experiments* explore a phenomenon through careful investigation and observation with the aim to generate assumptions or hypotheses about empirically testable relationships and regularities that may be generalised as laws or equations. At the beginning, there are no hypotheses yet for an unexplained, open situation. In contrast, *explanatory experiments* test theoretical statements. Their aim is to obtain an explanation for an observation (measurement). The procedure starts with a hypothesis or assumption based on empirical laws either from prior experiences or from theory. These different approaches are reflected in lab reports with different structures and other discourse rules.

As can be seen in Table 13.2, the lab report of an exploratory experiment, unlike an explanatory experiment, has no assumption or hypothesis. The hypothesis in the explanatory experiment is derived from the device theory for the electroscope that the deflection of the needle requires charges on the electroscope.

Another difference between these two types of experiments lies in the evaluation or discussion of the results. In the experimental report in Table 13.2, an assumed relation is justified with the observations. In this case, the newly assumed general relation is explained by the fact that charges on a conductor are displaced when electric charges are only brought close to the conductor. Since charge movements cannot be observed directly, the electroscope is needed as a detection device. In order for the device to indicate charges validly, there has to be a device theory. For this observation, it is most important that the electroscope is an electric conductor whose needle deflects when charges accumulate on the needle and the holder (the rule of physics states that identical electric charges repel each other). Only with this device theory can the claimed connection—from the observation that the needle of the electroscope deflects—be explained. In the explanatory assessment, the original hypothesis is first assessed and shown (justified) to be false due to the observation. Then, an explanation is given starting from the premise that the electroscope is an electric conductor on which charges can move. From this premise, in a chain of inferences, the observation that the needle deflects is derived under the condition that a charged rod is brought near the electroscope (cf. Tang 2015). Two more conditions and the resulting observations are explained further in the following sections.

Table 13.2 Example of a lab report for an exploratory and an explanatory experiment

Section of lab report	Exploratory experiment (discover a phenomenon)	Explanatory experiment (explain observations)
Research question	What happens when you only approach an electroscope with a rubbed rod without touching it?	
Assumption/hypothesis		If you don't transfer any charges to the electroscope, its needle won't deflect
Materials	Electroscope PVC rod Woollen cloth	
Procedure/methods	First, the PVC rod is rubbed over its entire length with a woollen cloth so that it is negatively charged. Then the charged PVC rod is brought close to the plate of the electroscope and the pointer of the electroscope is observed. After that the PVC rod is held near the plate for some time without touching it. Finally, the PVC rod is removed from the vicinity of the electroscope	
Observations/results	If the PVC rod is brought close to the plate of the electroscope, then its needle deflects from its rest position. As long as the charged PVC rod remains near the plate of the electroscope, the deflection remains. When the PVC rod is removed, the needle returns to its initial state	
Evaluation/discussion	*Justification*: If you bring electric charges only close to a conductor, then charges on it will be shifted The electroscope is an electric conductor whose needle deflects when charges accumulate on the needle If you bring a charged rod close to the electroscope, charges seem to accumulate on the needle because the electroscope deflects. This only results in a charge shift (and not a charging) because the deflection disappears as soon as the charged rod is removed (This shift of charge on electric conductors is called induction. The charged rod induces a charge shift on the electroscope)	*Checking of the hypothesis*: The assumption was wrong, because the needle of the electroscope deflects *Explanation*: The electroscope is an electric conductor in which charges can move freely If the negatively charged PVC rod is brought closer to the plate of the electroscope, the negative charges (electrons) on the plate of the electroscope are repelled by the negative charges (electrons) on the PVC rod, because charges of the same kind repel each other The negative charges accumulate around the needle of the electroscope, causing it to deflect As long as the charged PVC rod remains near the plate of the electroscope, the charge shift on the electroscope remains Only when the charged PVC rod is removed, do the negative charges (electrons) distribute evenly over the electroscope again and the deflection vanishes

Discourse Functions in the Lab Report

Each section of the lab report relates to one of the specific discourse functions. The discourse function DESCRIBE can be found in the procedure and observation sections of the lab report in Table 13.3. The discourse functions EVALUATE (here represented by the justification) and EXPLAIN are required in the discussion section depending on the kind of experiment. Describing is not simply reproducing superficial object features and procedures, but constituting them with reference to established, shared knowing by highlighting and marking relevant features (Feilke 2005). However, a closer look reveals that descriptive processes and methods in the procedure section requires different discursive actions than describing observations. This becomes clear as shown in Table 13.3, where the respective discursive activities are marked more precisely by additional interrogatives (cf. Krabbe et al. 2019).

As indicated above, by following Osborne and Patterson (2011), we distinguish explaining and justifying in the discussion section of the lab report as given below.

Explaining means that an observation surely exists and is accounted for by applying known laws, rules or causal relationships, whereas **justifying** is the reverse that a law, rule or causal relationship is generally asserted and made plausible with arguments or empirical data. According to Table 13.2, this applies specifically to explaining why an experimental procedure is done and justifying whether this can be done, but there are other possible discursive actions of explaining and justifying in other parts of the lab report, too. This shows that the more general discourse functions require subject-specific interpretation and must therefore be explicitly promoted in the instruction in physics lessons.

Table 13.3 Assignment of discursive actions to functional parts of the lab report

Discursive action	Discursive activity	Function in the lab report
Describe what with	*the experiment is realised*	Experimental setup (materials)
Describe how	*the experiment runs*	Processing of the experiment
Explain what for	*the electroscope is used*	Experimental setup and planning
Describe what happens if	*a charged rod is brought near*	Observation
Explain what	*an electroscope or induction is*	Definition of terms
Explain why	*the needle deflects*	Causal explanation (deductive)
Justify whether	*charges can be shifted on conductors*	Generalisation (inductive)
Justify which	*explanation is appropriate for the data*	Reasoning (for decision)

In the middle column, the activities are related to the induction experiment and should be read as a continuation of the first columns

Writing to Learn

For experimenting, different cognitive and procedural activities such as setting up the experiment, carrying it out, observing or recording and interpreting data can be trained separately. Similarly, for mastering of text production, the associated discursive activities can be practised separately for each section of the lab report. Combining both science learning and language practising is the idea of writing to learn approaches, which also reflect the conceptual writtenness of academic and technical language. According to Wallace et al. (2004), two perspectives on writing to learn practices can be distinguished. One perspective that follows Halliday and Martin (1993) assumes that "science language is a particular set of language practices, and that students need to master these ways of doing and recording science" (Wallace et al. 2004, p. 34). In the other perspective that follows Christie (1981), Wallace et al. viewed language as a learning resource when students use various writing types and alternative discursive resources to acquire physics concepts and scientific inquiry. Obviously, the example of the lab report presented here belongs to the first perspective, because it aims to generate a certain text and discourse competence (Dimension 4). Therefore, a limited repertoire of standard and technical language resources is required, for example, the use of passive constructions, temporal adverbs, nominalisations, causal, final or consecutive clauses and coherent textual references (Boubakri et al. 2017; Ricart Brede 2014; Beese and Roll 2013). To enhance students' ability to apply all rules adequately, they should be given explicit instruction in vocabulary, set phrases, grammar and pragmatics (Dimension 5).

In arguing for how students can improve their discourse competence by writing the lab report, we also refer to the Feilke's (2014) concept of text procedures, which are composed of a cognitive action scheme and associated linguistic means. For instance, the description of an experimental procedure requires a scheme of listing of the steps to be done in the correct order using temporal adverbs such as *first, then, next, after that* or *finally* as possible linguistic means. When describing observations, cause-and-effect relationships must be established between an independent and a dependent variable. This can be done using conditional sentences with the independent variable in the *if* part and the dependent variable in the *then* part. Such conditional means are also used for inferences in the explanation. The difference here is in the cognitive action scheme. Whereas relationships between observable variables are established in the results section, theoretical variables are addressed in the explanation. Another example of these linguistic means is the use of causal clauses as linguistic means in justification and explanation (see Table 13.4).

Once again, the difference lies in the cognitive action scheme. In the justification, a law or rule is stated in the main clause and supported with observations in the *because* subordinate clause. In the explanation, on the other hand, an observed phenomenon is stated in the main clause that is derived from general laws or theoretical concepts in the *because* subordinate clause. This shows that linguistic resources cannot simply be learned in language lessons and transferred to physics lessons, but must be clarified in the context of the subject, because language, reasoning and practices in physics are closely interlinked.

Table 13.4 Comparison of causal sentences in justification and explanation (cf. Osborne and Patterson 2011)

Justification	Explanation
A law, rule or causal relationship is generally asserted and made plausible with arguments (e.g. empirical data)	A phenomenon exists and is explained by the application of known laws, rules or relationships
Charges can be shifted on conductors, because the needle of the electroscope deflects if a charged rod approaches it	The needle of the electroscope deflects, because charges can be shifted on conductors if a charged rod approaches it

The analysis of the lab report using Voller and Thürmann's (2013) model has shown how closely text type competences are linked to discourse functions, cognitive action schemes and specific linguistic means. Language support can start with one of these dimensions. This enables a close integration of language and subject learning, and offers opportunities for specialisation and individual support. With novice language learners, one might focus on the linguistic means (e.g. sentence patterns), and promote physics thinking implicitly via the associated discourse functions and cognitive action schemes. Students who already have a good command of the language can be encouraged to use more scientific reasoning, by addressing the different structures of the exploratory and explanatory lab protocols with respect to the different epistemic purposes of exploratory and explanatory experiments.

13.4   Summary

The aim of this chapter is to highlight that, contrary to the opinions of many physics teachers (Airey 2012), the promotion of language is an important task of physics teaching. As shown in Fig. 13.3, there are basically two conceivable ways to promote language in physics classroom. One still common way would be to develop students' academic language skills in language lessons first and then to let them use these skills in physics lessons in order to present facts in a linguistically precise way with appropriate physics content. The other way is to develop students' language skills needed in the physics lessons, which can thus contribute to their general language development. The considerations of previous researchers discussed in this chapter should have made it clear that a transfer from language teaching is difficult, because the linguistic means must be used in a specific way that can only be learned in conjunction with physics learning. Therefore, reflecting on language and promoting science language must be an indispensable part of physics teaching.

The way language development is integrated into physics lessons depends on the language level of a particular group of students. For language learners, the attention may be more on the structure of the language and means of expression. For native speakers, on the other hand, attention may be more focused on the logical and coherent presentation of physics knowledge. In both cases, linguistic awareness that can be used to distinguish between different registers and their functional areas of use is required in order to be able to communicate appropriately to the addressee. The use of everyday and scientific concepts is also evident in the choice of language registers. In classroom instruction, therefore, the awareness of and contrast between, different registers should be made clear by the physics teacher. This also includes a meta-reflection on the respective function of the registers. Drawing on Lemke's (1990) ideas about talking science in the classroom, we consider that talking physics is not only talking about physics but also about the language of physics and its effective use in communicating physics concepts.

Acknowledgements We would like to thank Markus Sebastian Feser (Universität Hamburg) und Alexander Kauertz (Universität Koblenz-Landau) for carefully and critically reviewing this chapter.

References

Airey J (2012) I don't teach language. The linguistic attitudes of physics lectures in Sweden. AILA Rev 25(1):64–79

Anderson JR (2014) Cognitive psychology and its implications, 8th edn. Worth Publishers, New York, NY

Beese M, Roll H (2013) Versuchsprotokolle schreiben zur Förderung literaler Routinen bei mehrsprachigen SuS in der Sekundarstufe I. [Writing lab reports to promote literal routines for students in lower secondary school]. In: Decker-Ernst Y, Oomen-Welke I (eds) Deutsch als Zweitsprache: Beiträge zur durchgängigen Sprachbildung [German as a second language: contributions to continuous language education]. Fillibach bei Klett, Stuttgart, pp 213–230

Bernstein B (1999) Vertical and horizontal discourse: an essay. Br J Sociol Educ 20(2):157–173

Boubakri C, Beese M, Krabbe H, Fischer HE, Roll H (2017) Sprachsensibler Fachunterricht [Language-sensitive subject teaching]. In: Becker-Mrotzek M, Roth H-J (eds) Sprachliche Bildung-Grundlagen und Handlungsfelder [Language education—basics and fields of action]. Waxmann, Münster, New York, pp 335–350

Bullock OM, Colón Amill D, Shulman HC, Dixon GN (2019) Jargon as a barrier to effective science communication: evidence from metacognition. Public Underst Sci 28(7):845–853. https://doi.org/10.1177/0963662519865687

Cazden CB (2001) The language of teaching and learning, 2nd edn. Heinemann, Portsmouth, NH

Chomsky N (1980) Rules and representations. Behav Brain Sci 3:1–61

Christie F (1981) The language development project. Engl Aust 58:3–10

Cummins J (1981) The role of primary language development in promoting educational success for language minority students. In: California State Department of Education (ed) Schooling and language minority students: a theoretical framework, evaluation, dissemination and assessment center. California State University, Los Angeles, pp 3–62

Cummins J (2000) Language, power and pedagogy. Blue Ridge Summit: Multilingual Matters, Bristol. http://doi.org/10.21832/9781853596773

Cummins J (2008) BICS and CALP: empirical and theoretical status of the distinction. In: Hornberger NH (ed) Encyclopedia of language and education. Springer, Boston, MA. http://doi.org/10.1007/978-0-387-30424-3_36

Dalton-Puffer C (2013) A construct of cognitive discourse functions for conceptualising content-language integration in CLIL and multilingual education. Eur J Appl Linguist 1(2):216–253

Dalton-Puffer C (2016) Cognitive discourse functions: specifying and integrative interdisciplinary construct. In: Nikula T, Dafouz E, Moore P, Smit U (eds) Conceptualising integration in CLIL and multilingual education. Multilingual Matters, Bristol, Buffalo & Toronto, pp 29–54

Dalton-Puffer C, Bauer-Marschallinger S, Brückl-Mackey K, Hofmann V, Hopf J, Kröss L, Lechner L (2018) Cognitive discourse functions in Austrian CLIL lessons: towards an empirical validation of the CDF construct. Eur J Appl Linguist 6(1):5–29

Feilke H (2005) Beschreiben, erklären, argumentieren – Überlegungen zu einem pragmatischen Kontinuum [Describing, explaining, arguing—reflections on a pragmatic continuum]. In: Klotz P, Lubkoll C (eds) Beschreibend wahrnehmen – wahrnehmend beschreiben [Perceiving descriptively—describing perceptively]. Rombach, Freiburg i. Br., pp 45–60

Feilke H (2014) Argumente für eine Didaktik der Textprozeduren. [Arguments for a didactics of text procedures]. In: Bachmann T, Feilke H (eds) Werkzeuge des Schreibens. Beiträge zu einer Didaktik der Textprozeduren [Tools of writing. Contributions to an education of textual procedures]. Klett, Stuttgart, pp 11–34

Fodor JA (1983) The modularity of mind: an essay on faculty psychology. MIT Press, Cambridge, MA

Habermas J (1978) Umgangssprache, Wissenschaftssprache, Bildungssprache [Everyday language, scientific language, academic language]. Merkur 32(359):327–342

Halliday MAK (1978) Language and social semiotics: the social interpretation of language and meaning. Edward Arnold, London

Halliday MAK (1993) Towards a language-based theory of learning. Linguis Educ 5(2):93–116

Halliday MAK, Martin JR (1993) Writing science: literacy and discursive power. The University of Pittsburgh Press, Pittsburgh, PA

Keys CW (1997) An investigation of the relationship between scientific reasoning, conceptual knowledge and model formulation in a naturalistic setting. Int J Sci Educ 19:957–970

Koch P, Oesterreicher W (1985) Sprache der Nähe—Sprache der Distanz. Mündlichkeit und Schriftlichkeit im Spannungsfeld von Sprachtheorie und Sprachgeschichte [Language of proximity-language of distance. Orality and writing in the field of conflict between language theory and language history]. Rom Jahrb 36:15–43

Krabbe H (2015) Das Versuchsprotokoll als fachtypische Textsorte des Physikunterrichts [The lab report as a typical text type in physics lessons]. In: Schmölzer-Eibinger S, Thürmann E (eds) Schreiben als Medium des Lernens. Kompetenzentwicklung durch Schreiben im Fachunterricht [Writing as a medium of learning. Competence development through writing in the subject classroom]. Waxmann, Münster, pp 157–173

Krabbe H, Timmerman P, Boubakri C (2019) BESCHREIBEN, ERKLÄREN und BEGRÜNDEN im Physikunterricht [Describing, explaining and reasoning in physics lessons]. In: Maurer C (ed) Naturwissenschaftliche Bildung als Grundlage für berufliche und gesellschaftliche Teilhabe [Science education as a basic for professional and social participation]. Gesellschaft für Didaktik der Chemie und Physik, Jahrestagung in Kiel 2018. Universität Regensburg, pp 265–268

Langenmayr A (1997) Sprachpsychologie [Psychology of language]. Hogrefe, Göttingen, Bern, Toronto

Lemke J (1990) Talking science. Ablex Publishing Corporation, New York

Martin J, Rose D (2003) Working with discourse. Continuum, New York

Montgomery SL (1989) The cult of jargon: reflections on language in science. Sci Cult 1(6):42–77

Muckenfuß H (1995) Lernen im sinnstiftenden Kontext: Entwurf einer zeitgemäßen Didaktik des Physikunterrichts [Learning in a meaningful context: design of a contemporary didactics of physics education]. Cornelsen, Berlin

Norris SP, Phillips LM (2003) How literacy in its fundamental sense is central to scientific literacy. Sci Educ 87(2):224–240

Osborne JF, Patterson A (2011) Scientific argument and explanation: a necessary distinction? Sci Educ 95(4):627–638

Pineker-Fischer A (2017) Sprach- und Fachlernen im naturwissenschaftlichen Unterricht. Umgang von Lehrpersonen in soziokulturell heterogenen Klassen mit Bildungssprache [Language and subject learning in science teaching. How teachers in socioculturally heterogeneous classes deal with educational language]. Springer VS, Wiesbaden

Ricart Brede J (2014) Zur Didaktik des Versuchsprotokolls als Aufgabe eines sprachsensiblen Fachunterrichts und eines fachsensiblen Sprach(förder)unterrichts [On the didactics of the lab report as a task of a language-sensitive subject teaching and a subject-sensitive language (remedial) teaching]. In: Klages H, Pagonis G (eds) Linguistisch fundierte Sprachförderung und Sprachdidaktik: Grundlagen, Konzepte, Desiderate [Linguistically based language support and language education: Basic principles, concepts and desiderata]. Walter de Gruyter, Berlin, pp 173–191

Rincke K (2010) Alltagssprache, Fachsprache und ihre besonderen Bedeutungen für das Lernen [Everyday and special language and their role in science education]. Zeitschrift Für Didaktik Der Naturwissenschaften 16:235–260

Rincke K (2011) It's rather like learning a language: development of talk and conceptual understanding in mechanics lessons. Int J Sci Educ 33(2):229–258

Schleppegrell M (2004) The language of schooling. Lawrence Erlbaum Associates, Mahwah, NJ. https://www.coe.int/en/web/language-policy/languages-of-schooling

Selinker L (1969) Language transfer. Gen Linguist 9:67–92

Selinker L (1972) Interlanguage. Int Rev Appl Linguis Lang Teach (IRAL) 10(3):209–231

Tajmel T (2017) Die Bedeutung von Alltagssprache - eine physikdidaktische Betrachtung [The meaning of everyday language—a physics didactic consideration]. In: Lütke B, Petersen I, Tajmel T (eds) Fachintegrierte Sprachbildung [Subject-integrated language education]. De Gruyter Mouton, Berlin, Boston, pp 253–267

Tang KS (2015) The PRO instructional strategy in the construction of scientific explanations. Teach Sci 61(4):14–21

The Council of the European Union (2008) Resolution on a European strategy for multilingualism. Retrieved from https://eur-lex.europa.eu/LexUriServ/LexUriServ.do?uri=OJ:C:2008:320:0001:0003:EN:PDF

Thürmann E (2012) Subject literacies and the right to quality education for democratic citizenship and participation. Report on a seminar hold at Strasbourg, 27–28 September 2012. Council of Europe, Strasbourg. Retrieved from https://rm.coe.int/09000016805a18ee

Uesseler S, Runge A, Redder A (2013) "Bildungssprache" diagnostizieren: Entwicklung eines Instruments zur Erfassung von bildungssprachlichen Fähigkeiten bei Viert- und Fünftklässlern [Diagnosing "academic language": developing an instrument to assess educational language skills of fourth and fifth graders]. In: Redder A, Weinert S (eds) Sprachförderung und Sprachdiagnostik [Language support and language diagnostics]. Waxmann, Münster, pp 42–67

Vollmer HJ, Thürmann E (2013) Sprachbildung und Bildungssprache als Aufgabe aller Fächer der Regelschule [Language education and academic language as a task for all subjects of the regular school]. In: Becker-Mrotzek M, Schramm K, Thürmann E, Vollmer HJ (eds) Sprache im Fach [Language in the subject]. Waxmann, Münster, pp 41–57

von Glasersfeld E (1987) The construction of knowledge, contributions to conceptual semantics. Intersystems Publications, Salinas CA

von Polenz P (1981) Über die Jargonisierung von Wissenschaftssprache und wider die Deagentivierung [On the Jargonisation of scientific language and against the deactivation]. In: Bungarten T (ed) Wissenschaftssprache. Beiträge zur Methodologie, theoretischen Fundierung und Deskription [Language of science. Contributions to methodology, theoretical foundation and description]. Fink, München, pp 85–110

Vygotsky LS (2012) Thought and language (revised and extended edition). MIT Press, Cambridge, MA

Wallace CS, Hand BB, Prain V (2004) Writing and learning in the science classroom. Springer, Dorddrecht

Watson JB (1930) Behaviorism. Norton, New York

Wellington J, Osborne J (2001) Language and literacy in science education. Open University Press, Buckingham

Whorf BL (1963) Sprache, Denken, Wirklichkeit [Language, thought and reality]. Rowohlt Taschenbuch, Reinbek

Whorf BL, Carroll JB (1956) Language, thought, and reality: selected writings of Benjamin Lee Whorf. Technology Press of Massachusetts Institute of Technology, Cambridge

Chapter 14
Students' Conceptions

Michael M. Hull and Martin Hopf

Abstract This chapter aims to be a very practical and easy-to-read overview about students' conceptions with many examples to make the discussion concrete and relevant. The chapter consists of four parts. In part 1, we define students' conceptions with the goal of making readers aware of why it is important to be familiar with 1) the typical conceptions of students and 2) the theoretical views that describe those conceptions. In part 2, we discuss the different theoretical views that we allude to in Part 1 (the Theory Theory view, the Ontological view and the Knowledge in Pieces view). All discussions are clearly illuminated with concrete examples of students' conceptions. In part 3, we focus on the known strategies corresponding to each view for promoting student learning (respectively: cognitive conflict, ontological training and drawing on productive student resources). A main point is that cognitive conflict is generally limited in its effectiveness, and student naïve ideas tend to persist in some form even after research-based instruction. Finally, in the fourth part, we briefly describe how student conceptions are affected by emotional state and attitudes towards the content being learned.

14.1 Why Should You as a Physics Teacher Care About Students' Conceptions?

Imagine going out on the street and asking any passer-by, "Why is it hot in the summer and cold in the winter?" What answers would you expect to get? This scene depicts one of the famous moments in the history of physics education research. During the production of the film *A Private Universe* (Sadler et al. 1989), researchers from Harvard University did exactly that, only asking not strangers on the street but rather their students, who were proudly graduating on that day. As they found out, many of

M. M. Hull (✉)
Universität Wien, Vienna, Austria
e-mail: michael.malvern.hull@univie.ac.at

M. Hopf
Universität Wien, Vienna, Austria

© Springer Nature Switzerland AG 2021
H. E. Fischer and R. Girwidz (eds.), *Physics Education*, Challenges in Physics Education,
https://doi.org/10.1007/978-3-030-87391-2_14

the interviewees answered by claiming that the Earth is nearer to the Sun during the summer and further away during winter. This stereotypical—and wrong—answer is what you would expect to get with strangers on the street; this study from Harvard showed that you get the same answer despite the supposedly excellent education of the interviewees. Nor was this a fluke: other researchers have found the same result time and time again (for an overview see Sneider et al. 2011).

To science educators, it is horrifying to learn that many people after instruction seem not to have understood even the most basic principles of science. It is indeed a sad state of affairs that even graduates from renowned schools explain the seasons via the distance of the Earth to the Sun. From these studies, it becomes clear that society cannot blame the problem on dull teachers or unmotivated students. Furthermore, it becomes clear that it is ineffective to just tell the students that the seasons come from the tilt of the Earth's axis. All of the interviewees in Sadler et al.'s study, of course, had heard that in class before. Something more subtle must be going on.

So—what is going on? What happened to the correct answer about the Earth's tilt that the interviewees had heard in class? What we have described here is a prototypical example of a students' conception. Many years ago, science education researchers asked themselves why, when faced even with concepts as basic as, for example, *force*, *current* or *heat*, students often said, "just don't get it". Repeatedly, students did poorly even in simple conceptual tests. To address this question, researchers started to focus on the learning processes of students. And very soon, it was found that students bring to the classrooms ideas which are not compatible with scientific concepts (e.g. Duit and Treagust 2012). For example, students come into the classroom thinking that if they throw a ball up in the air, they give a force to the ball that the ball takes with it as it ascends. As soon as this force is used up, the ball will drop down again.

Although different education researchers use different technical jargon, we define students' *conceptions* as those ideas that students bring to the classrooms or sometimes also develop there. Research on students' conceptions has been done in science education research for more than 50 years, and many results are available for physics teachers to draw upon in their daily teaching practice. To be clear, most of the time, you, as a physics teacher, will not hear your students talk about their conceptions with the wording you will find in the following sections, even if you ask them to explicate what they are thinking. They might be completely unaware that they have these conceptions. The important point, however, is that many students will behave as if they believe these conceptions. So if you ask a middle school student during the teaching of electricity, "Will the current be used up in this circuit?", you will not often get a clear answer. But if you ask, "What will the current be like on this side of the light bulb?", many students (even after instruction) will claim that there is less current "after" a bulb than "before" (in fact, it is the same current on either side of a bulb). They argue as if they believe that the bulb will use up the current. That is enough to warrant actions from education researchers and physics teachers alike.

As the role of technology in people's everyday lives continues to grow, so too grows the danger associated with people not having a proper understanding of physics. People should aim for a society of voting citizens who will not just blindly believe the "experts", but who will have the knowledge necessary to grapple with issues deeply rooted in physics. In helping students develop such knowledge, it is crucial that physics teachers know about students' conceptions. To illustrate this, consider the following example: one of the typical topics in the high school curriculum is that visible light is only a small part of the spectrum. Usually, one starts to teach that by showing students that there really seems to be something more than visible light coming out of a light bulb. For this, a heat bulb is often used to introduce infrared radiation. However, there is a problem with this. Research has shown that many students in high school already have heard about infrared radiation and are familiar with heat lamps. In addition, they have come to the conclusion, before coming to school, that the red light, which (also) comes out of the heat lamp, is the infrared radiation everybody is talking about. For them, infrared radiation is red-coloured radiation (Libarkin et al. 2011; Neumann and Hopf 2011; Plotz 2017). Being aware of this conception, teachers can quickly see that it will not make much sense to use a heat lamp in the classroom; it is much better to use a terrarium lamp. Those are not too expensive, and they only emit infrared radiation (and not visible light). If a student puts a hand in front of the terrarium lamp, the student can easily feel that the hand is heated. But the student will not see any light. This is just one example to argue that if you know which conceptions your students will have when they come into your classroom, you will be much better able to plan your physics teaching in a way that students will be better able to learn.

14.1.1 Where Do Conceptions Come from?

To understand better, what leads to the formation of students' conceptions, it is helpful to understand how people learn. Research about learning is old and has produced many results to date. What comes to your mind when you think about how people learn? Perhaps you think about how, at some time during your studies, you heard something about Pavlov's dogs or Skinner's rats. Although these studies played an important role in the development of education theory, they themselves turn out to not be particularly helpful to understand physics learning. Research that is more modern has demonstrated that an individual's learning is heavily dependent on the richness of his or her internal mental life, and many education researchers utilize instead the constructivist theory. To get started in thinking about this theory, a fruitful starting point will be to reflect on your own experiences of learning physics.

Do you remember your first weeks in a large physics lecture? I hope that you remember greatly enjoying this time, but it is also likely that, at least sometimes, you felt frustrated. Most physics students have moments in their first weeks of study when they feel thoroughly confused regarding what the professor is talking about. This happens even if the lecturer does a great job of explaining the physics to you. The challenge you faced was that you had to make sense out of the lecture. You had to make sense out of the impressions you were getting on your retinae and eardrums. All the professor can do is activate some arrays of nervous responses of your brain without activating others. It is up to you to make sense of those patterns of responses.

It is not possible for the lecturer to "pour" knowledge into your brain as though it were an empty cup.

The fact that you got past that initial confusion and persisted in learning physics is evidence of how good your brain is at being able to make sense of the sensual impressions, which arrive. You are not unique in this amazing ability. A new-born baby learns to recognize his/her mother in a week, and young children learn to read in a few months. This same amazing ability is to blame for students entering the classroom with conceptions that are at odds with what you want them to learn. The brain is so good at making sense that it forcefully finds connections even in situations that actually have none. An example for this is what is known as "gamblers fallacy": if the ball at a roulette table fell for ten times on red, people think that surely the ball will now fall on black (in fact, the probability of the next spin is the same as it was for each of the preceding spins).

People make sense of new things by connecting them to what they already know, usually including what they have physically experienced in the world around them. Consider the earlier example of the seasons. If you ask someone to think about the origin of the seasons, that person will likely draw upon the experience that it is warm in summer and cold in winter. That person also has the experience that if they bring their finger nearer to a candle flame, it feels warmer. Therefore, it makes sense to assume that the origin of the seasons is the distance of the Earth to the Sun. To make sense of new information or to consider a new question, people draw upon knowledge they already have. This includes not only what they have physically experienced in the world around them, but also what they have heard from others outside and inside the classroom. Regarding the seasons example, many people will further justify their argument by stating that they remember their teacher telling them that the orbit of the Earth around the Sun is not a perfect circle. Indeed, the Earth is closer to the Sun at some points than at others, but that does not explain the seasons. In fact, the Earth is closest to the Sun in January.[1] That is the main idea of the constructivist theory: people use their existing ideas and experiences to *construct* their understanding of the unknown. Just as the second floor of a house can only be built upon the first floor, the things people already know serve as the base for the new knowledge they construct. Sometimes, one who first learns about constructivism thinks that it means doing something with one's hands. Hopefully, we have made it clear here that that is not, generally, what it means. "Construct" in this context refers to what happens in the knowledge networks of human brains.

Now that we have discussed what the constructivist theory is, let us use it to explain the development of students' conceptions. Students of all age groups have already had profound experiences in their daily lives with most of the topics that are discussed in the physics classroom. These students have already constructed an understanding of what they have observed in their world. They know how to hit balls, and they know that batteries run empty after a time of use. This "common sense" that students construct is often contradictory to the explanations the physics teacher will

---

[1] If you live in the southern hemisphere, you might want to withhold this fact from your students, as it might actually foster their incorrect conception.

offer. Since students have already spent their lives constructing their understanding of the world around them, they are likely to reject what you teach them when it contradicts their common sense. It is not necessarily the case that the "rejection" will take the form of openly complaining to you, but it often is the case that students do not make sense of what their physics instructor teaches, or that they conclude "well, this might be true in the physics classroom, but not in the real world". That is a problem that motivates education researchers and that should motivate you to keep reading to learn more about student conceptions and how to use this knowledge to inform your teaching.

Students' conceptions in physics have been extensively researched for more than 40 years now. In fact, the first Ph.D. in physics education research (Banholzer 1936) addressed students' conceptions. A search on Google Scholar for "student conceptions physics" found 358.000 entries in September 2020. So, there is a vast knowledge base of typical conceptions. To give even a first overview, much more than a book chapter would be needed. But we want to now present here some examples of students' conceptions from different fields of physics.

14.1.2 Examples of Students' Conceptions

The examples presented here can only give an extremely small glimpse into the many different conceptions, which have been documented by research. To learn more about concrete students' conceptions, we encourage you to have a look at the literature (e.g. Driver et al. 2005) or in the bibliographies of student conceptions that have been compiled (Duit 2009; Flores et al. 2014).

Mechanics

Mechanics is perhaps the most researched field regarding students' conceptions. A very good example to illustrate some typical problems is the following: students are asked which forces act on a ball during flight (see Fig. 14.1). Several versions of this question are around, e.g. in the testing instrument "Force Concept Inventory" (Hestenes et al. 1992) or in large-scale assessment studies (e.g. Neidorf et al. 2020). The correct answer is that, in every position of a ball in the air, the gravitational force will act on the ball in the direction of the earth. Generally, one must also consider a force from air resistance that acts in the direction opposite to the ball's motion, but these problems often tell students that this force is negligible.

This, however, is not the answer given by most students. On the plus side, students generally do include the gravitational force in their account. The problem, though, is that students of all age groups will say that there is also a force acting in the direction in which the ball is moving. They seem to think that force is directly connected to a motion and that a motion can only happen, if a force acts in the direction of the motion. Students also often assume that this force was given to the ball during the

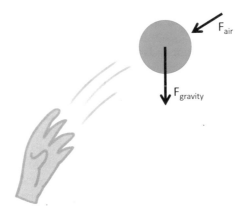

Fig. 14.1 Ball that has been thrown and is in the air will experience two forces and neither points in the direction of motion

throw, such that a "force" is put inside the ball. This "force" also will be used up during the flight. This becomes especially clear if the problem is stated with a ball thrown vertically in the air. Then, students will argue that an upward force is put in the ball. This "force" will be used up, and only after it has been used up, will the ball start to fall back down. The correct explanation that only a constant gravitational force accelerates the ball downward at all times is, for most students, extremely difficult to come to terms with.

Optics

One of the key ideas of optics is that one can see an object, if light from this object comes into one's eye. This seemingly simple idea contains some tricky aspects. First of all, observers' eyes sense light, but they do not "see" the light itself, as "seeing" includes cognitive processing of the stimuli to their retinae. Hence, what they "see" is actually the object where the light came from (see Fig. 14.2). Second, it is not

Fig. 14.2 Although an observer's eyes sense light, "seeing" involves cognitive processing of the stimuli: the observer "sees" the object and not the light itself

Fig. 14.3 Although the wall is not a light source, an observer can see it because it scatters light from a light source

clear at first how light can come from an object which is not a light source. The answer is that light from lamps and other light sources is reflected or scattered from (almost) all objects. During this, the characteristics of the reflected or scattered light can change (see Fig. 14.3). Finally, geometrical optics is a theory about processes. A person aiming to understand optical behaviour studies the light leaving the object, passing through or being reflected by optical objects like lenses or mirrors and finally reaching their eye. In this model, light rays are often used to depict those processes. All of this is happening with the speed of light, so the process does not seem like a process at all to an observer, but rather a simultaneous occurrence.

In all of these points, students' conceptions do not fit with scientific conceptions. Students think that they can see light itself. They also often think that they can see things without light being there. Many students find the idea that a non-luminous object also emits light to be absurd. They will argue, that they can clearly see, that a brick wall is not shining. Light propagates too quickly for us to see it going from the source to the wall, and so, rather than understanding optics as a process, students perceive it to be a static condition. So to learn the key idea of seeing things is, in many aspects, counterintuitive and hence quite a challenge.

Particulate Model of Matter

According to the particulate model of matter, every object consists of small, indivisible particles. Those particles are atoms (the standard model of particle physics plays no major role in the physics classroom in middle and high schools). Only about 100 different kinds of atoms exist. These different kinds of atoms are called elements, and they are organized in the periodic system of elements. In the physics classroom, the particulate model of matter is heavily drawn upon in the field of thermodynamics. With this model, one can explain the phases of matter or the behaviour of gases. One of the major challenges during the learning of the particulate model of matter is that of "emergence". "Emergence" is when systems consisting of many small parts show completely different behaviour than the single particles do. As an example

Fig. 14.4 Misleading figure that one might find in a textbook to show that water consists of particles

from thermodynamics, if one heats up a balloon, the volume of the gas in the balloon will increase. But that does not mean that the volume of the single gas atoms has increased.

Many students' conceptions regarding the particulate model of matter stem from a problem in understanding emergence. Students assume that the particles have the same properties as the system has. In addition to thinking that the volume of a gas atom can expand when the gas does, some students also think that the atoms of a gas will come to a rest once the gas has finished expanding, in the same way that a cart stops due to friction. In fact, although the gas itself is no longer moving, the individual atoms continue moving and colliding with each other at an average speed proportional to the temperature of the gas.

Another problem arises with the idea of vast emptiness described by the particulate model of matter. In the particulate model, everything consists of particles and emptiness between the particles. For students, this is a problematic idea. For them, it makes much more sense for there to be air (or water) between the particles. This incorrect conception is exacerbated by drawings in textbooks, where sometimes a water level is drawn into a depiction of a fluid consisting of particles (see Fig. 14.4).

Electricity

Think of the following circuit (see Fig. 14.5, a figure modified slightly from Shipstone et al. 1988) consisting of three identical light bulbs. Even after instruction, many students will answer that the current will divide in half at each connection, getting $I_1 = 6A$ and $I_2 = I_3 = 3$ A. They use a local argument and do not seem to have understood that an electrical circuit always has to be analysed as a complete system. Doing so yields the correct answer of $I_1 = I_2 = I_3 = 4$ A.

One can see another challenge to physics educators in this example. The rule of local argumentation is a good idea for many fields of physics, and one uses this all the time in, for example, mechanics. But electricity is a completely different kind of theory. For the analysis of simple circuits taught to middle and high school students, one relies upon so many idealizations in the model that sole reliance upon the rule of local argumentation gives wrong results. Much elaborated physics knowledge is needed to see such differences between the different fields and theories of physics.

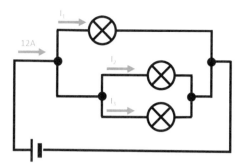

Fig. 14.5 Simple circuit problem that students answer incorrectly because of reliance on local reasoning: "first the current splits in half, so I_1 is 6 A, then the remaining 6 A splits in half again" (Shipstone et al. 1988)

14.2 What Is the Nature of Students' Conceptions?

There is disagreement in the field of physics education research regarding the structure of students' conceptions (diSessa 2009, 2017; Hammer 1996a, b; Vosniadou and Skopeliti 2014). One account considers conceptions to be specific to a given topic (e.g. "a conception about force") but to be stable and consistent across situations dealing with that topic. These researchers sometimes describe conceptions as being "theories" (Vosniadou and Skopeliti 2014), and they are thus sometimes referred to as using a *Theory Theory* view of student conceptions (diSessa 2017). A second group of researchers emphasizes how student conceptions are regulated by what the student perceives to be the nature of the object. Although you might imagine a party that is "three hours long", you surely cannot imagine a dog that is "three hours long". That is because a "party" is not merely different than a "dog"; it is a different *kind* of thing. That is, "dog" is categorized with a different "ontology" than "party". In the same way, it has been argued that students who think that current is used up in a light bulb do so because they are categorizing "current" with the ontology of "entity" when they should be categorizing it with the ontology of "process" (e.g. Chi 2013). In additional to the *Theory Theory* view and the Ontological view for thinking about student conceptions, there exists also the "Knowledge in Pieces" view (abbreviated often as "Pieces view"). In stark contrast to the other two views, the Pieces view attributes fluidity and flexibility to students' reasoning by perceiving their conceptions to be comprised of smaller knowledge pieces. Sometimes, you might hear a student say that an object stops moving on a horizontal surface because gravity slows it down. The *Theory Theory* view might be used to diagnose this student as having the theory that "gravity drains impetus". The Pieces view, in contrast, might describe the student as utilizing in this particular context two knowledge pieces, "dying away" (which is used for anything that decreases over time, e.g. the sound from a struck bell) and "interference" (which is used for something that impedes an effect and, in this case, is connected to gravity for the student). A crucial point about the Pieces view is that, whereas the theory "gravity drains impetus" is wrong, "dying away" and "interference" are not inherently correct or incorrect. Rather, they can either be useful or detrimental to student understanding, depending on the situation (Hammer 1996a).

There are other views in addition to the *Theory Theory* view, Ontological view and Pieces view, and to address all of them in a book chapter would not be feasible. We have chosen to focus only on these three because they are each well-developed, being established through decades of research findings, and, consequently, are well-known and respected by physics education researchers. Finally, they are views that you can draw upon in a useful way for your physics teaching. In this section, we elaborate upon each of these views in turn. In the next section, we discuss what implications this theoretical background has for your teaching in the classroom.

14.2.1 Theory Theory View

Researchers operating within the *Theory Theory* view attribute stability to student conceptions, and this characteristic is used to explain why students seem to "just not get it", even after a brilliant lecture. These researchers see incorrect conceptions (sometimes called "theories" by these researchers) as obstacles to learning that must be removed before effective learning can take place (Carey 1986, 2009; Strike and Posner 1982). With such a view, it follows that the most effective way to deal with incorrect theories is to have students confront them early on so that everyone can see that the idea is wrong and the theory can be made suspect. Doing this paves the way for successful learning to occur. Were this not to be done, then the incorrect theory might pop into the mind of the students in the middle of the lesson, making the lesson unintelligible. Confronting the theory early on thus acts as a sort of vaccine against misinterpreting the knowledge that will be presented in the ensuing lessons. The *Theory Theory* view sees students as having just one way of thinking about a conceptual topic (e.g. Newton's laws) at a given time. To be specific, students are often considered to have the incorrect theory prior to effective instruction (usually from everyday experiences in the world around them) and to have the scientific theory after conceptual change has occurred. Most literature documenting student conceptions implicitly treats them as though they are theories, for example, by concluding from survey responses that a student has a given (incorrect) way of thinking about the topic. Any of the examples discussed so far can serve as examples for theory-like conceptions, in the event that they do not readily change from one context to another.

14.2.2 Ontological View

In investigating the relative persistence of particular incorrect student conceptions, Chi and colleagues (e.g. Chi et al. 2012, 1994) found that, in the case of the more resilient conceptions, learners tend to characterize the relevant physics concept with the wrong *ontology* or mental category into which an individual divides the world (Keil 1979). The Ontological view uses a tree to organize these categories (see Fig. 14.6, which is modified from Chi 2013). The further down the tree you look, the more specific the category becomes and the more the category has in common with other categories at that level. "Birds", for example, are more similar to "Mammals" than "Animals" are to "Plants" (because both birds and mammals are animals). As such, many of the same words that are used to talk about birds can also be used to talk about mammals. One can say that either of the two animals is eating, resting or singing. These are not words that can be used to describe plants, unless it is understood in a figurative sense. The higher up the tree you are, the more difficult it becomes to convince someone to change how he or she categorizes the item. It is not particularly difficult to convince someone that a whale is a mammal and not a fish. But how would you ever convince someone that a bird is a "three-hour long process"?

According to the Ontological view, however, this is exactly what a teacher must accomplish when teaching people most topics of physics. Many concepts students encounter in physics such as force, electric current, heat and light are difficult for them to learn because, whereas the scientific understanding attributes to these concepts the ontology of "process", learners tend instead to see them as "substances" (Chi et al. 1994; Reiner et al. 2000; Slotta and Chi 2006; Slotta et al. 1995).

Even within the ontology of processes, Chi et al. found that students have difficulty replacing their incorrect conceptions related to diffusion, for example, of gas exchange in the lungs, because they are attributing a "sequential" ontology when the normative category is an "emergent" one (Chi 2005). That is, the scientifically accepted picture is that, on inhalation, the random collisions of oxygen and carbon dioxide molecules give rise to a net increase of oxygen and a net decrease of carbon dioxide in the bloodstream because of the initial concentration differences. This is in contrast to the theory of learners that oxygen is forced into the body and carbon

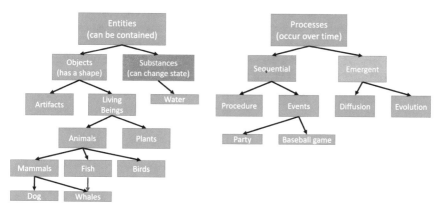

Fig. 14.6 Ontological view uses a tree to organize categories into which an individual divides the world

dioxide pushed out because of some driving agent. This ontological miscategorization in turn makes it difficult for students to remove the theory that, for example, after inhalation, *all* of the oxygen breathed in has gone into the bloodstream (Chi 2005).

The Ontological view of Chi et al. suggests that learners categorize objects or phenomena with a single ontology, which either is or is not the correct one. This is similar to the *Theory Theory* view, which treats learners as having only a single theory about a given topic that either is or is not correct.

Examples of student conceptions discussed by the Ontological view

The Ontological view generally does not focus on individual student conceptions, because the argument of the view is that *collections* of student conceptions arise in tandem as a result of categorizing a phenomena with the wrong ontology. Because students think of electrical current as being a "Substance" instead of a "Process", they are prone to believe the conceptions that current is stored in a battery and used up in a circuit, for example (Chi and Slotta 1993).

A popular example used by Chi et al. is that of diffusion. One incorrect student conception is that, when ink is diffusing, ink molecules do not move about in random directions, but rather in the direction of motion towards lower concentration, with intention. An additional conception is that, once the ink is uniformly distributed throughout the water, the molecules stop moving. Both of these conceptions plausibly originate from thinking about diffusion not as an emergent process, but rather as a sequential process (Chi et al. 2012).

In addition to explaining why student conceptions often exist in concert, the Ontological view also explains why many student conceptions are so resistant to change. Law and Ogborn (as cited in Chi 2013) discussed the progress of a student in writing a computer code to model the relationship of force and motion. This student wrote her code under the hypothesis that "Agents cause and control motion by acting as sources that supply force". When she did not get the expected results in the simulation, she tried patching her code in various ways, for example, by changing the agent causing the motion from gravity to air current and so on. She did not, however, change her underlying hypothesis, which corresponds to categorizing the motion of objects under the ontology of "sequential process". While it was easy for the student to try changing the agent responsible for the cause of motion, she did not succeed in changing the ontology with which she categorized the phenomena to the scientifically correct one of "emergent process" (Chi 2013).

14.2.3 Knowledge in Pieces View

A final view, known as "Knowledge in Pieces", is used to consider conceptions to generally be much more fluid than either the *Theory Theory* or Ontological views. The Pieces view treats students as having access to multiple ideas, which shift fluidly depending upon subtleties in the context surrounding the question at hand. This context-dependency is explained via attributing a different cognitive structure to conceptions than what the *Theory Theory* view does. Whereas in the *Theory Theory* view (usually implicitly), the conceptions themselves are considered to be the fundamental pieces of cognition, in the Pieces view, conceptions are considered to be assemblies of smaller pieces of knowledge that in turn come from physical experiences in the world (Smith et al. 1994) and that need not be tightly bound together (Hammer et al. 2006). It might seem difficult to understand the difference at first, but an analogy will help. Suppose a child builds a tower out of blocks. An observer can perceive what the child has as a single tower (a perspective representing the *Theory Theory* view in this analogy), but one can also perceive what the child has as a collection of blocks that, at the moment, *looks like* a single tower (a perspective representing the Pieces view in the analogy). According to the Pieces view, it surely is *possible* that the blocks become glued together into a single tower that can then not easily be changed (in this regard, the Pieces view subsumes the *Theory Theory* view), but that need not, in general, be the case. If one does not like the tower, it is not necessarily difficult to rearrange the blocks into, say, an elephant. These smaller knowledge pieces (the blocks in the analogy), which have been referred to as "resources" (Hammer 2000), "primitives" (diSessa 1993; Kapon and diSessa 2010), or "facets" (Minstrell 1992), are not inherently right or wrong (Hammer et al. 2006). For example, many students often erroneously think that a force is necessary for motion and perhaps go so far as to describe the "force of an object's motion". This is often accounted for as being an incorrect theory that "motion is caused by force". However, the Pieces view might account for this phenomenon by attributing its cause to, for example, an inappropriate usage of the primitive "maintaining agency". "Maintaining agency" is a knowledge element used to understand any continuing effect maintained by a cause, like how wood maintains a fire burning (Hammer 1996b).

Examples of Student Conceptions Described as Being Fluid in the Pieces View

Much support for the Pieces view comes in the form of interview and in-class recordings. (e.g. DiSessa 1993; Hammer 1996a, b, 2000; Hammer et al. 2006; Kapon and DiSessa 2010; Minstrell 1992; Smith et al. 1994).

DiSessa (2014) gave an example of students coming to understand Newton's law of cooling (that two bodies of different temperatures will reach an intermediate temperature at an exponentially decaying rate). One student (given the pseudonym "W") explained the observation that the liquid cools down quickly at first but then at a slower rate later with the following:

1. I think that the liquids like to be in an equilibrium.
2. So, when one is way off, they sort of freak out,
3. and work harder to reach equilibrium.
4. And when it is closer to equilibrium, they are more calm,
5. so they sort of drift slowly towards equilibrium.
6. So maybe that is why it moves fast at first, because it is like freaking out.
7. But then it just calms down as it approaches the right temperature (DiSessa 2014, p. 813).

The student invoked ideas about balancing in the explanation. One might encounter such an idea when balancing a balance scale when learning mechanics. Using abstract and intuitive knowledge pieces (in this case about balancing) across physics domains (mechanics and thermodynamics in this example) is one type of flexibility that one can observe in student reasoning. In this example, the agentive language (about the water "working harder") was only temporary. The students quickly and without any prompting removed this description and moved forward with a purely mathematical description. This is a second way student reasoning can be flexible. The strength of the Pieces view is that it provides a lens to physics teachers to perceive and capitalize on such flexibility.

As an additional example, let us return to the situation of throwing a ball upwards that we discuss at the start of this chapter. Many students, when faced with the situation of a ball being thrown upwards—and then asked what forces are acting on the ball—will say that there are two forces: gravity acting downwards on the ball and the force from the hand which stays with the ball and becomes smaller and smaller as the ball's speed decreases. This might at first suggest that students are reasoning with the "motion requires force" theory, with less motion, hence requiring less upward force from the throw. When thinking about this situation of the ball rising upwards, their reasoning strongly suggests that the "force from the hand" would reach zero when the ball comes to rest for a moment at the top of the throw. If this theory really were rigidly placed in the mind of students, one should expect them to give such an answer when next directly asked what forces act on the ball at the top of the trajectory. Many of these same students, however, reply in this case that the downward force of gravity must be balanced by an upward force, contradicting their prior reasoning (Hammer et al. 2006). The Pieces view can be used to account for this contradiction by describing the student response to each question as resulting from a temporary coherence of smaller knowledge pieces such as "force as mover" and "dying away" for the first question about the ascent and "dynamic balancing" for the apex of the throw (diSessa 1993). Although the particular combination of pieces might be incompatible with physics canonical knowledge, the pieces themselves are neither correct nor incorrect.

Hammer et al. (2006) discussed a student named Sherry who voiced the incorrect idea that one requires a full-length mirror to see one's full self in it. She gave this answer despite owning a half-length mirror in her room with which she saw her full self every day. Inside of her room, she would surely be aware that her mirror is only half her height and that she can see her full self in it. In the context of answering a physics question in class, however, the knowledge pieces that she draws upon for her reasoning fluidly shift. In another article, Hammer (2000) describes how students, when considering the collision of a moving truck with a stationary car, will tend to say that the truck exerts a larger force on the car than the car on the truck (in fact, the two forces are equal in magnitude, in accordance with Newton's third law). However, these same students, when first asked how the change in speed of the car will compare with that of the truck—and then asked what that means about the accelerations—are able to conclude (using $F = ma$) that the forces are, in fact, equal. Depending on how these students approach the problem, the knowledge pieces they draw upon change.

In an interview study conducted with high school students in Vienna, we found that student reasoning about the decay of an individual atom shifted fluidly, depending on the context. Bailey, for example, at first contended that an individual radioactive nucleus must fission at some point prior to the half-life. After considering some analogies involving flipping coins, Bailey seemed to recognize that half-life is not a useful concept for predicting when the fission will occur, as the fission can, in fact, take place at any time. However, in the context of a different interview prompt, Bailey reverted to using half-life to argue that the most likely time to see the atom fission would be on the day marking the half-life. (Hull and Hopf 2020).

So far, we have discussed the three most-established views for considering students' conceptions. At this point, you might be wondering, which view is most useful to bear in mind when planning your physics lessons? The answer is similar to the question of whether a photon is like a wave or a particle. Our answer is "it depends on the situation". It is therefore important for you to be aware of all three views and to consider their strengths as well as limitations. In the next section, we address how to put this theory into practice in the classroom.

14.3 What to Do in the Classroom

So far, we have compared three views for thinking about student conceptions and provided some examples that illustrate these views. This is relevant for you as a physics teacher because it can inform what you will do in response to student conceptions in the classroom. We must repeat that your students will not enter your classroom as blank slates that you can freely write upon; they will come with conceptions, many of which will be at odds with what you want to teach. You might find it surprising at first, but your students really will use arguments that reflect ideas that you will recognize from the research literature. So, it is not an option to just ignore students' conceptions if you want to be an effective teacher. One implication of this concerns carrying out experiments. According to the constructivist theory, students cannot observe without bias what is taking place; rather, they observe what they expect to occur. Many physics teachers assume that, if they show a demonstration to students, the students will be convinced of the truth that the demonstration shows. However, it is often the case that students see things that are not there. For example, consider a demonstration to show that current is a process and not a substance. You want students to see that the entire length of a wire gets hot and starts to glow at the same time when you connect it between two terminals of a high voltage source. Although that may in fact be what happens when the wire is connected, what students *see* is that the end of the wire closest to the positive end of the battery glows first if they think that current is a substance that travels from the positive to the negative terminals of the battery (Schlichting 1991). These students are building their new knowledge (their observation of glowing) onto their existing knowledge, and it results in strengthening their incorrect conception.

Just as there are different views about the nature of students' conceptions, there are different suggestions about how to help students undergo conceptual change. Depending on the nature of your students' conceptions, you can benefit from the suggestions of these education researchers. In this next section, we discuss the strategies put forth from practitioners of the *Theory Theory* view, the Ontological view and the Pieces view.

14.3.1 If Your Students' Conceptions Seem to Be Theory-Like

If you find that, time and time again, your students respond consistently with the same incorrect conceptions despite being faced with a variety of situations and despite you nudging them to think about the problem from a variety of perspectives, then it may be fruitful to think of your students' conceptions as being theory-like. In such a case, a general model for conceptual change is offered by Posner et al. (1982). According to this work, in order for a learner to replace existing ideas with canonical content knowledge, four criteria must generally be met as follows:

1. Students must recognize the inadequacy of their theory, for example, by being faced with a situation their theory cannot explain. This puts the learner in a state of cognitive conflict, without which, the learner will merely assimilate the new observation or fact into his or her existing theory without realizing that the theory was mistaken. Posner et al. pointed out that this first step might not take place if students compartmentalize the new knowledge. In the case of learning about the Earth being spherical, for example, students often think, "OK, so there are two Earths. One that we live on, which is flat, and another one up in the sky that goes around the Sun".

2. The new theory must be intelligible. This includes ensuring students know what variables in equations stand for, diagrams and graphs are clear and so on. Generally, students must be able to internalize the new theory. In the Earth example, if a teacher merely asserts that the ground beneath our feet is actually curved, that might be unintelligible to many children.

3. The new theory must appear plausible, in that (a) learners can imagine how the new theory can explain the situation their existing ideas could not explain, and (b) the new theory is consistent with other knowledge and fundamental assumptions that the learner has. In the case of a learner who is committed to the idea that the ground beneath his or her feet is flat, a hybrid model might emerge where the "ground beneath our feet" is curved—just not the part of the ground actually touching the feet (see Fig. 14.7). Such a hybrid model can hold traction for students because it is consistent with the ideas to which the learner is committed.

4. Learners must be able to imagine how the new theory can be used for additional situations as well. In the case of learning that the Earth is spherical, students will not likely see why the new theory is fruitful in their daily life without help

Fig. 14.7 Common hybrid model that learners develop when they hear that "the ground beneath our feet is curved", because the model is consistent with the ideas to which the learner is committed

from the teacher. If, on the other hand, the teacher leads them in seeing that the shape of the earth can be used for explaining day–night alternation, time zones, seasons and so on, then this final step could be fulfilled.

A general strategy which is consistent with these requirements is presented in the flow chart shown in Fig. 14.8. First, students are made aware that their existing conceptions are incorrect, resulting in students facing cognitive conflict and questioning their old ideas. Next, the correct conceptions are introduced to the students. The new ideas are explained to students, and they are defended with plausible arguments. Students then compare the new ideas with their former ideas to see the benefits afforded them by the new ideas. Finally, students apply the new ideas to new contexts. In the next section, we discuss some specific curriculum, which you can use if your students' ideas seem theory-like.

Fig. 14.8 Sequence of steps to meet the criteria that are generally necessary for conceptual change to occur if student conceptions are theory-like

Available Curricula to Help Bring About Conceptual Change (Theory-Like)

In practice, a useful technique to satisfy at least some of the steps described by Posner et al. (1982) is that of "Elicit, Confront, Resolve", which is utilized masterfully by curriculum from the University of Washington's Physics Education Group.[2]

In this approach, students consider questions that may elicit common incorrect theories and then are led to a situation in which their theories are shown to be incorrect or insufficient. They are then guided to "resolve" the conflict in favour of the correct theory (McDermott 1991). Figure 14.9 (modified from McDermott 1991) is from a learning sequence concerning electric circuits. At this point of the lesson, students are asked to discuss the arguments of Student #1 and Student #2 and to see which student they agree with. Many agree with Student #2—who reasons with the incorrect theory that current is used up in the circuit—and in so doing, have their incorrect theory "elicited". Students next set up the circuit and are "confronted" with the experimental result that, in fact, this theory is incorrect. At this point, students are guided to "reconcile" with the scientific theory that current is not used up.

This lesson comes from Tutorials in Introductory Physics (TIPs), a research-validated set of instructional materials. In TIPs, students work in small groups of three or four to complete guided worksheets, doing hands-on activities in some cases and having access to a shared writing space (such as a large sheet of paper or a white board in the middle of their table). TIPs has been shown to be effective in helping students undergo conceptual change across a wide variety of topics at different levels (Ambrose 2004; Ambrose et al. 1999; Lindsey et al. 2009; McDermott and Shaffer 1992; McDermott et al. 1994; Wosilait et al. 1998, 1999) with different populations of students (Sabella 2002; Steinberg and Donnelly 2002).

Student #1: All the current goes through A. Then it divides equally between B and C so they will both be equally dim compared to A. Then the current comes together again to go through D, which will be as bright as A.

Student #2: I think D will be much dimmer than A. There won't be much current left after it passes through the other three bulbs.

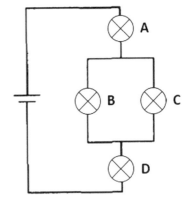

Fig. 14.9 Excerpt from a tutorial in introductory physics (reworded from Fig. 5 of McDermott 1991) designed to "elicit" the incorrect theory that current is used up as it passes through a bulb

---

[2] We do not mean to suggest that the University of Washington's Physics Education Group created their curriculum with the Theory Theory view in mind. Our point only is that the "Elicit, Confront, Resolve" approach, which is an approach well-aligned with the Theory Theory view, is frequently used in their materials.

An additional example comes from an intensive study on the construction of an optics curriculum for middle school students (Haagen-Schützenhöfer and Hopf 2020). One of the important aspects of optics is the spectral composition of white light. Usually, this is taught by teaching that white light can be split into different colours. But research has shown that this is very difficult for students to accept, especially since many seem to believe, that sunlight is yellow. Attempts to discover and utilize potentially productive knowledge pieces within this conception proved fruitless in the study, so it was decided to adapt a confrontational strategy instead. The optics curriculum demonstrates to students that sunlight and yellow light are different by showing them that red and green light, when mixed, is seen as yellow (Haagen-Schützenhöfer 2017). This demonstration serves as a confrontation after eliciting the incorrect theory from students. After this, students are ready to accept the resolution to the cognitive conflict when it is given by the teacher.

14.3.2 If Your Students' Conceptions Seem to Come from Ontological Mismatch

Do you find your students consistently thinking of force as something contained by an object? Do they consistently think that current is used up in going through a light bulb? And that heat is contained in hot molecules that is then given to cold molecules? Although these might at first appear to be unrelated incorrect theories, they might actually have something in common. They might all stem from an underlying difficulty of students categorizing physics phenomena using the "Substances" ontology rather than the "Processes" ontology.

According to the Ontological view, the approach to take in helping students successfully overcome resilient incorrect theories is to first teach them the correct ontology if it is not already present and to then help them build conceptual understanding of the topic from within that ontology (Slotta and Chi 2006). "Once students have successfully built such an alternative schema with its distinct set of properties, they can begin to assimilate new instruction (for example, about heat transfer) into the category" (Chi 2013, p. 68). According to the Ontological view, it is clear that the different ontologies are incompatible with each other, and, as such, teachers trying to help students learn a concept located in one ontology by drawing on their ideas in a different ontology would likely only lead to confusion (Chi and Slotta 1993; Slotta and Chi 2006).

According to the Ontological view, students have difficulty because they see much of physics content as "Entities", whereas they should be categorizing what they are learning as "Processes". When students do categorize with the "Process" ontology, they generally categorize the process as "Sequential", whereas, for most processes in physics, they should instead be categorizing the process as "Emergent". Whereas it may be relatively easy to tell students that a whale is a mammal and not a fish (because students have already compiled both the "mammal" and "fish" category),

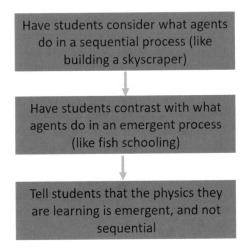

Fig. 14.10 Approach advocated by the ontological view in bringing about conceptual change

many students have not yet built a schema about emergence. This, then, is the first step proposed by the Ontological view in so-called ontology training (Chi et al. 2012; Slotta and Chi 2006). Once students have undergone the "ontology training" and successfully built the "emergent process" category, they can then be told that heat transfer is not a "sequential process", but rather an "emergent process". One thing teachers can do when pointing this out to students is to highlight the features which contrast the interactions between agents in emergent processes with those in sequential processes. For example, in emergent processes (like heat transfer), all agents behave more or less the same (molecules vibrate and bump into each other). In sequential processes (e.g. a baseball game), on the other hand, agents behave distinctly; for example, the pitcher does different things than the batter (Chi 2013). The flow chart in Fig. 14.10 summarizes the approach for conceptual change advocated by the Ontological view.

Available Curricula to Help Bring About Conceptual Change (Ontological Mismatch)

To the best of our knowledge, relatively few curricula have been created in physics that can readily be used for bringing about conceptual change when student conceptions seem to arise from ontological mismatch, as the work of Chi et al. has primarily been theoretical in nature. To teach students the "emergent process" ontology, Chi et al. created computer simulations on diffusion using NetLogo. These computer simulations are described in detail in Slotta and Chi (2006) and Chi et al. (2012). Text and questions are interspersed throughout these computer simulations to have students reflect on features contrasting the interactions between agents in emergent processes with those in sequential processes. A table of these interactions is available in Table 2 in Chi et al. (2012, p. 16).

14.3.3 If Your Students' Conceptions Seem to Be Pieces-Like

Do you find that your students demonstrate different understandings of the same concept across different contexts? Do they seem to have understood what you taught them in one moment, but then act as though they have an incorrect theory in the next? It may be the case that their ideas are not yet tightly woven together into theories, but rather that their conceptions are comprised of loosely bound knowledge pieces that can rearrange themselves depending on the context. If student conceptions were indeed context-dependent, effective instructional intervention would look quite different from that recommended by the *Theory Theory* view and the Ontological view described above.

From the perspective of the Pieces view, since the underlying knowledge, pieces are not inherently incorrect, and since their interaction with each other is relatively fluid, conceptual change is considered to be a change in which pieces are associated with which contexts; pieces are recombined, some become excluded, and others become included. With this perspective, the best way to help students learn involves capitalizing on the knowledge pieces embedded within their conceptions that are shared with the expert view of the material (e.g. Hammer 1996a). DiSessa maintains that it is important to build upon prior knowledge: "Students have a richness of conceptual resources to draw on. Attend to their ideas and help them build on the best of them" (DiSessa 2009, p. 41). According to the Pieces view, student conceptions contain knowledge pieces that can potentially be useful for reaching an expert understanding. With this view, the role of a teacher is not to structure their lessons to "dismantle and replace prior knowledge", but rather to "modify the organization and use of prior knowledge" (Hammer 1996a, p. 98). One strategy that encapsulates this philosophy of "modifying the organization and use of prior knowledge" is the "bridging strategy" (Brown and Clement 1989). In the bridging strategy, students are first asked a question that they will likely answer incorrectly; for example, "Does a table exert a force on a book at rest on it?" Many students answer incorrectly that there is no force, because "the table is just in the way" or "the table is not doing anything". This situation is referred to as the "target situation". Next, an analogous (from the perspective of the physics expert) situation that the student would most likely find intuitive and *would* answer correctly is subsequently posed: "suppose a book is sitting on a spring. Does the spring exert a force on the book?" This situation is referred to as an "anchoring example". In the case where the student, despite reasoning correctly about the anchoring example, continues to answer the target incorrectly, the teacher now modifies the anchoring example to be conceptually closer to the original target. With this example, the teacher might ask, "How about if the book is resting on a springy plank of wood that you see deform. Does the piece of wood exert a force on the book?" These situations that are intermediate between the original question and the anchoring example are known as "bridging analogies". This approach draws upon potentially productive knowledge pieces of students, in this case, the primitive of "springiness", to lead students to understand physics principles. These knowledge pieces are activated by students in some contexts (e.g. when thinking about pushing

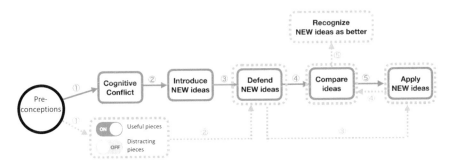

Fig. 14.11 Sequence of steps consistent with the pieces view (lower part of the flow chart) for bringing about conceptual change. In contrast, the sequence corresponding to the *Theory Theory* view is juxtaposed in the upper part of the flow chart

on a spring) but not in others (e.g. when pushing on a table). The goal is to create a bridge for students to activate these knowledge pieces in contexts where it leads to results that match the physics and not in contexts where they are at odds with the physics.

An alternative pedagogical approach consistent with the Pieces view is to pose new physics content to students in a way that will interact positively with the knowledge pieces they already have while avoiding triggering the knowledge pieces that would interfere with learning the scientific conceptions. Only once the new ideas have appeared sensible to students does the teacher then have them compare the new ideas with the typical incorrect conceptions. This process is represented in the lower part of the flow chart in Fig. 14.11.

Available Curricula to Help Bring About Conceptual Change (Pieces-Like)

A pre-print of the lesson plan described above that utilizes the bridging strategy to help students understand normal force is available freely on the internet (Camp and Clement 1994).

Open Source Tutorials, like Tutorials in Introductory Physics, emphasize conceptual understanding and consist of guided worksheets to be completed in groups. Whereas Tutorials in Introductory Physics heavily utilize the "Elicit, Confront, Resolve" strategy, Open Source Tutorials are characterized by having students build physics knowledge by "refining everyday thinking". Do you remember the example we describe earlier in this chapter of students first thinking that a moving truck exerts a larger force on a stationary car but then, when approaching the problem from a different perspective, are able to also find it intuitive that the forces are equal? That happens in the Newton's Third Law Open Source Tutorial. Open Source Tutorials and accompanying instructor's guides are available online at https://www.physport. org/curricula/MD_OST/.

An additional example of an educational approach that explicitly models student reasoning as being in "Pieces" that are not necessarily cohesively bound together is "Hypothesis—Experiment Class" (HEC). HEC curricular materials are comprised primarily of a "Classbook", which students receive from their instructor page by page. These pages contain multiple-choice questions asking students to make a prediction about an experiment. Students first make a personal prediction and then discuss in the whole-class setting. Once the discussion has come to a close, the experiment is carried out, and the next pages are distributed which contain the answer as well as the next question. HEC aims to have students consider a given problem from a variety of perspectives and to bring in various bits of knowledge in the discussions. For this reason, instructors are to encourage student ideas, for example, by saying: "Are there any other ideas that have not been heard yet? Maybe you chose A but you were kind of thinking B at first? What was attractive about B in the beginning?" Furthermore, to avoid "brutally forcing a theory on a student" (Itakura 2019, p. 11), minimal explanation, if any, is given between experiments in a Classbook (Tsukamoto 2004). Instead, the questions are carefully arranged such that students gradually come to a canonical understanding of the content material in a way that feels organic to them and that they are personally convinced of.

A final example of curriculum created in line with the Pieces View is the 2DD-mechanics curriculum (Spatz et al. 2020). This curriculum builds upon the knowledge pieces "Ohm's p-prim" (that more effort leads to a bigger result, mediated by a resistance) and "nothing comes from nothing". These ideas are used to show that the greater the force exerted on an object, the greater its change in velocity in that direction. This can be shown by hitting a rolling ball perpendicular to the direction of its velocity. The 2DD-curriculum was developed over a long time, and many research studies have been conducted to arrive at the final version of the curriculum. This version has been proven to enhance middle school students' learning of Newtonian mechanics significantly (Spatz et al. 2020).

14.4 Non-conceptual Factors that Affect Conceptual Change

Just as there is variability in how difficult it may be to have learners reach a scientific understanding of physics content, there is also variability in the reason why it is difficult. Sometimes, students fail to learn the physics content for conceptual reasons. Even if the content knowledge is crystal clear to the teacher, it might not make any sense to the students. Even if the new ideas are intelligible, students might still maintain their prior ways of thinking because it is not clear what additional explanatory power the new ideas provide. Often, however, the difficulty is not a purely conceptual one, but rather one interconnected with attitudinal factors, including interest and emotion. In a recent guest editorial in the *American Journal of Physics*, McIntyre (2019) described an exchange with a Flat Earth proponent. McIntyre suggested to the proponent to take a flight over Antarctica together. If the Earth is spherical, it should be possible to make the trip without refuelling. If, on the other hand, the Earth is flat (in which case one cannot "cross" Antarctica, but rather travel along it), it would be necessary to stop to refuel to reach the same destination. The flat-Earth proponent, however, argued that it might be the case that airplanes do not actually need to refuel to reach ANY destination on Earth, explicitly conceding that all of aviation history may have been a hoax (McIntyre 2019). Here, it does not seem to be the case that the flat-Earth proponent rejected the alternative theory because it was unintelligible or because it seemed to lack explanatory power. Rather, it seems plausible that a non-conceptual factor (in the sense that it does not directly relate to reasoning about the shape of Earth in this case) was to blame. Specifically, it seems plausible that a strongly held view that scientific consensus is a conspiracy theory could have enabled the flat-Earth proponent to defend ideas that flamboyantly defy science, turning a deaf ear to evidence and arguments that would show the learner's existing ideas to be inadequate (the first of four criteria we list above in the list from Posner et al. 1982). In their later work, Strike and Posner would describe this idea that non-conceptual factors can influence conceptual understanding in terms of a "conceptual ecology" (Strike and Posner 1992), and we encourage interested readers to consult their work to learn more about these factors.

14.4.1 Views About the Nature of Physics Knowledge and Learning

Hammer's (1991) Ph.D. dissertation pioneered a discussion of how an individual's approach to learning physics is affected by his or her beliefs about the nature of physics knowledge and about how physics should be learned. These student views might affect how a student prepares for class as well as his or her assessment of how well he or she "knows" the material being taught. A student who stably sees physics as a plethora of disconnected facts, for example, might be content to memorize equations without considering how those equations relate to each other or to the world outside the classroom (Hammer 1989; Lising and Elby 2005; Rosenberg et al. 2006). In interviews during his study, Hammer had students solve physics problems, explain equations and explain textbook passages. He found that students who think of physics knowledge as being made up of "formulas loosely associated with conceptual content" were prone to abandon their common sense and to conclude, for example, that two blocks connected by a cord will have different speeds despite travelling together or that a pendulum bob will have "zero velocity" despite continuing to move. On the other hand, Tony (one interviewee)—who felt that understanding physics means being able "to explain in qualitative terms" and that solving problems should involve conceptualizing the situation (instead of just plugging and chugging)—felt a need to find what mistake he had made in calculating that the two tethered blocks have different speeds (Hammer 1994). If students do not believe that learning physics involves learning concepts (and instead think of it just as a matter of plugging the right numbers into the right equation), then it is difficult to imagine how the process of conceptual change can succeed, since the concepts that are supposedly changing in conceptual change are irrelevant to the students. Similarly, if students do not feel that physics knowledge is coherent, and/or that learning physics does not involve cohering that knowledge together, then the step of "cognitive conflict" used so heavily by the *Theory Theory* view, for example, is unlikely to occur, because students will not mind that their ideas are inconsistent with what they observe in the classroom.

The importance of these attitudes on a longer time scale has been studied as well. For example, student responses to various attitudinal surveys administered prior to taking a college physics class have been shown to correlate with which course the students choose to take, with conceptual gains within that class (Halloun and Hestenes 1998; Perkins and Gratny 2010), with student interest in physics (Perkins et al. 2005) and with who decides to become a physics major (Perkins and Gratny 2010). In general, as a result of their beliefs about physics, students might have knowledge and abilities that they never use to their detriment both in terms of enjoying physics class and in terms of succeeding in it.

14.5 Conclusion

In this chapter, we have discussed students' conceptions and how awareness of them can influence your teaching. One has to understand that it never will be possible to eradicate students' conceptions. Many conceptions are deeply rooted in the system of human cognition and, as such, are extremely resistant to change. For example, everyone uses the word "sunrise" even though it is common knowledge that that it is not the sun rising but rather the Earth rotating that leads to the phenomenon. Other conceptions are flexible reactions to specific situations. Although physics teachers might succeed in suppressing some reactions of their students while promoting others in the context of the classroom, they must recognize with humility that the pattern of which conceptions activate will likely be different in a different context (e.g. once the students have graduated from the class). One realistic goal for a physics teacher, however, is to expose students to a new idea in a meaningful way, in the sense that the students can see that this new idea will prove better in some contexts. The more contexts in which teachers can get students to use the scientific ideas, the better.

For your instruction to be successful, it should be informed by awareness of the nature of your students' conceptions, whether they are theory-like, pieces-like or whether they seem to originate from miscategorizing physics phenomena. We encourage you, as a physics teacher, to listen carefully to your students, so that you can decide how best to characterize their ideas and respond accordingly. As you might expect, doing this is easier said than done. Indeed, it remains an important open question in education research how to facilitate teachers in making such decisions. Determining which view (or views!) is most productive in categorizing a students' understanding at a given moment requires practice, and we encourage readers to dedicate a small amount of time after teaching in the classroom to reflect on the lesson and to consider how that experience compares with what is described and encouraged by physics education researchers.

Acknowledgements We would like to thank Kitty Tang (associate professor in education at Southwest University in Chongqing), Eric Kuo (assistant professor in physics education, University of Illinois at Urbana Champaign), Ahmet Ilhan Sen (professor in physics education at Hacettepe University) and Luke Conlin (assistant professor in chemistry/physics at Salem State University) for their careful review and insightful comments on our chapter. We wish also to thank Shizuka Nakayama for the illustrations.

References

Ambrose BS (2004) Investigating student understanding in intermediate mechanics: identifying the need for a tutorial approach to instruction. Am J Phys 72:453

Ambrose BS, Shaffer PS, Steinberg RN, McDermott LC (1999) An investigation of student understanding of single-slit diffraction and double-slit interference. Am J Phys 67:146

Banholzer A (1936) Die Auffassung physikalischer Sachverhalte im Schulalter. Dissertation. Universität Tübingen

Brown DE, Clement J (1989) Overcoming misconceptions via analogical reasoning: abstract transfer versus explanatory model construction. Instr Sci 18(4):237–261. https://doi.org/10.1007/BF0011 8013

Camp C, Clement J (1994) Normal forces—lesson 2. In: Camp C, Clement J (eds) Preconceptions in mechanics: lessons dealing with students' conceptual difficulties, p 22. Retrieved from http:// citeseerx.ist.psu.edu/viewdoc/download?doi=10.1.1.673.8182&rep=rep1&type=pdf

Carey S (1986) Cognitive science and science education. Am Psychol 41(10):1123. Retrieved from http://psycnet.apa.org/fulltext/1987-08644-001.html

Carey S (2009) The origin of concepts. Oxford University Press, Oxford, UK

Chi MTH (2005) Commonsense conceptions of emergent processes: why some misconceptions are robust. J Learn Sci 14(2):161–199. https://doi.org/10.1207/s15327809jls1402_1

Chi MTH (2013) Two kinds and four sub-types of misconceived knowledge, ways to change it, and the learning outcomes. In: Vosniadou S (ed) International handbook of research on conceptual change, 2nd edn. Routledge, New York, pp 61–82

Chi MTH, Slotta JD (1993) The ontological coherence of intuitive physics. Cogn Instr 10(2–3):249–260. https://doi.org/10.1080/07370008.1985.9649011

Chi MTH, Slotta JD, De Leeuw N (1994) From things to processes: a theory of conceptual change for learning science concepts. Learn Instr 4(1):27–43. Retrieved from https://www.sciencedirect.com/science/article/pii/0959475294900175

Chi MTH, Roscoe RD, Slotta JD, Roy M, Chase CC (2012) Misconceived causal explanations for emergent processes. Cogn Sci 36(1):1–61

diSessa AA (1993) Toward an epistemology of physics. Cogn Instr 10(2–3):105–225. https://doi.org/10.1080/07370008.1985.9649008

diSessa AA (2009) A bird's-eye view of the "pieces" vs. "coherence" controversy (From the "pieces" side of the fence). In: Vosniadou S (ed) International handbook of research on conceptual change, p 35. Retrieved from https://books.google.at/books?hl=en&lr=&id=sdyOAgAAQBAJ&oi=fnd&pg=PP1&dq=A.+diSessa,+in+International+Handbook+of+Research+on+Conceptual+Change,+edited+by+S.+Vosniadou+(Routledge,+New+York,+2008&ots=qAW9Qp7Xqo&sig=ek10is6IrNONz8GE7XcL52IgTyU

diSessa AA (2014) The construction of causal schemes: learning mechanisms at the knowledge level. Cogn Sci 38(5):795–850

diSessa AA (2017) Conceptual change in a microcosm: comparative analysis of a learning event. Hum Dev 60(1):1–37

Driver R, Rushworth P, Squires A, Wood-Robinson V (2005) In: Driver R, Rushworth P, Squires A, Wood-Robinson V (ed) Making sense of secondary science: research into children's ideas. Routledge, London

Duit R (2009) Bibliography—STCSE students' and teachers' conceptions and science education. Retrieved July 31, 2019, from http://archiv.ipn.uni-kiel.de/stcse/

Duit R, Treagust D (2012) Conceptual change: still a powerful framework for improving science teaching and learning. Issues Challenges Sci Educ Res 1(1):43–55

Flores CF, Vega MEJ, Tovar MME, Bello GS, Gamboa F, Castañeda R et al (2014) Ideas previas. Retrieved July 31, 2019, from http://www.ideasprevias.ccadet.unam.mx:8080/ideasprevias/sea rching.htm

Haagen-Schützenhöfer C (2017) Students' conceptions on white light and implica-tions for teaching and learning about colour. Phys Educ 52:44003

Haagen-Schützenhöfer C, Hopf M (2020) Design-based research as a model for systematic curriculum development—the example of a curriculum for introductory optics. PRPER (accepted)

Halloun I, Hestenes D (1998) Interpreting VASS dimensions and profiles for physics students. Sci Educ 7:553

Hammer D (1989) Two approaches to learning physics. Phys Teacher 27:664

Hammer D (1991) Defying common sense: epistemological beliefs in an introductory physics course. University of California, Berkeley

Hammer D (1994) Epistemological beliefs in introductory physics. Cogn Instr 12(2):151–183. https://doi.org/10.1207/s1532690xci1202_4

Hammer D (1996a) Misconceptions or P-prims: how may alternative perspectives of cognitive structure influence instructional perceptions and intentions. J Learn Sci 5(2):97–127. https://doi. org/10.1207/s15327809jls0502_1

Hammer D (1996b) More than misconceptions: multiple perspectives on student knowledge and reasoning, and an appropriate role for education research. Am J Phys 64(10):1316–1325. https:// doi.org/10.1119/1.18376

Hammer D (2000) Student resources for learning introductory physics. Am J Phys 68(S1):S52–S59. https://doi.org/10.1119/1.19520

Hammer D, Elby A, Scherr RE, Redish EF (2006) Resources, framing, and transfer. In: Mestre JP (ed) Transfer of learning from a modern multidisciplinary perspective, vol 1, pp 89–121. Retrieved from http://umdperg.pbworks.com/w/file/fetch/51074580/Transfer_chapter_final.pdf

Hestenes D, Wells M, Swackhamer G (1992) Force concept inventory. Phys Teacher 30(3):141–158

Hull MM, Hopf M (2020) Student understanding of emergent aspects of radioactivity. Int J Phys Chem Educ 12(2):19–33

Itakura K (2019) In: Funahashi H (ed) Hypothesis-experiment class (Kasetsu). Kyoto Univ. Press, Kyoto and Trans Pacific Press, Melbourne

Kapon S, diSessa AA (2010) Instructional explanations as an interface—the role of explanatory primitives. In: AIP conference proceedings, pp 189–192. https://doi.org/10.1063/1.3515195

Keil FC (1979) Semantic and conceptual development: an ontological perspective. Harvard University Press, Cambridge, MA

Libarkin J, Asghar A, Crockett C, Sadler P (2011) Invisible misconceptions: student understanding of ultraviolet and infrared radiation. Astron Educ Rev 10(1)

Lindsey BA, Heron PR, Shaffer PS (2009) Student ability to apply the concepts of work and energy to extended systems. Am J Phys 77:999

Lising L, Elby A (2005) The impact of epistemology on learning: a case study from introductory physics. Am J Phys 73(372)

McDermott LC (1991) Millikan lecture 1990: what we teach and what is learned—closing the gap. Am J Phys 59(1):301

McDermott LC, Shaffer PS (1992) Research as a guide for curriculum development: an example from introductory electricity. Part I: investigation of student understanding. Am J Phys 60:994

McDermott LC, Shaffer PS, Somers MD (1994) Research as a guide for teaching introductory mechanics: an illustration in the context of the Atwood's machine. Am J Phys 62:46

McIntyre L (2019) Calling all physicists. Am J Phys 694–695

Minstrell J (1992) Facets of students' knowledge and relevant instruction. In: Research in physics learning: theoretical issues and empirical studies. Proceedings of an internatnl workshop, pp 110–128. Retrieved from http://www.citeulike.org/group/9538/article/4445910

Neidorf T, Arora A, Erberber E, Tsokodayi Y, Mai T (2020) Student misconceptions and errors in physics and mathematics: exploring data from TIMSS and TIMSS advanced. Springer, Berlin

Neumann S, Hopf M (2011) Was verbinden Schülerinnen und Schüler mit dem Begriff 'Strahlung.' Zeitschrift Für Didaktik Der Naturwissenschaften 17:157–176

Perkins KK, Gratny M (2010) Proceedings of the 2010 physics education research conference. AIP, Portland, OR, pp 253–256

Perkins KK, Gratny MM, Adams WK, Finkelstein ND, Wieman CE (2005) Proceedings of the 2005 physics education research conference. AIP, Salt Lake City, UT, pp 137–140

Plotz T (2017) Lernprozesse zu nicht-sichtbarer Strahlung. In: Empirische Untersuchungen in der Sekundarstufe 2, vol 240. Logos Verlag GmbH, Berlin

Posner GJ, Strike KA, Hewson PW, Gertzog WA (1982) Accommodation of a scientific conception: toward a theory of conceptual change. Sci Educ 66(2):211–227

Reiner M, Slotta JD, Chi MTH, Resnick LB (2000) Naive physics reasoning: a commitment to substance-based conceptions. Cogn Instr 18(1):1–34. https://doi.org/10.1207/S1532690XCI1 801_01

Rosenberg S, Hammer D, Phelan J (2006) Multiple epistemological coherences in an eighth-grade discussion of the rock cycle. J Learn Sci 15:261

Sabella MS (2002) Proceedings of the 2002 physics education research conference. AIP, Boise, ID

Sadler PM, Schneps MH, Woll S (1989) A private universe. Pyramid Film and Video, Santa Monica, CA

Schlichting HJ (1991) Zwischen common sense und physikalischer Theorie - wissenschafts-theoretische Probleme beim Physiklernen. Der Mathematische Und Naturwissenschaftliche Unterricht 44(2):74

Shipstone DM, Rhöneck CV, Jung W, Kärrqvist C, Dupin JJ, Johsua SE, Licht P (1988) A study of students' understanding of electricity in five European countries. Int J Sci Educ 10(3):303–316

Slotta JD, Chi MTH (2006) Helping students understand challenging topics in science through ontology training. Cogn Instr 24(2):261–289. https://doi.org/10.1207/s1532690xci2402_3

Slotta JD, Chi MTH, Joram E (1995) Assessing students' misclassifications of physics concepts: an ontological basis for conceptual change. Cogn Instr 13(3):373–400. https://doi.org/10.1207/ s1532690xci1303_2

Smith JPI, diSessa AA, Roschelle J (1994) Misconceptions reconceived: a constructivist analysis of knowledge in transition. J Learn Sci 3:115–163. https://doi.org/10.2307/1466679

Sneider C, Bar V, Kavanagh C (2011) Learning about seasons: a guide for teachers and curriculum developers. Astron Educ Rev 10(1):3847

Spatz V, Hopf M, Wilhelm T, Waltner C, Wiesner H (2020) Introduction to Newtonian mechanics via two-dimensional dynamics—the effects of a newly developed content structure on German middle school students. Eur J Sci Math Educ 8(2):76–91

Steinberg RN, Donnelly K (2002) PER-based reform at a multicultural institution. Phys Teacher 40:108

Strike KA, Posner GJ (1982) Conceptual change and science teaching. Eur J Sci Educ 4(3):231–240. https://doi.org/10.1080/0140528820040302

Strike KA, Posner GJ (1992) A revisionist theory of conceptual change. Philos Sci Cogn Psychol Educ Theor Pract 176

Tsukamoto K (2004) The theory of hypothesis—experiment class and 'new physics education research' developed since 1980 in Britain and the US. Butsuri Kyouiku 52(2) (in Japanese)

Vosniadou S, Skopeliti I (2014) Conceptual change from the framework theory side of the fence. Sci Educ 23:1427–1445

Wosilait K, Heron PR, Shaffer PS, McDermott LC (1998) Development and assessment of a research based tutorial on light and shadow. Am J Phys 66:906

Wosilait K, Heron PR, Shaffer PS, McDermott LC (1999) Addressing student difficulties in applying a wave model to the interference and diffraction of light. Am J Phys 67:S5

Chapter 15
Formative Assessment

Mathias Ropohl and Claudia von Aufschnaiter

Abstract Students who enter the physics classroom hold a wide range of ideas about physics phenomena that they have established from their everyday experiences: Forces can be transferred to an object and are consumed, while the object is moving, current is used up in a bulb, and atoms enlarge when an object is heated. Research has demonstrated that students are more likely to learn if teachers elicit their ideas about these phenomena and adapt instruction accordingly. Teachers should furthermore monitor students' learning during, and as a result of, instruction to promote further learning. Assessment is, therefore, an important teacher activity aiming at improving student understanding of physics. Within this chapter, formative assessment—also called assessment for learning—how such assessment can inform teaching and learning processes. Based on a literature review, the chapter offers an overview about the different theoretical perspectives on assessment, focusing particularly on key characteristics of formative assessment. The process of formative assessment and its components are presented. Furthermore, feedback as a means for promoting students to make further progress is unpacked. Within the chapter, theoretical concepts and empirical results are related to examples from the physics classroom and some practical guidelines are also offered.

Keywords Formative assessment · Assessment for learning · Assessment practices · Feedback

15.1 Setting the Scene—A Lens on Physics Classrooms

Imagine that a teacher has used experiments and some information to help his or her students to understand that light is not just "there" but travels from a source to an

M. Ropohl (✉)
Universität Duisburg-Essen, Essen, Germany
e-mail: mathias.ropohl@uni-due.de

C. von Aufschnaiter
Justus-Liebig-Universität Gießen, Gießen, Germany
e-mail: Claudia.von-Aufschnaiter@didaktik.physik.uni-giessen.de

© Springer Nature Switzerland AG 2021
H. E. Fischer and R. Girwidz (eds.), *Physics Education*, Challenges in Physics Education,
https://doi.org/10.1007/978-3-030-87391-2_15

object from which it is scattered or reflected. The experiments are used to demonstrate that a detector determines light that is directly or indirectly (via an object) hitting the sensor. Parallels between a detector and the human eye are discussed by the students. The students are then asked to complete the task depicted in Example 1 (see Fig. 15.1) that focuses on the process of seeing. Even though being exposed to the same instruction, the students solve the task in very different ways, indicating that their understanding of the topic varies.

Whereas Nena in Example 1 seems to understand that light travels from a source to an object and from there to the eye in order for her to see, Tolga acknowledges that light is relevant to seeing, but does not consider any direction. Lars and Dana both seem to have understood that light travels to an object, but may not yet regard seeing as a process that involves light entering the eye. However, Dana may have some idea of this process, as she seems to consider a direction "looking at the object." How might the teacher proceed now? It does not seem very likely that the same explanation would fit the process of seeing for all four students. What kind of individual feedback and instruction for further learning would you as a teacher offer to each student?

Without a task similar to the one used in Example 1, the teacher may not even have had the chance to notice that each student may differ in their individual understanding of the topic. Without further monitoring the students, the teacher would also not be able to tell whether her or his feedback and related instruction helps the students to develop their understanding further. The example, therefore, demonstrates that a particular process, which is called *formative assessment*, is an important instructional approach. Formative assessment is an attempt to elicit students' competences: their knowledge and understanding, their cognitive skills, abilities, and their

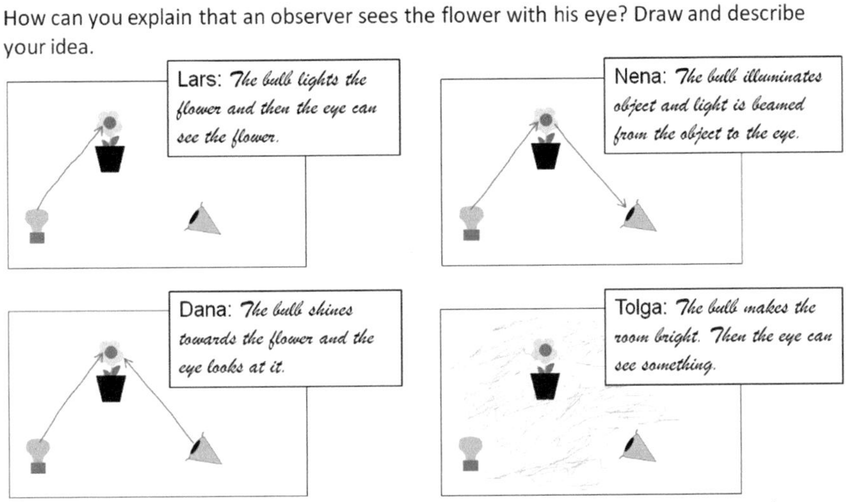

How can you explain that an observer sees the flower with his eye? Draw and describe your idea.

Lars: The bulb lights the flower and then the eye can see the flower.

Nena: The bulb illuminates object and light is beamed from the object to the eye.

Dana: The bulb shines towards the flower and the eye looks at it.

Tolga: The bulb makes the room bright. Then the eye can see something.

Fig. 15.1 Students' explanations of the process of seeing (Example 1; translated from von Aufschnaiter et al. 2020, p. 534)

motivation and interest to engage with physics activities, phenomena, and explanations (Cizek 2010; Liu 2010). Formative assessment is essentially based on the idea that learning is most likely to happen if instruction matches students' current competences and their learning processes. Already in the late 1960s, Ausubel (1968) mentioned this connection: "The most important single factor influencing learning is what the learner already knows. Ascertain this and teach [...] accordingly" (p. vi). This quote indicates that formative assessment is more than simply saying that a student has or has not mastered a particular task, does (not) know or understand a particular concept, can(not) enact a particular strategy, or is (not) willing to engage with physics. Unpacking what a particular student understands and is able to do from the student's (verbal) activities and products can be challenging to teachers but has also great potential to enrich teachers' professional knowledge and can result in more satisfying learning support. Formative assessment is essentially oriented toward eliciting evidence that helps a teacher to promote further learning. Therefore, formative assessment is central at the beginning and during a teaching unit, in order to assess which competences students bring to the classroom. It is also known as *diagnostic assessment* (Liu 2010), which is an essential ongoing activity to identify students' progress during instruction. Within the next sections, we conceptualize formative assessment and its components and elaborate how to enact formative assessment in the classroom.

15.2 Formative Assessment—What Is It?

Assessment in general and formative assessment in particular is considered as a key to improve teaching and learning. A first milestone in this field was a review by Black and Wiliam (1998) that highlights the promising relation between an ongoing assessment and effective classroom learning. Such an ongoing assessment with a strong focus on students' needs is called formative assessment. For two decades, science education research has focused on this kind of assessment. It is often referred to as *assessment for learning* in contrast to summative assessment that addresses *assessment of learning*. Two decades later, it seems to be confirmed that formative assessment has a strong positive influence on teaching and learning. Especially, Hattie (2009) summarized factors that positively affect learning processes, some of which are key features of formative assessment such as feedback (Hattie and Timperley 2007).

In recent years, researchers have tried to find a comprehensive definition of formative assessment. Until today, different extant definitions have typically specified the construct by describing single characteristics. One example is the definition by Cizek (2010):

> Broadly conceived, formative assessment refers to the collaborative processes engaged in by educators and students for the purpose of understanding the students' learning and conceptual organization, identification of strengths, diagnosis of weaknesses, areas for improvement,

and as a source of information that teachers can use in instructional planning and students can use in deepening their understandings and improving their achievement. (pp. 6–7)

Harlen and James (1997), as well as Bell and Cowie (2001), describe basic principles that underlie the construct of formative assessment and that become visible in the definition cited above.

1. **Fostering individual learning**. Formative assessment is an ongoing process aiming for further development of learning opportunities in order to better support individual student learning. It does not only identify students' current competences (e.g., their knowledge and understanding) but also assesses students' progress during learning. The support offered on the basis of the assessment can be provided at different levels: New advanced questions and tasks can be issued or differentiated explanations and tasks can be formulated according to students' current understanding and abilities. With its focus on development, the notion *assessment for learning* is often used instead of the term *formative assessment*. As summative assessment is basically not aiming at fostering learning, it is also called *assessment of learning*. In order to support further development of learners, formative assessment holds a positive stance toward what students know and are able and willing to do. Summative assessment, in contrast, focuses on whether or not students' performance meets particular standards.

2. **Incorporating various resources**. Formative assessment is typically conducted at the beginning of and during instruction and can incorporate different resources of evidence. Teachers offer tasks, ask questions, or offer prompts that help to elicit student thinking. These tasks, questions, or prompts lead to data that serve as resources for further interpretation. Resources are, for example, students' written answers on worksheets or observations while students work on instruction, interact with learning material or with each other. Each resource has different potential and limitations for interpreting student understanding and abilities. Ideally, teachers collect different resources to obtain an authentic and comprehensive picture of student thinking (Cowie et al. 2011).

3. **Weighing efficiency against purpose**. Not only are the resources relevant for the quality of the interpretation but also how they are collected. On the one hand, formative assessment can be realized spontaneously and on the fly, for instance, by observing students' interactions with particular learning material. This kind of formative assessment is less standardized and leads to unsystematic and undifferentiated data that may limit the interpretation. At the same time, this approach is more open to the dynamic situation of teaching and learning in the classroom. This means that this kind of formative assessment can be used on a more frequent, everyday basis to get a sense of students' learning. On the other hand, formative assessment can be standardized and accurate by, for instance, incorporating rubric-based analysis of students' written answers. Such approach requires to plan the assessment more carefully and needs expanded knowledge about how to interpret the data, but it can result in more sound interpretations. It has to be noted that summative assessment is often characterized in

similar way—the necessity to plan and standardize the assessment—which can be confusing. It is therefore important to distinguish both types of assessment by the purpose for which data is collected, assessment for learning vs. assessment of learning, and not by the level of standardization.

4. **Considering errors as resources for learning.** In summative assessment, weaknesses would be detected as errors, but for formative assessment, they provide *diagnostic information* that can be used to infer the next steps in learning. For instance, in the example in Fig. 15.1, Tolga's drawing in particular is incorrect from a scientific perspective, yet plausible: We do not see any light traveling from a source to somewhere; rather, if we turn on the light, the room is typically instantly bright. With her mistake in answering the task, Tolga offers information on his understanding. He can be involved in comparing his drawings with the others, thinking about an experiment that can help to understand that light is not automatically everywhere.

5. **Engaging teachers *and* students.** Formative assessment is interactive in the sense that teachers *and* students enact it. It can involve individual students as well as the whole class. This is the case with self- and peer-assessment. For instance, students could evaluate each other's written texts and give each other feedback that highlights what is elaborated well and what may need to be improved. In contrast, summative assessment is not interactive, it is planned and enacted by the teacher. Besides, in formative assessment, students have to be active in their own learning. They have to understand their own strengths and weaknesses; in order to achieve this, they must be actively engaged in analyzing their own activities and learning products. Therefore, a basic principle of formative assessment is students' responsibility for their own learning.

The definition and the description of the underlying principles 1–5 above indicate that formative assessment is a complex process, not only because it includes different sources of evidence and is enacted by all members of a classroom, but also because it can focus on different aspects of student competence. Example 1 in Fig. 15.1, for instance, depicts students' drawings in completing a task on the process of seeing; thus, assessment focuses on students' understanding of a particular physics principle. The next example, Example 2 (see Fig. 15.2), offers a transcription of how students conduct an experiment. Here, a teacher could formatively assess students' abilities to manipulate experimental material and their cognitive skills of controlling variables; by hanging the masses below each other, the students move the center of mass a bit, which affects the length of the pendulum. The example also demonstrates that a teacher might focus on students' motivation to engage with physics. Finally, Example 3 (see Fig. 15.3) can be used to identify not only the students' understanding of Newton's First and Second Law but also how they experience physics and how confident they are about their competences.

The three examples presented so far not only demonstrate that teachers can focus on different aspects of students' competences (knowledge and understanding, skill, ability, and motivation and interest) but also that teachers can use different kinds of data as sources of evidence. They may refer to students' drawings (e.g., Example 1

Task 1

Investigate which factors have an impact on the periodic time of a pendulum. You can use different masses and strings (material, length). Remember to use small elongations only and to measure the time for ten oscillations which you then divide by ten in order to get the time for one period.

Students' conversation

S1: Ok, should we start?

S2: What do we have next, Sports? Oh, yes. Are we going to play basketball again?

S1: Come on, Mr. B is watching us. Let's really start. (takes a string and attaches it to the set-up; hangs one piece of mass at the end of string; pulls pendulum a little bit to one side and lets it go) One, two, three, did you start the stopwatch?

S2: Oh, forgotten. How do I get it started? (presses several buttons)

S1: (takes stopwatch) Stop (presses right hand button), reset (presses left hand button), and start (shows right hand button, passes watch to S2).

S1: Ok, again. (pulls pendulum). Now! (lets pendulum go) One, two, three, four five, six, seven, eight nine, ten-stop!

S2: (has started and stopped watch, looks at watch) 22.9 seconds. Divided by ten. 2.29 seconds.

S1: (tries to attach the second piece of mass below the first piece; the second piece falls to the ground; tries again, set-up wobbles, both pieces of mass fall on the ground)

S2: (talks with other students about basketball)

S3 from another group joins in: Can I help you? (attaches both pieces of mass to the string, pieces are hanging below each other)

S1: Cheers mate. Hey [to S2] would you mind coming over or are you discussing something super important? [ironically]

S2: Come on. (takes stopwatch)

S1: (pulls pendulum) Now! (lets pendulum go) One, two, three, four, five, six, seven, eight, nine, ten-stop!

S2: (looks at stopwatch) 23.2 seconds. Longer, isn't it?

S1: Dunno. Let's try five pieces of mass. (hangs pieces below each other, takes three to four trials for each additional piece) That's so stupid fiddly.

S2: (talks to another group about a movie)

S1: Ready?

S2: What? (turns around). Yes, sure.

S1: (pulls pendulum) Now! (lets pendulum go) One, two, three, four, five, six, seven, eight, nine, ten-stop!

S2: 23.5 seconds. Longer.

S1: Ok. Let's write "The more mass is attached to the pendulum, the longer one period takes." Well, not quite a surprise though.

Fig. 15.2 Task and students' conversation while enacting scientific inquiry (Example 2; adapted from von Aufschnaiter et al. 2020, p. 547)

Task 2

A gardener pulls a big flower bucket through his market garden. He exerts a constant horizontal force; therefore, the bucket moves with constant velocity along the floor.

The constant horizontal force exerted by the gardener is

(A)... as great as the bucket's weight.

(B)... greater than the bucket's weight.

(C)... as great as the sum of all forces acting against the bucket's movement.

(D)... greater than the sum of all forces acting against the bucket's movement.

Students' conversation

S2: *[...] Well, I would say greater, isn't it?*

S1: *Greater? (reads aloud) "than the sum of all forces acting against the bucket's movement. [D]" (reads) "greater than the bucket's weight. [B]" I don't understand...*

S2: *(interrupts) But no, wait. Hold it. Same magnitude, because the bucket is moving already. We don't have to accelerate it. It says "the bucket moves with constant velocity", that is, it moves. (indicates movement on the table) And we are right in the middle of the movement. Therefore, they have to be the same [C].*

S1: *Oh, physics can be so confusing! I hardly understand anything.*

S2: *Well, I don't know either. If they are the same, then the bucket would stand still, wouldn't it? (indicates stopping with his hands)*

S1: *Gosh, what a mess!*

S2: *Let's tick D.*

S1: *Ok. (writes on worksheet)*

Fig. 15.3 Task and students' conversation while solving the task (Example 3; translated and adapted from von Aufschnaiter et al. 2020, p. 545)

in Fig. 15.1), multiple-choice or written answers or answers they offer verbally (e.g., decision to tick answer D at the end of Example 2 in Fig. 15.3). Thus, they focus on students' *solutions* of particular tasks. These solutions can be used to infer students' competences but are also limited as information is missing on the *processes* by which students arrived at a particular solution. With this information, teachers get a more comprehensive idea of students' competences. For instance, Example 3 in Fig. 15.3 reveals that S2 seems to have some understanding of Newton's First and Second Law. However, it flips between an appropriate physics concept (solution C of Task 2, if the velocity is constant, then the net force is zero) and one that is plausible from her everyday experiences: In order to keep an object moving, you always have to exert a force (solution D of Task 2). If a teacher would only analyze the students' final answer—expressed with checking box D—he or she would lose important information, demonstrating that S2 has some appropriate understanding, even though this is not yet stable. Collecting data on solution processes offers not only more comprehensive information about a particular facet of students' competences (e.g., their understanding of Newton's Laws), it also has the chance to offer more information on other facets such as whether students feel confident with what they know (see Fig. 15.3) or are willing to engage in particular tasks (see Fig. 15.2). If students know Newton's First Law already but seem to have some problems identifying where it applies, they would need different types of assistance as compared to those who have not yet understood the law. If students do not engage in physics tasks, promoting their learning would need different contexts or tasks that are more challenging.

Acknowledging that formative assessment can use students' *solutions to a task* only or can more holistically (also) consider the *processes by which a particular task is solved*, two different types of formative assessment can be distinguished: *solution-oriented formative assessment* and *process-oriented formative assessment* (von Aufschnaiter et al. 2020). Collecting evidence from solutions only within solution-oriented formative assessment makes this type of assessment more convenient within a classroom as teachers cannot easily assess the solution processes of various students or groups (e.g., by observing what they do while working on a particular task). Solutions can also very clearly indicate particular learning difficulties (such as those depicted in Example 1 in Fig. 15.1). However, teachers should be aware that they might lack some important information on students' current competences. Solution-oriented formative assessment may not be further applied to identify particular competences; for instance, competences related to scientific inquiry (see Example 2 in Fig. 15.2 that indicates, among other aspects, students' limited understanding and/or consideration of measurement uncertainty). Thus, not only can process-oriented formative assessment expand information on students' competences, but it can also identify *specific* competences.

When reflecting on formative assessment practices in classrooms, experts conclude that formative assessment requires a certain environment or culture (OECD 2005). It is stated that formative assessment should be based on the establishment of a classroom culture that promotes interaction between the teachers and the students as well as between the students. Also, it should encourage the use of assessment tools. This sounds trivial, but surveys revealed that tasks used and discussions occurring are often not conducive for formative assessment (e.g., Schiepe-Tiska et al. 2016). One important characteristic of such a conducive classroom culture is the fact that students can make mistakes without fearing negative consequences and that they do not attempt to anticipate a particular answer or reaction that might be expected by the teacher or their peers (Cowie et al. 2011). Anticipated expectations and reactions might influence students' actions strongly. For example, students sometimes desist from seeking help by asking questions and from discussing their ideas in plenary because they fear that teachers and students will find them less capable. Furthermore, in order to make students' competences and their further learning needs *visible* to both teachers and students, a conducive classroom—and tasks and questions— should encourage students to elaborate their ideas instead of offering one-word or half-sentence answers (Cowie et al. 2011). This leads to a larger data basis that can be used for solution- and process-oriented formative assessment.

15.3 Unpacking Formative Assessment—What Are Its Key Components and How to Enact Them?

Formative assessment itself is a process in which evidence has to be collected, interpreted, and conclusions have to be inferred and implemented. In order to enact formative assessment, research has disentangled this complex process into components that have some similarities with how experiments are conducted in science. Components of formative assessment have been described in the science education literature as *formative assessment practices* or *strategies*. Furtak et al. (2016) highlighted four practices teachers have to apply when enacting formative assessment: (1) designing formative assessment tasks, (2) asking questions to elicit student ideas, (3) interpreting student ideas, and (4) providing feedback. In a similar way, Wiliam and Leahy (2015) described five practices: (1) clarifying, sharing, and understanding learning intentions; (2) engineering effective discussions, tasks, and activities that elicit evidence of learning; (3) providing feedback that moves learners forward; (4) activating students as learning resources for one another; and (5) activating students as owners of their own learning. The last two practices relate more to the role of learners in formative assessment. In the cyclical formative assessment process described by Dolin et al. (2018), teachers have to (1) establish clear goals, (2) design or chose a method for collecting data about student performance, (3) implement this method to collect evidence relating to goals, (4) interpret evidence, and (5) decide about the next steps in learning.

Even though the models about practices to enact formative assessment vary slightly, a process with seven components can be derived from the different conceptualizations (see Fig. 15.4). Being clear about the learning goals, Component 1 is not only necessary to design instruction according to these goals, but is also relevant to identifying whether status-oriented formative assessment can offer enough information about students' progress according to the goal (e.g., how they consider the process of seeing; see the example in Fig. 15.1). Alternatively, whether a deeper insight into students' processes of arriving at particular solutions is needed (e.g., to understand better how students investigate physics phenomena; see the example in Fig. 15.2). By deciding which type of formative assessment suits the learning goal best, teachers also have to decide on which method they want to use for collecting the data (paper–pencil, observation, video recording, etc.) and which task(s) they want to use in Component 2. Teachers then collect the data in Component 3, for example, by collecting students' drawings (see the example in Fig. 15.1) or by collecting an audio recording of their discussion (see the example in Fig. 15.3). They may highlight or describe relevant information in Component 4, and try to make sense of this information in Component 5, according to the learning goal. *Making sense* not only includes teachers' action to *infer* students' competences but also for them to *discuss*, maybe with themselves, alternative interpretations and reasons why the students may (not) demonstrate a particular competence or do (not) progress further. From their analysis, teachers would try to conclude what to do next to foster student learning in Component 6. This may include that teachers revise the learning goals or think about

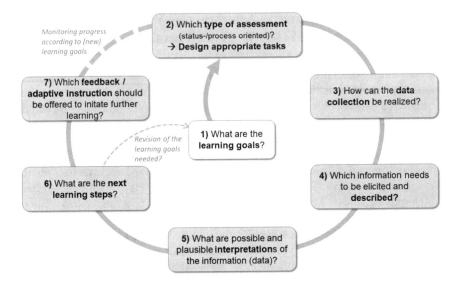

Fig. 15.4 Seven components of formative assessment organized in a cyclical process

adaptive learning goals for particular students. Finally, teachers would concretize their ideas by giving feedback to their students to engage them in their learning and/or design adaptive instruction in Component 7.

The sequence of components in Fig. 15.4 is ideal–typical; in most cases, the process of formative assessment will not follow the strict order given, but will require recourse and branching. For example, if a student's answer is ambiguous, the teacher may want to collect further data or analyze the available data from a different perspective in order to refine the feedback/the instruction. Furthermore, components may be missing, that is, if teachers have to deal with evidence collected by other means.

The question that guides teachers when to implement formative assessment should be: How can I elicit student thinking and how can I react to their thinking in order to improve their learning processes? The seven components described in Fig. 15.4 are tools to answer this question. They are explained in detail in the following sections that also offer guidelines on how to enact them. The components are grouped in three sections: collecting evidence (Components 1–4), interpreting student solutions and their activity (Component 5), and establishing and implementing decisions (Components 6 and 7).

15.3.1 Collecting Evidence—What Tools Are Appropriate?

Formative assessment encompasses the collection of data that either stem from solutions, such as verbal or written answers to questions, final graphical representations, or a final experimental setup (for solution-oriented formative assessment), or from the processes leading to those solutions (for process-oriented formative assessment). The questions or tasks can be formulated in advance or they can be developed on the fly. The quality of the data and thus of the questions or tasks has a significant influence on how meaningful the assessment can be in view of its influence on prospective teaching and learning. Thus, it is important to have appropriate tools to collect data (i.e., questions and tasks). However, which tools are appropriate?

It is important to start with thinking about what students should learn in science (see Fig. 15.4, Component 1). A look at curricula and standards like the Next Generation Science Standards (NGSS Lead States 2013) shows that in general two areas are defined in addition to the explication of crosscutting concepts: (1) disciplinary core ideas and (2) science practices (NGSS Lead States 2013). The first area encompasses knowledge about the basic concepts in science such as *structure of matter, energy,* or *chemical reaction* (American Association for the Advancement of Science—Project 2061 2001, 2007). The second area refers to the competences students need when conducting scientific inquiry. Without being mentioned explicitly, further learning goals in the second area—or expectations toward the competences students should bring to the classroom—may refer to students' social skills, their willingness to engage with science, or other more general competences such as reading and mathematical skills.

Being clear about learning goals—of physics education in general and for the current classroom setting in particular—is an important guideline for enacting formative assessment. The questions are: Where is a teacher's focus for data collection and analysis? Are students' conceptions of a particular core idea in the focus or a particular practice? Does a teacher want to assess students' social skills (in order to identify whether they can engage fruitfully in group work)? Among the variety of classroom events, learning goals can offer the lens with which a teacher decides on what data to collect and/or on which particular aspects of student activity to look at. Some learning goals may only require the teacher to collect evidence from students' solutions, whereas others may require him or her to observe them carefully (maybe by collecting audio- or video-data if permitted) while they are working on particular tasks. Thus, the type of assessment, solution- or process-oriented formative assessment, needs to be chosen carefully according to the learning goal and the information a teacher wants to collect with respect to that goal.

In addition, the tasks themselves need to be chosen according to the learning goals: Which thinking, which performance should the chosen tasks elicit? How can the tasks support the collection of as much information as possible? Evidence about students' competences can be collected by very different tasks, ranging from asking students to write, draw, calculate, or set up experiments, depending on the learning goal (see Fig. 15.4, Component 2). As solution-oriented formative assessment is restricted to reconstructing students' competences from their solutions only, special attention should be paid to choosing tasks and questions carefully, so that they can elicit student understanding and knowledge. Various examples have been described in the literature to elicit student understanding including prompting students to search for and correct errors in texts, drawings, concept cartoons, write texts, or create drawings (see examples in Keely 2015). Even though assessing solutions only, tasks can ask for at least some information on why students think their solution is appropriate or how they have arrived at a particular solution. The examples in Fig. 15.5 (from Urban-Woldron and Hopf 2012) demonstrate that even multiple-choice formats can be combined in questions or tasks, so that teachers get more information about students' thinking that may have resulted in choosing a particular solution. For both

The bulb in the circuit shown below flashes.

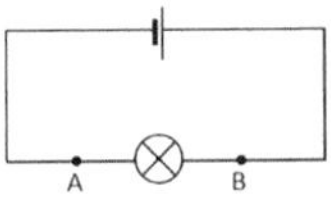

What can you say about the current in points A and B?
- ☐ The current in A is greater than in B.
- ☐ The current in B is greater than in A.
- ☒ The current is the same in A and B.

How do you explain your decision?
- ☒ The current is the same in the entire circuit.
- ☐ The current is partly consumed by the bulb.
- ☐ The current is completely consumed by the bulb.

The resistor R_1 in the circuit shown in the upper diagram below is small.
It is replaced by a greater resistor R2 as shown in the lower diagram.

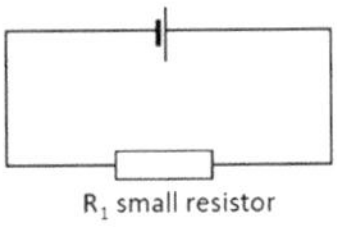

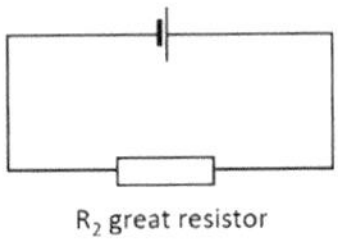

What happens to the current in the circuit in the lower diagram compared to that in the upper one?
- ☐ The current is greater.
- ☒ The current is lesser but not zero.
- ☐ The current is the same.
- ☐ There is no current any more.

How do you explain your decision?
- ☐ The battery is not strong enough to push any current through the greater resistor.
- ☒ The battery cannot push current as great as before through a greater resistor.
- ☐ A greater resistor requires more current than a smaller one.
- ☐ It is the same battery; therefore, the current remains the same.

Fig. 15.5 Assessment of students' understanding of electric circuits (adapted and translated from Urban-Woldron and Hopf 2012, pp. 208–209, correct solutions ticked)

solution- and process-oriented formative assessment, teachers should have in mind that questions or tasks that only require a "yes" or a "no" answer will not provide sufficient information to infer next steps for students' learning. Thus, appropriateness means that an assessment delivers as much meaningful data as possible and necessary.

For the actual collection of data (see Fig. 15.4, Component 3), it is essential to keep in mind that formative assessment has an explicit focus on fostering learning. It is therefore important to establish a culture in which formative assessment is part of learning and any answers—even vague or "wrong" ideas from students' point of view—are very welcome as they offer information relevant to supporting further learning (OECD 2005). This is especially the case, if formative assessment is not planned and students may not even notice that the teacher is currently collecting data (is observing their solutions and discussions). Still, students need to experience a classroom culture in which the teacher is not "controlling" them but rather is interested in their thinking, is trying to understand why their ideas and activities are plausible from their point of view, and is aiming at fostering their learning. For more planned formative assessment, students should regard data collection not as a measure of their capabilities but as an opportunity for themselves and the teacher to identify the strength of their ideas and where their understanding might need to be improved.

In order to being aware of what can be observed in the data—in contrast to interpreting this observation—it can be useful to explicitly (mentally) describe or highlight the observation (see Fig. 15.4, Component 4). As data can be very rich, especially in process-oriented formative assessment, allowing different observations to be made, the desired learning goals (Component 1) can serve as a lens to extract observation that is *relevant* according to the goal. This prevents other aspects that are less important for the learning goal from coming into the focus of the interpretation. However, at the same time, the teacher should remain open to critically check whether other observations can be implicitly relevant, even though they are not related directly to the learning goal. In Fig. 15.2, the off-task activities do not seem to be relevant to a learning goal that is related to conducting experiments. However, they may have relevance as they can indicate that for the particular student the assessment task does not fit well (is not engaging the student to demonstrate his experimental skills). His teacher might then need to construct another task to verify this assumption (see Fig. 15.4, Component 2). The off-task activities may also be considered as possible indications that the learning goal is too challenging because competences relevant to meaningful engagement with according learning activities are not observed. Though not being directly associated with the aspect of competence to be assessed, data might encompass information about barriers to engage with the task as planned by the teacher. Here, the teacher can strike a balance between different interpretations regarding the relevance of the same information, which also may result in different decisions about how to proceed with the formative assessment. The need to decide on what information is relevant to a particular learning goal demonstrates that even though the description of observations should avoid any one single interpretation (so that different interpretations are possible and can be discussed against each other), the orientation toward what to describe is always influenced by what the assessment is about to reveal.

15.3.2 Interpreting Student Ideas—What Are Students Thinking?

Based on his or her observations, a teacher can think about different interpretations of the same observation he or she has made (see Fig. 15.4, Component 5). Vice versa, that is, if a teacher immediately arrives at an interpretation, he or she can think about the observations that support that interpretation and those that may not, so that the teacher can make another interpretation more, or equally, likely. Moving back and forth between observations and interpretations is an essential process when enacting formative assessment, and is one reason why we stress earlier in this chapter that formative assessment does not necessarily follow the components or steps in Fig. 15.4 in the strict order given.

When interpreting, teachers should bear in mind that students actual thinking can never be measured directly. Again, focusing on learning goals when collecting and interpreting data can help teachers to arrive at sound interpretations. *Interpretation* is then the essential process in which a teacher attempts to understand the meaning of the student's utterances and activities *from the student's point of view*: What does the activity indicate about the student's understanding and how does this relate to the learning goal? What meaning has the concept to be mastered for the student? Which aspect of the targeted learning goal is not yet met? Interpretations should unpack what students do seem to understand, be able and are willing to do, and what might be difficult for them or missing. Only if interpretations are "more" than simply recordings of whether a student is "wrong" or "right," they can assist in concluding which further learning steps are needed. Unpacking interpretations also support the identification of students' potentials instead of focusing exclusively on their deficits.

The process of unpacking can be guided by empirical evidence and theoretical consideration from science education research. In science education, students' conceptual understanding has always played an important role (see Chaps. 3 and 14). When establishing an understanding of science concepts, students develop concepts from their everyday experiences (and any informal instruction, for example, from their parents) that might not match the scientific concepts. For movement of objects (see the example in Fig. 15.3), it has long been documented that students often think that any movement requires a force in the direction of the movement (e.g., Halloun and Hestens 1985). Such an idea is plausible from students' everyday experiences in which they always have to exert a force in order to keep an object in its state of motion. However, students may not be aware of the compensating effects of friction. What may appear as a "misconception" (force is proportional to velocity) is actually an appropriate conclusion from everyday experiences indicating missing conceptions about friction (e.g., von Aufschnaiter and Rogge 2010). Some conceptions are plausible from the students' point of view but are in contrast with scientific ideas, other conceptions would easily be considered as correct, because they make sense in a particular context: We often say that objects are at rest (without acknowledging that these move within space and we would need to express the reference system). At cloudy days and during nights, we would say that the sun is not shining (even though the sun is always shining). Therefore, interpretations should not only analyze students' competences but also relate this analysis to the particular situation attempting to understand students' point of view.

For teachers, it is crucial to know as much as possible about students' conceptions in order to reconstruct these from what they have observed and to think about why these may be plausible and what the next learning steps could be. According to a constructivist view on learning, new learning opportunities should build on existing competences. This is why science education should start to collect and structure students' understanding of disciplinary core ideas. For example, for the concept of matter, students' conceptions can be classified according to four aspects: (1) structure and composition, (2) physics properties and change, (3) chemical properties and change, and (4) conservation. Progress of conceptual understanding within each aspect can furthermore be distinguished using levels. Hadenfeldt et al. (2014) defined five different levels which can be used to interpret students' utterances and activities. Table 15.1 briefly describes the levels for the aspect *structure and composition*. Aspects and levels result in a rubric that can help teachers to reconstruct students' conceptual understanding from what they express and do: Imagine, a student would say "If you heat metal, the atoms enlarge as well as the air between the atoms." With the rubric, his or her teacher can first think about which aspect is addressed in the student's utterance (structure of matter). Using levels as an interpretational tool, a teacher can interpret that this student considers particles as building blocks and assigns them macroscopic properties ("atoms enlarge"). The student does not seem to distinguish clearly between the components of matter and the matter itself or air that exists between the components. The student would seem to hold ideas described at Levels 3 and 2, which are reasonable as it is difficult to imagine that within a substance, there is nothing but its components and that these behave in a completely different manner than the substance itself.

From research on conceptual change and so-called learning progressions, various detailed descriptions of the development of students' conceptual understanding are available which are either unidimensional or multidimensional (e.g., Neumann et al. 2013 for energy; e.g., Alonzo and von Aufschnaiter 2018 for forces and motion; e.g., Plummer and Krajcik 2010 for celestial motion). From this area of research, rubrics can be inferred that do not only define the milestones for the development of scientific concepts such as the examples in Hadenfeldt et al.'s (2014) study, but also articulate and specify the expectations for an assignment, or a set of assignments, in a more detailed way (Panadero and Jonsson 2013). As shown in the above example about students' understanding of composition of matter, typically, specific criteria are listed and levels of quality in relation to each of these criteria are described. More generally speaking, the aim is that a rubric should help to analyze what someone can do and

Table 15.1 Levels of students' understanding about the structure and composition of matter (Hadenfeldt et al. 2014, pp. 193–194)

Level	Name of level	Description of students' understanding for the aspect *structure and composition*
5	Systemic particle concepts	Students are able to describe and to explain the structure of complex molecules […]. They are able to explain why specific interactions in a system of particles occur […]
4	Differentiated particle concepts	Students are able to describe particles with the use of a differentiated atom model (e.g., nucleus-shell, shell model) […]. They differentiate between atoms and molecules and can distinguish between different bond types […]
3	Simple particle concepts	Students understand particles as a building block of matter […]. There is nothing between the particles. These particles are often described as the "last divisible unit" which is why they are often described with macroscopic properties as the particles inherit these properties through this division process […]
2	Hybrid concepts	Students understand particles as entities embedded in matter […]. Between the particles is the actual substance […]. Students are not able to use their perception of particles to explain the structure of matter […]
1	Naïve concepts	Students describe structures without the use of the particle concept […]. They view matter as dividable but continuous […]

what someone cannot do yet. As such, these rubrics are not only an important tool to interpret student utterances and activities—especially, when they do not perfectly express their ideas—they also help to infer what needs to be developed next or what needs to be stabilized further (see Fig. 15.4, Component 6; see also the next Sect. 15.3.3 and Alonzo 2018).

Within the discussions of the examples taken from the figures above and of the student idea that atoms enlarge when metal is heated, not only possible interpretations of students' utterances and activities were presented, but also reasons why students may behave (observation) and think (interpretation) about the way they do. We argue earlier in this chapter that some student ideas are plausible from their everyday experiences: In a world with friction, you have to exert a force to maintain a steady speed (without noticing that the net force is zero, example in Fig. 15.3). As objects enlarge, why should atoms not behave the same way? And students usually cannot see that light travels from a source to an object. Such kinds of reasons are based on (individual) students' life experience and can also include their prior learning experiences in physics. For instance, students who have not established sufficient knowledge about the concept of matter (ideas described at Levels 1 and 2 in Table

15.1) cannot explain density in an appropriate way. Lacking prior knowledge may therefore be a reason why students do not understand content currently taught. The usage of everyday language is another reason referring to the student life experience. Typically, students use their everyday language to describe and explain scientific phenomena. Forces are, for instance, something that you have (being strong) and not necessarily something that you exert. Thus, the meaning and use of everyday language is sometimes in conflict with scientifically correct descriptions of scientific concepts. Therefore, students should develop the competence to describe phenomena and respective concepts scientifically correct. The desired goal is that students are able to use science to explain why their everyday ideas and experiences happen.

In addition to the reasons that are grounded in students' life experiences, the nature of physics concepts can make it difficult for students to understand these concepts. In particular, physics concepts of which students cannot have first-hand experiences (such as energy, current, and voltage, or particles of matter) are more likely to result in conceptions that are not appropriate. Students cannot "see" voltage and current, so they either confuse these two concepts or assume that current is something that is consumed (which is also supported by their everyday language of "current consumption"). A third possible group of reasons is the assessment task itself or the context in which the assessment takes place. It may provoke specific student behaviors or ideas because, for instance, the students misinterpret what the task is about. The fact that the situation is a possible cause should always be considered, but can often only be confirmed by observations and interpretations of different students. This could reveal that the same task causes similar difficulties for different learners.

Attempting to infer plausible reasons is important because it helps teachers to positively engage with student thinking; very often, students do have good reasons, even if their result of thinking does not match teachers' expectations. In addition, reasons can offer ideas on how to promote further learning. If, for instance, the everyday use of language may be a reason, then assessment tasks that employ drawings or contrast everyday language with scientific language may be good tools to promote further learning and to assess student understanding (Hardy and Stern 2011). If students assume that any constant velocity of an object requires a positive net force because this is their everyday experiences, then it is important to investigate how varying friction is correlated with the amount of force needed to maintain the object's steady speed (Alonzo and von Aufschnaiter 2018). Despite the relevance of inferring reasons, it should be noted that sometimes it is impossible to construct plausible reasons. Therefore, we encourage teachers to think about plausible reasons even if they may arrive at the conclusion that they cannot identify any.

15.3.3 Making Decisions About the Next Steps in Learning—What Should Be Done Next?

When thinking about the next steps in learning (see Fig. 15.4, Component 6) and deriving a decision (Component 7), a teacher should allow for the preceding components (from 1 to 5) of the formative assessment process and their outcomes to form the basis for decisions on fostering further learning of students. In other words, the different components are linked together in order to offer what is often called *formative feedback* (Shute 2008). The outcomes of each component deliver information that could be used as part of the feedback given to the students or used by the teacher to adapt his or her teaching activities. Feedback in this sense offers some kind of a link between previous learning activities and prospective activities that should be based on the identified success of the previous ones.

The starting point of any decision about the next steps in learning is a comparison of the actual student competence (i.e., interpretation; see Fig. 15.4, Component 5) and the intended learning goal as a reference (see Fig. 15.3, Component 1; Black and Wiliam 1998). The aim of formative assessment is not just to *measure* the distance between actual and reference competence but to conclude the next steps in learning in order to reduce the discrepancies. Thus, this comparison yields information, which then could be used to alter the gap between the actual and the reference competence. There are different scenarios that could be true: The students have arrived at the learning goal, they have exceeded the learning goal, or they have not yet reached the learning goal. For the first two cases, the teacher might need to consider a new learning goal either to foster further learning beyond the initial goal, for the latter case the teacher might need to establish intermediate learning goals to reach the intended learning goal step by step. Rubrics that stem from research on students' conceptual progress as described above (see Table 15.1) are a helpful tool to identify what the next learning goals and the corresponding steps of learning might be. In the example with the student who assumes that particles enlarge when being heated, his or her understanding stands in contrast to a learning goal that stresses that interaction between particles changes with temperature. Though it might be straightforward "just" to explain the physics concept(s), to establish intermediate learning goals is often a more appropriate way to foster further learning: A first goal can include instruction to stabilize the student's better understanding of the (counter-intuitive) idea that there is nothing in between particles of matter (see Table 15.1, Level 3). After that, the student can be supported to establish the idea that the particles and their arrangement determine the macroscopic features and that their features are very different from macroscopic features of the same system (Level 4): Macroscopic features like color and temperature cannot be applied to the microscopic level. This can serve as a basis to understand better that particles can behave in a different way than the corresponding substance (Level 5).

Hattie and Timperley (2007) do not refer to components as described in Fig. 15.4 but, rather, to three questions that guide effective formative feedback. The three questions are: "Where am I going? …How am I going? …and Where to next?"

(p. 86). Even though these questions should lead to answers similar to those derived for the Components 1 and 5 in Fig. 15.4, the different framing can assist teachers in establishing feedback for their students and instruction to foster their further learning.

The answer to the first question "Where am I going?" more or less repeats the learning goal the students should achieve and that was defined by the teacher (see Fig. 15.4, Component 1). In other words, one could also ask: "What are the goals?" However, goals without clarity as to when and how a student (and teacher) would know they were successfully met are often too vague to serve the purpose of enhancing learning. Feedback or adaptive instruction cannot lead to the reduction of the gap between the status of learning and the desired goal if the goal is poorly defined. In this case, the gap between the status of learning and the intended learning goal is unlikely to be sufficiently clear for students to see a need to reduce it. Therefore, it is important that teachers are explicit when they communicate to their students the learning goals and highlight the positive and negative aspects of their solutions related to these goals. Usually, rubrics help teachers to answer this question (see Sect. 15.3.2 in this chapter). These aspects bridge the first question to the second one.

Answering the second question "How am I going?" usually involves the teacher providing information relative to a goal or expected standard, to prior performance, or to the success of other students. The question could also be: "What progress is being made toward the goal?" Within formative assessment literature, this aspect is called *feedback* and could also be given by peers or by the student him- or herself if peer- or self-assessment are applied (Hattie and Timperley 2007). Students are often looking for information on "How they are going" even if they may not welcome the answers. Too often, responses would only state what is wrong and what is right. In this case, information is missing to tell the student what exactly is right or wrong and why it is right or wrong. The information given in this second part of feedback is taken from the Components 2 to 5 of the formative assessment process (see Fig. 15.4).

For the third question ("Where to next?"), the answers should guide students in their future learning. From a teacher's perspective, one could rephrase it and ask: "What do I need to do to get this particular student or students with a similar profile to the next or the reference level?" From students' perspective, the question would be: "What do I need to do to get there?" (Black and Wiliam 1998; Hattie and Timperley 2007). Consequently, the teacher must find answers in terms of teaching and learning approaches, tools, or materials that push students' learning forward. For Example 1 in Fig. 15.1, the teacher should (again) elaborate that the learning goal is to describe the process of seeing and maybe how the task relates to the learning goal ("Where is the student going?"). Ideally, the teacher would then discuss each solution with the individual student, elaborating with the drawings what the teacher has diagnosed about the individual student understanding ("How is the student going?"). For Lars, the teacher may express that the drawing indicates that he already considers light propagation from the source to the object but would need to figure out what happens with the light afterward. However, it is important for Lars to learn first something about light being scattered from objects—even those that do not emit light on their own or are mirrors. The teacher may support Dana to think about reasons for the arrow between the flower and the eye to point to the other direction. Dana may be prompted by her teacher to work on that question on her own, being (more) responsible for her own learning. Lars, in contrast, may need more teacher-organized instruction, that is, real and virtual experiments in which he can investigate how light is scattered and reflected ("Where to next/What needs to be done to get there?"). Ultimately, all initiated changes in the activities should result in the expansion of knowledge and the further development of particular skills.

Responding to the questions "Where to next?" and "What do I need to do to get there?" as part of feedback is also called *feed forward* and refers to the Components 6 and 7 of the formative assessment process (see Fig. 15.4). Potentially, new goals could be also part of the information, in order to initiate a new formative assessment process. Feed forward in that sense also allows students and/or their teachers to set further challenging goals if the previous ones are attained. By this, conditions for ongoing learning are established. Besides, goals are more effective when students share a commitment to attaining them, because they are more likely to seek and receive feedback. The establishment of goals is closely linked to students' prerequisites, as both must be coherent.

Although the three questions constituting feedback are presented separately, they of course do not work in isolation when enacting formative feedback. They typically work together. Only by providing students with information on all the three questions the power of feedback can have an effect on student learning.

15.3.4 On Which Facets Can Formative Feedback Have an Effect?

In addition to the descriptions of formative assessment and feedback so far, it is furthermore helpful to unpack the three facets feedback can address. Feedback information can focus on (a) the solution of the task (solution-oriented feedback), (b) the activities undertaken to solve the task (process-oriented feedback), and (c) the self-regulation influencing the process of solving the task (self-regulation-oriented feedback; e.g., Hattie and Timperley 2007). Being very specific about the facets that can be addressed with feedback concurs with the idea of unpacking the interpretation as described above (see Fig. 15.4, Component 5). Teachers and students can derive their interpretation on "How am I going?" by (a) analyzing the solution of the task, (b) analyzing whether the process leading to the solution is appropriate, and (c) analyzing whether the learner employs appropriate strategies of self-regulation.

(a)   Analyzing the solution of the task

Based on solution-oriented formative assessments (see Fig. 15.4, Component 2), feedback can provide information about the solution of a task or the task itself. This feedback tells students whether their answer is correct or incorrect. Consequently, it is called "corrective feedback or knowledge of results" (Hattie and Timperley 2007,

p. 91). However, *corrective feedback* is not very useful to foster students' learning processes, because many students cannot conclude how they could improve their learning based on the information provided. For the students, it is important to know what exactly is correct, what is not correct, and what constitutes correct/incorrect solutions. Thus, the teacher has to interpret why something is correct or incorrect in order to provide feedback that fosters students' reasoning about a particular issue. It is less effective if the students just "know" the correct solution without having an idea how they know that a particular solution is correct. Corrective feedback usually cannot be transferred to other tasks as it is closely connected to a specific task or task types. Without receiving such an expanded corrective feedback, the generalizability is limited and the students cannot learn much that they can also use in other learning situations. Furthermore, too much corrective feedback may encourage students to focus on particular answers and not on strategies to evaluate whether their answer is correct or not.

(b) Analyzing whether the process leading to the solution is appropriate

Like the entire assessment itself, feedback can also focus on the process enacted when completing a task or creating a product (see Fig. 15.4, Component 2). In this case, the feedback usually informs students about typical mistakes they have made during the process. In general, feedback focusing on solution processes appears to be more effective than corrective feedback for enhancing deeper learning processes. Hattie and Timperley (2007) concluded that feedback information is most useful for students when it supports them in rejecting wrong hypotheses and provides direction for searching. Both would have an effect on the process. Besides, they argued that feedback at the process level and feedback at the task level can reinforce each other.

(c) Analyzing whether the learner employs appropriate strategies of self-regulation

In addition, feedback to students can refer to students' self-regulation. Self-regulation addresses the way students monitor, direct, and regulate their activities toward the learning goal (e.g., Pintrich 2004). The regulation implies that they adapt their activities during the solution process against the learning goal. Hattie and Timperley (2007) emphasized that there is a number of aspects mediating the effect of feedback about self-regulation. One aspect is students' ability to generate feedback themselves. Effective learners generate internal feedback, whereas less effective learners depend more on external feedback given by the teacher or their peers. Another aspect of students' self-regulation is their self-assessment. Effective learners are powerful in selecting and interpreting information in ways that provide internal feedback. Students' ability to review and evaluate their competences through a variety of self-monitoring processes is part of the self-assessment as well as the ability to monitor and regulate their behavior through planning, correcting mistakes, and using solution strategies. When students are successful learners, they have the metacognitive skills to evaluate their levels of understanding, their effort and strategies used on tasks, their attributions and opinions of others about their performance, and their improvement in relation to their goals and expectations. We consider feedback about self-regulation and processing a task from Hattie and Timperley's point of view of the most powerful feedback in terms of students' deep processing and mastering of tasks.

15.4 What Are Teachers' Roles and Students' Roles in the Process of Formative Assessment?

In view of the process of formative assessment (Fig. 15.4) and of the consequences drawn from this process, it is useful to reflect on teachers' roles and students' roles, this might help beginning teachers to become aware of their own role and of the importance of formative assessment for their profession. Usually, two perspectives are differentiated when thinking about consequences drawn from the process: (1) what could the *teacher* do in terms of adapting his or her instruction to students' needs and (2) what should the *student* do in terms of initiating individual learning processes?

The first perspective means that the teacher is the person who plans and realizes science lessons; he or she can use the evidence generated by the process of formative assessment to adapt instruction to students' needs. The second perspective means that the student not only stands in the center of each learning process but is also responsible for his or her learning. Therefore, the student has to engage actively in a lesson in order to foster his or her own learning—no one would expect that students can learn to solve physics problems if they only sit and wait for learning to occur *naturally*. This does not explicitly mean that students can or should infer the next steps in learning on their own. The teacher should estimate the effect of possible consequences in order to determine which possibility works best in a specific situation and which works best in view of the learning goals. However, students could also collect and interpret data representing their learning efforts and they need to engage in the process of enacting the feed forward. Both would help to develop students' self-regulation skills. Of course, these two perspectives are closely linked. Actually, they should highlight that both teachers and students benefit from the enactment of formative assessment and that both are active players in the classroom with different roles (Hattie and Timperley 2007; OECD 2005).

It has to be stressed that formative assessment in general and feedback in particular are often only seen as information for the students to improve his or her learning. In this case, feedback is seen as a tool for individual support for single students and then has consequences for individual learning processes. On a time scale, the effect of feedback could be classified as a short-term effect that directly influences the next learning processes. Usually, it is neglected that *feedback is also a tool for teachers* to further develop and improve their instruction. In this case, it might be difficult to notice a direct effect on student learning as the effect of feedback is delayed and related to future learning processes. The teacher can use diagnostic information to rethink his or her way of teaching and planning instruction. In this case, the teacher needs self-regulation skills that help to change his or her own teaching activities. A possible consequence with a short-term effect could be that the teacher reconsiders the tasks and examples so that they fit better with the competences that should be developed by the students. A possible consequence with a delayed effect could be to take a closer look at the subject matter to be learned and to examine how reduced requirements can be planned for this purpose. This could eventually lead to completely different instruction and thus have a different effect on the development of students' competences.

The separation between the role of the teacher and the role of the students cannot be maintained in interactions between students. In peer-assessment, for example, students can take on the role of the assessor and/or the assessed (Topping 2009). It is possible that learners either take both roles alternately or only one of both. This means that students can take over the role of the teacher and assess their peers' competences. Peer-assessment in a formative sense can follow the process of formative assessment and encompass the same components; especially the components 4–7 (cf. Fig. 15.4). Assessor and assessed can profit by this kind of assessment in different ways. Findings indicate that peer-assessment mainly has an effect on the assessor as he or she needs to assess a competence of his peer (Cho and Cho 2011). To do so, the assessing student has to understand the science concept. Besides, by giving feedback, in-depth thinking and self-reflection processes are initiated.

15.5 Summary

The goal of formative assessment is to monitor student learning to provide ongoing formative feedback that can be used by the teacher to improve his or her teaching and by the students to improve their learning. More specifically, formative assessment helps students to identify their strengths and weaknesses and the teacher to recognize where his or her students are struggling in their learning and address their problems immediately. The process of formative assessment can be described by seven components:

(1) Being clear about the learning goals, (2) designing appropriate tasks meeting the learning goal, (3) collecting data with the tasks, (4) making sense of the information collected, (5) interpreting the data according to the learning goal, (6) concluding the next steps in learning, and (7) giving feedback/adapting instruction (Fig. 15.4).

Formative assessment provides the teacher with a way to align learning goals, physics concepts, and assessment, allows for the purposeful selection of teaching and learning strategies, embeds assessment in instruction, and guides instructional decisions. Overall, the desired goal is that students' learning is optimized by formative assessment.

Acknowledgements We would like to thank Prof. Erin Marie Furtak (University of Colorado Boulder) and Prof. Heike Theyßen (University Duisburg-Essen) for carefully and critically reviewing this chapter.

References

Alonzo AC (2018) An argument for formative assessment with science learning progressions. Appl Measur Educ 31(2):104–112

Alonzo AC, von Aufschnaiter C (2018) Moving beyond misconceptions: learning progressions as a lens for seeing progress in student thinking. Phys Teach 56(7):470–473. https://doi.org/10.1119/1.5055332

American Association for the Advancement of Science—Project 2061 (2001) Atlas of science literacy, vol 1. American Association for the Advancement of Science; National Science Teacher Association, Washington, DC

American Association for the Advancement of Science—Project 2061 (2007) Atlas of science literacy, vol 2. American Association for the Advancement of Science; National Science Teacher Association, Washington, DC

Ausubel DP (1968) Educational psychology: a cognitive view. Holt, Rinehart, and Winston, New York

Bell B, Cowie B (2001) The characteristics of formative assessment in science education. Sci Educ 85(5):536–553. https://doi.org/10.1002/sce.1022

Black P, Wiliam D (1998) Assessment and classroom learning. Assess Educ Principles Policy Pract 5(1):7–74. https://doi.org/10.1080/0969595980050102

Cho YH, Cho K (2011) Peer reviewers learn from giving comments. Instr Sci 39(5):629–643. https://doi.org/10.1007/s11251-010-9146-1

Cizek GJ (2010) An introduction to formative assessment: history, characteristics, and challenges. In: Andrade HL, Cizek GJ (eds) Handbook of formative assessment. Routledge, New York & London, pp 3–17

Cowie B, Jones A, Otrel-Cass K (2011) Re-engaging students in science: issues of assessment, funds of knowledge and sites for learning. Int J Sci Math Educ 9(2):347–366. https://doi.org/10.1007/s10763-010-9229-0

Dolin J, Black P, Harlen W, Tiberghien A (2018) Exploring relations between formative and summative assessment. In: Dolin J, Evans R (eds) Transforming assessment: through an interplay between practice, research and policy. Springer International Publishing, Cham, pp 53–80

Furtak EM, Kiemer K, Circi RK, Swanson R, de León V, Morrison D, Heredia SC (2016) Teachers' formative assessment abilities and their relationship to student learning: findings from a four-year intervention study. Instr Sci 44(3):267–291. https://doi.org/10.1007/s11251-016-9371-3

Hadenfeldt JC, Liu X, Neumann K (2014) Framing students' progression in understanding matter: a review of previous research. Stud Sci Educ 50(2):181–208. https://doi.org/10.1080/03057267.2014.945829

Halloun IA, Hestens D (1985) Common sense concepts about motion. Am J Phys 53(11):1056–1065

Hardy I, Stern E (2011) Visuelle Repräsentationen der Dichte: Auswirkungen auf die konzeptuelle Umstrukturierung bei Grundschulkindern [Visual representations of density: effects on conceptual restructuring of primary school children]. Unterrichtswissenschaft 39(1):35–48

Harlen W, James M (1997) Assessment and learning: differences and relationships between formative and summative assessment. Assess Educ Principles Policy Pract 4(3):365–379. https://doi.org/10.1080/0969594970040304

Hattie JAC (2009) Visible learning: a synthesis of over 800 meta-analysis relating to achievement. Routledge, London & New York

Hattie JAC, Timperley H (2007) The power of feedback. Rev Educ Res 77(1):81–112. https://doi.org/10.3102/003465430298487

Keely P (2015) Science formative assessment: 75 practical strategies for linking assessment, instruction, and learning. Corwin, Thousand Oaks, CA

Liu X (2010) Essentials of science classroom assessment. Sage, Thousand Oaks, CA

Neumann K, Viering T, Boone WJ, Fischer HE (2013) Towards a learning progression of energy. J Res Sci Teach 50(2):162–188. https://doi.org/10.1002/tea.21061

NGSS Lead States (ed) (2013) Next generation science standards: for states, by states. National Academies Press, Washington, D.C. http://doi.org/10.17226/18290

OECD (2005) Formative assessment: improving learning in secondary classrooms. OECD Publishing, Paris

Panadero E, Jonsson A (2013) The use of scoring rubrics for formative assessment purposes revisited: a review. Educ Res Rev 9:129–144. https://doi.org/10.1016/j.edurev.2013.01.002

Pintrich PR (2004) A conceptual framework for assessing motivation and self-regulated learning in college students. Educ Psychol Rev 16(4):385–407

Plummer JD, Krajcik J (2010) Building a learning progression for celestial motion: elementary levels from an earth-based perspective. J Res Sci Teach 47(7):768–787. https://doi.org/10.1002/tea.20355

Schiepe-Tiska A, Schmidtner S, Müller K, Heineörg-Henrik, Neumann K, Lüdtke O (2016) Natur-wissenschaftlicher Unterricht in Deutschland in PISA 2015 im internationalen Vergleich [Science education in Germany in PISA 2015 in international comparison]. In: Reiss K, Sälzer C, Schiepe-Tiska A, Klieme E, Köller O (eds) PISA 2015: Eine Studie zwischen Kontinuität und Innovation [A study between continuity and innovation]. Waxmann, Münster, New York, pp 133–175

Shute VJ (2008) Focus on formative feedback. Rev Educ Res 78(1):153–189. https://doi.org/10.3102/0034654307313795

Topping KJ (2009) Peer assessment. Theor Pract 48(1):20–27. https://doi.org/10.1080/00405840802577569

Urban-Woldron H, Hopf M (2012) Entwicklung eines Testinstruments zum Verständnis in der Elektrizitätslehre [Development of a diagnostic instrument for testing student understanding of basic electricity concepts]. Zeitschrift Für Didaktik Der Naturwissenschaften 18:201–227

von Aufschnaiter C, Rogge C (2010) Wie lassen sich Verläufe der Entwicklung von Kompetenz modellieren? [How can the development of competences be described?]. Zeitschrift Für Didaktik Der Naturwissenschaften 16:95–114

von Aufschnaiter C, Theyssen H, Krabbe H (2020) Diagnostik und Leistungsbeurteilung im Unter-richt. In: Kircher E, Girwidz R, Fischer HE (Hrsg) Physikdidaktik I Grundlagen. Springer Spektrum, Berlin, Heidelberg, pp 529–571. http://doi.org/10.1007/978-3-662-59490-2_14

Wiliam D, Leahy S (2015) Embedding formative assessment: practical techniques for K-12 classrooms. Learning Sciences International, West Palm Beach, Florida

Literature for Further Reading

Hattie J, Clarke S (2018) Visible learning: feedback. Routledge, London, New York. https://doi.org/10.4324/9780203887332

Liu X (2010) Essentials of science classroom assessment. Sage, Thousand Oaks, CA. https://dx.doi.org/10.4135/9781483349442

Chapter 16
Methodical Basics of Empirical Research

Hans E. Fischer, William Boone, and Heiko Krabbe

Abstract To be up-to date, teachers should be able to follow current research on teaching and learning in physics. Therefore, they have to be able to assess if the results presented in publications are meaningful and trustworthy. In this chapter, we detail requirements research studies should comply with in order that measuring results, conclusions and generalisations can be trusted. In the case of the calculation of means for example, the requirements of such calculations are familiar to physics teachers. However, when considering more complicated statistics, for example the meaning of correlations or Rasch analysis, physics teachers may be less familiar with such statistical topics. To identify the relations between numerous variables, for example, the effect of migration background, social status and cognitive abilities on school success, more complex mathematical models are necessary. The starting point for all empirical investigations must be a valid theoretical model. Such a model should include both the variables considered important and the design of a study.

16.1 Introduction

Trustworthy research findings are needed for researchers to develop new theoretical models and research questions and for science educators at universities and teacher training institutions to substantiate curricular content of teacher education. In addition, students in teacher education need to know what to look for when choosing study literature in their studies, teachers in schools need to know how to optimise

H. E. Fischer (✉)
Universität Duisburg-Essen, Essen, Germany
e-mail: hans.fischer@uni-due.de

W. Boone
Miami University, Oxford, OH, USA
e-mail: boonewj@miamioh.edu

H. Krabbe
Ruhr-Universität Bochum, Bochum, Germany
e-mail: heiko.krabbe@rub.de

© Springer Nature Switzerland AG 2021
H. E. Fischer and R. Girwidz (eds.), *Physics Education*, Challenges in Physics Education,
https://doi.org/10.1007/978-3-030-87391-2_16

their teaching, and policy makers need to know what they can base their decisions on if they want to change education systems or just parts of them.

Trustworthiness of findings always requires sound research methods. To propose trustworthy and forward-looking statements about effects of teaching and learning physics and respective learning environments, research methods are primarily grounded in psychology and sociology and specified from inherent aims of physics education. Such statements should be understood as hints because empirical research can in principle only optimise the fit between theoretical models and empirical findings and can never result in definitive statements (Patten and Newhart 2018). According to Popper (2002 [1959]) the empirical sciences are never are able to derive general statements from theoretical assumptions but can falsify statements with only one reproducible contradicting experiment. This is also true for methods of empirical science education, which mostly have been developed in empirical sociology and psychology. Furthermore, the object of investigation in the field of teaching and learning is usually sufficiently complex that the set of information is incomplete. Therefore, researchers have to consider many influencing variables and never are able to consider all potential variables. Therefore, findings of research in physics education always produce statements to increase the probability of teachers' success in the classroom. Nevertheless, a particular teaching unit can be inefficient even when research results are included because some influencing factors were not or could not be taken into account. Some factors cannot be fully considered for various reasons. For example, ethical constraints can hinder the types of information one can take into account. This means that some variables and their relationships to other variables cannot be fully considered.

Such a scenario means these are differences between empirical research in physics and physics education. In physics, the number of dependent and independent variables often can be reduced in a laboratory setting or through assumptions (hypotheses) about the intended finding to create controllable predictions. In contrast, research on teaching and learning generally has to consider systems with many different individuals and many different characteristics. Those systems are much more complex than most physics systems. The complexity is comparable when physicists investigate new fields, such as dark matter or black holes. As in physics education, the theories of a new field are not elaborated in as much detail as is necessary, possible variables are perhaps not detected and researchers may not know if the measuring devices measure what they should. For example, astronomical research of the last decades found that the matter of the known universe covers only 5% of energy density of the universe. The remaining 95% is called dark matter, an umbrella term for structures, functional relations, and matter characteristics about which we do not know the details. It is even unclear how this dark side of physics can be integrated into already developed and validated theories and how dark matter can be measured. In those situations, a disadvantage of physics compared to physics education is that physicists cannot directly ask their research objects to gain a first impression; they always have to use devices and complex experimental settings. However, empirical research on physics education has to consider that the answers of human individuals always contain subjective components which can result in biased findings. As it is often not possible to eliminate the bias, the goal in educational research is to understand and account for the bias.

All empirical sciences use descriptions and measures for collecting and processing data, and for producing trustworthy statements. Therefore, investigations must meet the scientific criteria of the community of responsible scientists (Patten and Newhart 2018). Those criteria are not always obligatory, and they are open for negotiation, but, they always follow core certain principles of objectivity, validity, reliability and significance (Fischer et al. 2014a). In the text, which follows, these principles are explained in more detail. These principles are directly dependent on the *field of research,* the *theoretical approach,* the *design of the study* including the choice of *sample* and *measuring methods* and the use of adequate *instruments* (e.g. chosen or self-developed tests, rating scales, questionnaires).

16.2 The Field of Research

The general aim of physics education research concerns the description and the improvement of teaching and learning physics in all types of settings. Therefore, the main conditions and the general and specific boundary conditions of teaching and learning physics must be described. Some educators have developed models considering the interrelations between social, educational and methodical preconditions of teaching and learning in general and for teaching and learning physics specifically (see Fig. 16.1) (Fischer et al. 2005).

To assess the quality of physics lessons, a general model for teaching and learning sciences is needed. As described in Chap. 1, until the 1970s, teaching and learning was conceptualised as a transfer of information, and the interaction between teacher and learner was described with a "transmitter–receiver-model". To improve teaching and learning, the idea was to optimise the decoding and encoding processes of information. Today researchers assume that the actors in teaching–learning processes act independently from each other, and the cognitive development is considered as substantially being at the individual level. Merrill (1991) has suggested that this theoretical model could be viewed as second-generation instructional design. Learning processes are described as the relation between cognitive constructs and teaching activities which are seen as an adaptive design of learning environments as a balance between outside control and self-determination of the learner. As indicated in Fig. 16.1, learning in a school context is considered as dependent on a goal orientation of the provided learning environments and offered as opportunities for students to use it for physics learning. The theoretical background considers instruction and learning environments as a teacher's offer to the learner and as utilisation of that offer by the students. The offer, e.g. to approach the concept of speed with an experiment, can be optimised by controlling the achievement of goals and by moderating the ongoing teaching–learning process or by investigating and changing the conditions. As described in Chap. 4, the concept of goal orientation includes cognition, motivation, (self-regulated) learning and performance. The influences of these goals need

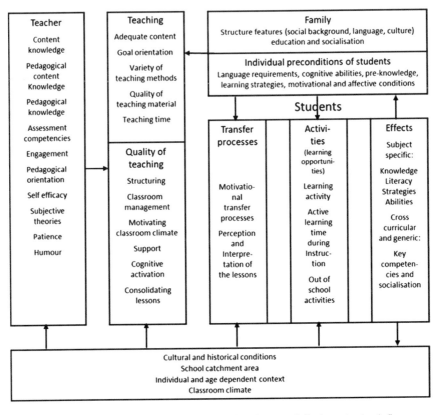

Fig. 16.1 Learning opportunities model as a theoretical framework for investigating influences on physics instruction (translated and adapted from Helmke 2009, p. 73)

to be empirically investigated to increase the quality of physics teaching in general (Pintrich 2000).

According to Fischer et al. (2005), three levels of research can be identified for structuring research programs.

1. The education system, school conditions, location of the school and family conditions at the system level (e.g. cultural conditions, curricula, school equipment, learning conditions at home)
2. The connection between teaching and learning as the level of instruction (e.g. offer, utilisation, performance and effect)
3. The cognitive and emotional conditions of students and teachers during instruction processes (e.g. pre-knowledge, learning gains, motivation, beliefs, professional knowledge).

The three main domains of teacher knowledge are called pedagogical content knowledge (PCK), content knowledge (CK) and pedagogical knowledge (PK). All three knowledge areas are necessary to study to become a good teacher as Evens et al. (2018) have pointed out. They found that in teacher education, the explicit and integrated study of all three domains PCK, CK and PK is a necessary precondition for becoming a good teacher. Only studying two domains such as CK and PK or PCK and CK is not sufficient to develop the missing domain.

Although all three domains are critical, CK and PCK require special attention. Usually student teachers of physics have to learn physics on an academic level. Sorge et al. (2019, p. 5) point out that physics teachers "develop their CK mostly in formal learning opportunities such as lectures (e.g. introductory physics, theoretical physics), seminars (e.g. discussing tasks from lectures) or lab work…". According to Eichenlaub and Redish (2019) for physicists, the content of physics is strongly related to mathematical structures and the physics interpretation and meaning of those mathematical structures which is fundamental for understanding physics. However, on all school levels, the necessary mathematical aspects of academic physics cannot be presented in full to understand physics at this level. It has been suggested that PK and PCK be adjusted to the appropriate mathematical sophistication as a function of the age or grade level in which the physics content developed in the university is taught. To show student teachers of physics and in-service teachers how to transfer physics concepts from academic settings to the high school or even the elementary school level, more than academic knowledge in physics is needed. For this transfer, we need to consider PK and PCK, and take into account findings from teaching and learning psychology, which, for example, can provide details regarding the chronological development of students' cognitive abilities. Therefore, we need particular efforts on the part of researchers of physics education to transform the *academic physics content* structure into an *instruction structure of physics content*.

16.3 Theory and Evidence

How do people proceed when they need an explanation for something that has taken place in everyday life but do not trust their own perception? The three main steps to answer such questions on a scientific level are outlined below:

1. The obvious approach is to ask experts, or search for expert opinion, in a journal or the Internet. However, as often occurs, opinions of a number of experts are not consistent. Especially, if experts publish new findings, a standard is needed to compare and assess different and sometimes contradicting statements. The standards have to be accepted by the community of empirical researchers. Accordingly, new research studies consider already known, and yet better, already successfully applied theoretical models. These models take into account already accepted research findings and apply already accepted research methods. When an investigation is conducted, the goal of the researchers should be to publish project findings in scientific journals for discussion and replication. An important result of publishing findings is to stimulate scientific conversations regarding the study conclusions. Trustworthiness of findings and replication of

investigations are inseparably connected. For example, in physics or psychology, findings are only trustworthy if they are repeated many times in independent studies.

2. According to Popper (2002 [1959]) empirical research, as done in biology, chemistry, physics, sociology and psychology, can never prove something as is possible in mathematics. At the conclusion of an empirical study, a researcher should be able to assess if a specific learning environment, or intervention, reaches the theoretically assumed learning goal with a higher probability than is possible using other strategies. For example, McLure et al. (2020) found that students' initial ideas about heat could be developed by applying a so-called thinking frames approach. The approach is a sequence of teaching and learning interactions to encourage students to use their own ideas to form scientific explanations. The sequence contains five key sections: observe, activate your thinking, formulate a (research) question, compare your experience with your ideas and describe and discuss what you found. In an experimental comparison group design, the comparison of the learning outcomes of an intervention group and a comparison group showed that the thinking frames approach led to significantly better test results in the intervention group compared to learning with traditional methods in the control group.

3. It is critical that a newly developed learning environment or learning material should be able to influence students' learning processes, behaviour, and motivation to learn physics or their competences as intended. Such new learning environments or learning materials should be derived from previous findings and theoretical models. It is critical that if a theoretical model used is not tested accordingly, it is not allowed to be used in general and the new finding must be restricted to the tested case (Olson 2004). This is a principle demand of empirical research, which is directly connected with the trustworthiness of the produced findings. This precondition should also be applied to research utilising small samples and to research using case studies, which have to be restricted to specific small samples and cases because a scientifically proofed generalisation is not possible. In small samples and especially in studies with only few cases, large differences between the small sample and the much larger population can arise simply by chance. This is important when policy makers, teachers and researchers are interested in the effect of a particular intervention for a larger population such as all students of a certain grade in a particular type of school (Watt and van den Berg 2002).

4. In an ideal research effort, empirical research on physics education should be able to provide teachers with clear and consistent guidance as how to increase teaching quality. However, due to the complexity of most of the theoretical constructs and the situations found in the field, not all research questions can be answered reliably. For example, a lot of teaching time is spent addressing so-called students' conceptions (also named misconceptions) although evidence is lacking that teacher knowledge about those conceptions increases the quality of teaching and learning physics. For the teaching of mathematics, Hill and Chin (2018) found inconsistency in the predictive power of knowledge on students'

misconceptions (KOSM). Moreover, KOSM is not a stable construct, as studies on mathematics learning have confirmed. In the field of physics education, a comparison of Finnish and German physics lessons on electrodynamics showed that Finnish students performed much better than German students, although Finnish teachers knew nearly nothing about KOSM compared to their German colleagues (Olszewski 2010).

5. Sometimes a problem for teacher education utilising research results is the poor quality of some research in (physics) education. In a meta-analysis on scientific inquiry, Furtak et al. (2012) found that only 59 out of 5800 peer-reviewed publications met their selection criteria for a rigorous investigation. Another important study is a meta-analysis by Ruiz-Primo et al. (2008). They analysed the effect of innovative learning environments on physics lessons. Out of 400 reviewed articles, 51 met the quality criteria of a quantitative synthesis of the measured effects.

6. Theories and models should guide empirical investigations and direct researchers' choice and application of adequate research methods. The key is that empirical research is principally engaged in confirming or refuting theoretical attempts, theoretical models or findings of former research to develop theory and to draw conclusions for larger samples (generalisation) and quality teaching (practical consequences) (Bahr and Mellor 2016). The simultaneous advancement of theoretical models and quality teaching is partly a result of a corresponding development of theory and research methods. In the 1960s, the applicability of models for quality teaching were difficult to confirm by observations of lessons. At that time, observers had to fill in questionnaires, utilising theory conform categories, while sitting in the classroom. For each category, the number of times a topic was mentioned and how many times a category was observed were the basis of the analysis. Only in the late 1980s were video techniques available to repeat the analysis of each lesson, this reduced coding mistakes and allowed one to answer new research questions. Also through the development of better test instruments and category systems, the complex structures of physics lessons could be described and analysed in more detail (Neumann et al. 2012).

7. An investigation should always be guided by an accepted theory or theoretical model. To measure, for example, a change of students' abilities to solve physics tasks after introducing a new learning environment, a researcher needs a theoretical definition of abilities and tasks. *Abilities* should be differentiated from *knowledge* and *competencies*, and physics *tasks* should also be differentiated from physics *problems* (see Chap. 9). The findings should confirm that the theoretical models used are able to describe the expected effect. If the observations, the findings, of a study match what is predicted from a theoretical model, the conclusions of a study may be a confirmation of the usability of the newly developed learning environment. The connection between theory and analysis is established by postulating a certain behaviour of the students theoretically (learning outcomes such as abilities, motivation and learning activities) as result of an intervention. In this case, it might be postulating the change of students'

ability to solve physics tasks (using tests, questionnaires). As physics teaching is always very complex and ordered hierarchically (school system-school-teacher-student), the theoretical model must consider as many independent variables as necessary for mapping the hierarchical structure. In this case, the hierarchy may be described as interdependencies between curriculum, the technical capacity of the school and teachers' professional knowledge including his or her ability to prepare and utilise the new learning environment. The learning environment must consider students' cognitive abilities, social background and the offered opportunities to learn. The analysis must discuss internal relations between the assumed dependent variables (e.g. ability, motivation, learning activities of the students) and the independent variables (e.g. professional knowledge and activities of the teacher). Furthermore, the relations between the variables of both groups must be considered. The relations can be described as causal and/or probabilistic.

8. According to Clausen (2002), empirical investigations on quality teaching need to address three important considerations:

(a) The variables of the theoretical model should validly map quality teaching (validity is described in Sect. 16.4).

(b) Hypothetical relations between the variables should be described in a structural model.

(c) A measuring model should be developed which contains measurable factors to describe the theoretical model as comprehensively and validly as possible (Fischer and Neumann 2012) (see Fig. 16.2).

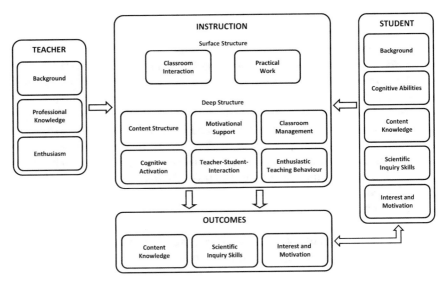

Fig. 16.2 Structure and measuring model to guide the evaluation of the quality of physics teaching (Fischer et al. 2014a)

Each *grey box* of Fig. 16.2 represents one construct (one variable, one trait), which can be measured with developed or adapted instruments. *Instruction* provides examples of key variables that can be investigated through the video observation of lessons. The constructs are measured using respective categories for an analysis at a surface level (which provides results of low inference) and interpretive (high inference) level. The boxes labelled *Teacher* and *Student* contain cognitive, demographic, social and motivational (independent) low inference variables. The *Outcomes* box presents the (dependent) high inference variables which provide quality measures of physics teaching. Each concept in the grey boxes of Fig. 16.2 is theoretically deduced and empirically reasoned from past investigations in physics education, psychology and/or pedagogy. Already existing or newly developed tests, questionnaires or categories for the video analysis operationalise the concepts. The complete measuring model is described and applied by Fischer et al. (2014a).

Among many researchers McMillan and Schumacher (2010) as well as Shavelson and Towne (2002) mentioned the following key principles to guide the conducting of deductive empirical research in education to confirm already existing theories:

1. A newly posed research question should be relevant for learning and teaching.
2. The question should be empirically accessible, that means it should be based on testable and refutable hypotheses.
3. A hypothesis always includes empirical research and theoretical attempts to formulate expected outcomes of the new investigation. In this way, research is linked to relevant theory.
4. The research designs and methods (including the samples) used must allow the answering of the research questions. Therefore, researchers should chose suitable methods to answer the research questions and researchers should develop the necessary design for answering the questions.
5. All predictions, explanations and conclusions must be derived logically from evidence, provided by previous research and theory and supported by coherent reasoning. Research studies should also detail study limitations, potential shortfalls in studies, as well as counterarguments and biases. By considering such components of a study, future research can be suggested
6. To allow a study to be further validated, critiqued, and replicated, it must be published in as much detail as possible. To publish findings in a scientific journal or to present at a scientific conference, there should first be peer review. The professional community of scientists in each field sets the standards of this process.
7. To provide researchers, teachers and policy makers a chance to assess the findings with respect to their own work and to discuss the generalisability of the study in view of other findings.

Inductive research concerns making a general statement with the help of observations with only very vague theoretical assumptions. In this type of research, one tries to derive hypotheses and general conclusions. Inductive research is done in the following steps:

1. The research topic must be described and already existing assumptions must be checked for the fit to the observed object.
2. Already made observations of the object or similar objects must be checked.
3. Already existing research methods have to be checked for adequacy and, if possible, extended and adapted to the new observation task.
4. The observation must be carried out according to the rules of empirical studies. Especially important is providing an exact description of the procedure, so that the observation can be repeated in the scientific community.
5. The data collection must be checked for reliability and validity utilising already existing theoretical approaches.
6. If possible, hypotheses can be formulated on the basis of the measurement results and the theoretical approaches already available.
7. The hypotheses may lead to new research questions, or even to a new theoretical model. Such new models should be validated by further investigations.
8. If a theoretical model can be formulated, the inductive approach transforms into a deductive one. The transition is usually hybrid and therefore cannot be determined exactly.

To be trustworthy, both approaches are subject to certain rules and requirements, which are described in more detail below.

16.4 Elements of Trustworthiness

In principle and relevant for both quantitative and qualitative research, there are four criteria to evaluate the trustworthiness of empirical investigations: *objectivity, reliability, validity* and *significance*. It is assumed that each variable (e.g. physics competence, motivation to learn physics) is distributed normally (e.g. there is a Gaussian distribution) of motivation if the (sub-)sample, such as selected students of grade 9 in a country, is big enough for ensuring precision and represents all students of grade 9 in the same country. Therefore, also the study data, such as the results of a competence test, should be normally distributed (Verma 2019). If the study data are not normally distributed, then either the measuring instrument is not reliable, and/or the sample is not representative for the target group. Non-normally distributed data will not be considered in this chapter.

16.4.1 Objectivity

Sir Francis Bacon (1561–1626) was an English philosopher and political leader, who formulated principles of objective research which he named idols (idola) (Bacon 2017).

Idols of the Tribe (*Idola tribus*) refer to human nature, this idea points out that a researcher must consider that all perception is made by the human mind and not by the objective universe.

Idols of the Cave (*Idola specus*) refers to individual experience of nature and its educational development. This process is responsible for the scientist's particular preference for specific theoretical approaches and research methods. Researchers should be aware of their individual cave (today we call it a bubble), which influences both perception and awareness.

Idols of the Market (*Idola fori*) refers to the development of a person mainly through social interaction. Meaning of words and related and underlying scientific theories should be used as precisely as possible. This means that scientists should be aware of the changing meaning of words and the related confusion from the changing meaning of words, which can take place in a communicative situation. Physics teachers for example should be aware of the difference between every day and scientific language use of physics concepts such as force or energy.

Idols of the Theatre (*Idola theatri*) are a matter of being influenced by dogmata and misleading principles of presentation. Scientists should be encouraged to question what has been proposed previously by philosophers and scientists, and they should distinguish between dogma (such as theological principles) and theoretical attempts (science communication should be evidence based).

Bacon's idols are still present in modern research guidelines utilised by research associations all over the world (e.g. DFG 2019). Accordingly, objectivity of empirical investigations is an important consideration when empirical studies are conducted. It should be that all individuals involved in studies attempt to insure objectivity. The aim being to minimise subjective influences on measurements and goal-oriented observations. To enable other researchers to understand and replicate a study, the conditions of sampling, testing, interviewing or video recording and the whole process of data processing and interpreting the data should be as consistent and precise as possible.

16.4.2 Reliability

Reliability is the consistency, or the amount of error, of the data produced in repeated measurements under consistent conditions (objectivity). There are many kinds of reliability, often reliability is expressed using values between 0.00 (much error) and 1.00 (no error), this is very similar to how error is measured in physics. "Stability" refers to the measuring error or the proportion of the variance of one measuring value related to the total variance of a measure. There should be no or only little change of this proportion if the measurement is repeated. To accurately understand the *true value T* of a measure, we have to consider statistical and systematic errors. The *random error R* is caused by variances of the behaviour of the test participants and the measuring devices (e.g. test instruments for quantitative data or rating people and categories for qualitative data). In educational research random errors are, for example, produced by differing attention of students and teachers, variations of students' discipline in the classroom, disruptions from outside, such as temperatures or noise, or mistakes in the evaluation of the tests or questionnaires. A *systematic error S (bias)* has no influence on reliability, but bias does affect validity (see Sect. 16.4.3). For example, it can occur because the students are afraid that the teacher might find out the test results. Therefore, one needs to be sure that the results are for use by the researchers only. Social desirability impacting student responses is another source of systematic errors. Teachers, for example, normally overestimate the time needed by students for an experiment during a lesson. Social desirability must be avoided or controlled, e.g. by measuring the time of experimental work in a lesson directly.

If x is the measured value of a physics test, T the true or expected value, R the random error and S the shift of T caused by a systematic error (bias), the expected value is expressed as $T = x + R + S$. Random errors often have a normal distribution, and statistical methods can be used to analyse the data. The mean M of a number of measurements should be located near the expected value, whereas the standard deviation s (or sd) of the measurements expresses the variability in the data. If the number of measures converges towards infinity, the mean converges towards the expected value. In a normal distribution, about 68% of all values are located in an interval of +/1 1 SD around the expected value T. s is a value for the distribution of the measured values around T or $T + S$ (Fig. 16.3):

$$s = \sqrt{s^2 = \frac{\sum_{i=1}^{n} (x_i = \overline{x})^2}{n - 1}}$$

There are different types of reliability; the three most commonly used are described in the following (McLeod 2013):

Test–retest reliability indicates that substantially the same results should be produced if the measurement is repeated with the same group of test takers over time. Often a repetition of the measurement is not possible. However, the reliability can be assessed for example by splitting the data of one test into half and comparing the consistency/error of both parts *(split-half reliability)*.

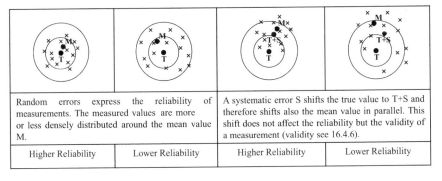

Random errors express the reliability of measurements. The measured values are more or less densely distributed around the mean value M.		A systematic error S shifts the true value to T+S and therefore shifts also the mean value in parallel. This shift does not affect the reliability but the validity of a measurement (validity see 16.4.6).	
Higher Reliability	Lower Reliability	Higher Reliability	Lower Reliability

Fig. 16.3 Cases of reliability

Interrater reliability measures the agreement between two or more raters, for example in their rating of categories in a video analysis of physics lessons.

Internal consistency reliability refers to the correlations between different items of a used test. For quantitative measurement in physics education, the used tests and questionnaires have to be tested for consistency, one component of reliability. The items of one theoretical construct such as physics knowledge should constantly measure in the same way (e.g. easy and difficult items of the test are easy or difficult for each test participant and independent of other knowledge areas, such as reading ability). The consistency of dichotomous items (can be answered with yes/no or correct/incorrect) are commonly calculated with the Kuder–Richardson formula and the consistency of interval scaled items is commonly measured with Cronbach's α. Interval scaled items (e.g. rating scale items) allow us to assess the mathematical difference between items but there is no real zero. Many of the test scales in physics education research, for example knowledge scales or motivation scales, are interval scales and not ratio scales. The Kelvin temperature scale is a ratio scale. It has a theoretically predicted absolute zero point. As there is no absolute zero point for living human beings, for example for knowledge or motivation, these scales are interval and not ratio scales.

For someone reviewing the science education literature, Cronbach's α (alpha) would appear to be very important as it is an often reported value in many publications. However, it is important to bear in mind that α is not the reliability of a test instrument but the internal consistency of its items and only a component of evaluating reliability. For a specific test, the value of α will fall in an interval between 0 and 1 [0, 1]. This value is a measure of the average correlation between the items of a test. α reflects how consistent the items, as an operationalisation of the construct such as motivation or knowledge, measure these constructs. If, for example, physics knowledge of students is measured before and after (pre and post) a newly developed unit in mechanics is implemented, the pre-value of α must be lower than the post-value of x. This is because in a pre-test, students often answer by guessing. To investigate the construct of the test using α, it is important that x changes from pre-test to post-test. One would target for an α (post-test) to be larger than 0.70. α values in an interval [0, 1] are written without zero before the decimal point, but all other indices summarising have to include zero, such as 0.57 for Cohen's κ (kappa) (see below). Recently, Cronbach's α has been critically discussed. Now sometimes researchers report a ρC (rho-C) as congeneric reliability. ρC, takes into account that a complex construct, such as force in physics, is not one absolute homogeneous trait but that the trait consists of different factors with different contributions to the construct. ρC helps provide an assessment if the different factors, or components, of the construct are measuring something similar (Cho 2016).

In addition, it is important to note that qualitative methods, such as category-based interpretation of lessons using video recordings, should be used to interpret the analysed classroom situation (the classroom environment) as precise as possible. The analysis must be based on a theoretical model and an operationalisation of the model (such as categories of behaviour or the number of specific events per time). If, for example, the quality of lessons is assessed, a model for quality teaching is needed, and that model must contain the necessary variables (see Fig. 16.2) to evaluate the model. According to Neumann et al. (2012), quality teaching can be described by *sequencing of content, time on task, classroom management* and *cognitive activation. Sequencing of content* describes the orientation of a lesson to the planned goal, *time on task* is the time during a lesson when students have focused on tasks directed towards the lesson goal, *classroom management* describes activities of the teacher to allow a maximum of time on task, and *cognitive activation* refers to elements of the lesson which encourages students to conduct goal-oriented thinking. The above categories belong to the *deep structure* of a lesson because these categories need high inference interpretation. The description of students' activities per time is needed to locate the events of the theoretical model (e.g. disturbances or form of teaching), are called *surface structure*, those events are low inference interpretations (see Fig. 16.2).

In a study, it is important that persons who have been trained for the analysis, rate or code events and categories. To assess the reliability of the rating interpretations provided by different raters, ratings of different coders have to be compared. For video analyses, for example, at least 10% of all ratings have to be provided by at least two different raters. One metric of interrater reliability can be expressed as a "percent agreement". However, this metric does not consider the accidental agreement of a rating. To compare two raters, most often researchers calculate a reliability using Cohens κ. To compare more than two raters, Fleiss κ is the method of choice. The κ values lie between < 0 (lower than randomly expected) and $+1.0$ (full agreement). The rule of thumb in social science research is a κ value of 0.6 or higher should be observed. Generally, the interrater κ reliability of deep structure events is above 0.6, whereas surface structures are mostly coded with a very high interrater reliability κ near 1.0.

As described in Fischer et al. (2014b), the concept of quality teaching is very complex. Studies evaluating quality teaching should apply quantitative and qualitative methods in parallel. To evaluate the quality of a lesson, we need to analyse both types of data, and also the activities of teachers and students in the classroom must be evaluated. For example, how cognitive variables such as students' physics knowledge and motivation are correlated. For a detailed description of empirical research and methods for different data settings (e.g. with an ordinal scale), refer to Patten and Newhart (2018).

16.4.3 Correlation

Correlations are very important for interpreting the relevance of empirical findings. A correlation is the degree of linear (Pearson) and monotonic relation (Spearman) or of a concordance of categories (Kendall) between two variables. Correlations (r) are expressed from -1 to 1. Data are positively correlated if the values of both variables increase or decrease, they are negatively correlated if the value of one variable increases while those of the others decrease (Fig. 16.4). There is no correlation if no relationship can be identified (Fig. 16.5).

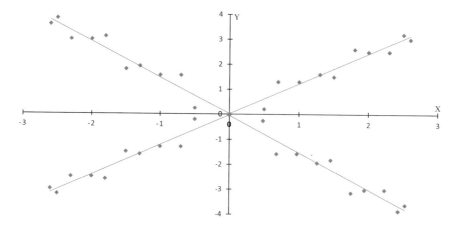

Fig. 16.4 High correlations of the variables X and Y

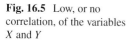

Fig. 16.5 Low, or no correlation, of the variables X and Y

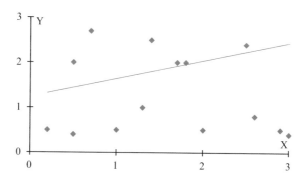

When correlation is used to evaluate data, it is important to note that variables are not considered dependent and independent.

Swimming in a lake causes sunburn does not express a causal relation, although sunburn often appears when many people swim in lakes, swimming in the lake did not cause the sunburn. In some regions of the world, a correlation between a *decreasing number of storks and a decreasing number of baby humans* may be noted. However, it is incorrect to assume that storks increase the number of babies. Obviously, the relationship between number of storks and babies is indirect. Correlations of this kind are called *spurious correlation* because there are other variables responsible for the observed correlation. For example, ultraviolet light was the reason for sunburn. It is important to note even in cases of real correlations, such as the correlation between motivation and physics competence, the causal direction cannot be analysed simply by calculating the correlation. To estimate the causal relationship between dependent and independent variables, one needs an analysis of data utilising statistical methods such as *regression analysis* or *structural equation models*. For example, using a complex model of motivation and self-concept, Ardura and Pérez-Bitrián (2019) were able to predict that students with low motivation and low self-determination opt out of physics and chemistry at the end of secondary level and their performance was lower than for those who did not opt out. The authors used a complex theoretical model to guide an analysis of the influences of motivation and self-determination on student performance. The authors used cluster analysis and structural equation modelling for their analysis in an effort to identify the magnitude and direction of influences.

In physics education research, the Pearson product-moment correlation coefficient is the most common method for analysing the distance between measured and expected values by a line of best fit (see Figs. 16.4 and 16.5). Values of the correlation coefficients r or ρ (rho) are located in an interval $[-1 \leq r \leq 1]$. Although reference values for interpreting values of high correlations are discussed and published, there are no generally accepted values.

It should be noted that a mathematical (statistical) analysis of a correlations is always crucial because the character of a correlation cannot be identified from a plot alone. There are plots that look correlated and the correlated variables turn out to be weakly correlated when calculated, and perhaps most importantly, there are plots that look correlated and turn out to be uncorrelated.

16.4.4 Statistical Significance

Testing for statistical significance is necessary in order to generalise the results occurring in a sub-sample of a basic population. For example, a study might be conducted with a sample of students (e.g. 500 students) in order to generalise more broadly but statistical tests must be conducted. It is hypothetically expected that a specific correlation between two variables does not appear by chance. Significance indicates a more than accidental connection between variables. The confidence level (p value) will lie in the interval $[0 \leq \alpha \leq 1]$.

Before a measurement (a priori) is conducted, a confidence level (p value) is determined, for example, researchers may choose an α of 0.01 as the criteria in order to decide when the null hypothesis (the null hypothesis being there is no effect) is to be rejected. A significance of $\alpha = 0.05$ is accepted for most studies (the confidence level for an effect is set 95%, meaning there is less than a 5% chance it is random).

The method to calculate significance depends on what is measured and which kind of data is being evaluated. If the data are discrete (the number of choices to answer a question is finite), a Chi-Square test is often applied.

A research question and a presumed hypothesis for testing physics competence of university physics students is given in Table 16.1.

The test Trends in International Mathematics and Science Study (TIMSS) is an existing test of physics knowledge (Mullis and Martin 2017). If the correlation in Hypothesis 1 is statistically significant with an $\alpha = 0.05$, it is confirmed with 95% certainty. A type 2 error (β error) occurs if the null hypothesis is accepted, but in reality, the null hypothesis is actually false.

Table 16.1 Research question and theoretically derived hypotheses to measure physics competency

Research question:
To which extent does the model of competence used in the study measure the physics competencies of physics students?
Hypothesis 1:
The correlation between the estimated person ability using the competence test and the TIMSS-test is above $\alpha = 0.7$
Hypothesis 2:
The correlation between the estimated personal ability as a result of the competence test and other references of physics competence (grades in physics, grades in mathematics, measures of cognitive abilities) is higher than the correlation of grades in German language and the personal ability calculated by the competence test

The probability (likelihood) that the significant correlation of variables really exists as a characteristic of the sample is expressed by the *power* of a measurement. Power is determined by the type 2 error and not only dependent on error probability but also on effect size, experimental design, and sample size (see Sect. 16.4.5).

16.4.5 Relevance and Effect Size

Effect size allows researchers to assess the practical relevance of empirical findings. It is a sample-dependent standardised measure for comparing differences between groups in one study (e.g. intervention and control groups) and for comparing the findings of studies in order to conduct a meta-analysis (Hattie 2016). As already described, physics teaching and learning at school is a very complex process. Many results of the processes analysed as dependent variables such as knowledge, competence, motivation and others are influenced by many independent variables such as age, gender, type of school, social background, professional knowledge of teachers, pre-knowledge of students as well as other factors. It can be the case that the relative effects of the independent variables on many dependent variables are not very high. Nevertheless, the theoretical model should consider all known variables. Therefore, to compare results of studies, it is important to compute and utilise the effect sizes of different variables. By doing so, one can gain an impression of how relevant the values of an interesting variable related to other variables might be. However, it is important to note, when samples are small in size, only large effect sizes turn out to be significant. To detect small effect sizes, big sample sizes are necessary. According to Bakker et al. (2019) measures of effect size refers to either measures of *differences* or *correlations* as one definition of the effect size:

$$\text{effect size} = \frac{\text{mean of expimental group} - \text{mean of control group}}{\text{standard deviation of experimental group}}$$

Division by the standard deviation makes the measure independent of units (standardised). *Cohen's d* is a common measure of effect size. Other (e.g. correlational) effect sizes often utilise the percentage of explained variance to express effect size. Effect sizes d in educational publications typically range from $]{-}2 < d < 2[$. Cohen (1988) has suggested that effect sizes of 0.2 be viewed as small, 0.50 as medium, and 0.8 as large. However, Cohen (1988) cautions that effect sizes should always be interpreted in light of 0.2 as small, 0.50 as medium, and 0.8 as large but recommends discussing the values always in the frame of the findings and the theoretical approach of the study. Taylor et al. (2018) explain how to use effect size in a meta-analysis and how to interpret effect size regarding interventions in science education.

16.4.6 Validity

Validity refers to the application of a scientific theory of a theoretical model for methods and instruments of measurements. Methods and instruments should be able to represent the investigated theoretical model (e.g. of motivation or competence) as precisely as possible. Therefore, a necessary requirement to evaluate validity is the fit between theory (or theoretical model), design and measuring instruments. For example, when directly measuring current in physics, we need to utilise an ampèremeter, a voltmeter would not be appropriate. The theoretical model of current as a flow of electrical charges means that the effect of moving charges must be detectable by the instrument.

To measure physics "competence" requires a test instrument for "competences" and not for intelligence. Therefore, we need to define what "competence" is, we do not want to define intelligence, knowledge or ability. In addition, a theoretical model is needed for competency, but there are theoretical characteristics, which are also needed. For example, the model should help one distinguish between high and low demands to measure high or low competences. The test items therefore should represent the different demands as precisely as possible and the instrument must contain an adequate number of difficult, medium and easy tasks to distinguish between low and high performing students. In Leutner et al. (2017), theoretical and practical aspects of modelling and measuring competence are described in detail for different subjects. Weßnigk et al. (2017) consider this issue for physics. The robustness of a theoretical model can be assessed from different perspectives. Therefore, different types of validity have been proposed, some of these validity types are described below. It is important to note that we do not attempt to summarise all types of validity.

Face validity, or *logical validity*, is a type of validity in which one evaluates an instrument or measuring process appears to be plausible for reaching the goal of the measurement. For example, there is no statistical procedure for evaluating face validity as face validity is evaluated just through a document review. It is preferable to have a face validity provided by an expert in the field. Face validity is the weakest form of validity because it does not allow a researcher to make substantive statements if a test measures what it is hypothesised to measure (Nevo 1985).

Content validity refers to the content of the modelled construct and relates to accuracy from an accuracy/precision perspective. For example, a test on professional knowledge should include the three hypothesised components of professional knowledge: content knowledge (CK), pedagogical content knowledge (PCK), and pedagogical knowledge (PK). If a competence test for secondary one level students has content validity, the test must meet all competencies outlined in the official curricula of the area of interest for different grades. Content validity can be assessed by asking experts such as physics professors and physics teachers if the test instrument covers the content to be tested by the instrument.

Criterion validity contains different subtypes of validity. *Concurrent validity* refers to comparing results of a test with the results of a test purported to measure the same topic. *Prognostic validity* compares the results of a test with an event in the future, such as the results of an entrance test for a university with the academic success of the same students at the end of their studies. Prognostic validity can be tested with longitudinal or quasi-longitudinal designs.

Internal validity considers whether there is a causal relationship between independent and dependent variables of the construct. Also there should be no influence of control variable or confounding variable. Those variables must be controlled or eliminated to describe the relation between independent and dependent variables. For example, in most physics lesson studies, the motivation of students and their knowledge of physics correlate. However, it is not clear whether students are motivated because of their good performance or whether their motivation leads to good performance. Therefore, it is not possible to decide which is the dependent and which the independent variable. It is also possible that the motivation of the students (and thus perhaps their performance) correlates with the pedagogical content knowledge (PCK) of the teacher and that the relationship between motivation and student performance cannot be fully explained. It is also possible that the motivation of the students (and thus perhaps their performance) correlates with the didactic knowledge of the teacher. As a result, the correlation between motivation and student performance cannot be fully elucidated. **Construct Validity** can be viewed as a component of *internal validity*. A test has construct validity if the ordering of item difficulty of a test and the items' ability to discriminate between high and low achieving students matches that predicted from theory. When the results of a test correlate with an already existing test one can say the test has **convergent validity**. To decide if a test for physics competence measures physics and, e.g., not reading or problem-solving competence, the results of the related tests with the same sample have to be correlated (**discriminant validity**, *low correlation expected*). If the correlations are high, we would say that provides evidence of convergent validity.

External validity, or *generalisability*, must be discussed and taken into account at the very beginning of a study. To generalise, the results to other populations of the sample must be representative of the population and the sampling must be random and not accidental. To apply the test in other settings, the planned setting should be described as precisely as possible (this is also a matter of objectivity). For example, the results of a test on physics competency at the end of grade 10 should be generalisable to all students in the educational system of the respective country, and after adapting content and difficulty of the items, the study should be applicable to grade 10 of other countries. Therefore, the composition of the sample is very important, because, in our example, it is not possible to test all students of grade 10 in one country. Because the diversity of students can be quite high in schools, the validity of a test must also be examined with regard to issues such as race and gender. In some cases the difficulty rankings of the items may differ as a function of subgroups. Such issues can be investigated and interpreted using differential item functioning (DIF) using a latent class (LC) analysis approach (Tsaousis et al. 2020). To produce generalisable results of a sub-sample for an entire population, the chosen sample must represent the population with all its characteristics. As it is known that also different schools can represent different characteristics of students, schools can be a basis for sampling. For example, in a study, the chosen schools must represent (1) all school types with physics as taught subject, (2) all regions of the country, (3) all types of schools with grade 10 representing all differences of teaching and social and cognitive differences of students. It is more important to note that the more often the results of a study are replicated and confirmed with different sub-samples, the higher is its external validity. (Tsaousis et al. 2020).

Construct validity is a partial aspect of validity. It is present when the measurement of a construct, such as knowledge of physics or problem-solving competence, is not influenced by systematic errors or other factors. The criteria chosen for constructing the construct contribute to a homogeneous measurement model. Construct validity of a physics competence test at university can be clarified, among other measures, by having biology and physics students of a university write the same test after the first semester. It can be expected that the test results of the two groups have a low but statistically significant correlation (e.g. $r = 0.35$). On average, both groups can be expected to have high cognitive abilities; after all, they have passed all tests up to university enrolment. After the first semester, however, the physics students are already on their way to becoming physics experts. The test therefore measures more than cognitive abilities but in addition, something that physics students may have learned in the first semester (but not the biology students). For example, the correlation (although low) may match the expectation of the researchers, and thus, this result may be of importance. If this is the case, it is more likely that the test measures physics competence. If biology students perform as well as physics students the test more likely measures competences which are not learned in physics lessons, for example in everyday situations to solve everyday problems. For more details, see for example, Cohen et al. (2013) and Borsboom et al. (2004).

If the population being studied is small, such as all beginners of physics studies at universities in one country, external validation can be performed with contrasting samples. For example, two samples of physics and biology students are compared with the same physics competence test, and it should be expected that biology students perform not as well as physics students.

It should be noted that a mathematical (statistical) analysis of a correlations is always crucial because the character of a correlation cannot be identified from a plot alone. There are plots that look correlated and the correlated variables turn out to be weakly correlated when calculated, and perhaps most importantly, there are plots that look correlated and turn out to be uncorrelated.

16.5 Analysis of Lessons

As stated at the beginning of this chapter, research in physics education is essentially concerned with clarifying the conditions for high-quality teaching (for an overview of quality teaching see Neumann et al. 2012). We currently know, for example, little about whether pedagogical knowledge (PK), pedagogical content knowledge (PCK) and content knowledge (CK) that is taught in physics teacher education at the university which has a significant impact on the quality of pre-service teachers later teaching at school. Investigating this issue will lead to more precisely formulated goals and content in teacher education at the university as well as other institutions.

In order to draw conclusions about the effects of teachers' professional knowledge on instructional quality, it is first necessary to develop theoretical assumptions about how professional knowledge contributes to instructional quality. Naturally some characteristics of good teaching are of course well known (Seidel and Shavelson 2007). For example, effective classroom management correlates with learning success (Fricke 2016) and cognitive activation correlates with being able to solve increasingly complex tasks (Wischgoll et al. 2019). It has been found that appropriate feedback, the timing of feedback, as well as the use of positive and negative feedback can have different effects (Hattie and Timperley 2007, p. 81). In addition, long-term stability of the instructional concept (Seidel and Prenzel 2006, p. 228) contributes to the "(1) organisation of activities, (2) quality of teacher-student interactions and (3) students' perception of learning conditions".

With this empirical guideline, a measurement model (see Fig. 16.2) can be constructed for investigations that can differentiate between school classes that show a distribution of learning outcomes. With such a model, it should be possible to establish at least an average correlation between the test results of students and the test results of their teachers. If this is not the case, consideration must be given to other influences that have not yet been taken into account or the model should be modified. Even if there is a correlation of the test results of students and teachers, it is important for teacher education to know exactly which teacher activities can be linked to individual characteristics of professional knowledge. An analysis of teaching utilising the criteria of professional knowledge is the method of choice in such a case.

The variables displayed in Fig. 16.2 can be measured with test instruments and questionnaires as well as with an analysis of videos recorded, for example, videos recorded during an intervention study in physics lessons.

Figure 16.6 refers to Chap. 1, Fig. 1.1 and the utilisation of learning opportunities model. It indicates which influences of teaching and learning (teacher–lesson–students) and control variables (personal and material conditions, social background and demography) should be considered for investigating instruction quality under

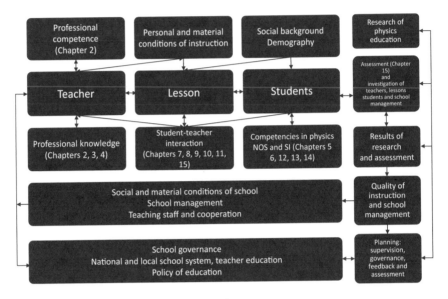

Fig. 16.6 Variables of physics education research

field conditions. The results of research on physics education should be included in the university curricula as professional knowledge of teachers (see Chap. 2). Knowledge about student–teacher interaction, students' physics competencies and motivation are the result of teacher education at the university, and therefore, all over the world are a precondition for quality teaching at schools and universities.

Depending on the goals of respective research studies, designs, samples, instruments and empirical methods have to be chosen.

16.5.1 Design and Samples

Today, the development of mathematics and science education research allows researchers to investigate *individual learning processes and learning in groups, general classroom activities* as well as *school systems*. Concepts about *teaching* and *learning* mostly provide the basis upon which the goals for investigating all kind of educational situations are investigated. In this book, these environments are restricted to institutional situations such as in schools or universities.

Accordingly, the first concept, *teaching,* is an activity, which requires a person to perform in a certain way. The activity of the teachers should be planned and guided by the expected learning processes of the students. However, the content and the structure of respective learning environments depends on the teachers' professional knowledge including his or her theoretical background. For example, a teacher or researcher who refers to behaviourism will mostly produce a teacher and not a student oriented and student-centred approach for organising their lessons (see Chap. 1).

The second central concept, *learning,* is described as individual cognitive processes, dependent on specifically arranged situations of teaching. Due to basic cognitive facilities, learning describes a process which might arise when a learner attentively observes the environment. Through related interactions with the environment, it is expected that changes of the learner's knowledge, beliefs, motivation and emotions may lead to changes of the individual cognitive repertoire of behaviour and ultimately to changes in the person's behaviour. Fiorella and Mayer (2015, p. 5) have considered the general definition to instructional situations in institutions.

Learning is a generative activity. This statement embodies a vision of learning in which learners actively try to make sense of the instructional material presented to them. Learners accomplish this goal by actively engaging in generative processing during learning, including paying attention to the relevant aspects of incoming material (which we call selecting), organising it into a coherent cognitive structure in working memory (which we call organising), and integrating cognitive structures with relevant prior knowledge activated from long-term memory (which we call integrating).

The teaching–learning process just described may require long periods to achieve the desired goals in the learning process. To explore the achievement of the goals and to describe both the process and the differences between the beginning, intermediate and end of the process, we therefore need research that can measure over this time span.

16.5.2 Longitudinal Design of Comparative Studies

Longitudinal research, in general, is research for detecting changes over time for a sample or an individual. For example, in a research study using longitudinal techniques, one might follow the development of functioning of school systems or the motivation of individuals. With longitudinal research, the same variables are observed for two or more distinct and consecutive periods. However, it is important to note that the design and sample of a longitudinal research study can only be determined when the goals of an investigation have been developed utilising a theoretical model.

In longitudinal studies, a distinction is made between trend studies and panel studies. In a trend study, the same study is conducted at several points in time, each with a different sample. With the help of trend studies, it is possible to trace changes at the level of an entire sample. However, no changes can be derived at the individual level. The PISA study is an example at the entire sample level study.

In a panel study, the same sample is examined at several points in time. In this way, individual changes can also be recorded over a certain period of time. As with a trend study, changes affecting the entire sample can also be measured. Examples range from experimental studies of the effect of a physics lesson to national educational panels investigation educational processes (competence development, educational processes, educational decisions and returns to education) in formal, non-formal and informal contexts over the entire life span (Rieger et al. 2018).

1.   If, for example, the difference of students' physics (science) competences in the school systems of different countries is to be described, physics competence has to have been theoretically modelled and a measurement instrument utilising a theoretical model must be developed that can measure this competence. The Programme for International Student Assessment (PISA) (OECD 2019) is an example of a study which measures changes in mathematics and science competencies. The goal of the PISA effort is to inform participating countries' education systems. Since 2000, PISA collects data every three years from a representative sample of 15-year-old students. The samples differ (different students are tested every three years), but samples are comparable over time. The PISA tests must be piloted several times and modified as needed to meet the above-mentioned quality criteria. In the process of developing the test instrument, cultural differences, in particular as a function of a country, are a major consideration. Therefore, the officials of PISA ask selected individuals from all participating countries to evaluate the tasks with regard to their cultural suitability for the respective country. Often tasks have to be changed or deleted based on a deficiency in validity. Also, teachers, physics educators and physicists from the participating countries check the validity of the contents of the tasks regarding correctness, appropriateness to age and cultural background, difficulty of the tasks and curricular validity. If the test meets all quality criteria, it is applied according to strict rules (objectivity) to a representative sample (generalisability).

For such studies, an important consideration is also how the socio-economic status (an independent variable) correlates with students' physics competence (a dependent variable). This step is necessary in order to identify and evaluate differences and, if the measurement is repeated after three years, to identify and describe any change of the country performance in the PISA test. However, it is important to bear in mind that correlations are generally not causal. The relationship between cause and effect, or the direction of the effect, is not described by a correlation. If, for example, a high correlation between motivation in physics lessons and physics competence is found in a country, this does not indicate whether the competence is high because motivation is high or whether high physics competence leads to higher motivation or if the high correlation is due to a different variable.

In the case of country comparisons, the sample must be drawn at random from a group of schools ordered according to defined selection criteria. In school systems such as in Canada, Germany, India or Switzerland all federal states, or cantons, must be represented in the sample because they represent different educational subsystems. However, including federally organised countries in a comparison is only possible if the same selection criteria for the sample of the whole country are applied to each of the subunits (states, provinces, cantons), which considerably enlarges the sample (Beaton and Barone 2017).

Since only average competences are needed for comparisons, the test only has to be administered once at each test date. The results can be ordered, which means quality can be compared through ranking. However, conclusions regarding the causes of teaching quality are not possible without additional investigations. Only by comparing the correlations between test results and specific country characteristics, such as, for example, per-student expenditure for education can recommendations for changes in the system be proposed.

The PISA design does allow conclusions regarding the development of the respective educational systems of the participating counties, because the test is repeated every three years. However, no conclusion can be drawn about the development of the physics abilities of students, as each test in the respective year is only carried out among 16-year-old.

2. If it is necessary to follow for example the development of physics competencies of students during their time at school, a different design must be utilised. Recording the development of student competencies is only possible in a longitudinal study. In such a study, the same students would be interviewed at least twice after a period. One problem of long-term longitudinal studies is the time span needed to collect data (often a time span of years), and sometimes, a significant numbers of students have to be dropped from the study if data are not collected from a student at every time point. Yuan et al. (2020) and many other researchers propose procedures to insert missing data without major errors. But even with the possibility of compensating for missing data, the samples must be large enough, since the error of the investigation increases with the amount of missing data.

By testing the same students in successive grades, in different schools, at the same time, it would be expected that the development of personality can be measured and can be compared vertically (Luan et al. 2019) (e.g. PISA; OECD, 2019 as described above) [e.g. learning progression; see Neumann et al. (2013)]. Vertical measures can be used to describe the development of physics concepts and competencies of individual students. In the horizontal case (e.g. comparing students' outcome on certain grades in different countries; Geller et al. (2014)], samples and time periods are the same, and the conditions (e.g. time on task) of each grade can be controlled, such as demographics or professional knowledge of teachers, and therefore compared among different grades. The results of both attempts can be used to adapt physics teaching to individual learning processes and teachers, and student teachers can be guided as how to design their lessons more successfully (Allmendinger et al. 2011; White and Arzi 2005).

3. If students of different class levels are tested at the same time, the study is named a *quasi-longitudinal study*. Such studies allow only a limited number of conclusions regarding the individual development of competence. For example, the development of physics competences from grade 5 to grade 9 (Neumann et al. 2013) can describe school-related changes of competencies, but the measured competence differences may not be causally related to the change in individual students' abilities. For example, the school system may have changed in different ways for different age groups (e.g. examination regulations, curricula, school

equipment, etc.). Students' interest in physics may have changed due to events that have a different effect on different age groups of students. For example, there may have been programmes run by the ministry of education to promote science education for grade 9, but not for grade 7. The measured difference in competence of students may therefore have been influenced by many age-related events that cannot be included in the measurement as control variables (factors that are difficult or impossible to control). As a result, guidance, based upon such studies, can be provided regarding student related designs of learning environments, or the construction of curricula, but not on the development of physics concepts at school.

16.5.3   Intervention and Causality—Experimental and Quasi-experimental Research

Physics education research is interested in exploring conditions for improving student performance, motivation or interest. Such designs are called *experimental* or *quasi-experimental*.

1. An investigation is called *experimental* when it is possible to distinguish between dependent and independent variables. Thus, the investigation goes far beyond a mere description. Expected correlations are investigated in at least two groups (e.g. two comparable classes of a school) and the data of the groups are compared. Although multiple groups are allowed, typically two groups are compared, typically one group received a "treatment" while the other group does not. This group is often named the "treatment" group, and the other is named the "control" group. If the learning gain in the treatment group is actually greater in comparison with the control group, there may be many reasons for the gain. For example, if there have been different teachers in both groups or the cognitive abilities of the students of both groups were different, the effect might have been caused by the students' abilities or by the teacher. Therefore, in experimental settings, those variables have to be controlled.

2. Hypotheses derived from a theoretical model and past empirical studies can be further evaluated by changing the learning outcomes (dependent variable), and by changing the intervention in a school class (independent variable). Confounding variables, such as different teacher experience in both groups, additional tutoring in the intervention group or different time on task in both groups must be excluded because they make it impossible to identify whether the intervention, or that confounding variables are responsible for the difference in outcomes.

3. All other variables should be controlled (control variables): the groups to be studied are either selected randomly or in parallel, i.e. they are assembled in such a way that they have comparable conditions. For example, the students in the groups must have a similar average knowledge of the test content (e.g. physics competence of mechanics), a similar average intelligence and similar

average scores on other control variables (e.g. age, gender, reading ability, social background). However, the similarity of the corresponding mean values is not sufficient, since equal mean values may have arisen from very different distributions (variance) of the individual values (e.g. competence of individual students). For example, the sets of values (3, 4, 5, 6, 7) and (0.1, 0.5, 1.2, 1.5, 4, 4, 8.5, 20.2) result in the same mean value $m = 5$, although the distributions are different. Therefore, different personal characteristics of the tested students can be assumed, and such differences compared. For example, the variance of the groups must be compared, and in most cases, the variances should be similar. Experiments can be carried out both in the laboratory and in the field. In laboratory experiments, confounding variables can be more easily excluded and control variables more easily controlled than in the field (e.g. in the classroom).

4. An investigation is called *quasi-experimental* if the sample is not (or cannot be) put together randomly or in parallel. Physics education research is therefore often quasi-experimental. In principle, with such designs, it is often not possible to clarify causal relationships, since it can never be ruled out that the sample composition was responsible for the measured effects.

An intervention in a physics classroom should be able to track the effect of a teaching method or a newly developed learning environment back to specific causes. It must therefore be possible to attribute teaching success to the intervention and not to coincidence or conflicting variables caused by the intervention itself. If, for example, in a new learning environment, the influence of the subject structure of Newtonian mechanics on students' concept development is examined, it must be determined, after the random selection of the sample, which results are expected and which results could be used to describe teaching as successful. For example, the new learning environment should, compared to the previous, produce a similar (or even better) a greater learning increase. In addition, there should not be a decrease in the motivation of students impacted by the intervention. At the same time, it must be clarified theoretically and from previous empirical studies, which variables might be responsible for the success of the learning environment. For this purpose, a comparison group is always needed. The comparison group receives instruction on the topic, but does not receive the treatment.

In order to confidently state that the results were a product of the intervention, at least the following criteria should be met to control confounding variables and to control for all expected variables:

(a) The success of the intervention group may indeed be due to the new structuring of the subject matter. In order to adequately support this, the control group must receive instruction with a different structure. This control group instruction is planned just as carefully as that presented to the intervention group. The implementation of the control group instruction is supervised just as intensively as was the case for the intervention group.

(b) The success of the intervention group may be due to the material made available to the teachers. To control for the material, the material for the intervention and for the control group must be produced with similar care and explained to the teachers in both groups with the same level of detail.

(c)   The materials provided to teachers may also differ in features that are not directly related to the physics structure of the subject matter, e.g. the requirements for reading comprehension, the graphic representations or the required mathematical skills. Such variables must therefore also be controlled or parallelised in both groups. Controlling is usually achieved through the administration of tests evaluating reading comprehension and for evaluating mathematical skills. The correlation with the measures of these test results and the results of the measures for physics competence should be low. Otherwise, an influence of the control variables on the measures for physics competence can be expected.

(d)   If learners' competence in a particular physics subject (e.g. mechanics or thermodynamics) is to be a distinguishing feature of the two groups, a statistically significantly higher level of knowledge must be measured in one of the groups. In determining the difference, the differences between the mean competences before and after the intervention must be taken as a basis, since each achievement on a particular subject after a certain (intervention) learning period depends on the previous achievement (state) level. General interest in physics, on the other hand, needs to be measured only once in an intervention, since interest is seen as a disposition (trait) that does not change rapidly as the result of teaching (Krapp 2007).

(e)   The test that is used to measure students' performance before and after the intervention must be appropriate for both the treatment and the control group with regard to the content of the intervention. The test must therefore not measure content that has not been taught to both groups (Borsboom et al. 2004).

(f)   All teachers participating in the study should have comparable professional knowledge and teaching experience. There are tests for the study teachers that can be used to compare the treatment teachers and control teachers (Fischer et al. 2012).

(g)   The socio-economic status of students should be comparable in comparison groups, as socio-economic status has been shown to be an important factor, impacting student learning. The impact of socio-economic status can be seen by reviewing OECD indices.

5.   In experimental designs, data analysis can be carried out using a wide range of statistical procedures. One common procedure is an analysis of variance (ANOVA). The variance (σ^2) is a measurement of the spread between the values of a data set, in our example the results of the measurement of students' competencies. The variance compares the distribution of the values of one data set against each other and against the mean. The variance of one or more dependent variables is explained by one or more independent variables. The simplest

form of analysis of variance tests the influence of a nominally scaled independent variable on a metrically scaled dependent variable. The scale can be interval, ratio, or absolute with equal distances between the values.

In an ANOVA with grouping (intervention and control group) as an independent variable and students' competence as a dependent variable, it is possible to investigate the extent of differences between the groups. In the case of competency being the dependent variable, it is possible to investigate the mean changes in the dependent variable (competence) for different groups, an effect of the independent variable (intervention). Such an analysis is not possible with a simple correlation. Another common statistical technique utilised is a multi-variate analysis of variance (MANOVA). This statistic considers the simultaneous influence of several independent variables on a dependent variable (for details of multi-variate statistics see Mertler and Vannatta Reinhart (2017).

Choice of the Sample

The most important questions that must be answered in order to determine the study sample are: for which group should the result be generalised (e.g. certain schools, schools in a certain town or schools in a certain country) and which groups should be distinguished (e.g. students or teachers). There are some basic rules for the choice and assembly of a sample.

As is the case of a physics laboratory, we need at least 20 independently performed measurements to determine the mean, statistical error and variance of the relationships between the variables to be investigated. The larger the sample, the more accurate these calculations. If the results in physics are to be generalised, it is important to have independent events or to control the variables as precisely as possible. Independent events means that one event should not be affected by previous events or by the observation. For example, the probability that a flipped coin will remain lying with the number facing up is always 1/2, as with every coin toss. The tosses are independent of the each other (the measurement) and independent of the observer. In physics, the object to be measured is often changed by the measurement. It is usually not possible to measure the pressure on a car tyre without changing the pressure when the filler neck is put on the valve. It is also not possible to measure the distance with a laser distance meter from the hardware store without changing the object that reflects and absorbs the laser light and therefore changes its energy state. In this case, the effects of the measurement are negligible, but the object nevertheless changes. In contrast, we experience dramatic influence of measurements on what is to be measured in particle physics. The position of an electron on its way from the cathode to a grid can be determined independently only once, because the position measurement changes the track of the electron in an irreversible way. In order to produce generalisable results, many measurements must therefore be performed.

For research in education, dependencies between variables and between observer and variables are unavoidable, and therefore, variables must be controlled and the sample must be randomised.

If the average physics competence of a particular age group is to be determined, to be on the safe side, 20 students should complete a corresponding test. Otherwise, great efforts would have to be made to compensate for the error. However, this selection criterion is not sufficient to make a statement about a specific age group. The sample must also represent the group for which the statement is to be made, perhaps for all 16-year-old physics students in a country. In order to compare 16-year-old boys and girls with regard to a specific characteristic (e.g. physics competence or motivation), at least 20 male and 20 female 16-year-olds must be tested. Since the sample is intended to represent the 16-year-old students in one specific country, all school types of the country must also be included, since learning opportunities might differ as a function of school type. If, for example, five school types are identified, at least 40 students of each type are needed. If 16 different countries shall be compared, we determine the need for a sample of 3200 if we want to determine, among other things, whether 16-year-old boys and girls differ in their physics competence or motivation. If differences between urban and rural areas are taken into account, we determine that 6400 16-year-old students who have to be tested, and even more students will have to be tested if there are substantially different regions (e.g. states, departments, ethical differences, language preferences). Depending on the theoretical model of the expected relationships between the characteristics of the individuals to be tested (e.g. social status, cognitive abilities or ethnic background) even larger sample sizes will be necessary.

If gender differences between teachers are to be investigated. For example, the correlation between teachers' pedagogical content knowledge (PCK) and the overall physics competence of their classes is to be established, at least 20 male and 20 female teachers of the same school type are required. If a distinction is also to be made between teachers with regular education as physics teachers and teachers from other disciplines who teach physics, we need 80 teachers in each of the groups, all in all 4000 teachers. To investigate students, assuming an average class size of at least 25, the physics competence of the students must be assessed using a sample of 100,000 students from throughout the country. The sample would have to be even larger when rural and urban populations are considered. This means that complex models that want to describe interdependencies between several levels (e.g. the levels of school type, class and student) require large samples.

It is important to note that when there is a low number of teachers per sample, large effects must be observed in order to be able to make any statements at all about differences between these groups, e.g. between male and female teachers. In order to be able to detect even medium effect sizes, $N = 40$ is required as an initial sample of teachers. This means about 200,000 students and 8000 teachers are needed to investigate a correlation between teachers' PCK and students' physics competence. However, the sample size can be reduced by reducing the demand for information. Maybe a large sample size is not necessary to distinguish between male and female teachers because studies in the past showed that the difference is not statistically significant.

To summarise, in order to plan the sample, precisely formulated research questions and details about the group to which the results are to be generalised are needed. For more details of basic statistics, see for example Gravetter et al. (2020).

Analysis of Classroom Videos

Originally, classroom analyses most commonly consisted of the qualitative description of social interaction. Inductive methods, such as the search for behavioural patterns, are suitable for such qualitative analysis. Such methods were developed by Ulich Oevermann in 1969. Most commonly social situations were observed directly (in a laboratory) or from analogue film recordings. A component of the analysis is discussing the event with a group of experts and reaching a consensus as to what was observed. Because such investigations are labour and time intensive, usually only small sections of a longer behavioural scene can be investigated. Such analysis leads to findings, which then have to be confirmed with quantitative methods. In order to better analyse interactions between teachers and students in classrooms, researchers started using videotaping around 1976, when the first viable video devices were available. Repeated ratings of entire physics lessons and individual activities of students and teacher started in the 1980s. Later on, results from such video studies became important for assessing the quality of teaching and learning (Fischer and Neumann 2012). Among others, Stigler and Hiebert (1997) and Hiebert et al. (2003) using video data have found in international comparative studies of mathematics teaching that there are specific teaching patterns for individual countries, so-called cultural scripts, whereas in the individual countries, the structure of instruction was highly similar. In many other video studies, the technical requirements of the process of recording in the classroom, and the video evaluation methods were standardised.

The development of deductive, category-based analysis methods has made video-supported classroom analysis accessible for quantitative methods. According to Mayring (2007), a deductive approach can be used to generate quantitative data from qualitative interpretations (e.g. time-related student activities in a group experiment are correlated with teacher activities for cognitive activation). This has standardised the interpretation process and made it possible to control the reliability and validity of the category system required for the analysis, e.g. for assessing classroom management or for assessing the attention of students. Deductive methods can be used to analyse teaching processes and teaching structures.

To analyse ideographs of lessons, researchers need theoretically derived categories, e.g. of behaviour and interaction, to describe what happens in a classroom and to identify patterns concerning the topics of classroom management, cognitive activation, motivational support or enthusiastic teacher behaviour, time on task and the analysis of content structure of the lesson (see Fig. 16.2). Depending on the category, the analysis is carried out using "so-called" events (*turns* or *intervals*) that are defined before the analysis. Turns always comprise completed activities, such as a certain phase in students' experiments. Depending on the student group or lesson, events can vary in length and are defined and evaluated according to an operationalised category. With interval-based coding, one has to decide whether a certain event (e.g. a lesson disturbance) has occurred in an interval. The interval lengths are determined in such a way that a clear decision can be made regarding the category in question.

The development of a category system, e.g. for sequencing lessons, determining the cognitive activation of students, evaluating how teachers work with physics experiments or describing teachers' ability to lead lessons, requires a theoretical model. First, based on theory, it must be explained what is meant by sequencing of teaching etc.

In such analyses, a coding book is developed to define how certain turns or intervals are to be interpreted. For example, how many different disturbances, of which type, have taken place in a certain part of the lesson or how many actions of which type for the cognitive support of the students have been initiated by the teacher. When the coding instructions are based on a theoretical model, coders are trained according to the established coding rules until an acceptable interrater reliability is achieved. Recall that interrater reliability is indicia to evaluate the similarity of raters. In order to achieve this, it may be necessary to differentiate the coding instructions further in order to detect systematic differences in understanding between coders.

Before video recording of lessons, the number and positions of cameras and microphones must be determined, and camera operators must be trained, so that the recordings conform to the category system. If, for example, student experiments are to be analysed, more than two cameras are needed (usually to follow the teacher and the students) and the camera positions must be adapted to the goal of the analysis. Fischer and Neumann (2012) describe different conditions and corresponding arrangements of cameras. Often transcripts of videos are needed for the analysis. Depending on the aim of the analysis, transcripts might contain the writing of statements, further information about lesson details, or even information regarding gestures and facial expressions of the people involved. Mayring (2007) describes the theoretical basis of qualitative research and provides examples of categorising, transcription rules and coding of lesson videos that should be considered in order to set the stage for the reliable analysis of physics lessons.

At the end of a video analysis, the quality of the data is evaluated. Interrater reliability (also known as interrater agreement or interrater concordance) can in its simplest form, be evaluated as the percentage of agreement among raters. According to Tong et al. (2020), 10% of the recorded teaching time should be randomly selected and interpreted by two independent coders to ensure sample representativeness. Interrater reliability provides a measure of the homogeneity in the ratings provided by independent raters.

The validity of the measured construct is determined by correlations with test and/or questionnaire data that verify the claimed relationships. If, for example, the consistent sequencing of instruction is to be a measure of instructional quality, the learning outcomes, including physics competence and motivation, must correlate satisfactorily with the number and quality of the teacher's sequencing activities. If this is not the case, or if the interrater reliability is not satisfactory, the theoretical assumptions must be checked and/or the category systems must be modified. If modification in the coding system are needed, then naturally there will be a change, in the coding manual and in the training of coders (Fischer et al. 2005; Mayring 2014).

16.6 Conclusion

This chapter is intended as an introduction to empirical research in physics education. The methods of empirical research on teaching and learning today use probabilistic methods (item response theory), including the Rasch model as an important method for test development evaluation (Boone and Noltemeyer 2017). Rasch is also used for survey development and analysis in physics education research. In addition, multi-level models (hierarchical linear modelling, HLM) are applied, in order to elucidate relationships between variables which, as is common in teaching systems, are located at different but related levels (educational governance, school system, school, classroom, individual) (Feldstain et al. 2012).

In addition, linear structural equation models (LISREL) and confirmatory factor analysis (CFA) play a role in empirically testing theoretical assumptions about complex cause–effect relationships (Westland 2015). These methods of analysis have an impact on research on physics teaching and learning, but they go beyond the scope of an introductory chapter.

Whether a newly developed teaching unit on a modern field of physics or new material on Newtonian mechanics actually shows the desired effects, certainly depends on the structure and the complexity of the content taught. However, it may also depend on personal characteristics of the teacher, such as his or her professional competencies (see Chap. 2) or other characteristics that influence his or her relationship to the class. In addition, the socio-economic background of the students, or their previous knowledge and cognitive abilities are important factors to consider in studies.

In order to be able to consider all these influences in physics education research, we first need a theoretical model that takes all these factors into account and describes their relationship (see Figs. 16.1, 16.2 and 16.6). In the end, our hope is to provide statements that increase the probability of good teaching (if teachers follow such guidance). The critical point is only with validated theoretical models, can statements be made about the quality of the newly introduced learning environment or the needed instructional structure of the new content.

The implementation of empirical findings in teaching practice or in teacher training (see Chap. 3) has not been discussed in this chapter. Since physics education and educational research were not empirically oriented for a very long time, there are instructional structures, teaching methods, teaching contents at universities and teacher seminars and recommendations for teachers for which we are not sure of their intended effects. This situation gives rise to myths about the effect of teaching measures and intuitive beliefs that are difficult to refute. Such myths are often based on personal experience, which, from the perspective of empirical research, must be limited to these persons. The situation is just now changing in favour of trustworthy results regarding teaching quality. Therefore, it is important for teachers and student teachers alike to understand research results in order to be able to assess the relevance and quality of empirical research for their own teaching.

Acknowledgements We would like to thank Robert Evans (University of Copenhagen) and William Romine (Wright State University, Dayton) for carefully and critically reviewing this chapter.

References

Allmendinger J, Kleinert C, Antoni M, Christoph B, Drasch K, Janik F et al (2011) Adult education and life-long learning. In: Blossfeld H-P, Roßbach H-G, Maurice JV (eds) Education as a lifelong process—the German national educational panel study (NEPS), vol 283–299. VS Verlag für Sozialwissenschaften, Heidelberg

Ardura D, Pérez-Bitrián A (2019) Motivational pathways towards academic achievement in physics & chemistry: a comparison between students who opt out and those who persist. Chem Educ Res Pract 20(3):618–632. https://doi.org/10.1039/C9RP00073A

Bacon F (2017) In: Bennett ABJ (ed) The new organon: or true directions concerning the interpretation of nature

Bahr N, Mellor S (eds) (2016) Building quality in teaching and teacher education, vol 61. ACER Press, Camberwell, Victoria

Bakker A, Cai J, English L, Kaiser G, Mesa V, Van Dooren W (2019) Beyond small, medium, or large: points of consideration when interpreting effect sizes. Educ Stud Math 102(1):1–8. https://doi.org/10.1007/s10649-019-09908-4

Beaton AE, Barone JL (2017) Large-scale group-score assessment. In: Bennett RE, von Davier M (eds) Advancing human assessment: the methodological, psychological and policy contributions of ETS. Springer International Publishing, Cham, pp 233–284

Boone WJ, Noltemeyer A (2017) Rasch analysis: a primer for school psychology researchers and practitioners. Cogent Educ 4(1):1416898. https://doi.org/10.1080/2331186X.2017.1416898

Borsboom D, Mellenbergh G, Heerden J (2004) The concept of validity. Psychol Rev 111:1061–1071. https://doi.org/10.1037/0033-295X.111.4.1061

Cho E (2016) Making reliability reliable: a systematic approach to reliability coefficients. Organ Res Methods 19(4):651–682. https://doi.org/10.1177/1094428116656239

Clausen M (2002) Qualität von Unterricht – Eine Frage der Perspektive? [Quality teaching—a question of perspective?]. Waxmann, Münster

Cohen J (1988) Statistical power analysis for the behavioral sciences, 2nd edn. Lawrence Erlbaum Associates, Hillsdale

Cohen J, Cohen P, West SG, Aiken LS (2013) Applied multiple regression. Correlation analysis for the behavioral sciences, 3rd edn. Routledge, London

Eichenlaub M, Redish EF (2019) Blending physical knowledge with mathematical form in physics problem solving. In: Pospiech G, Michelini M, Eylon B-S (eds) Mathematics in physics education. Springer International Publishing, Cham, pp 127–151

Evens M, Elen J, Larmuseau C, Depaepe F (2018) Promoting the development of teacher professional knowledge: integrating content and pedagogy in teacher education. Teach Teach Educ 75:244–258. https://doi.org/10.1016/j.tate.2018.07.001

Feldstain A, Woltman H, MacKay J, Rocci M (2012) Introduction to hierarchical linear modeling. Tutorials Quant Methods Psychol 8:62–69. https://doi.org/10.20982/tqmp.08.1.p052

Fiorella L, Mayer RE (2015) Learning as a generative activity: eight learning strategies that promote understanding. Cambridge University Press, Cambridge

Fischer HE, Neumann K (2012) Video analysis as a tool for understanding science instruction. In: Jorde D, Dillon J (eds) Science education research and practice in Europe: retrospective and prospective. Sense Publishers, Rotterdam, pp 115–139

Fischer HE, Klemm K, Leutner D, Sumfleth E, Tiemann R, Wirth J (2005) Framework for empirical research on science teaching and learning. J Sci Teacher Educ 16(4):309–349

Fischer HE, Borowski A, Tepner O (2012) Professional knowledge of science teachers. In: Fraser B, Tobin K, McRobbie C (eds) Second international handbook of science education. Springer, New York, pp 435–448

Fischer HE, Boone WJ, Neumann K (2014a) Quantitative research designs and approaches. In: Lederman NG, Abell SK (eds) Handbook of research on science education, vol II. Taylor and Francis (Routledge), New York, pp 18–37

Fischer HE, Labudde P, Neumann K, Viiri J (eds) (2014b) Quality of instruction in physics—comparing Finland, Germany and Switzerland. Waxmann, Münster, New York

Fricke K (2016) Classroom management and its impact on lesson outcomes in physics. A multi-perspective comparison of teaching practices in primary and secondary schools. Logos Verlag Berlin GmbH, Berlin

Furtak EM, Seidel T, Iverson H, Briggs D (2012) Experimental and quasi-experimental studies of inquiry-based science teaching: a meta-analysis. Rev Educ Res 82(3):300–329

Geller C, Neumann K, Boone WJ, Fischer HE (2014) What makes the finnish different in science? Assessing and comparing students' science learning in three countries. Int J Sci Educ 36(18):3042–3066. https://doi.org/10.1080/09500693.2014.950185

Guidelines for safeguarding good research practice—code of conduct (2019)

Hattie J (2016) Visible learning for literacy, grades K-12: implementing the practices that work best to accelerate student learning. Corwin Press, Thousand Oaks, USA

Hattie J, Timperley H (2007) The power of feedback. Rev Educ Res 77(1):81–112. https://doi.org/10.3102/003465430298487

Helmke A (2009) Unterrichtsqualität und Lehrerprofessionalität. Diagnose, evaluation und Verbesserung des Unterrichts [Instruction quality and teacher professionality. Diagnose, evaluation and improvement]. Kallmeyer, Seelze

Hiebert J, Gallimore R, Garnier H, Bogard Givvin K, Hollingsworth H, Jacobs J et al (2003) Teaching mathematics in seven countries: results from the TIMSS 1999 video study

Hill HC, Chin M (2018) Connections between teachers' knowledge of students, instruction, and achievement outcomes. Am Educ Res J. Advance online publication. http://doi.org/10.3102/0002831218769614

Krapp A (2007) An educational–psychological conceptualisation of interest. Int J Educ Vocat Guidance 7(1):5–21. https://doi.org/10.1007/s10775-007-9113-9

Leutner D, Fleischer J, Grünkorn J, Klieme E (eds) (2017) Competence assessment in education: research, models and instruments. Springer International Publishing, Cham

Luan Z, Poorthuis AMG, Hutteman R, Denissen JJA, Asendorpf JB, van Aken MAG (2019) Unique predictive power of other-rated personality: an 18-year longitudinal study. J Pers 87(3):532–545. https://doi.org/10.1111/jopy.12413

Mayring P (2007) Mixing qualitative and quantitative methods. In: Mayring P, Huber GL, Gürtler L, Kiegelmann M (eds) Mixed methodology in psychological research. Sense Publishers, Rotterdam, pp 27–36

Mayring P (2014) Qualitative content analysis: theoretical foundation, basic procedures and software solution. Klagenfurt

McLeod SA (2013) What is reliability? Simply psychology. Retrieved from https://www.simplypsychology.org/reliability.html

McLure F, Won M, Treagust DF (2020) Teaching thermal physics to year 9 students: the thinking frames approach. Phys Educ 55(3):035007. http://doi.org/10.1088/1361-6552/ab6c3c

McMillan JH, Schumacher S (2010) Research in education: evidence-based inquiry, 7th edn. Pearson Education Ltd., Essex

Merrill MD (1991) Constructivism and instructional design. Educ Technol 31(5):45–53. Retrieved from http://www.jstor.org/stable/44427520

Mertler C, Vannatta Reinhart R (2017) Advanced and multivariate statistical methods. Routledge, New York

Mullis IVS, Martin MO (eds) (2017) TIMSS 2019 assessment framework. Boston College. Retrieved from TIMSS & PIRLS International Study Center website

Neumann K, Kauertz A, Fischer HE (2012) Quality of instruction in science education. In: Fraser B, Tobin K, McRobbie C (eds) Second international handbook of science education. Springer, New York, pp 247–258

Neumann K, Viering T, Boone WJ, Fischer HE (2013) Towards a learning progression of energy. J Res Sci Teach 50(2):162–188. https://doi.org/10.1002/tea.21061

Nevo B (1985) Face validity revisited. J Educ Meas 22(4):287–293

Olson DR (2004) The triumph of hope over experience in the search for "what works": a response to Slavin. Educ Res 33(1):24–26

Olszewski J (2010) The impact of physics teachers' pedagogical content knowledge on teacher action and student outcomes, vol 109. Logos, Berlin

Patten ML, Newhart M (2018) Understanding research methods: an overview of the essentials, 10th edn. Routledge, New York

Pintrich PR (2000) Chapter 14—the role of goal orientation in self-regulated learning. In: Boekaerts M, Pintrich PR, Zeidner M (eds) Handbook of self-regulation. Academic Press, San Diego, pp 451–502

Popper K (2002 [1959]) The logic of scientific discovery. Routledge, Abingdon-on-Thames

Rieger S, Hübner N, Wagner W (2018) NEPS technical report for physics competence: scaling results for the additional study Thuringia. Retrieved from Bamberg: https://www.neps-data.de/Portals/0/SurveyPapers/SP_XL.pdf

Ruiz-Primo MA, Briggs D, Shepard L, Iverson H, Huchton M (2008) Evaluating the impact of instructional innovations in engineering education. In: Duque M (ed) Engineering education for the XXI Century: foundations, strategies and cases. ACOFI Publications, Bogotá, Colombia, pp 241–274

Seidel T, Prenzel M (2006) Stability of teaching patterns in physics instruction: findings from a video study. Learn Instr 16(3):228–240. https://doi.org/10.1016/j.learninstruc.2006.03.002

Seidel T, Shavelson RJ (2007) Teaching effectiveness research in the past decade: the role of theory and research design in disentangling meta-analysis results. Rev Educ Res 77(4):454–499. https://doi.org/10.3102/0034654307310317

Shavelson RJ, Towne L (eds) (2002) Scientific research in education. National Academy Press, Washington, DC

Sorge S, Kröger J, Petersen S, Neumann K (2019) Structure and development of pre-service physics teachers' professional knowledge. Int J Sci Educ 41:862–889. https://doi.org/10.1080/09500693.2017.1346326

Stigler JW, Hiebert J (1997) Understanding and improving classroom mathematics instruction: an overview of the TIMSS video study. The Phi Delta Kappan 79(1):14–21. Retrieved from www.jstor.org/stable/20405948

Taylor JA, Kowalski SM, Polanin JR, Askinas K, Stuhlsatz MAM, Wilson CD et al (2018) Investigating science education effect sizes: implications for power analyses and programmatic decisions. AERA Open 4(3):2332858418791991. http://doi.org/10.1177/2332858418791991

Tong F, Tang S, Irby BJ, Lara-Alecio R, Guerrero C (2020) Inter-rater reliability data of classroom observation: fidelity in large-scale randomized research in education. Data Brief 29:105303. http://doi.org/10.1016/j.dib.2020.105303

Tsaousis I, Sideridis GD, AlGhamdi HM (2020) Measurement invariance and differential item functioning across gender within a latent class analysis framework: evidence from a high-stakes test for university admission in Saudi Arabia. Front Psychol 11(622). http://doi.org/10.3389/fpsyg.2020.00622

Verma JP (2019) Normal distribution and its application. In: Statistics and research methods in psychology with excel. Springer Singapore, Singapore, pp 201–235

Watt JH, van den Berg S (2002) Populations and samples: the principle of generalization. In: Research methods for communication science, pp 50–61. http://www.cios.org/: CIOS Open Text Project

Weßnigk S, Neumann K, Viering T, Hadinek D, Fischer HE (2017) The development of students' physics competence in middle school. In: Leutner D, Fleischer J, Grünkorn J, Klieme E (eds) Competence assessment in education: research, models and instruments. Springer International Publishing, Cham, pp 247–262

Westland JC (2015) An introduction to structural equation models. Structural equation models: from paths to networks. Springer International Publishing, Cham, pp 1–8

White RT, Arzi HJ (2005) Longitudinal studies: designs, validity, practicality, and value. Res Sci Educ 35(1):137–149. https://doi.org/10.1007/s11165-004-3437-y

Wischgoll A, Pauli C, Reusser K (2019) High levels of cognitive and motivational contingency with increasing task complexity results in higher performance. Instr Sci 47(3):319–352. https://doi.org/10.1007/s11251-019-09485-2

Yuan C, Hedeker D, Mermelstein R, Xie H (2020) A tractable method to account for high-dimensional nonignorable missing data in intensive longitudinal data. Stat Med 39(20):2589–2605. https://doi.org/10.1002/sim.8560

DFG (2019) Guidelines for safeguarding good research practice - code of conduct. https://www.dfg.de/en

OECD (2019) PISA 2018 Assessment and analytical framework. https://doi.org/10.1787/b25efab8-en

Gravetter FJ, Wallnau LB, Forzano LAB, Witnauer JE (eds) (2020) Essentials of statistics for the behavioral sciences (10 ed). Cengage, Boston

Chapter 17
Qualitative Research on Science Education in Schools

Michaela Vogt and Katja N. Andersen

Abstract Specific features in the three steps of theoretical framing, data collection and data analysis characterise qualitative research on science education. As a general tendency, the qualitative paradigm contributes to research results that are gained by the interpretation of non-numerical data collected through a rather open, not hypothesis-driven, process-like research (Bortz and Döring in Research methods and evaluation: for human and social scientists. Springer Medizin, Heidelberg, 2016; Lamnek and Krell in Qualitative social research: With online material. Beltz, Weinheim, 2016). Beyond this pragmatic shortcut to the paradigmatic perspective, it should be emphasised that the following contribution is based on a fundamental understanding of qualitative research in the sense of a multidimensional modular system. The individual components of this system can be used and combined flexibly. However, this must happen based on the solid foundation of theory and the principled orientation towards the object of research or research questions. This contribution presents and discusses current trends in qualitative research on science education in schools. The chapter focusses on the four steps (a) theoretical groundwork for a research project in didactics, (b) data collection implying sampling, methods and technical support, (c) data analysis with its diverse methods and criteria of quality and (d) the interpretation of the analysed data related to the theoretical framework as well as to the research field.

17.1 Introduction

Planning a Qualitative Research Project: An Initial Overview

Planning a qualitative project generally begins with a subject-didactic problem or challenge and related research questions. Suitable research questions arise in relation

M. Vogt (✉)
University of Bielefeld, Bielefeld, Germany
e-mail: michaela.vogt@uni-bielefeld.de

K. N. Andersen
University of Luxembourg, Esch-sur-Alzette, Luxembourg

© Springer Nature Switzerland AG 2021
H. E. Fischer and R. Girwidz (eds.), *Physics Education*, Challenges in Physics Education,
https://doi.org/10.1007/978-3-030-87391-2_17

to the state of research or state-of-the-art identified and may lead to reflecting on definitions and defining key terms. Moreover, decisions related to *scientific theory and methodology* need to be made in accordance with the research interest and in view of deciding the concrete specific methods to be used (see Fig. 17.2, Sect. 17.2). This kind of theoretical foundation later serves to explain steps of *data collection* (Sect. 17.3), *data analysis* (Sect. 17.4) and *interpretation* of the results (Sect. 17.5) (Diekmann 2018; Flick et al. 2004). Additionally, the validation of quality criteria is of enormous significance for raising the quality and transparency of one's own research process (Sect. 17.5). Whether and in how far the subject-didactic natural sciences have already integrated qualitative research with its standards into their research repertoire will be elucidated in a *conclusion* (Sect. 17.6).

The steps of conducting a qualitative research project listed in Fig. 17.1—which will be elaborated on and illustrated by means of example projects from the field of natural science subject didactics—strongly depend on each other. In addition, qualitative research, in particular, is distinguished—due to the emphasis it places on the *adequate approach to its subject*—by the fact that it is the norm rather than the exception that no existing approach can be simply applied. Indeed, in most cases, existing methods and approaches need to be adapted and, during the actual research process, adjusted further, if any discrepancies arise. In this sense, the approach to the individual steps of qualitative research detailed in the following represents a point of reference for conceptualising qualitative research projects that is, in principle, adjustable.

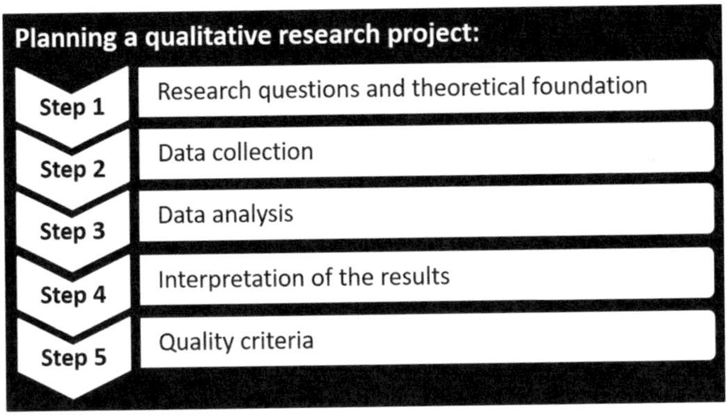

Planning a qualitative research project:

Step 1 — Research questions and theoretical foundation

Step 2 — Data collection

Step 3 — Data analysis

Step 4 — Interpretation of the results

Step 5 — Quality criteria

Fig. 17.1 Five steps of planning a qualitative research project

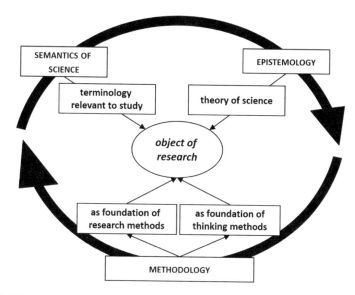

Fig. 17.2 Theoretical groundwork for a qualitative research project

17.2 Step 1: Research Questions and Theoretical Foundation for a Research Project in Didactics

Discovering and demonstrating a research gap, first and foremost, necessitates an intense engagement with the existing studies relevant to the topic (see Fig. 17.2). This is done in the form of researching and identifying the *state of research*, often also referred to as the state of the art. In this, it is important to maintain an interdisciplinary perspective, as a potential object of research will, as a rule, have been studied and researched in several distinct disciplines.

17.2.1 Object of Research

Once a gap has been identified through researching the state of the art and preliminary *research questions* (not hypotheses, as in the quantitative paradigm) have been developed, a *theoretical engagement with the object of research* must precede the decision for a specific approach regarding data collection, analysis and interpretation. Although this theoretical groundwork also plays a role in quantitative research, it is especially crucial in qualitative research for finding an approach that is adequate to the object of research and maintaining principled openness in methodology are the key. Generally speaking, this theoretical reflection lays the groundwork for subsequent decisions in the research and comprises the following aspects, to be detailed further in the subsequent sections of this chapter:

1. the clarification of crucial terminology relevant to the study
2. the methodological groundwork as foundation for subsequent decisions regarding data collection and analysis
3. the epistemological and theory-of-science foundation for concluding interpretation of data.

17.2.2 Semantics of Science

Clarifying key terminology relevant to the project may require a twofold approach to definitions. On the one hand, terminology as such must be defined in reference to the project using a *semantics-of-science approach*. On the other hand, the structures of other, related scientific theories are relevant insofar as they are linked to such terminology (cf. Harris 2005). In this regard, context theories, semantic or structuralist approaches to theory, among others, can be helpful (cf. Carrier 2004).

Methodology

Furthermore, a *methodological basis* must be found to correspond to the research questions and the clarified definitions. This basis, in principle, frames both the specific research method and the related method of thinking, thus providing the point of departure for concrete steps of data collection and analysis.

Options for the methodological basis for the specific research method:

- *Hermeneutics*
- *Phenomenology* (e.g., *Kakkori* 2009)
- *Discourse theory according to Foucault* (1989)
- *Ethnomethodology according to Garfinkel* (1991)
- *Symbolic Interactionism according to Blumer* (1969).

Viable methods of thinking (Mannheim 1982; Reichertz 2004a) are:

- *Deduction*
- *Induction*
- *Abduction.*

The methodological approaches listed here are central concepts. However, above and beyond this, there are more approaches existing.

Epistemology

Besides the methodological basis for data collection and analysis, the circumstances may require establishing the science-theoretical foundation for data interpretation in terms of a theory of science. This is based on an *epistemological foundation* that conceptualises the elementary process of human insight. The theory of science can, in terms of its subject matter, be regarded as the extension of epistemology. Thus, theory-of-science approaches strive to focus on scientific insight and thus on possibilities and foundations of scientific research (e.g., Lamnek and Krell 2016). In the strict sense, theory-of-science approaches according to Tschamler (1996) include the logic of insight as the clarification of terminological structures. Such logic is transferred to the constitutive problems of the respective object domain in research. Thus, it functions as the explanatory foundation for the interpretation of scientific findings on the object of research in the specific research project.

Central concepts for the epistemological basis of a research project (e.g., *Bhaskar* 2008; *Matthews* 1993; *Tschamler* 1996):

- *Rationalism*
- *Empiricism*
- *Constructivism*
- *Structuralism.*

Central concepts for theory-of-science foundations for data analysis and interpretation:

- *Social Constructivism* (*Berger and Luckmann* 1966)
- *Systems Theory* (*Luhmann* 2012)
- *Critical Rationalism* (*Popper* 1998).

Initially, both a methodological and a theory-of-science foundation should be employed for the specific approach of a research project. After a thorough assessment, these levels may, as circumstances allow, be combined. An example of such a theoretical foundation for a project is provided by Reinhoffer (2000). On subject domains of the natural sciences, Reinfoffer conducted a diachronic analysis of curricula for the teaching of the subject "local history" or "Sachkunde" in Baden-Württemberg. Another example is Porter and Córdoba's (2009) study in the UK, in which they extended the notion of *Systems Theory* as it pertains to sustainability science by using systems thinking as a practical and pedagogical framework (see also Waddock 2006). In general, it must also be noted that the theoretical foundation for a research project is meaningful only if one continuously refers back to these basic assumptions during data collection, analysis and interpretation. Moreover, the argumentative structure of the analysis and interpretation should be built on these theoretical considerations. Should this not be done, the theoretical foundation of the research loses its purpose and the research conducted loses both scientific quality and internal consistency.

Based on the decisions made in step 1, the design of the study is created. In this process of design development, such decisions have to be made as, for example, whether to choose a single case design, a descriptive design, an evaluation design or an intervention study. While the single case design is based upon the idea of evaluating one individual instead of generalising, the descriptive design builds on the approach of describing a phenomenon of the total population examined. The evaluation design again requires the development of a structure that provides the information needed to answer each of the formulated assessment questions. Finally, the intervention study provides opportunities to determine the effectiveness of an intervention, of which the investigator assigns the exposure. This differs from the observational studies (see also Sect. 17.3.2), in which the researcher focuses mainly on the authentic learning environment without influencing it. Especially in qualitative research around the didactics of natural sciences, the concept of intervention studies is of high importance.

17.3   Step 2: Data Collection

In principle, the steps of data collection should be regarded separately from those of data analysis, as there is no fixed correspondence between specific methods of data collection and analysis in qualitative research. However, it is necessary to connect these meaningfully to each other. Means of data collection comprise, in principle, all steps for the collection but also delineation of a pool of data related to the research question (Bortz and Döring 2016; Flick 2016).

In terms of data collection, important decisions on the specific approach to be taken relate to deciding the sample, the choice of a specific method of data collection and, as circumstances may require, planning the use of technical means or tools for collection.

17.3.1   Deciding the Sample

By way of examples, this section presents a selection of common sampling strategies of the qualitative paradigm but also forms of representative sampling (see Fig. 17.3). Traditionally, the latter are more often relevant to quantitative approaches. However, given a pertinent research question and rationale, they can also be useful in qualitative research projects. In this case, however, the aim of representativeness is usually not

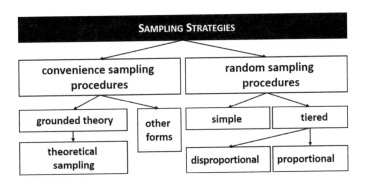

Fig. 17.3 Overview of sampling strategies

crucial (e.g., Flick 2016). In addition to the question of setting a sample, contextualising data by means of complementary material is important. This is also addressed here briefly in this chapter.

The best-known strategy in qualitative research is the step-by-step delineation of the sample by *theoretical sampling* according to Glaser and Strauss (2017), which is connected to *grounded theory* (Sect. 17.4). This approach describes a step-by-step sequence of selection decisions aligned with a theory developed in parallel with the research. In this, a continuous comparison between the newly gained insights and the already collected data is crucial (Strauss and Corbin 1997). In addition, there are *further, deliberate and thus convenience sampling strategies* (e.g., analytical induction, selection of typical cases; Kromrey et al. 2016; Lamnek and Krell 2016). Given a pertinent rationale that is adequate to the object of research, *random sampling strategies* can also be employed productively in qualitative research. These are carried out in simple or tiered fashion and, in the latter case, to relate proportionally or disproportionally to the original distribution in the parent population according to specific criteria (Kromrey et al. 2016). Among studies from the subject didactics of the natural sciences, Dunker (2016) used *theoretical sampling* to select contrastive cases in studying teachers' views on students' experiments in natural science classes. In contrast, Günther et al. (2004) used a disproportionally tiered sample to analyse conceptualisations of science held by teachers in training and teachers in the natural sciences. In another example, Zinn (2008) devised a proportionally tiered sample in relation to the parent population of the project THINK ING in order to determine the degree of correspondence between the experience of physics classes and the interests of students in 11th-grade. The summary of Zinn's project is given below.

Example project Zinn (2008):

As Zinn's project is part of the larger project THINK ING, the selection of case studies is guided by the parent population defined there. To do so, key features of the population's composition (e.g., gender, age, preference for specific forms of teaching among the teachers) were taken into account and used to compile a feature-specific, representative sample for the study at hand. This sample consisted of 10 male students, 14 female students, 4 female teachers and 7 male teachers.

It must also be noted that, in principle, a combination of various sampling strategies may be appropriate in a given case in order to successively narrow down the sample.

As a rule, the data collected for a qualitative study also come with complementary contextual information that must be defined in its range and explored in its structure. This is especially so in historically oriented projects in subject didactics, where context plays an important role. In other topic areas, too, context must be meaningfully integrated into the sample or connected to it (e.g., social data or other materials that complement interviews, influencing factors on observation or interviews). As historical research, the study conducted by Lind (1999) can be named here: It analysed the teaching of physics in German grammar schools from the start of the eighteenth to the start of the twentieth century by way of examples. Aiming for an international comparison, the project *Innovation Naturwissenschaftlich-technischer*

Bildung in Grundschulen der Region Bodensee (InTeB) studied the school contexts for the successful use of "learning boxes" for teaching aspects of the topic "flying" related to physics and technology (Wagner 2016a). A structured engagement with contextual factors at school and national levels plays an important role in Wagner's project.

17.3.2 Deciding the Method of Data Collection

A detailed look at the individual options for data collection is not possible within the scope of this chapter. Instead, the three main approaches of qualitative data collection according to Flick (2016) and Diekmann (2018) are briefly explicated and differentiated here both from each other and from their quantitative counterparts.

Interviews and Open Questionnaires

The first option for qualitative data collection is *different forms of interviews and questionnaires*, which, in contrast to their quantitative counterparts, generally uses open or at most semi-standardised forms. In this method of data collection, the subjective views of the social actors interviewed are crucial, including, for example, students' subjective explanations of observable physics processes, experiments and phenomena. In such method, the interviewers themselves act as "instruments" of data collection (Bortz and Döring 2016, p. 309) and include into the analysis, where relevant, their own thoughts, emotions and reactions during the interviews. Examples for such qualitative data collection are, among others, guideline-based interviews, narrative interviews, focused interviews, ethnographical interviews, expert interviews or, in the broadest sense, also group discussions (e.g., Flick 2016).

When conducting an interview or open-questionnaire, the following aspects should be known:

- *Arguments for the choice of a specific form of interview or open-questionnaire that must, above all, focus on the epistemological interest of the research*
- *Differences between distinct types of question*
- *Techniques of different forms of interview, open-questionnaire and documentation*
- *Potential sources of errors in interviews or open-questionnaires*
- *Adequate use of transcription conventions* (e.g., *Diekmann* 2018).

A concrete example for a qualitative interview is the study conducted by Grygier (2008), who used such kinds of interviews, among other methods, to analyse primary school children's understanding of science in "Sachunterricht" classes. Menger (2011) dealt with students' conceptions of simple mechanical machines, which were reconstructed after a phase of practical activities by means of problem-centred group interviews (see also Leavy 2016). Schick (2000) used semi-standardised interviews, among other methods, to collect data on the physics-related self-constructions of grammar school children in 8th-grade as well as the connections between these constructions and the students' behaviours in their physics classes. The summary of Schnick's project is given below.

Example project Schick (2000):

   In the course of the study—using various methods of data analysis—which focused on the teaching unit "electricity and water", two semi-standardised interviews with students were conducted and video-recorded for documentation. One of these focused mainly on the participants' prior knowledge of the physics of electricity and parts of their self-construction, whereas the other focused primarily on self-related cognitions. In addition, the following methods of data collection were also applied:

- *Questionnaire to document the student's self-assessment at the beginning of the teaching unit*
- *Video documentation of the two participating student groups' activities during the teaching unit*
- *Collecting written work produced (e.g., class work, work sheets) by the students*
- *Opinion-focused questionnaire on teaching practice to be completed by students after the teaching unit.*

Observations

In addition, qualitative data can also be collected by means of *observations* that can relate to the direct analysis of human actions, linguistic utterances, nonverbal responses as well as other social traits (Diekmann 2018; Flick 2016). In contrast to quantitative alternatives, qualitative observations focus mainly on the authentic learning environment, the researcher's principled openness to new insights as well as the additional exploration of latent meaning structures (which, in part, can be observed only indirectly) during the research. The aim is to investigate how something really works or happens. For such studies, the various positions the researcher can take are highly significant (ranging from full participation to full pure observation). Ethnographic approaches, in particular, have gained significance regarding this method of data collection over the recent years.

When conducting observations, the following aspects, among others, should be known:

- *Differences between various techniques of observation*
- *Advantages and disadvantages of the various methods of observation*
- *As required by the circumstances, knowledge of transcription conventions* (e.g., Flick 2016).

An exemplary use of this method of data collection in contexts of teaching or teaching-and-learning situations can be seen in the study by Kaiser and Dreber (2010) who used participant observation to study nursery school children's understanding of natural sciences and technology. The summary of Kaiser and Dreber's project is given below.

Example project Kaiser and Dreber (2010):

In the context of the project Wissenschaft im Kindergarten (Science in Nursery School), nursery school-children were observed during their experimental activities, using a semi-standardised technique of observation. The data thus collected by the researchers were recorded in research journals. In addition, conversation logs were created for the conversations between the children regarding their experiments. The analysis of the data collected used the method of scaling-structuring content analysis (see Sect. 17.4.1) and the QDA software MAXQDA (see Sect. 17.4.2).

Video cameras and visual representations were used by Azevedo (2018) in a US study on changing students' learning behaviours by introducing them to complex topics in STEM disciplines to trigger their situational interest. The summary of Azevedo's project is given below.

Example project Azevedo (2018):

By providing students access to visual representations (e.g., diagrams, Cartesian graphs, pictures, tables, drawings), Azevedo sought to expose the students to core aspects of professional scientific work and material culture while also drawing on their competence with design-based activities. In total, the researcher team collected about 30 h of videotapes and research assistants took extensive field notes on classroom events. Analysis was conducted by identifying a triggered situational interest through (i) a shift in a student's mode of participation, as seen in his/her (ii) increased attention, focus, and engagement.

Non-reactive Methods

As the third option in qualitative data collection, *non-reactive methods* play an important role as well. These approaches are distinguished by the fact that the researcher cannot influence the persons or events studied, since the researcher and the object of research do not come in contact with each other. Instead, the researcher encounters the object(s) of research in their finished state. The key basis of non-reactive data collection methods are documents that were either created for the express purpose of the study (e.g., journals or children's drawings) or already existed (e.g., archive files, registers, meeting minutes, but also curricular documents, textbooks or other teaching materials). Generally speaking, non-reactive methods can be distinguished by the various types of documents that serve as source materials. These documents can be characterised, first and foremost, according to authorship and accessibility (e.g., Carrington et al. 2005; Flick 2016; Scott 1990, 2017).

When using non-reactive methods, the following aspects should be known, among others:

- *Steps to building a text corpus that corresponds to the research questions*
- *Deciding adequate selection criteria for scaling down a text corpus*
- *Options for combining various data sources.*

Document analyses in subject didactics of the natural sciences are often linked to a historical perspective, as in the case of Sauer's (1992) exemplary analysis of the paradigm shifts in the representation of the phenomenon "thunderstorm" in nineteenth-century science textbooks (see also Cech and Giest 2005). Focusing on contemporary data, in another example, Altenburger and Starauschek (2012) studied the class registers of 30 teachers with respect to, inter alia, notes on the range of physics topics taught over two school years in 3rd- and 4th-grades in Baden-Württemberg.

17.3.3 Deciding Technical Means of Research

The potential use of various technological means to support data collection varies according to the approach chosen. In the case of *interviews*, audio-recording devices are used most frequently, and subsequently digital transcripts are created in adherence with specific transcription conventions (as necessary, supported by transcription software). Beyond this, complementary notes taken by the interviewer and, if necessary, other materials can play a role. This includes, for example, photographs, video files or drawings, which may be handed over by the interviewee or created by the interviewer (Bortz and Döring 2016). *Observations*, as a rule, are documented in the form of structured notes in a notebook or field journal. Here, too, photographs, video files and drawings or sketches may be relevant as complementary data—especially in case of second-hand observations (Flick 2016). Since, in the case of *non-reactive methods*, the data already exists in the form of text or image, the task at hand tends to be their digitalisation and preparation. Here, bookeye-scanners and OCR software can play an important role, as can databank or reference management software for the purpose of well-structured file management.

17.4 Step 3: Data Analysis

The steps of data analysis may either follow those of data collection or alternate with them in a more or less continuous back-and-forth fashion (see Sect. 17.5). Although there are forms of data collection and analysis that are frequently combined and match well, there is—as a rule—no fixed link between these two areas of the research process (Fig. 17.4).

Decisions on the specific approach for analysing the collected data should—as with data collection—depend *only on the research interest or adequacy to the object of research*, and thus also on the fit with one's basic methodological assumptions. This ultimately also entails the possibility of varying existing methods, as demonstrated, for example, by Lechte (2008). Moreover, the steps of data analysis must be conceptualised in conjunction with approaches to generalisation right from the beginning, when planning the project, since this is ultimately the aim of all qualitative research (Bortz and Döring 2016; Flick 2016). Besides the method chosen (or developed), data analysis also includes the corresponding use of technology.

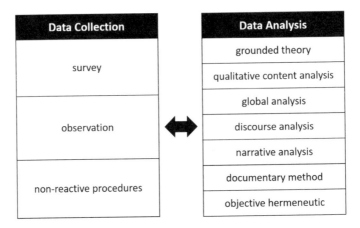

Fig. 17.4 Overview of methods of data collection and analysis

17.4.1 Deciding the Method of Data Analysis

As when presenting various forms of data collection, the options for data analysis can only be discussed briefly here. This overview is also not comprehensive due to the diversity of methods in qualitative research and its variability in adapting approaches to the specific research interests. The following discussion thus focuses on such methods of data analysis as have been used in the field of qualitative research on natural science didactics or are, in principle, suitable for projects in this field.

Grounded Theory

One form of data analysis is *grounded theory*, an approach at first jointly published by Glaser and Strauss (2017) and then separately developed into a rather theoretical and practical direction (e.g., Corbin and Strauss 2015; Glaser 1996, 2005; Strauss and Corbin 1997). This approach is carried out in parallel to data collection and is distinguished, first and foremost, by distinct forms of coding.

Glaser and Strauss (2017)*distinguished the following forms of coding:*

- *Initial, open coding that follows units of meaning in the textual data*
- *Selective coding for the selection and enrichment of promising codes*
- *Axial coding for the clarification of the core category and thus the phenomenon, followed by relating this category to the other codes.*

Grounded theory is used, for example, in the US context by Dennis (2018) to investigate the emerging conceptualisation of neuroscience practices and implications for promoting a positive classroom climate. In German research contexts, Kaiser and Schönknecht (2016) used this approach to analyse think-aloud protocols, thereby investigating students' understanding of visualised process representations in, inter alia, school textbooks. Similarly, Landwehr (2002) used grounded theory to research, drawing on episodic interviews, the root causes of primary teacher's disproportionally infrequent choice to specialise in physics. The summary of Landwehr's project is given below:

Project example Landwehr (2002):

In order to approach the relationship that teachers and university students in teacher training programs have to the teaching subject of physics, Landwehr used a multistep coding system in the framework of grounded theory to analyse her data: First, in the "open coding" step, line-by-line paraphrases were extracted from the transcribed material and then condensed into a hierarchical coding system. Next, "axial coding" was carried out to reorganise the codes according to topic areas. This was followed by "selective coding", which mainly focused on the core category of "educational relevance" and implied corresponding subcategories in the sense of enrichment.

Qualitative Content Analysis

As a further form of *qualitative data analysis*, qualitative content analysis provides comparatively precise guidelines for the step-by-step procedure of investigating the data collected (e.g., Mayring 2014; Schreier 2012). For this method, data collection is carried out prior to the actual analysis. As with *grounded theory*, this approach is also based on the principled orientation towards a coding scheme developed for the analysis. However, this is complemented by a general model that encompasses the entire process of data analysis.

Mayring (2014) *divided his version of qualitative content analysis into three subforms*:

- *Summarising content analysis, which paraphrases and condenses material in order to allow generalisations on a higher level of abstraction,*
- *Explicating content analysis, which generates explicating paraphrases through enriching ambiguous text passages with the help of context analyses, and*
- *Structuring content analysis, which—according to the variant chosen—filters out formal, thematic, triangulating or scaling structures from the material.*

Besides Mayring (2014), there are also more open forms of qualitative content analysis (e.g., Cho and Lee 2014). In the context of the subject didactics of the natural sciences, however, qualitative research—which, overall, frequently uses Content Analysis—tends to employ the version developed by Mayring (2014). This can be seen, for example, in Wagner (2016b), who analysed school-level contextual conditions for conveying contents of physics curricula in primary school based on a deductive approach to structuring content analysis complemented by inductive insights. Similarly, Hempel (2008) analysed guideline-based interviews with primary school children with respect to their understanding of science (regarding natural sciences or social sciences), and, among others, also in Zinn (2008). The summary of Hempel's project is given below.

Example project Hempel (2008):

The results of the guideline-based, problem-centred interviews with two children each from 4th- and 2nd-grade classes from schools in Vechta were transcribed verbatim and analysed by using a version of qualitative content analysis. This implicated a categorical-hierarchical exploration of the data, accompanied by steps of condensation and reduction to key aspects. This facilitated understanding of the children's lived experiences and cognitions. However, the actual analytical procedure that was followed in Hempel's project was only an approximation of Mayring's (2014) method.

Global Analysis

In contrast, *global analysis* according to Legewie (1994) focused mainly on gaining a broad overview of the thematic range of the text to be interpreted. This approach comprises ten distinct steps of structuring and reducing the data in order to ultimately focus on key terms and statements.

Legewie (1994) *described the following ten steps*:

1. *Gain initial orientation within the text*
2. *Activate the text's context (of production)*
3. *Work through the text by marking key passages*
4. *Develop ideas on the text*
5. *Write an index of key topics in the text*
6. *Summarise the text analytically-thematically or sequentially*
7. *Assess the text (e.g., credibility)*
8. *Decide on key terms for analysis in order to grade the text regarding its relevance to the research question*
9. *Assess the text's relevance for subsequent analysis*
10. *Presenting key results of the analysis (e.g., an assessment or evaluative statement, index of topics).*

In the course of this approach, texts of up to approximately 20 pages are assessed and, as the circumstances may require, evaluated regarding their suitability for inclusion in the sample. Thus, this method often complements other methods of data analysis (e.g., grounded theory or qualitative content analysis) and is rarely used as a stand-alone method for data analysis within a research project.

Discourse Analysis

As being strongly focused on contents and topics, *discourse analysis* is used to analyse discursive phenomena and thus, first and foremost, to identify the construction of versions of events in reports and representations (Flick 2016). Data analysed can include everyday conversations as well as interviews, media reports or, in particular, school textbooks. The context in which these are produced also constitutes part of the data basis. For discourse analysis in particular, this synthesis of textual and contextual data is highly relevant. This accords an important role to the connections between linguistic action and linguist forms as well as between linguistic action and social structures. However, discourse analysis has been received and developed in slightly different ways in various disciplines.

Some examples for discipline-specific variants of discourse analysis are as follows:

- *Critical discourse analysis according to van Dijk* (1993)
- *Social sciences discourse analysis according to Wodak and Krzyzanowski* (2008)
- *Historical discourse analysis according to Brinton* (2001).

In the subject didactics of the natural sciences, discourse analysis has so far not received much attention, despite the fact that it would be well-suited, for example, to tracing the communicative and cooperative processes that underlie the emergence and change of topical trends in natural science curricula or textbooks. In the US research context, Brown and Spang (2008) built on discourse analysis to investigate how science language was used to accomplish the normative practices of the classroom. Following the aim of understanding the dialogical meaning of a science text and how a science text acts to socially construct the identity and experience of the reader, Chambers (2008) applied the discourse analysis approach in his Canadian study to analyse the semantic content and rhetorical interaction in relation to language and other symbolic systems in a social context. The summary of Chambers's project is given below.

Example project Brown and Spang (2008):
In their study on synthesising everyday and science language in the classroom, the researchers engaged in an 8-month ethnographic study of the language of a 5th-grade classroom, involving 27 African-American students (17 girls and 10 boys), while doing science activities. In a discourse analysis, message units were used to document the fundamental message in each statement. These segments were accompanied by analytical units that defined what was accomplished by using a particular tone, volume, or pace of talk. These units document the embedded meaning associated with each utterance that helped to identify how classroom interactions were shaped by different styles of classroom talk.

Narrative Analysis

This method, originally developed by Schütze (2005), focuses on the social reality perceived by the informant. The aim of data analysis in this approach can vary due to potential objectives of reconstructing factual processes as well as the analysis of the respective processes of construction leading to the narratives (e.g., Rosenthal and Fischer-Rosenthal 2004). More specifically, this approach connects various focal points and procedures. Nevertheless, they all follow a sequence-analytical approach to the data—typically collected as narrative interviews—according to key units and a synthesis of these data with social data. As a rule, this is complemented by a comparison of multiple case stories. Biographical research, in particular, often uses *narrative analysis* to study life histories (Wengraf and Chamberlayne 2006). An example of this can be seen in Lechte (2008), who, drawing mainly on interviews she conducted, used elements of narrative analysis in her data analysis in order to investigate 11th-grade students' experiences with physics and physics classes as well as the relationship between these experiences and the students' attitude to this field of knowledge. The summary of Lechte's project is given below.

Example project Lechte (2008):

Drawing on narrative analysis, but complementing this with references to the documentary method, the interview data collected were paraphrased, analysed structurally, and condensed. The aim of this approach was to identify, within the chronological flow of narration, the individual physics stories of the students in a condensed manner—without losing sight of moods, the turns of the conversation or the views of the narrating individuals. Finally, these condensations had to be authorised by the interviewees in the sense of a communicative validation (see Sect. 17.6). In particular, during the subsequent interpretation of the thus condensed data, the focus on the subjective life-practice as well as the objective structures underlying the actual statements brought the documentary method into play. Complementary material for this method consisted of pictures drawn by the interviewees showing their understanding of physics.

Documentary Method

The use of *documentary method* aims to overcome the divide between subjectivism and objectivism, since the informants' knowledge is used as the empirical basis of the analysis. Nevertheless, the practices of action are explored through the underlying process structures, which elude the perspective of the informants themselves (Bohnsack et al. 2008). As basic principles of this method, Bohnsack (2013) referred to type formation, generalisation and comparative analysis. In the formation of types, he discussed those of common sense, the praxeological, sense-genetic and socio-genetic, and ascribed a generalisation of type to the praxeological formation of types through their multidimensionality. Nentwig-Gesemann (2013) emphasised the "methodological foundation, its analytical approach, the methodological steps [...] and the multidimensional structure" (p. 295) as characteristics of these types.

Analytical steps of the documentary method are as follows (Bohnsack 2001, 2008):

Formulating interpretation in order to explore social reality from the perspective of the individual actors and thus the exploration of the thematic structure of texts, that is, the underlying contents to be analysed

Reflecting interpretation analyses as the regularity of empirically observable reactions in connection with statements to be triggered in order to obtain a classification of the reactions and to reconstruct the production of this practice and thus the underlying patterns of orientation

Comparative analysis to contrast different cases in order to more clearly identify the frames of orientation

As necessary, analysis of the discourse's trajectory in order to differentiate the distinct realisations of the topic.

Given the documentary method's now established subforms and variations, it can be used not only for group discussions and observations but often also for open biographical interviews, field research protocols and analysis of historical texts (Bohnsack et al. 2008), as well as for image and video interpretation (Knoblauch 2006; Wagner-Willi 2006). This method is relevant, for instance, to the study of Lechte (2008), who combined narrative analysis and documentary method in her data analysis. This opens up possibilities to gain knowledge and conclusions about action practice (Bohnsack et al. 2008). The study of Feierabend and Eilks (2011) on curriculum innovation in science education can be used as an example of *participatory action research* (PAR). Aiming at curriculum development in science teaching with implications for the development of strategies and materials to potentially improve science practices, Feierabend and Eilks chose an interpretation of action research that is more researcher-centred in that the practitioners, for example, participated in the development of concrete teaching practices.

Objective Hermeneutics

This method focuses on identifying "latent meaning structures" as objective meanings of an utterance or action. However, this does not correspond to the subjective meaning of the acting subject (Becker-Lenz et al. 2016; Lueger and Hoffmeyer-Zlotnik 1994). Exploring these meaning structures is done sequentially by groups of interpreters, following a fixed procedure.

Analytical steps of objective hermeneutics are as follows:

Sequential preliminary analysis of the outside contexts embedded in an utterance
Sequential detailed analysis in multiple steps by means of procedures of formulating hypotheses and falsification in order to identify structures of the interaction
Validating the identified interactional structures in further material.

In addition to written documents, the data to be analysed can also include images and photographs. In contrast, decoding the subjective meanings of utterances and actions plays a marginal role in *objective hermeneutics* (Flick 2016). Other hermeneutic approaches such as *social sciences hermeneutics* or *hermeneutical sociology of knowledge* later place their emphasis on the social construction of knowledge and also vary in terms of methodological procedures (Reichertz 2004b). With regard to the natural science content of "Sachunterricht" in primary school, hermeneutical approaches can, for example, be found in Vogt et al.'s (2011) analysis of drawings—created by children in 1st- and 4th-grades—of the hands of people of various ages (baby to 65 years old). Their aim was to identify concepts of ageing in primary school children (see also James and James 2012). Biester (1991) offered a discussion of drawing as a helpful tool for understanding in "Sachunterricht".

17.4.2 Deciding Technical Support

The use of software that supports data analysis in qualitative research has been gaining increasing importance in recent years. Today, such applications have become standards of good research for many variations of methods. However, before researchers undertaking a qualitative project decide to use a specific software, they should reflect on the adequate software options available (Flick 2016; Kuckartz 2014). Three steps are recommended for choosing an adequate technical means to support data analysis. *First*, the basic decision to use or not use software to support the process of data analysis must be taken according to the following aspects:

Some software is too expensive or requires too much time to learn relative to its benefit for the project
The use of software does not necessarily improve the quality of the study
The decision should be guided not by any previous experience with a specific software but by the object of research itself.

If the basic decision is to use software to support analysis, the *second step* is to find the appropriate level of software. Here, one option is using the available functionality of text processing software to support a particular method (e.g., comments, search, extending functionality through the use of macros); a second option is to work with databank software that can organise codings outside the text in separate databanks (and can usually do so without further, sophisticated functions); and the third option is to use specialised QDA software developed for the analysis of qualitative data.

Advantages of using QDA software include the following (John and Johnson 2000):

Organising and managing the text corpus
Using various options for coding
Direct links between corpus of sources and the categories developed
Option to establish hierarchies and other forms of linkage within category systems developed
Various forms of visualising the data
Diverse options for search and management functionality.

If the use of QDA software has been found to be appropriate in a particular research project, the *third step* is to choose a specific software package. In this regard, it must be noted that QDA software has been developed with different methods as well as fundamentally different methodological assumptions in mind.

The substantial discrepancies between various software options necessitate an initial learning period to become acquainted with the respective trial versions, thoroughly reading the relevant literature and possibly attending workshops. Currently, widely used software includes MAXQDA, Atlas.ti and NVivo (Bazeley and Jackson 2015; Friese 2018; VERBI Software 2012). These can be broadly characterised as follows:

MAXQDA is seen as user-friendly, easily understood software, but offering fewer functions than other QDA software
Atlas.ti is a relatively variable and adaptable software with strong network functions
NVivo is seen as supporting a wide range of methods and as applicable in various ways for diverse file formats.

Examples for the use of MAXQDA or its predecessor WINMAX can be found, for example, in Kaiser and Dreber (2010) or in Landwehr (2002). Similarly, Coryn et al. (2014) used MAXQDA in their explorative study to analyse factors associated with successful school strategies for native Hawaiian students in the State of Hawaii.

17.5 Steps 4 and 5: Interpretation of the Results and Quality Criteria

The interpretation of the results makes up a large part of qualitative research and is, therefore, built extensively on the fulfilment of quality criteria. Compliance with quality criteria is of particular importance in the context of qualitative research, since the possibilities of interpretation in this area are extremely diverse.

17.5.1 Interpretation of the Results

As the interpretation of the results is closely connected to data analysis, builds directly upon it, and thus differs according to the method of analysis chosen, this section will only cover basic considerations that are of particular relevance to the concluding interpretation of data in research projects. These are the following:

The interpretation of the results must be grounded in the theory-of-science considerations that inform the direction of interpreting data as well as the boundaries of interpretation.

The fundamental perspective of the interpretation of the results must necessarily follow from the research questions posed initially, all of which must be directly addressed in the course of generalisation.

It is crucial that the interpretation of the results builds on the data collected and analysed in a transparent fashion; it is equally important that appropriate evidence is given and that cross-references to the data are made to allow readers to follow and understand the interpretative perspectives gained at any point.

17.5.2 Quality Criteria

In good qualitative research projects, validating the criteria of quality is as important as in quantitative research. Such projects are equally guided by the criteria of *validity*, *objectivity* and *reliability*; however, within this triad, they include more varied options than the quantitative criteria of "test quality" and are, furthermore, supported by the necessary proximity of all research activities to its object (see Fig. 17.5). With regard to quality criteria of qualitative research, Steinke (2004) recommended above all the consideration of intersubjective traceability, indication of the research process, empirical anchoring, limitation, coherence, relevance and reflected subjectivity.

Validity

In order to test the validity of a research project, a number of options are viable. *Triangulating measures* (triangulation through data, researcher, theory or method) allows counter-balancing the weaknesses of a method or approach by means of the strengths of another. This allows a comparison between different results (Lamnek and Krell 2016). In the case of reactive approaches (observations, interviews and open-questionnaires), the procedure of *communicative validation* can serve to validate the insights gained through conversations with the people involved. *Argumentative validation*, in contrast, aims to achieve intersubjectivity of the interpretative results by means of a dialogue with other researchers (Mayring 2016).

Objectivity

In the context of qualitative research projects, too, objectivity can be achieved by means of measures appropriate to the research paradigm. This includes, for example, *emergentist objectivity* according to Kleining (1982), which in the context of constructivism emphasises *intersubjective plausibility* as a crucial criterion. In addition, validating two forms of consistency—*external consistency,* which implies matching with knowledge external to the respective study, and *internal consistency,*

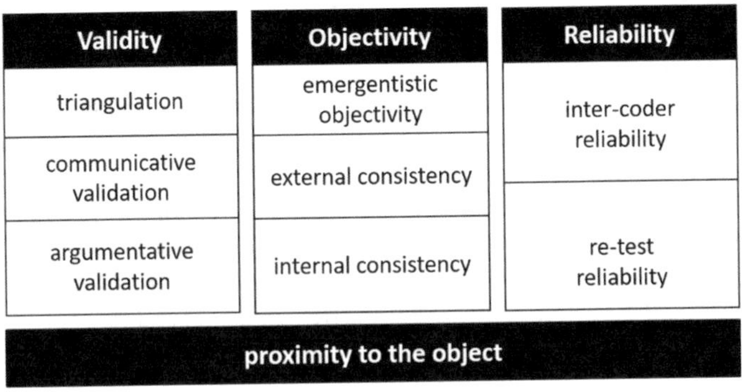

Validity	Objectivity	Reliability
triangulation	emergentistic objectivity	inter-coder reliability
communicative validation	external consistency	
argumentative validation	internal consistency	re-test reliability
proximity to the object		

Fig. 17.5 Overview of criteria of quality in qualitative research (e.g., Lamnek and Krell 2016; Mayring 2016)

which identifies the consistency within data collection, analysis and interpretation—also constitutes criteria of objectivity in qualitative research (Lamnek and Krell 2016; Mayring 2016).

Reliability

As for determining reliability, the validation of *intercoder reliability* can be carried out with the calculation of several coefficients. In case of Krippendorff's alpha, which is the most frequently used coefficient where correct agreements are offset by random ones (Mayring 2014). Parallel to intercoder reliability, the same coefficients can be used in case of a single researcher to determine *test–retest reliability*.

Despite the relevance of the three criteria: validity, objectivity and reliability, the ultimate aim—and thus crucial point of reference which informs all criteria of quality or complements these as the most relevant—is that the *proximity to the object of research,* which is achieved by qualitative research, must be maintained throughout research (Mayring 2016).

17.6 Conclusion

In summary, it can be stated that the five steps of planning a qualitative research project are differentiated into various facets and decision-making processes. Since the formulation of research questions and the theoretical foundation as the *first step* of the planning process are based on the elaboration of the state of research and the theoretical engagement with the object of research, a number of central decisions have to be made already in this introductory step of the research project. First and foremost, the semantics of science, the methodology and epistemology must be defined at this point in the research project.

The *second step* in the planning of a qualitative research project, the data collection, involves decision-making processes with regard to the selection of the sample, the methods for data collection and technical implementation. In this step, the choice between convenience sampling versus random sampling, above all, must be made. It must be decided whether qualitative data are generated by the use of interviews, observations or non-reactive procedures, for example, and it must be decided on the choice of the technical implementation of the data collection, taking into account the variation in the context of the fit between methods for data collection and data analysis.

Data analysis is the *third step* in the planning of a qualitative research project. In addition to the decision regarding technical support, the focus is primarily on the choice of the method of data analysis. In general, the methods of grounded theory, qualitative content analysis, global analysis, discourse analysis, narrative analysis, documentary method and objective hermeneutics can be separated from each other. Selecting one of these data analysis methods is closely dependent on the choice of data collection, so that steps 2 and 3 of the planning process are mutually dependent. Also, the *fourth step* (interpretation of the results) and the *fifth step* (quality criteria) are in an intermediate relationship to each other. The interpretation of the results follows, on the one hand, the research questions and, on the other hand, builds on the collected data. The criteria of validity, objectivity and reliability are, after all, guiding factors for the quality of research.

The different steps are generally closely interdependent; a change in one step will also result in changes in the others. Thus, qualitative research is ultimately a circular process with circular adjustments and readjustments in all areas of the research project until everything fits together. The central criterion, especially in qualitative research, is always that the various steps are oriented in their form to the object of research and are appropriate to it. By way of conclusion, it can be noted that qualitative research in the subject didactics of the natural sciences is currently very limited and, in most cases, takes the form of triangulating research designs. Purely qualitative approaches, in contrast, tend to be rare. A possible cause for this may be recognised in the as-yet-unused advantages of using QDA software. Indeed, such advantages could well significantly extend current capabilities of qualitative research to gain insights into the subject didactics of the natural sciences (Garz 1997).

Acknowledgements We would like to thank Philipp Mayring (Alpen-Adria Universität Klagenfurt, Austria) and Astrid Huber (Private Pädagogische Hochschule der Diözese Linz, Austria) for carefully and critically reviewing this chapter.

References

Altenburger P, Starauschek E (2012) Physikalische Themen im Sachunterricht Baden-Württembergs in den Jahrgangsstufen 3 und 4 [Physics topics in Baden-Württemberg's science education in grades 3 and 4]. In: Giest H, Heran-Dörr E, Archie C (eds) Lernen und Lehren im Sachunterricht: Zum Verhältnis von Konstruktion und Instruktion [Learning and teaching in the classroom: on the relationship between construction and instruction], Bd. 22. Klinkhardt, Bad Heilbrunn, pp 71–78

Azevedo FS (2018) An inquiry into the structure of situational interests. Sci Educ 102:108–127

Bazeley P, Jackson K (2015) Qualitative data analyses with NVivo. Qual Res Psychol 12(4):492–494

Becker-Lenz R, Franzmann A, Jansen A, Jung M (eds) (2016) Die Methodenschule der objektiven Hermeneutik: Eine Bestandsaufnahme [The methodological school of objective hermeneutics: an inventory]. Springer VS, Wiesbaden

Berger P, Luckmann P (1966) The social construction of reality. Doubleday, New York

Bhaskar R (2008) A realist theory of science. Routlegde, London

Biester W (1991) Zeichnen als Hilfe zum Verstehen im Sachunterricht der Grundschule [Drawing as an aid to comprehension in primary science education]. In: Lauterbach R, Köhnlein W, Spreckelsen K, Bauer H (eds) Wie Kinder erkennen. Probleme und Perspektiven des Sachunterrichts [As children recognise. Problems and perspectives of science education], Bd. 1. IPN, Kiel, pp 82–97

Blumer H (1969) Symbolic interactionism: perspective and method. Englewood Cliffs, New Jersey

Bohnsack R (2001) Dokumentarische Methode: Theorie und Praxis wissenssoziologischer Interpretation [Documentary method: theory and practice of sociological interpretation of knowledge]. In: Hug T (eds) Wie kommt Wissenschaft zu Wissen? Einführung in die Methodologie der Sozial- und Kulturwissenschaften [How does science come to knowledge? Introduction to the methodology of social and cultural sciences], Bd. 3. Schneider, Baltmannsweiler, pp 326–345

Bohnsack R (2008) The interpretation of pictures and the documentary method. Forum Qual Soc Res 9(3):Art. 26

Bohnsack R (2013) Typenbildung, Generalisierung und komparative Analyse: Grundprinzipien der dokumentarischen Methode [Type formation, generalisation and comparative analysis: basic principles of the documentary method]. In: Bohnsack R, Nentwig-Gesemann I, Nohl A-M (eds) Die dokumentarische Methode und ihre Forschungspraxis: Grundlagen qualitativer Sozialforschung [The documentary method and its research practice: the foundations of qualitative social research], 3rd edn. VS-Verlag, Wiesbaden, pp 241–270

Bohnsack R, Pfaff N, Weller W (eds) (2008) Qualitative analysis and documentary method in international educational research. Barbara Budrich, Opladen & Farmington Hills

Bortz J, Döring N (2016) Forschungsmethoden und Evaluation: Für Human- und Sozialwissenschaftler [Research methods and evaluation: for human and social scientists], 5th edn. Springer Medizin, Heidelberg

Brinton LJ (2001) Historical discourse analysis. In: Schiffrin D, Tannen D, Hamilton HE (eds) The handbook of discourse analysis. Blackwell, Malden, pp 138–160

Brown BA, Spang E (2008) Double talk: synthesizing everyday and science language in the classroom. Sci Educ 92:708–732

Carrier M (2004) Theoriesprache [Theoretical language]. In: Mittelstraß J (ed) Enzyklopädie Philosophie und Wissenschaftstheorie [Encyclopedia of philosophy and theory of science], Bd. 4: Sp-Z. J.B. Metzler, Stuttgart, pp 283–289

Carrington PJ, Scott J, Wasserman S (2005) Models and methods in social network analysis (Structural analysis in the social sciences book 28). Cambridge University Press, Cambridge

Cech D, Giest H (2005) Sachunterricht in Praxis und Forschung: Erwartungen an die Didaktik des Sachunterrichts [Science education in practice and research: expectations on the didactics of science education], Bd. 15. Klinkhardt, Bad Heilbrunn

Chambers JM (2008) Human/nature discourse in environmental science education resources. Can J Environ Educ 13(1):107–121

Cho JY, Lee E-H (2014) Reducing confusion about grounded theory and qualitative content analysis: similarities and differences. Qual Rep 19(64):1–20

Corbin J, Strauss AL (2015) Basics of qualitative research: techniques and procedures for developing grounded theory, 4th edn. Sage, Los Angeles

Coryn CLS, Schröter DC, McCowen RH (2014) A mixed methods study of some of the factors associated with successful school strategies for native Hawaiian students in the State of Hawaii. J Mixed Methods Res 8(4):377–395

Dennis SR (2018) Applied educational neuroscience in elementary classrooms: a grounded theory study. Doctoral thesis, submitted to the Faculty of the University Graduate School. Indiana University

Diekmann A (2018) Empirische Sozialforschung: Grundlagen, Methoden, Anwendungen [Empirical social research: principles, methods, applications], 12th edn. Rowohlt Verlag, Reinbek b. Hamburg

Dunker NK (2016) Überzeugungen von Sachunterrichtslehrkräften zum Experimentieren im Unterricht [Convincing science education teachers to experiment in the classroom]. In: Giest H, Goll T, Hartinger A (eds) Probleme und Perspektiven des Sachunterrichts: Sachunterricht zwischen Kompetenzorientierung, Persönlichkeitsentwicklung, Lebenswelt und Fachbezug [Problems and perspectives of science education: science education between competence orientation, personality development, life world and subject reference]. Klinkhardt, Bad Heilbrunn, pp 107–115

Feierabend T, Eilks I (2011) Innovating science teaching by participatory action research: reflections from an interdisciplinary project of curriculum innovation on teaching about climate change. Center Educ Policy Stud J 1(1):93112

Flick U (2016) Qualitative Sozialforschung: Eine Einführung [Qualitative social research: an introduction], 7th edn. Rowohlt, Reinbek b. Hamburg

Flick U, von Kardorff E, Steinke I (eds) (2004) A companion to qualitative research. Sage, London

Foucault M (1989) The order of things: an archaeology of the human sciences. Routledge, London

Friese S (2018) Atlas.ti 7: user guide and reference. ATLAS.ti Scientific Software Development GmbH, Berlin

Garfinkel H (1991) Studies in ethnomethodology. Polity, Cambridge

Garz D (1997) Qualitative Forschungsmethoden für die Sachunterrichtsdidaktik [Qualitative research methods for the didactics of subject teaching]. In: Marquardt-Mau B, Köhnlein W, Lauterbach R (eds) Forschung zum Sachunterricht: Probleme und Perspektiven des Sachunterrichts [Research on subject teaching: problems and perspectives of subject teaching], vol 7. Klinkhardt, Bad Heilbrunn, pp 43–60

Glaser BG (1996) Theoretical sensitivity. Advances in the methodology of grounded theory. Sociology Press, Mill Valley

Glaser BG (2005) The grounded theory perspective III: theoretical coding. Sociology Press, Mill Valley

Glaser BG, Strauss AL (2017) The discovery of grounded theory: strategies for qualitative research. Routledge, London

Grygier P (2008) Wissenschaftsverständnis von Grundschülern im Sachunterricht [Understanding of science by primary schoolchildren in the classroom]. Klinkhardt, Bad Heilbrunn

Günther J, Grygier P, Kircher E, Sodian B, Thoermer B (2004) Studien zum Wissenschaftsverständnis von Grundschullehrkräften [Studies on the understanding of science by primary school teachers]. In: Doll J, Prenzel M (eds) Bildungsqualität von Schule: Lehrerprofessionalisierung, Unterrichtsentwicklung und Schülerförderung als Strategien der Qualitätsverbesserung [Educational quality of schools: teacher professionalisation, classroom development and student support as strategies for quality improvement]. Waxmann, Münster, pp 93–113

Harris R (2005) The semantics of science. A&C Black, London

Hempel M (2008) Zum (Vor)Wissen von Wissenschaft bei Grundschulkindern [On (pre-)knowledge of science in primary schoolchildren]. In: Giest H, Wiesemann J (eds) Kind und Wissenschaft [Child and science]. Klinkhardt, Bad Heilbrunn, pp 83–95

James A, James AL (2012) Key concepts in childhood studies, 2nd edn. Sage, London

John SW, Johnson P (2000) The pros and cons of data analysis software: a review. J Nurs Sch 32(4):393–397

Kaiser A, Dreber I (2010) Empirische Effizienzüberprüfung eines Kindergartenprojektes zum elementaren naturwissenschaftlich-technischen Lernen [Empirical efficiency review of a kindergarten project for elementary scientific and technical learning]. In: Fischer H-J, Gansen P, Michalik K (eds) Sachunterricht und frühe Bildung: Forschungen zur Didaktik des Sachunterrichts [Subject teaching and early education: research on the didactics of subject teaching], Bd. 9. Klinkhardt, Bad Heilbrunn, pp 81–92

Kaiser L, Schönknecht G (2016) Lernhilfe oder Hindernis? Visualisierungen im Sachunterricht [Learning aid or obstacle? Visualisations in subject lessons]. In: Giest H, Goll T, Hartinger A (eds) Sachunterricht – zwischen Kompetenzorientierung, Persönlichkeitsentwicklung, Lebenswelt und Fachbezug [Science education—between competence orientation, personal development, life world and subject reference], Bd. 26. Klinkhardt, Bad Heilbrunn, pp 49–57

Kakkori L (2009) Hermeneutics and phenomenology problems when applying hermeneutic phenomenological method in educational qualitative research. Paideusis 18(2):19–27

Kleining G (1982) Umriss zu einer Methodologie qualitativer Sozialforschung [Outline of a methodology of qualitative social research]. Kölner Zeitschrift Für Soziologie Und Sozialpsychologie 34(2):224–253

Knoblauch H (2006) Videography. Focused ethnography and videoanalysis. In: Knoblauch H, Schnettler B, Raab J, Soeffner H-G (eds) Video analysis: methodology and methods—qualitative audiovisual data analysis in sociology. Peter Lang, Frankfurt a.M., pp 69–83

Kromrey H, Roose J, Strübing J (2016) Empirische Sozialforschung: Modelle und Methoden der standardisierten Datenerhebung und Datenauswertung [Empirical social research: models and methods of standardised data collection and data evaluation], 13th edn. UVK Verlagsgesellschaft, Konstanz

Kuckartz U (2014) Qualitative text analysis. Sage, London

Lamnek S, Krell C (2016) Qualitative Sozialforschung: Mit Online-Material [Qualitative social research: with online material], 6th edn. Beltz, Weinheim

Landwehr B (2002) Distanzen von Lehrkräften und Studierenden des Sachunterrichts zur Physik: Eine qualitativ-empirische Studie zu den Ursachen [Distances of teachers and students of subject teaching to physics: a qualitative-empirical study on the causes]. Logos, Berlin

Leavy P (2016) Essentials of transdisciplinary research: using problem-centered methodologies. Routledge, London

Lechte M-A (2008) Sinnesbezüge, Interesse und Physik: Eine empirische Untersuchung zum Erleben von Physik aus Sicht von Schülerinnen und Schülern. Studien zur Bildungsgangforschung [Sensory references, interest and physics: an empirical study of the experience of physics from the perspective of pupils. Studies on educational pathways], Bd. 23. Budrich, Opladen

Legewie H (1994) Globalauswertung [Global evaluation]. In: Böhm A, Mengel A, Muhr T (eds) Texte verstehen: Konzepte, Methoden, Werkzeuge [Understanding texts: concepts, methods, tools]. Universitätsverlag, Konstanz, pp 100–114

Lind G (1999) Der Physikunterricht an den deutschen Gymnasien vom Beginn des 18. Jahrhunderts bis zum Beginn des 20. Jahrhunderts [The physics lessons at German grammar schools from the beginning of the 18th century until the beginning of the 20th century]. In: Goodson IF, Hopmann S, Riquarts K (eds) Das Schulfach als Handlungsrahmen: Vergleichende Untersuchung zur Geschichte und Funktion der Schulfächer [The school subject as a framework for action: comparative study of the history and function of school subjects]. Böhlau, Köln, pp 109–150

Lueger M, Hoffmeyer-Zlotnik JHP (1994) Hermeneutic interpretation in qualitative research: between art and rules. In: Borg I, Mohler PP (eds) Trends and perspectives in empirical social research. de Gruyter, Berlin, pp 294–307

Luhmann N (2012) Introduction to systems theory. Wiley, Hoboken

Mannheim K (1982) Structures of thinking. Routledge & Kegan Paul, London

Matthews MR (1993) Constructivism and science education: some epistemological problems. J Sci Educ Technol 2:359–370

Mayring P (2014) Qualitative content analysis: theoretical foundation, basic procedures and software solution. Beltz, Klagenfurt

Mayring P (2016) Einführung in die qualitative Sozialforschung [Introduction to qualitative social research], 6th edn. Beltz, Weinheim

Menger J (2011) Das Modell der zirkulären Entfaltung von Denkwegen als Basis technischer Verstehensprozesse [The model of the circular unfolding of ways of thinking as a basis for technical understanding processes]. In: Giest H, Kaiser A, Schomaker C (eds) Sachunterricht: Auf dem Weg zur Inklusion [Science education: on the way to inclusion], Bd. 21. Klinkhardt, Bad Heilbrunn, pp 163–167

Nentwig-Gesemann I (2013) Die Typenbildung der dokumentarischen Methode [The type formation of the documentary method]. In: Bohnsack R, Nentwig-Gesemann I, Nohl A-M (eds) Die dokumentarische Methode und ihre Forschungspraxis: Grundlagen qualitativer Sozialforschung [The documentary method and its research practice: the foundations of qualitative social research], 3rd edn. VS-Verlag, Wiesbaden, pp 295–324

Popper KR (1998) Objektive Erkenntnis: Ein evolutionärer Entwurf [Objective knowledge: an evolutionary design], 4th edn. Hofmann und Campe, Hamburg

Porter T, Córdoba J (2009) Three views of systems theories and their implications for sustainability education. J Manag Educ 33(3):323–347

Reichertz J (2004a) Abduction, deduction and induction in qualitative research. In: Flick U, von Kardorff E, Steinke I (eds) A companion to qualitative research. Sage, London, pp 159–164

Reichertz J (2004b) Objective hermeneutics and hermeneutic sociology of knowledge. In: Flick U, von Kardorff E, Steinke I (eds) A companion to qualitative research. Sage, London, pp 290–295

Reinhoffer B (2000) Heimatkunde und Sachunterricht im Anfangsunterricht: Entwicklungen, Stellenwert, Tendenzen [Local history and science education in initial lessons: developments, significance, trends]. Klinkhardt, Bad Heilbrunn

Rosenthal G, Fischer-Rosenthal W (2004) Objective hermeneutics and hermeneutic sociology of knowledge. In: Flick U, von Kardorff E, Steinke I (eds) A companion to qualitative research. Sage, London, pp 259–265

Sauer M (1992) „Vom Nutzen des Gewitters": Paradigmen elementarer naturkundlicher Unterweisung im 19. Jahrhundert [„The benefits of thunderstorms": paradigms of elementary natural history instruction in the 19th century]. Neue Sammlung 32(1):134–153

Schick A (2000) Der Einfluss von Interesse und anderen selbstbezogenen Kognitionen auf Handlungen im Physikunterricht [The influence of interest and other self-related cognitions on actions in physics lessons]. Logos, Berlin

Schreier M (2012) Qualitative content analysis in practice. Sage, London

Schütze F (2005) Biography analysis on the empirical base of autobiographical narratives: how to analyse autobiographical narrative interviews? Curriculum development funded by the EU Leonardo da Vinci programme. http://www.zsm.ovgu.de/zsm_media/Das+Zentrum/Forschung sprojekte/INVITE/B2_1-p-140.pdf (21.04.2020)

Scott J (1990) A matter of record: documentary sources in social research. Polity, Cambridge

Scott J (2017) Social network analysis, 7th edn. Sage, London

Steinke I (2004) Quality criteria in qualitative research. In: Flick U, von Kardorff E, Steinke I (eds) A companion to qualitative research. Sage, London, pp 184–190

Strauss AL, Corbin J (eds) (1997) Grounded theory in practice. Sage, London

Tschamler H (1996) Wissenschaftstheorie: Eine Einführung für Pädagogen [Theory of science: an introduction for educators]. Klinkhardt, Bad Heilbrunn

Van Dijk TA (1993) Principles of critical discourse analysis. Discourse Soc 4(2):249–283

VERBI Software (2012) MAXQDA the art of data analysis: Einführung. http://www.maxqda.de/download/manuals/MAX11_intro_ger.pdf (24.04.2013)

Vogt H, Mogge S, Wolfram A (2011) „Oma hat Falten, ich nicht": Konzepte von Grundschulkindern über das Altern des Menschen ["Grandma has wrinkles, I don't": concepts of primary schoolchildren about ageing]. In: Heinzel F (ed) Generationenvermittlung in der Grundschule: Ende der Kindgemäßheit [Generational mediation in primary school: the end of child conformity]? Klinkhardt, Bad Heilbrunn, pp 223–238

Waddock S (2006) Leading corporate citizens: vision, values, value-added, 2nd edn. McGraw Hill, New York

Wagner K (2016a) Gelingensbedingungen des Einsatzes eines mobilen Lernarrangements zum Thema „Fliegen" im Grundschulunterricht [Successful conditions for the use of a mobile learning arrangement on the subject of "flying" in primary school lessons]. In: Liebers K, Landwehr B, Reinhold S, Riegler S, Schmidt R (eds) Facetten grundschulpädagogischer und -didaktischer Forschung: Jahrbuch Grundschulforschung [Facets of research in primary education and didactics: yearbook Primary School Research], 20. Springer VS, Wiesbaden

Wagner K (2016b) Schulische Rahmenbedingungen aus der Sicht von Sachunterrichtslehrkräften: Ein empirischer Beitrag zur Identifikation von Gelingensbedingungen von Unterrichtsentwicklung [The school environment from the perspective of subject teachers: an empirical contribution to the identification of conditions for successful instruction development]. In: Giest H, Goll T, Hartinger A (eds) Sachunterricht – zwischen Kompetenzorientierung, Persönlichkeitsentwicklung, Lebenswelt und Fachbezug [Subject teaching: between competence orientation, personal development, life world and subject reference], Bd. 26. Klinkhardt, Bad Heilbrunn, pp 167–174

Wagner-Willi M (2006) On the multidimensional analysis of video-data: documentary interpretation of interaction in schools. In: Knoblauch H, Schnettler B, Raab J, Soeffner H-G (eds) Video analysis: methodology and methods–qualitative audiovisual data analysis in sociology. Peter Lang, Frankfurt a.M., pp 143–153

Wengraf T, Chamberlayne P (2006) Interviewing for life-histories, lived situations and personal experience: the biographic-narrative interpretive method (BNIM). London

Wodak R, Krzyzanowski M (eds) (2008) Qualitative discourse analysis in the social sciences. Red Globe Press, London

Zinn B (2008) Physik lernen, um Physik zu lehren: Eine Möglichkeit für interessanteren Physikunterricht. Studien zum Physik- und Chemielernen [Learning physics to teach physics: an opportunity for more interesting physics lessons—studies on physics and chemistry learning], Bd. 85. Logos, Kassel